Design for Advanced Manufacturing

Technologies and Processes

LaRoux K. Gillespie Editor

McGraw Hill Education

New York Chicago San Francisco
Athens London Madrid
Mexico City Milan New Delhi
Singapore Sydney Toronto

Library of Congress Control Number: 2017930265

Design for Advanced Manufacturing: Technologies and Processes

1 2 3 4 5 6 7 8 9 QFR 21 20 19 18 17

ISBN 978-1-259-58745-0
MHID 1-259-58745-2

This book is printed on acid-free paper.

Sponsoring Editor
Robert Argentieri

Editorial Supervisor
Stephen M. Smith

Production Supervisor
Lynn M. Messina

Acquisitions Coordinator
Lauren Rogers

Project Managers
Nitesh Sharma and Gaurav Prabhu, MPS Limited

Copy Editor
Bindu Singh, MPS Limited

Proofreader
Kirti Dogra, MPS Limited

Indexer
Edwin Durbin

Art Director, Cover
Jeff Weeks

Composition
MPS Limited

Design for Advanced Manufacturing

About the Editor

LaRoux K. Gillespie is past president of the Society of Manufacturing Engineers. Before retiring with 40 years in the industry, he was Manager of Quality Assurance for a major Honeywell facility. Dr. Gillespie has edited or written 42 books and over 260 other publications, has B.S., M.S., and Dr. Eng. degrees in mechanical engineering, and is a Certified Manufacturing Engineer and a registered Professional Engineer.

About Tooling U-SME

Tooling U-SME, a division of SME (sme.org), delivers versatile, competency-based learning and development solutions to the manufacturing community, working with more than half of all *Fortune* 500® manufacturing companies, as well as 600 educational institutions across the country. Tooling U-SME partners with customers to build high performers who help their companies drive quality, profitability, productivity, innovation and employee satisfaction. Working directly with hundreds of high schools, community colleges and universities, Tooling U-SME is able to help prepare the next-generation workforce by providing industry-driven curriculum. Tooling U-SME can be found at toolingu.com or on Facebook (facebook.com/toolingu) and Twitter (twitter.com/toolingu).

Visit the Tooling U-SME website to download a copy of our latest white papers:

- Proving the ROI of Training: Moving from Expense to Necessity
- Embracing Millennials: Closing the Manufacturing Skills Gap and Gaining a Competitive Advantage
- Manufacturing Insights Report: Winning Practices of World-Class Companies

Contents

Contributors

David M. Allen *Cranfield University, Bedford, United Kingdom* (Chap. 3.21)

Jim Boldig *Custom Wire Technologies, Inc., Port Washington, Wisconsin* (Chap. 3.16)

Stanley Bovid *LSP Technologies, Inc., Dublin, Ohio* (Chap. 1.12)

Jack Burley *KAISER Precision Tooling, Inc., Hoffman Estates, Illinois* (Chap. 3.2)

Richard T. Chen *Microfabrica Inc., Van Nuys, California* (Chap. 3.44)

Marc Chooljian *University of California, Berkeley, California* (Chap. 3.25)

Masato Fukushima *Sodick, Inc., Schaumburg, Illinois* (Chap. 3.13)

Andrew S. Geiger *Leister Technologies, Itasca, Illinois* (Chap. 1.7)

LaRoux K. Gillespie *Manufacturing Consultant, Andover, Kansas* (Chapters not otherwise listed in this Contributors section)

Andrew Graves *Stratasys Direct Manufacturing, Valencia, California* (Chap. 2.11)

L. Mike Heglin *Rofin-Baasel, Devens, Massachusetts* (Chap. 1.10)

Chandra Shekhar Jawalkar *PEC University of Technology, Chandigarh, India* (Chap. 3.10)

Aaron Johnson *Accumold, Ankeny, Iowa* (Chap. 3.22)

Dorian Liepmann *University of California, Berkeley, California* (Chap. 3.25)

Rick Lucas *The ExOne Company, North Huntingdon, Pennsylvania* (Chap. 2.2)

Simone Maccagnan *Gimac Microextruders, Castronno (VA), Lombardia, Italy* (Chap. 3.28)

Scott Malkasian *MicroArc Welding, Worcester, Massachusetts* (Chap. 3.39)

Kenneth J. Mandile *Swissturn/USA, Inc., Oxford, Massachusetts* (Chap. 3.4)

Jason McNary *Comco Inc., Burbank, California* (Chap. 3.42)

Nicholas A. Meisel *The Pennsylvania State University, University Park, Pennsylvania* (Chap. 2.5)

Marty Mewborne *Amada Miyachi America, Monrovia, California* (Chap. 3.36)

Don Miller *Finepart Sweden AB, Bollebygd, Sweden* (Chap. 3.20)

Justin Morrow *University of Wisconsin, Madison, Wisconsin* (Chap. 1.3)

Stanislav Němeček *RAPTECH, U Vodárny, Czech Republic* (Chap. 1.5)

Mark I. Norfolk *Fabrisonic LLC, Columbus, Ohio* (Chap. 2.10)

Kenneth Norsworthy *Owens Corning Ridgeview, Duncan, South Carolina* (Chap. 3.12)

Jacobo Paredes *CEIT and Tecnun (University of Navarra), San Sebastián, Spain* (Chap. 3.25)

Wayne M. Penn *Alabama Laser, Munford, Alabama* (Chap. 1.9)

Frank Pfefferkorn *University of Wisconsin, Madison, Wisconsin* (Chap. 1.3)

Don Risko *DGR Consulting, Jamestown, Pennsylvania* (Chaps. 3.8, 3.9)

Jan Romberg *BIOTRONIK SE & Co. KG, Berlin, Germany* (Chap. 2.9)

Jim Russell *General Lasertronics Corporation, San Jose, California* (Chap. 1.4)

Ronald D. Schaeffer *PhotoMachining, Inc., Pelham, New Hampshire* (Chap. 3.6)

Rick Schiffer *Apollo Seiko Ltd., Bridgman, Michigan* (Chap. 1.8)

Carolyn Conner Seepersad *University of Texas, Austin, Texas* (Chaps. 2.1, 2.6)

Mitsushige Seto *Apollo Seiko Ltd., Japan* (Chap. 1.8)

Geoff Shannon *Amada Miyachi America, Monrovia, California* (Chap. 3.37)

Apurbba Kumar Sharma *Indian Institute of Technology, Roorkee, India* (Chap. 3.10)

Jeffry J. Sniegowski *Sandia National Laboratories, Albuquerque, New Mexico* (Chap. 3.14)

Colin Weightman *Comco Inc., Burbank, California* (Chap. 3.42)

Robin C. Whitmore *Medical Micro Machining, Inc., Colfax, Washington* (Chap. 3.5)

Christopher B. Williams *Virginia Tech, Blacksburg, Virginia* (Chap. 2.5)

Paul Witherell *National Institute of Standards and Technology, Gaithersburg, Maryland* (Chap. 2.7)

Hitomi Yamaguchi *University of Florida, Gainesville, Florida* (Chap. 3.19)

Preface

This handbook focuses on what designers need to know about the *advanced manufacturing processes* in use today. By "advanced" we mean both new and distinctively different. In some instances technology has progressed to the point that much closer tolerances can be held, a whole new population of sizes is now possible, or a totally new family of materials now has opened up new possibilities. The technology is moving so quickly in some manufacturing fields that while yesterday it was in the development phase this year it is actually producing production parts. A few of these processes appear so "gee whiz" that readers will wonder if they are real. But they are. The world of nano, micro, lasers, and additive processes is vibrantly successful today, but we will still see tremendous growth in these processes in the next few years.

There have been major process advances in three critical arenas—laser processing, micro manufacturing, and additive manufacturing. These cover almost every material used in industry, and their use or the anticipated use of them requires good insight into their capabilities and their limits. The chapters in this book are designed to be a straightforward, crisp read that provides enough insight that designers can readily tell whether that technology is applicable to their designs and how to best design to that technology. The chapters are not a history, not a theory, not a potential; they are based on actual physical capabilities today.

The intent of this work is to provide product designers with enough knowledge of specific processes and materials to intelligently design products and to work with those who will probably use those processes. This handbook does not detail all that a manufacturer needs to know; it presents only what a designer needs. At the same time, this is an instructive book for those new to advanced manufacturing. As such, it is useful for understanding potential new processes, and for expanding current knowledge no matter what the job title or function.

Beyond "advanced," we limit the processes in this book to those that mechanical and electrical manufacturers may expect to use. Chemical advances, cloth and fiber manufacture, and semiconductor manufacturing are in general excluded here, but are discussed in other McGraw-Hill handbooks. We approach nano manufacturing only from the larger micro world that in some cases reaches down to those minute dimensions. As a result, the manufacture of nanotubes is not included here, but processes that use nano items to make micro features are included. Similarly, the world of bio product manufacturing, while additive, micro, and nano, is not included in this handbook. Surprisingly, however, that field also includes mechanical and electrical processing, but in a biological environment. The additive processes described in this handbook are already being used by doctors, radiologists, researchers, and pharmacy representatives as standard processes for model making.

A typical chapter includes: (1) a brief overview of *how the process works*, illustrated with one or two graphics (flow diagrams, charts, images); (2) *product characteristics* and applications, such as shapes, surface finishes, tolerances, and known limitations; (3) typical production *quantities*, cycle times, or aspects that influence cost, including advantages of the process; (4) *materials* suitable to this process; (5) *specific design recommendations*, including shapes, materials, tolerances, and related aspects for specific shapes or applications—in short, any details a designer should know.

Enough information is provided for most of the processes that designers can design and expect that their design can be made with the process. There are, however, some processes that are so complex that, even though a product designer will define the need in physical terms, he or she will still need to turn the final design over to specialists, because only the producing company and their specialized computer programs can provide the fine detail needed to successfully define the product in a manner that allows it to perform as needed.

An example of this is MEMS technology. It employs all the processes used for integrated chips (ICs), which involves a proprietary material properties library at the atomic level, micro structure level, and micro properties level. These are interwoven with manufacturing knowledge into an integrated design/build package that is not typically available to general product designers, but is exclusive to IC and MEMS designers. Other processes having similar unique design constraints include dip-pen and polymer-pen nanolithography and capillary forming of structures. Thankfully there are only a few of these highly unique processes.

As editor, I am indebted to each of our contributors for his or her work, and to many others. I have spent thousands of hours identifying pertinent technology, finding leading authorities in their fields, and working with them to develop this handbook. They, in turn, have also spent thousands of hours in their fields to bring these pages to you the reader in both a convenient and an authoritative manner.

I encourage anyone who finds errors or omissions in this work to contact me. Similarly, as new processes come online I would love to hear about them; I will find a way to include them in a future edition of this handbook. Last, editing any publication is a work of love that expands the editor's knowledge immensely; I sincerely hope that the resulting book greatly expands each reader's knowledge, too.

LaRoux K. Gillespie

Introduction

This is a book for designers of products, designers who want to get the design right from the start in a manner that can be manufactured economically. It is a handbook, a reference for advanced processes—the processes and technologies recently developed for commercial use. It covers only what can be reasonably described as advanced 21st-century processes. A related handbook with a wealth of useful information on the processes in use for decades and centuries already exists (James G. Bralla, *Design for Manufacturability Handbook*, 2nd ed., McGraw-Hill, New York, 1999).

The technologies, processes, and materials are advancing so quickly that practiced designers and those just entering the field cannot be expected to keep up with all the design issues in play. This handbook has been carefully crafted to give designers and those who work with them the insight to stay clear of the majority of design issues that slow progress.

The many chapters on individual topics present size, shape, and surface issues—illustrating both poor design and the better approach. Dimensional tolerances, edge configurations, and surface finish capabilities that are typical or possible are defined. Appropriate materials, and the reasons they are good choices, are highlighted. Typical applications for the manufacturing process are given as well as special design features that can be produced.

This handbook is arranged to provide the maximum useful information to the designer in the most straightforward way with a minimum of extraneous material. It presents information in a way that is immediately usable. Design tips are presented in list format as well as many "do, don't," "poor, better," "this, not this" illustrations. This book is both a primer and a reference that will be used for years.

Because so many of these technologies are relatively new, the companies that employ them are often resistant to revealing all the design clues that a designer would like to have. In part, the processes have come into being so fast that companies have scurried to just keep up and have not themselves written down what they know. For competitive reasons, they are willing to share with direct customers who come to them with firm projects, but not with the general public for concern that competitors will quickly use that knowledge and the originating company will lose its edge. That edge can be very significant.

This handbook is the result of the work of hundreds of people and review of literally over a thousand sources of information. It is a compilation for designers like no other.

In brief, this is a reference dedicated to designers of all levels of experience, but it is also useful as a training tool for new designers. At the same time, it can be a cost-saving tool for individuals who are not designers by function, but who still need to find more effective manufacturing solutions. These individuals have such titles as:

Cost analyst

Value engineer

Lean manufacturing specialist

Tooling engineer

Manufacturing engineer

Manufacturing team manager

Research and development engineer

Educator

Consultant

Quality assurance engineer

Industrial engineer

Manufacturing team leader

The designer needs to know key elements of the manufacturing process, but he or she does not need to know all the nuances and details of the process. Manufacturing engineers and shop personnel will take care of those issues. The emphasis here is on the product and the designer, but the book also raises issues that the manufacturing engineer and shop face when they manufacture the product.

This handbook on advanced manufacturing processes is a companion to the above-mentioned handbook by Bralla, which goes into great detail on processes common in the 1990s. That 1300-page book covers basic design for manufacturability principles that are common to all products regardless of size, tolerance, or material, as well as ideas for all processes. We do not repeat the material in that foundational book, but do want to call attention to it for the many designers who must design for both advanced processes and historical manufacturing processes. Together the two handbooks form a unique knowledge base for all persons involved in manufacturing.

PART 1

Manufacturing with Lasers

CHAPTER 1.1

Overview of Laser Manufacturing Processes

LaRoux K. Gillespie

Manufacturing Consultant
Andover, Kansas

While using lasers in manufacturing is not new, the advances made in recent years truly speak of new technology. Femtosecond lasers have been able to cleave stainless steel grains in half as sharply and cleanly as a knife through butter, and their economics and new applications have made them recently commercially viable for many applications (Fig. 1.1.1). Today lasers can form both thin and thick parts, bending them without physically any tool touching the part. Lasers can correct part shapes and straightness, and coat parts with a variety of chemicals and coatings. Lasers can peen surfaces without touching them, smooth surfaces, add desirable residual stresses, and provide unusual textures that repel liquids. They are used to solder, weld, and cut with unprecedented accuracy and minuteness as well as with exceptional speed. Today lasers can drill minute holes at a rate of 5000 holes per second, and such applications are in common use in aerospace engines and automotive fuel systems. Lasers can make round holes or nozzles of various shapes to direct flows economically. They can make square holes, produce flat bottom cavities, slice brittle materials, and cut grooves or channels in a variety of materials with consistent accuracy and form. In some emerging applications, lasers can even move minute items, cause fluorescence, clean and decorate surfaces, and mark parts with extensive almost unseen characters, symbols, and information blocks.

The introduction of several new laser types in the past two decades, together with new materials, truly extensive research in laser design, operation, and processes, and innovative applications to production, has been the source of the new developments in laser-utilizing processes. While a single advance opened doors, it was the combination of all these issues that made today's production advances possible. A scan of today's

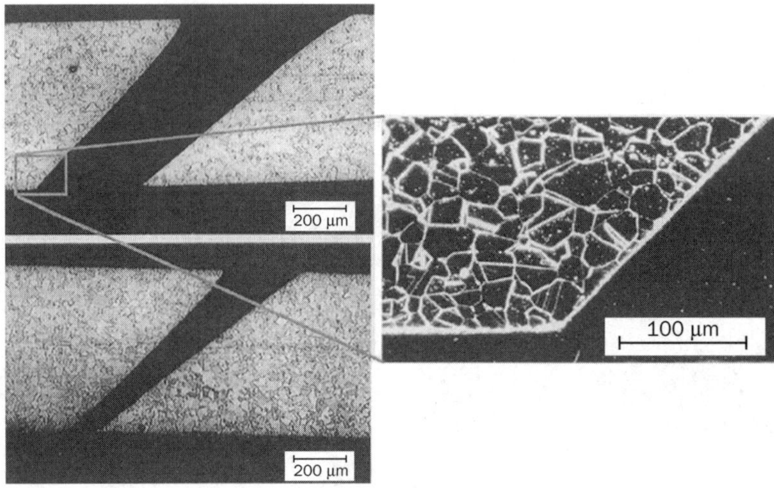

FIGURE 1.1.1 Femtosecond laser cuts through stainless steel. The grains are cleaved rather than cut or melted.

research journals involving lasers in general as well as processes takes weeks, months, and even years for those wanting to apprise themselves of all the new advances.

This Part 1 of the handbook describes the advanced processes utilizing lasers. The traditional or more foundational processes are well covered already in a variety of books and other formats. While a number of additional applications and process variations are under intense research and development, the chapters in this part cover only those that have reached industrial maturity and processes that involve historically typical part dimensions. Micro applications are presented in Part 3 and uses in additive processes are described in Part 2 rather than here.

The impact of lasers is perhaps best illustrated by the fact that they are described heavily in the three parts of this handbook, which describes what designers need to know about all major advanced processes. Lasers are everywhere, and in many cases the laser is the underlying aspect that makes many of these processes truly advanced.

In addition to the topics presented in the following chapters, the latest variation of new processes involves hybrid processes, processes such as laser-assisted turning, laser-assisted milling, laser-assisted grinding, laser-assisted EDM, laser-assisted ECM, ultrasonic-assisted laser machining, and laser-assisted chemical etching. Laser-assisted prepreg tape winding combines winding equipment with focused diode laser heat to tack it down, preventing wrinkles and gaps. Laser-assisted friction stir welding, laser-assisted purification (of EB-deposited platinum), and laser-assisted assembly are yet other examples of these hybrid processes. Each of these, except for assembly and purification, is commercially in use, and the laser either softens the material while being machined, accelerates the reaction, or removes something from a surface that allows better machining. In these cases, the laser is not the primary process; it assists the process. These laser-assisted processes are not included here since the basic process is already well defined, but in these and in all other cases the designer needs to clearly identify any restraints on the material characteristics in the finished product—not just the starting material. "Restraints" as used here means metallurgy, appearance, conductivity, water repellence, corrosion resistance, since all of these can and will be affected to some extent by the laser processes. They can be improved or they can be denigrated.

Laser forming is one of the least advanced of the laser processes, and requires an in-depth understanding of metallurgy by its users, but it has its own niche in which it provides unique benefits. Readers will find that there are three subcategories of laser forming, and that will often be true of the other processes as well.

In addition to laser advances, several machine tool manufacturers have combined lasers with the conventional processing equipment so that either or both may be used at one time. For the machining processes as noted elsewhere, "[the] laser provides intense localized heating to the workpiece ahead of the cutting region. By lowering the material strength in the cutting area at a certain elevated temperature, LAM [Laser Assisted Machining] can achieve lower cutting force, slower tool wear progression rate, higher material removal rate and better surface quality."[1] By this approach it is possible to machine very hard metals, metal matrix composites, and even ceramics such as silicon nitride that are not possible by methods other than grinding. Milling with micro tools [say those under 1 mm (0.040 in.)] is extremely difficult because of tool breakage. But in materials including aluminum, steel, stainless steel, Inconel 718, and Ti6Al4V, laser-assisted micro milling can prevent breakage and provide higher quality surfaces, as well as fewer burr formations.

Reference

1. "Laser Assisted Machining," Laser Materials Processing Lab (University of Iowa), webpage research.engineering.uiowa.edu/ding/LAM, 2015.

CHAPTER 1.2

Laser Cutting

LaRoux K. Gillespie

Manufacturing Consultant
Andover, Kansas

While lasers have been used for cutting metal for decades, today's capabilities far exceed the abilities defined in past records. The beam can be focused as finely as 25 μm (0.001 in.) and the cut that is made (kerf width) can be controlled for metals as narrow as 100 μm (0.004 in.). For most cutting the beam does not have to be so finely focused and for many applications it is less than 0.32 mm (0.0125 in.) in diameter.

Laser cutting provides very rapid design to product abilities with CNC programming and that translates also to immediate production with design changes. It simply results in new parts within minutes and to high production as well as one off production. This provides a relatively high production rate and an unattended process that accommodates many parts from the same sheet, as well as many different parts from the same sheet.

For very thick metals, plasma cutting is generally preferred for its lower cost and thicker cutting capabilities.

A water jet is used by some machines to guide the laser beam. This approach removes debris and cools the material while guiding the beam. It also provides high dicing speeds, parallel kerf, and omnidirectional cutting for dicing applications. It can cut with 1-μm (39-μin.) accuracy, but the kerf will be as wide as the water jet diameter. That diameter can be as narrow as 50 μm (0.002 in.), but more often is 80 μm (0.003 in.) or larger. Wafers thicker than 1 mm (0.040 in.) are typically better cut via conventional mechanical dicing processes. The advantage of this variation is that localized heating is minimized, and no molten material is ejected so that the problem of redeposition of material on the wafer is prevented. In addition, because it is a cool process local heating of the bulk material is also minimized.

Five different phenomena are used in cutting, depending on application.

- In *vaporization cutting*, the heat creates vapor that erodes the molten walls and blows out or ejects debris, thus enlarging the hole. This approach is often used on materials that do not melt (wood, carbon, and thermoset plastics). More energy is needed to vaporize metal than to melt it. Therefore, sublimation cutting requires high laser power and is slower than other cutting processes.

However, it produces high-quality cuts. It is used for delicate work like stent cutting. Plastic sheeting and textiles vaporize with the application of only a small amount of energy.

- *Melt and blow*, also known as fusion cutting, relies on high-pressure gas to blow molten material from the cutting area. The material is heated until it melts and then a gas jet blows the molten material out of the cut. Metals are normally cut in this approach. It is the most widely used process variation.
- *Thermal stress cracking* is used on brittle materials to create a crack that is then directed by the laser beam to make a complete fracture in a relatively straight line and smooth surface. Glass can be cut at rates of several meters per second.
- *Stealth dicing* is used to separate micro electronic chips on silicon wafers. This process may also be called *laser wafer dicing*.
- *Reactive cutting*, burning stabilized laser gas cutting, or flame cutting is used to cut very thick steel plates in a manner similar to oxygen torch cutting.

Materials

Almost all metals can be laser cut. Mild steel, stainless steel, and aluminum are the most common applications, but copper is another application, which is more challenging. Lasers cut wood, plastics, glass, and ceramics as well, although different lasers and different cutting modes may be used for them. Gold, silver, and other precious metals can also be cut as well as titanium and its alloys [up to 2 mm (0.080 in.) in thickness].

CO_2 lasers cut diamonds, silicon wafers, Plexiglas, and other thermoplastic and thermosets. Paper is cut at a speed of 2000 mm/s (79 in./s) and wood at speeds of 1.5 to 4 m/min (59 to 157 in./min). Ceramics are often cut at single-pass depths of 10 μm (>0.001 in.). Thicker ceramics are cut in multiple passes. Textiles are cut at speeds of up to 5 m/min (200 in./min). An advantage of using lasers for cutting textiles is that the thermal cutting does not cause fiber frazzle as mechanical cutting does. It "cauterizes" the cut edges. Lasers can also cut holes of any size in fabrics for artistic and performance purposes. Foils, labels, and self-adhesive materials are all cut with CO_2 lasers.

Tables 1.2.1 and 1.2.2 illustrate the abilities of laser cutting for various common materials.

Depending on the laser frequency, speed of cut, use of gas jets to eject material, and material processed, the bottom edges of parts may have some amount of dross (molten resolidified metal) on them. Some splatter also may exist on surfaces, requiring sanding or other processes to remove.

Equipment Capabilities

Fiber laser systems can cut 25.4-mm-thick (1-in.) mild steel, and some can handle 3 × 1.5 m (120 × 60 in.) sheets. More common upper limits of sheet thickness for powerful fiber lasers are approximately 22 mm (0.87 in.) for mild steel, 18 mm (0.7 in.) for stainless steel, 15 mm (0.6 in.) for aluminum, and 7.6 mm (0.3 in.) for copper and brass. Less powerful ones (less than 4000 W, for example) may be able to cut only half these thicknesses. Fiber lasers can cut at up to almost 127 mm/min (5 in./min) depending on sheet thickness and material being cut.

Material	Cut	Drill	Comments
Stainless steel	G–E	G–E	
Carbon steel	G	G	
Free machining steel	G	G	
Titanium	G–E	G–E	Inert atmosphere required
Copper	F–G	F–G	High reflectivity causes uneven results
Zirconium	G	G	
Brass	P–F	P–F	Zing depletion
Nickel	G	G	
Inconel 625	G–E	G–E	
Gold	G	G	
Silver	G	G	
Aluminum	G–E	G–E	Depending on thickness
Ceramic	F	F	
Plastic	P–F	P–F	Drilling poor on thermoplastics, cutting poor on most

P, poor; F, fair; G, good; E, excellent.

Table 1.2.1 Laser Processing of Materials with Solid-State Lasers (Courtesy Ebtec, www.ebteccorp.com)

A commercial 8-kW solid-state laser cutting machine can cut 1-mm-thick (0.039-in.) steel at a speed of 51 m/min (2000 in./min). Solid-state lasers can also cut 25-mm (1-in.) mild steel at speeds greater than 889 mm/min (35 in./min).

Laser cut equipment has positioning accuracies of up to ±5 μm (±0.0002 in.) and a repeatability of ±5 μm (±0.0002 in.). Most machines have capabilities somewhat less than that.

Surface finishes produced in laser cutting may range from 3.2 to 6.4 μm (125 to 250 μin.) Ra. Roughness increases as sheet thickness increases, and going from a 1- to a 6-mm-thick (0.039- to 0.236-in.) cut on the same material may result in the finish increasing by a factor of 2 to 3.

A low-power (30- to 120-W) desktop CO_2 laser is available to cut wood, acrylic, plastic, Delrin, cloth, leather, melamine, paper, rubber, veneer, cork, and other materials. That same machine is used to engrave wood, acrylic, plastic, glass, leather, Corian, fabric, coated metals, anodized aluminum, stone, marble, ceramics, Mylar, pressboard, and similar materials.

Removing rust, paint, or lacquer is a requirement for making fast, high-quality laser cuts, but not for less demanding cuts. Some machines are designed to blast the surface

Material	Cut	Drill	Comments
Stainless steel	G–E	G	
300 series	F–G	F–G	
400 series	P–G	G	
Galvanized steel	G	G	Zinc depletion during weld process
Carbon steel	E	E	
Free machining steel	E	E	Presence of sulfur can cause porosity
Coated steels	F–G	F–G	Coating degradation t cut edge
Spring steel	E	E	
Titanium	G–E	G–E	
Zircalloy	G–E	G–E	
Hastelloy X	G	G	
Copper	P	P	
Zirconium	G	G	
Brass	P–F	P–F	
Aluminum	P	P	Reflectivity and thermal conductivity restrain process
Ceramic	G–E	G–E	
Plastic	G–E	G–E	
Polyethylene	G	G	
Polycarbonate	P	P	
Acrylic	G	G	
Delrin	G	G	
Nylon	E	E	
Wood	E	E	
Plywood	E	G	
Fiberglass	F–G	F–G	
Laminates	F–G	F–G	Depends on epoxies used
Glass	G–E	G–E	Depends on coefficient of thermal expansion
Rubber	F–G	F–G	Best result in sulfur or green state
Leather	E	E	
Paper	G–E	G–E	Control of by-products necessary in high-speed processing

P, poor; F, fair; G, good; E, excellent.

Table 1.2.2 Laser Processing of Materials with CO_2 Lasers (Courtesy Ebtec, www.ebteccorp.com)

FIGURE 1.2.1 Six-axis laser cutting system's cut features at any position on formed parts. (Courtesy Mazak Optonics Corp.)

with a short laser pulse that vaporizes the corrosion. That is followed with an automatic switch to laser cutting conditions. While most laser cutting is done on sheet materials, lasers also cut, pierce, and drill holes in metal tubing, and can cut through a variety of 3D shapes. At the micro level, as described in Chap. 3.6, lasers are used to machine medical stents from tubing into mesh-like products that are inserted into human bloodstreams to reinforce the blood vessel walls. Cutting in three dimensions requires movable laser heads, or rotating chucks or rotary tables, all of which are readily available as add-ons or as OEM equipment.

Laser cutting systems can be provided with 6 axes of motion (Fig. 1.2.1), which allows almost any 3D configuration to be cut. This allows parts to be cut after forming rather than requiring that to be done in flat sheets.

Design Considerations

Sandwich-like composites that consist of several different materials are not a good choice for lasers since the laser frequency is tuned to a frequency for a specific type of material. Each material has a different reaction to the beam, and the beam can accommodate only one material at an optimum level. As a result some materials experience poor cutting conditions that may be unacceptable. When those layers are deep in a narrow cut, the effect may not be readily noticed.

The laser can only cut areas it can directly "see." Undercuts must be made with other processes. Tapers can be accommodated in any direction with multi-axis machines, but clear undercuts cannot.

Some metal alloys may produce toxic gases and some plastics may also produce harmful gases. All manufacturers can provide the necessary health and safety controls, but it is an issue that designers need to be aware of. Gallium arsenide, indium phosphide, and beryllium copper are three materials that some commercial houses will not laser cut because of the toxic gases or health issues.

Laser cuts produce a heat-affected zone (HAZ) along the edge of the cut. That zone will exist for 25 to 102 μm (0.001 to 0.004 in.) from the edge into the parent metal. For most parts that is not an issue. In certain applications it is. For critical parts, designers need to assess the risk that the metal micro fissures might exist and that the metallurgy in the HAZ is different from the parent metal. Many aerospace and medical applications have regulations limiting that HAZ. Extensive literature (hundreds of studies) exists on HAZ issues.

Laser cutters have to decelerate when they cut sharp 90-degree corners. That increases cut time and can reduce part quality in that area. That slowdown can overburn the corners, causing dross (or more dross) and it can burn corners away totally. Provide a large radius whenever possible to avoid these issues.

Additional design considerations include:

- Laser-drilled holes will be tapered approximately 1% of the drill depth, unless they are trepanned with 4- or 5-axis machines. Taper is particularly visible in material 1.27 mm (0.50 in.) and thicker.
- Lasers cannot drill a blind hole to a precise depth.
- Adherent metal must be removed from the exit side of the hole.
- Large-diameter holes can be trepanned or helical drilled rather than single shot or pulsed through cutting.
- If the part is subsequently to be bent, designers need to provide bend relief. Dross can get in the way of effective bends when relief is too small.
- Lasers can machine before a part is bent or after, although it will be faster doing it as sheet and that allows less expensive equipment to do it.

Chapter 3.6 provides details more attuned to micro cutting capabilities.

CHAPTER 1.3

Laser Surface Texturing

Frank Pfefferkorn and Justin Morrow

University of Wisconsin
Madison, Wisconsin

Laser surface texturing is a broad umbrella term for any process using a laser beam to change surface geometry by moving or removing surface material. This can enable the designer to make a rough surface smoother, a smooth surface rougher, selectively destroy unwanted features, or add desirable features by removing material (e.g., removing material to "add" a dimple feature). Surface texturing is considered to be a finishing operation and is used to tailor the surface after the main manufacturing processes (e.g., machining, grinding) have created the final part geometry. This is an important distinction as laser-based processes can also be used for both additive and subtractive primary manufacturing processes such as directed energy deposition (see Chap. 2.3), laser cutting (see Chap. 1.2), and laser micro machining (see Chap. 3.6). Only surface modification uses are discussed here.

The first distinguishing feature in laser-based process is selectivity: It is easy to add or remove features in one area of a surface and not another. The other distinguishing feature is that it is an energy-based process. This means that the surface will very briefly have much more thermal energy (higher temperature) than the interior and this can lead to both desirable and undesirable results. If there is a reason why thermal energy should not be introduced to the surface, then other methods such as mechanical (e.g., micro mechanical milling) or chemical (e.g., chemical etching, electrochemical polishing) removal methods should be considered.

The common thread among all of the methods discussed here is that energy in the form of intense coherent light (i.e., a laser) hits a surface and adds energy to change the surface geometry. The processes are distinguished by where, how, and when this energy is deposited during processing and what response is elicited from the surface. These characteristics of the energy transfer and material reaction are both a function of the laser *and* the surface material properties (optical and thermal). Therefore, it is important that a designer starting with a surface material and a desired final texture identify what energy transfer method can best create these features *in the given material* and what laser type is capable of executing this mode of operation.

The goal of this chapter is to help the designer navigate these decisions to two ends: First, to determine if one of the processes falling under "laser surface texturing" is a good candidate for a given application and second, to provide some general guidance in "design for manufacturing" that can help a designer reflect on whether any changes should be made in the design to make laser-based texturing simpler or more feasible.

Process Physics

Figure 1.3.1 shows a qualitative view of the physical mechanisms involved with laser surface texturing. On the far left, laser heating uses the least intense laser spots with largest surface area coverage and does *not* result in any surface changes, but is often used for laser hardening (see Chap. 1.5). Using more highly focused laser spots and higher power to increase the laser *intensity*, the laser power divided by the heated surface area leads to laser surface melting. This allows the surface topography to change inside the laser spot, which can be used for both surface smoothing and texturing. If the laser intensity is further increased, the surface begins to vaporize (ablate) leaving behind a crater or dimple. This process is typically used for texturing of smooth surfaces, but can also be used for selectively removing unwanted features on an otherwise smooth surface. The last process, laser-induced plasma machining, requires extremely high laser intensity and operates on a different physical principle. Instead of absorbing laser energy directly into the surface to cause heating, the laser beam is focused just above the surface so tightly that a plasma briefly forms and removes surface material before dissipating. The crater size and shape is directly related to the shape of the plasma created. This process typically requires immersing the surface to be machined in a fluid (liquid or specialty gases) that aids in plasma formation, so the reactivity of the surface material to this fluid (typically water) should be considered. The other processes *can* be done without any external gas, but typically use a shielding inert gas, such as argon, to prevent unwanted surface reactions with air (oxidation).

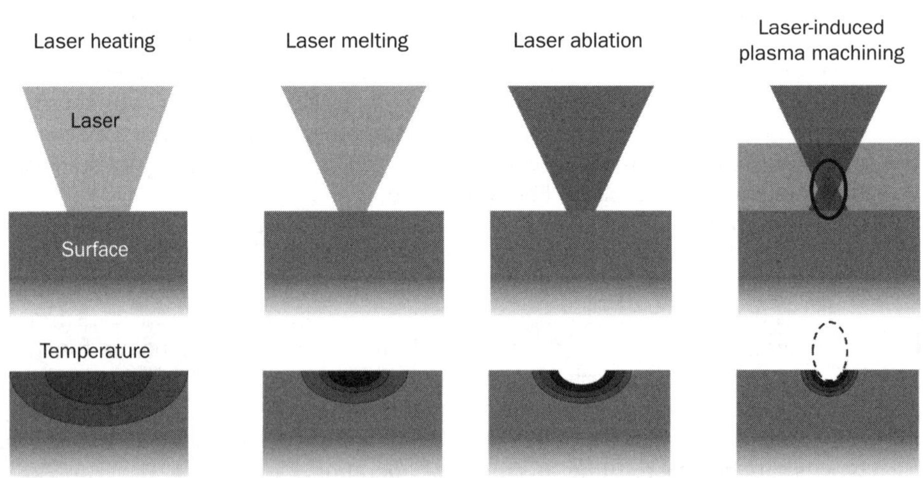

FIGURE 1.3.1 A basic physical explanation of laser heating, laser melting, laser ablation, and laser-induced plasma machining. From left to right in the figure the laser intensity is increased, resulting in higher peak temperatures and more localized heating.

Why Use Laser Texturing?

Texturing allows the tailoring of the surface to exhibit certain properties that are desirable to the designer. Specifically both adding (structuring) and subtracting (smoothing) surface roughness features can be beneficial in several ways:

Smoothing

- Improved reflectivity (functional and aesthetic)
- Reduced surface contact stress concentrations
- Improved fatigue performance on structural parts
- Improved surface mechanical properties (similar to solid-state laser hardening, Chap. 1.5)

Structuring

- Reduced or altered reflectivity (functional and aesthetic)
- Reduced surface contact area
- Altered friction and wear properties (e.g., lubrication pockets)
- Altered surface wetting (e.g., hydrophobicity)
- Altered microbial interaction

Several of these properties such as friction, wetting, and microbial interactions can potentially be made both better or worse through the choice of texture feature geometry, and significant academic literature is available on these topics. Several other methods of surface texturing are available, such as micro milling (see Chap. 3.3), micro stamping (see Chap. 3.24), chemical mechanical polishing (see Chap. 3.14), etc. The distinguishing features of laser texturing are that it is:

- Noncontact
- Non-consumable (no chemicals)
- Selective
- Flexible (multiple laser types and pulsing modes)

This makes laser texturing potentially attractive from a technical perspective for difficult problems such as polishing or structuring the surface of micro pins without modifying the base material and potentially from an economic perspective by eliminating the need for recurring mechanical media purchases. Some samples of laser-structured surfaces are shown in Fig. 1.3.2. Additional information can be found in the large body of research work currently being done in both surface smoothing [1–3] and surface structuring [4].

Design Considerations

Material Suitability

Laser texturing is very dependent on the properties of the surface material. Therefore, the designer should consider all suitable materials for the potential application as some may be more suitable for laser texturing than others. Metals are generally the most

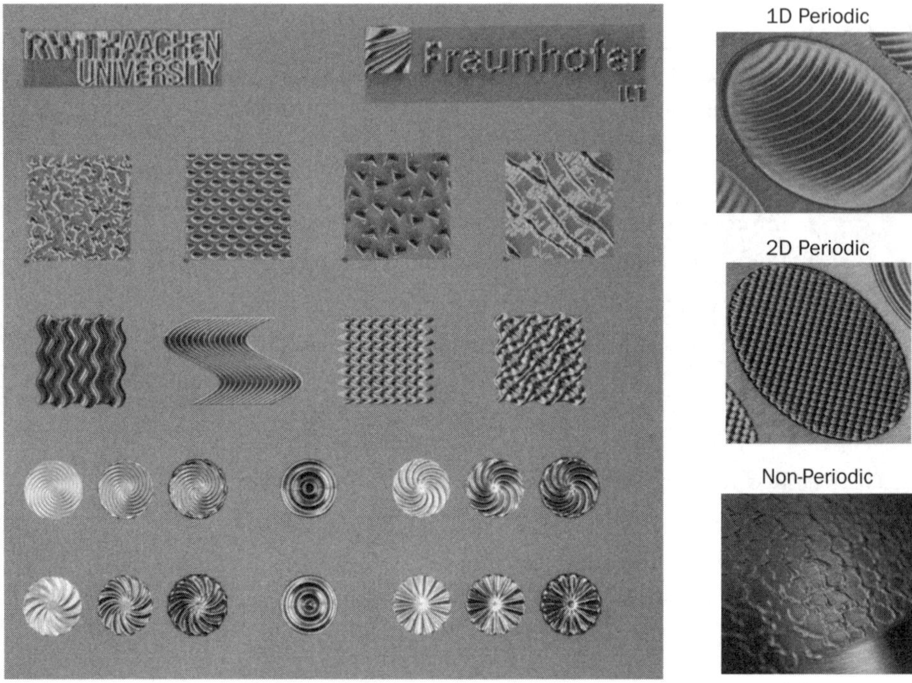

Figure 1.3.2 Examples of laser-structured surfaces created with continuous-wave laser melting on H13 steel showing the versatility of the process. Areas can be circular, square, or freeform, and the features can be periodic in one or two directions or non-periodic. (Images courtesy of Fraunhofer ILT and RWTH-Aachen.)

suitable for laser texturing, but some plastics and ceramics can also be successfully laser melted and ablated. Some issues that can arise in different types of materials are:

Metals

- Low absorptivity at certain wavelengths (reflective losses)
- High thermal conductivity (making melting difficult)

Polymers

- Prone to thermal distortions
- May discolor
- Thermoset polymers tend to burn instead of melt
- Low absorptivity at certain wavelengths (reflective and transmissive losses)
- Low boiling point (easy to overshoot melting)

Ceramics

- Low absorptivity at certain wavelengths (reflective and transmissive losses)
- Prone to thermal cracking
- High melting point

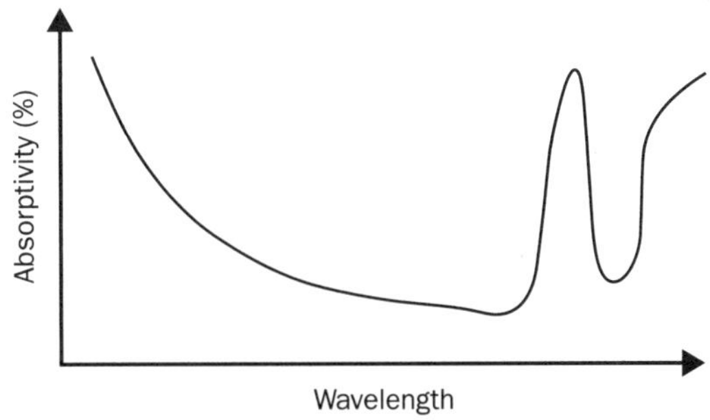

FIGURE 1.3.3 Figurative absorption spectrum.

The easiest recurring issue to solve is choosing the right wavelength for processing. A figurative example of an *absorptivity spectrum* is given in Fig. 1.3.3. This shows how much energy will be absorbed at a given energy wavelength and can be used to help choose which laser type will be most effective for processing a given material. In this example, the highest absorptivity would be achieved at either very low wavelength, very high wavelength, or by matching the absorptivity peak. Other issues such as thermal distortion or thermal cracking can be tougher to eliminate, but can sometimes be dealt with by using secondary processes such as workpiece preheating.

Laser versus Electron Beam

The heating, melting, and ablation mechanisms shown in Fig. 1.3.1 can also be achieved using an electron beam instead of a laser beam. This can be thought of as replacing an intense stream of photons with an intense stream of electrons. Significantly more research has been done in laser-based surface modification techniques to date, but both laser polishing and electron beam polishing machines are currently commercially available. Electron beam polishing typically takes place in a vacuum since electrons interact strongly with gas whereas lasers will not. Functionally, this is not a large distinction as laser beam polishing is usually done in a controlled atmosphere as well to prevent surface oxidation of metals.

Both lasers and electron beams can be used for metallic and nonmetallic surface modification and both laser and electron beam-based systems are commercially available for polishing metallic surfaces. However, these systems use different modes of operation that lend them to different applications:

Laser-based systems

- Use a small circular spot of energy [30-μm to 5-mm (0.0012- to 0.2-in.) diameter]
- Quickly move the focused spot across the surface at high speed [100 to 1000 mm/s (3.9 to 39 in./s)] without revisiting any positions
- Use tight line overlaps [10 to 100 μm (0.0004 to 0.0040 in.)]
- Achieve roughly 1 to 10 mm^2/s (0.0016 to 0.016 $in.^2/s$) area processing speed
- Require inert gas atmosphere

Electron beam systems

- Use a large irradiation spot [60 mm (2.36 in.) diameter]
- Irradiate same area multiple times (~30 times)
- Use minimum spot overlap for full coverage
- Achieve roughly 3.6 to 180 mm^2/s (0.006 to 0.28 in.^{2}/s) area processing speed
- Require low-pressure atmosphere (no fill gas)

Based on these capabilities, electron beam polishing currently has a higher typical processing speed, but is not marketed for selective polishing due to the large irradiated area. Laser-based systems are also capable of being more flexible for both surface smoothing (eliminating surface features) and surface structuring (creating surface features). The limiting factor in laser polishing speed is mainly the tight overlap and in electron polishing is the number of times a given area is irradiated.

Continuous versus Pulsed Operation

Laser systems can be operated in continuous output and pulsed output modes and pulse modes can vary widely in the pulse duration, intensity, and pulsing rate. Different modes of operation can often be used to achieve the same result, so the most important factor for the designer to stipulate is the desired final outcome to determine how much flexibility exists in the texturing process design. Some examples of questions that can help a designer determine whether laser texturing is suitable (after having identified an appropriate laser-material pairing) are:

1. How sensitive is the bulk material to heat? (e.g., deformation, cracking, phase transition)
2. Are the desired surface features smooth or sharp?
3. How small are the desired features?

These questions can help identify what laser mode should be used. *Pulsed* operation typically:

- Minimizes heat input to the bulk
- Can create high aspect ratio, sharp-edged, and micro scale features
- Opens up many options for pulse shape, frequency, and intensity
- Is used in both smoothing and roughening

Continuous operation is:

- Simpler: defined by laser power, spot size, movement speed, and movement path
- Typically used for hardening and polishing operations
- Less selective than pulsed operation

Surface Topography

If laser texturing is being done on a flat surface, then little thought needs to be given to the effect of the surface topography on the process, other than fitting the part inside the laser x-y scanning range. However, since laser processing typically uses a focused spot, any height variations on the surface to be textured will change the laser spot size at the surface and can have a large impact on the process. For example, height steps can cause the laser spot to suddenly go from in-focus to out-of-focus and move from a melting regime into a heating regime or from ablation to melting. Sloped surfaces will cause the laser spot to change from circular to elliptical, reducing the power input and potentially changing the texture geometry. The designer should therefore keep in mind that flat geometries are easier to laser process than complex surface geometries, but both electron beam and laser polishing are regularly done on curved and sloped surfaces.

Polishing Limitations

Extensive research has shown that the surface roughness is usually reduced by 50% to 80% when using laser melting to do surface polishing, depending on the starting roughness and material. A few limiting factors govern how much laser polishing can improve a surface:

1. *Low-frequency roughness:* Roughness features must be captured inside the laser spot to allow polishing. Therefore, long-wavelength roughness features are often left-over when using laser polishing. The beam spot can be increased to capture these roughness features, but this also introduces the potential for large unwanted material flows during the laser melting process and changing part geometry. Developing strategies for eliminating these roughness features is a hot area of current laser polishing research.
2. *Deep scratches:* Particularly deep scratches or gouges can also be particularly difficult to polish out since the melted layer must be deeper than the surface roughness for effective polishing. If possible, deep gouges or scratches [tens to hundreds of micrometers (thousandths to hundredths of an inch) deep] should be avoided.
3. *Edge effects:* The majority of a laser polished area often appears quite homogeneous, but secondary effects such as material buildup and overheating can sometimes occur at the edges. If this effect is unacceptable in the final application, then strategies such as adaptive power control can "fade-out" the polishing near the edges.

Structuring Limitations

Both laser melting and ablation can be used for creating surface grooves, dimples, and complex geometries. A few limitations should be kept in mind when considering either laser surface structuring route:

1. *Laser spot size:* The smallest unit in laser texturing is the size of the focused laser spot and this is physically limited based on the focusing lens, laser wavelength, and beam quality. As a rule of thumb, it is fairly straightforward to get a focused spot in the tens to hundreds of micrometers (thousandths to hundredths of an inch) in diameter, but difficult to get to the nanometer scale.
2. *Maximum laser output:* It is fairly easy to get even poorly absorbing or highly thermally conductive materials to melt, vaporize, or burn under a laser, given enough time. However, in laser structuring it is often desirable to functionalize the surface geometry while minimizing thermal impact on the bulk or to

vaporize without significant melting. This is highly dependent on the material properties and the laser pulse energy and may require a thoughtful material-laser pairing in consultation with manufacturing engineers.

Cost

Laser texturing has a relatively high up-front capital cost, ranging from a few thousand dollars for low-power lasers (<1 kW) to over a million dollars for high-power integrated laser systems with computer control and maintenance support service. Laser expertise is also often needed to choose the proper operation mode for a given application. Therefore, it may be advisable to begin by contracting with laser experts in the short term to pilot a laser-texturing process, then decide whether to contract out over the long term or to bring the process in-house if the pilot is successful. Modern lasers are extremely user-friendly, being fairly easy to use and maintain, so purchasing a laser system for in-house use is often achievable and economical over the long term.

As mentioned earlier, laser processing speed is typically in the range of 1 to 10 mm^2/s (0.0016 to 0.016 $in.^2/s$) for polishing applications in research applications. Laser polishing typically uses very tight spot overlaps, so texturing may be done much more quickly depending on the feature spacing. This is extremely economical compared to mechanical polishing for small areas, but may come to a cost trade-off with traditional mechanical polishing factoring in the availability and cost of mechanical polishing expertise as well as recurring purchases of mechanical polishing media.

References

1. Wang, Q., Morrow, J. D., Ma, C., Duffie, N. A., and Pfefferkorn, F. E., 2015, "Surface prediction model for thermocapillary regime pulsed laser micro polishing of metals," *J. Manuf. Process.*, **20**(1), pp. 340–348.
2. Vadali, M., Ma, C., Duffie, N. A., Li, X., and Pfefferkorn, F. E., 2012, "Pulsed laser micro polishing: Surface prediction model," *J. Manuf. Process.*, **14**(3), pp. 307–315.
3. Temmler, A., Graichen, K., and Donath, J., 2010, "Laser polishing in medical engineering," *Laser Tech. J.*, **7**(2), pp. 53–57.
4. Temmler, A., Willenborg, E., and Wissenbach, K., 2015, "Design surfaces by laser remelting," *Mater. Werkst.*, **46**(7), pp. 692–703.

CHAPTER 1.4

Laser Ablation for Cleaning, Decoating, and Surface Preparation

Jim Russell

General Lasertronics Corporation
San Jose, California

Basic Science of Laser Ablation

Ever since lasers were invented, one of their most promising applications has always been cleaning and stripping coatings from surfaces.

Concentrated laser energy does a terrific job of "ablating" coatings and surface contaminants, removing them cleanly by instantaneously heating the coating material, ideally to the point of chemical dissociation. If the laser can "see" a surface, it can strip virtually any coating or contamination, regardless of its composition or the complexity of the surface. Lasers are especially effective on irregular surfaces.

Figure 1.4.1 shows ablation in simplified terms.

The obstacle to adoption of laser ablation has always been the likelihood of damage to the substrate. Once the coating or contaminants have been removed, which can happen quickly, the substrate is exposed to laser energy. Any laser powerful enough to strip coatings quickly is also powerful enough to damage many substrates. Attempts to teach human operators to quickly turn off the laser as the coating disappears have all failed, because human reactions are far too slow. Primarily for this reason, laser ablation has not displaced conventional methods of stripping or cleaning surfaces.

One approach that solves this problem takes a design approach specifically aimed at protecting the substrate. The approach has two fundamental elements that differentiate it from other laser systems.

- It uses a pulsed laser rather than a continuous wave (CW) laser. This pulsed laser produces very high irradiance over very short pulse duration. It directs enough photonic energy into the coating to ablate it, but through too short a duration for dangerous amounts of thermal energy to be conducted to the substrate during ablation.

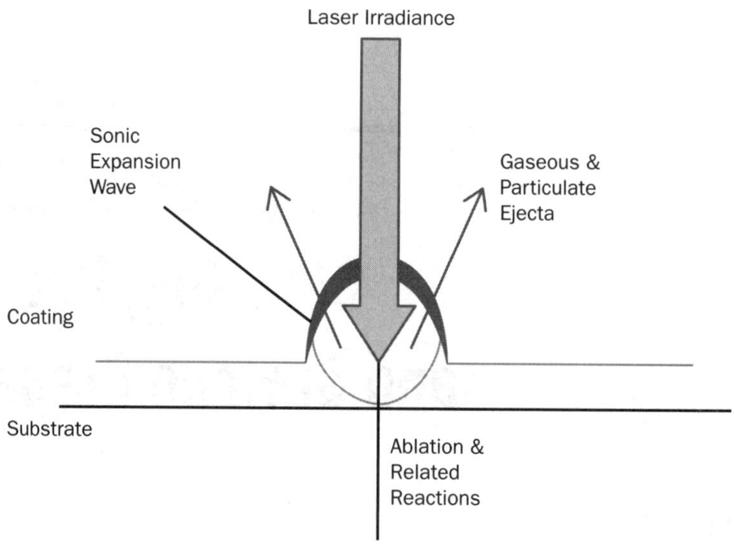

Figure 1.4.1 Laser ablation diagram.

- It uses an optical surface-sensing workhead that selectively ablates only the material specified by the operator. The workhead continuously scans the surface to detect presence or absence of the material to be removed. The electronically controlled laser fires only when the coating or contaminant is detected, which prevents the laser from ever striking the bare substrate.

The protection provided by this system allows its use on delicate materials where high-temperature excursions are unacceptable, such as aircraft skins, structural composites, and warship hulls. This approach works so well that it is approved by the FAA for use on flight-critical surfaces of commercial transport aircraft.

This system quickly strips all kinds of coatings and contaminants, including paints, sealants, anti-erosion coatings, corrosion, and common surface contaminants. The laser is indifferent to the specifics of the coating or the substrate.

When removing corrosion, if the corrosion has not already damaged the substrate, ablation returns the surface to its virgin condition. Regardless of the contaminant or coating, ablation leaves a completely clean surface, ready to accept coatings such as adhesives and paints without further processing.

The laser can be precisely controlled to produce a wide range of surface effects. For example, it can prepare an intermediate layer of a multilevel paint system for better adhesion or superior appearance of a finish top coat. It lends itself readily to CNC or robotic operations for high throughput and consistent application over large and varied surfaces.

In addition to removing coatings, laser ablation can also cause their chemical dissociation. Both effects dramatically reduce the volume of waste products generated in cleaning or stripping operations. With dissociation, components of stripped materials that are inherently gaseous separate from the solids, leaving a much diminished mass and volume of material requiring disposal. Figure 1.4.2 illustrates this comparison.

This dramatic reduction in waste volume along with the complete elimination of solvents and blast media can often, by themselves, pay for adoption of the laser.

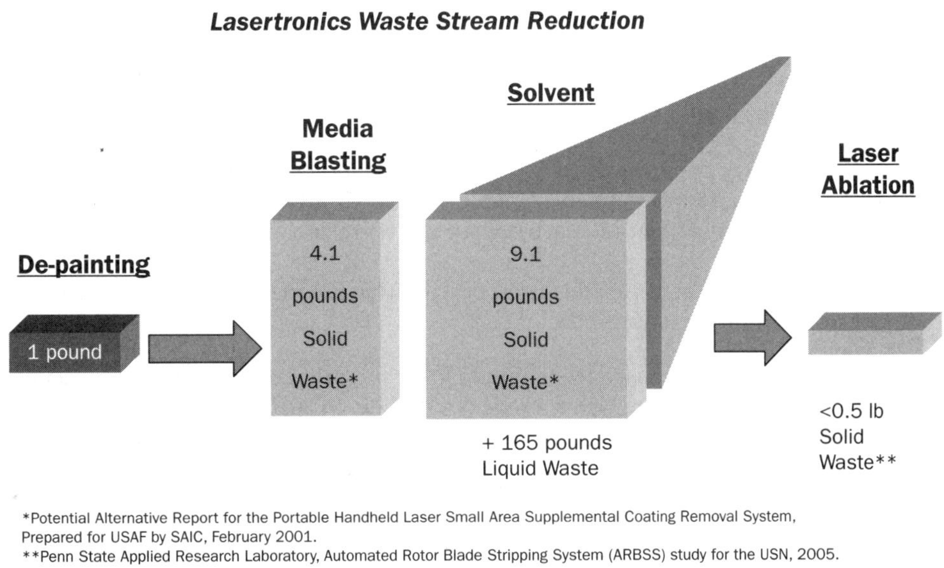

FIGURE 1.4.2 Waste stream reduction possible with laser ablation.

Another benefit of laser ablation in preparing metal surfaces for repainting is that the laser removes chlorides that would otherwise remain behind and attract water through layers of paint. This mechanism is frequently the cause of the bubbles that form under paint and can ultimately cause the coating to fail. Elimination of these chlorides can be expected to lengthen the life of any paint coating over metal.

Other advantages of laser ablation include elimination of repetitive motion injuries, more thorough cleaning of surfaces, and complete elimination of damage risk to substrates.

Surface Preparation in Manufacturing

In addition to coating removal, a well-designed system as described here can be used for preparation of surfaces prior to application of coatings. The initial application for this pretreatment is composites, where selective laser ablation produces measurable improvement in bond toughness, and more than 100% improvement in bond consistency. This degree of enhancement in bonding consistency obviously can bring important benefits in the manufacture of composite structures. This technique also enhances adhesion of other coatings applied to composites, such as paints and sealants.

Implementation Considerations

- Speed: Laser ablation (coating removal) rates are proportional to the power of the laser, and inversely proportional to the thickness of the coating to be removed. Both relationships are linear. The standard formula for ablation rate is

$$A = 2P/t \tag{1.4.1}$$

where A is the ablation rate in square feet per minute, P is the laser power in kW, and t is the coating thickness in mils (thousandths of an inch).

For example, for a 0.5 kW laser stripping a coating 3 mils thick:

$$A = 2 \times 0.5 \text{ kW}/3 \text{ mils} = 0.33 \text{ square feet per minute} \tag{1.4.2}$$

For a 76-μm coating that equates to 0.03 m^2/min.

In practice, paints applied to aircraft and in other applications where minimizing weight is important are in the range of 3 mils (76 μm) thick. Common contaminants that accumulate of their own accord, such as corrosion and oil mist, are typically much thinner.

To go faster, simply add laser power. The physics of ablation and closed-loop control remain the same.

- Costs: Laser costs are falling fast, on the order of 10% to 20% annually, so the prices cited here will soon be out of date. In 2016, a 500-W pulsed laser system equipped with the closed-loop control described here, manual (non-robotic) workhead, and vacuum waste capture system costs about US$300K. Two years ago, a similar system cost US$400K.

For applications where cleaning, rather than coating removal, is the objective, a less powerful laser could suffice. Depending on desired capabilities, these systems would start at US$125K.

Robotic systems that automatically control application of the laser's directed energy are more expensive by virtue of the robotics involved, but the laser costs are about the same. A novel capability of (computer-controlled) robotic laser ablation systems employing the described closed-loop control is their ability to control the robot in real time, based on the laser control system's continuous surface observation. This capability dramatically simplifies robot programming and often provides incremental speed and/or effectiveness that would otherwise be impossible.

- Approvals:
  - Metals: The closed-loop control beam delivery system described here is approved by the U.S. Federal Aviation Administration (FAA) for use on aluminum commercial transport aircraft. Such approval involved extensive before-and-after fatigue testing and took over a year. By 2016 only one laser vendor had earned this approval. Because aircraft skins are more delicate than most other metals and less tolerant of damage, this approval by the FAA is generally taken to also confirm the safety of these systems for use on other metal surfaces.
  - Composites: The closed-loop control beam delivery system is approved for use in paint removal from Sikorsky CH-53E helicopter's main rotor blades. These rotor blades are made from fiberglass composite and have a T_G (glass transition temperature) around 80°C (176°F). Extensive testing confirmed that the described system never raised the composite temperature above 40°C (~104°F). Based on these findings, Sikorsky and Naval Air Systems Command (NAVAIR) approved use of this robotic system for stripping these flight-critical surfaces. As with the FAA approval for metal aircraft, this approval for fiberglass composite rotor blades confirms the safety of such technology for use on structural composites.

While other laser systems can remove coatings, they may not protect substrates as well or as consistently as the closed-loop system described here.

Applications Where Laser Ablation Works Best

- Stripping or cleaning irregular surfaces, where nooks, crannies, crevices, fasteners, threads, grooves, and irregularities defeat other techniques. If the laser can "see" the surface, it can strip or clean it. For example, the laser can perfectly clean the threads of a screw, the links of a chain, and other highly irregular surfaces quickly and completely, leaving them ready to accept new coatings.
- Stripping coatings or removing contaminants that are chemically challenging for conventional techniques. For example, CARC paint, CMAS vitrified sand, and Hontek anti-erosion coating all come off quickly.
- Stripping coatings or removing contaminants from substrates that can easily be damaged by conventional techniques. For example, the closed-loop controlled laser can remove paints from composite helicopter rotor blades and accumulated mold release from composite molds without damaging the surface underneath. The laser eliminates any need for scraping or media blast, which can easily damage these substrates.
- Removing deliberately applied protective coatings. For example, the laser can remove thick polyurethane coatings from composite propeller blades by quickly defeating the adhesive that attaches them to the substrate, enabling the coating to peel off as a sheet. This is not ablation of the coating, but rather of the adhesive.
- Removing coatings that generate a costly waste disposal stream when removed by blasting, solvents, or abrasion. Laser ablation can chemically dissociate the coating into the elements making it up. The attached vacuum capture system sequesters the toxic particulate material in a paper filter. Laser ablation also converts some toxic materials into benign valences of the same material, making disposal and personnel protection dramatically simpler.
- Removing a top coating and preserving the primer underneath. This capability can save the cost of reapplying a primer, when only the top coat needs to be replaced or repaired.
- Removing radiological contamination from the surface of irradiated materials. In most cases, the laser can clean to free release.
- Removing coatings or preparing surfaces where conventional techniques cause repetitive motion injuries. The laser requires neither manual pressure nor specific motion patterns.
- Removing coatings underwater. The photonic energy produced by the pulsed laser does not suffer significant degradation by operating through water.
- Reducing the manpower required to accomplish specific cleaning or decoating objectives. Laser ablation generally requires fewer workers than legacy methods.
- With CNC or robotic control, dramatically enhancing stripping or cleaning throughput of similar items.

Applications Where Laser Ablation May or May Not Bring Important Benefits

- Stripping thick coatings from large, flat areas where high throughput is critical, such as paint from the hull of a ship. The common thickness of these coatings makes the laser no faster, and possibly slower, than legacy techniques.
- Preserving a very thin prime or Alodine coating under a much thicker paint that is being ablated. Depending on relative thicknesses, it can be impossible to prevent the thinner sub-layer from coming off along with the thicker upper layer.
- Stripping a top metal coating from a lower metal. Especially where colors are similar, the laser control system cannot always discriminate between the layers.
- Situations where it is important to retain legacy manpower in a coating removal work force, because laser ablation generally requires fewer workers than the methods it replaces.
- Organizations where the culture is unwilling to consider new approaches.

Manual versus Robotic Laser Ablation

Robotic ablation can be cost-effective where the need is to process a large number of similar surfaces in the same way. For example, the Navy's Automated Rotor Blade Stripping System (ARBSS) is designed to strip hundreds of nearly identical helicopter rotor blades each year (Fig. 1.4.3).

FIGURE 1.4.3 Robotic Navy laser stripping system.

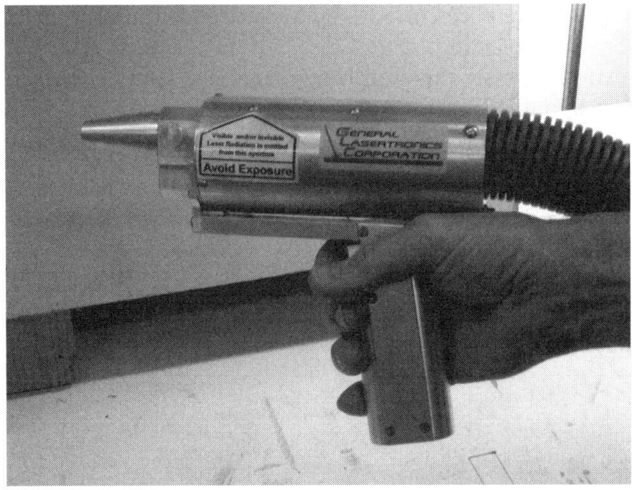

FIGURE 1.4.4 Laser manual cleaning head.

The efficiencies achievable by robotically applying laser energy to the surfaces of these rotor blades are significant, and merited the capital investment required to perfect this application. Work is under way on systems that will use several robotically controlled lasers to strip an entire aircraft.

An easily justified manufacturing robotic application is pretreatment of composite parts prior to bonding them together into larger structures. Each part is virtually identical, and a high throughput is required. A robotic system can deliver this high throughput and ensure that the faying surfaces receive identical pretreatment overall.

CNC applications can also be effective in this mode, at lower cost than robotic systems.

In contrast, where the application calls for cleaning or removing coatings from a variety of surfaces with different shapes and sizes, a manual system is the more logical choice. The most commonly chosen system delivers 500 W to the work surface through a workhead shaped like a pistol and weighing about a kilogram (2.2 lb) (Fig. 1.4.4).

Laser ablation produces audible and visible signatures that enable the operator to easily maintain the workhead at the optimal distance and orientation. Workers accustomed to using equipment that requires treating work surfaces with tools, for example painters and welders, can learn to use manual laser ablation equipment in under an hour.

Precautions and Safety

Industrial lasers are widely available, and their cost is falling rapidly. For applications where preserving the substrate is not important, many lasers could conceivably do an acceptable job of cleaning or stripping. But where substrate preservation is important, the manufacturing engineer must evaluate the laser's interaction with the intended surfaces, and confirm that no unacceptable damage occurs. Many cleaning and stripping applications will require closed-loop surface recognition control technology for adequate substrate protection.

It is always necessary to wear laser-protective safety glasses when using or working near an industrial laser. These cost about $200 per pair and are available at industrial supply houses. This is critical. Only these glasses reliably prevent irreversible eye damage, which could happen to a naked eye.

CHAPTER 1.5

Laser Hardening

Stanislav Němeček

RAPTECH
U Vodárny, Czech Republic

Process

Laser hardening offers short processing times and appreciable cost savings, as well as capabilities to accommodate almost any geometry, and control the depths and widths of the treated surfaces. It rarely requires the use of any special tools. In addition, laser-hardened surfaces exhibit additional beneficial properties when compared to conventionally hardened surfaces. These include differences in tribological, fatigue, and corrosion behavior, specifically an improved wear life, fatigue, and corrosion resistance.

Applicable Lasers

Continuous wave laser sources are preferred for laser hardening applications. The most widely used systems rely on diode or fiber lasers. One of the reasons is that the requirements for the beam quality (i.e., the beam divergence) are less strict for hardening than for cutting and welding. In addition, lasers offer the benefit of the lowest operating costs which has enabled them to compete with the widely used induction and flame hardening processes.

As the laser beam provides the energy, the beam-delivery optics control the beam size. The power delivered by industrial continuous-wave lasers is normally between 3 and 10 kW. The energy density (the amount of energy per unit of area) is controlled through the beam spot size. In cutting applications, the beam diameter is approximately 0.1 mm (0.004 in.), whereas in welding, it is in the range of 0.5 to 1 mm (0.020 to 0.040 in.). By contrast, surfaces to be hardened are typically rather large and the spot used is therefore tens of square millimeters in area. Using lasers, layers with special properties can be weld deposited on workpiece surfaces. In laser hardening, the power density reaches hundreds of watts per square millimeter, whereas the levels used for laser welding are on the order of thousands of watts per square millimeter.

Laser Hardening Materials

The main criterion for selecting steels for hardening is their carbon content. The relationship between carbon level and hardness is practically linear. The higher the carbon content, the higher the hardness. Steels for surface hardening typically contain 0.4% to 0.5% carbon in order to obtain a hardness of more than 55 Rockwell C hardness (HRC).

Transformation hardening of steels is based on iron's polymorphism characteristic. At low temperatures, atoms in steel are arranged in a cubic crystal structure with one additional atom in the center of the cube (body-centered cubic lattice). After heating above the Ac3 temperature, the value of which depends on chemical composition, the crystal structure transforms to the face-centered cubic lattice which has different distances between neighboring atoms. This transformation is known as austenitization. Concurrently, the carbides present in the steel dissolve and the released carbon atoms migrate to the larger voids in this lattice, which allows approximately 100 times more carbon to dissolve than the body-centered cubic crystal structure. Without austenitization, no hardening would be possible. Subsequent quenching transforms the austenitic structure back to the initial body-centered cubic structure but the short time available causes the carbon atoms to become locked in nonequilibrium and not quite regular locations. As a result, the spacing between atoms in one direction increases and the cubic lattice becomes a tetragonal lattice. This is directly proportional to the amount of carbon in the steel. The transformations which take place during heating and cooling are diffusion-controlled processes, which means they depend on temperature and time. In 1% carbon steel, the volume upon transformation to martensite increases by 4 volume percent. The slower transformation to pearlite expands the volume by a "mere" 2.4%. Such expansions cause internal stresses which, on a micro scale, are responsible for surface hardness and, on a macroscale, for the distortion in hardened parts. Typically, the goal is to maximize hardness and minimize distortion.

Grain Size

Rapid heating has both favorable and adverse impacts. First, high heating rates shift the transformation temperatures to higher values. Therefore, conventional transformation diagrams are not valid for laser hardening and the appropriate values of quenching temperatures become strictly protected knowledge of individual companies.

Hardenability of steels increases with grain size. When the grain size changes from 0.02 to 0.12 mm, hardenability increases by approximately 50%. However, rapid heating cannot provide adequate homogenization and leads to a variation in hardness because larger carbides take longer to dissolve. This may cause problems, for instance, in massive castings and machine parts, in which diffusion requires longer times. If such carbides fail to dissolve, the matrix will not contain enough carbon and the process will in fact turn into the hardening of low-carbon matrix. Hence, tool steels which contain primary carbides require different hardening parameters than structural steels with fine carbide precipitates. Illustrative examples are the widely used 1.2311 and 1.2379 steels. On the other hand, rapid heating helps to retain fine grain micro structure. Fine austenite grains transform to fine martensite, which is less susceptible to nucleation and propagation of cracks.

This means that in hardening, one should distinguish between cast parts (which are likely to contain coarse grains) and formed parts (typically with finer grains). The heat treatment history of the part should be taken into account as well. Equilibrium micro structures formed by slow transformation contain large carbides which take longer to dissolve. In normalized steels, cementite is spheroidized. By contrast, hardened materials with martensite or bainite micro structure contain very fine and easy-to-dissolve carbides. There are differences between Ac3 transformation temperatures as well. In annealed materials, it may be 200°C (392°F) higher than in the hardened ones. The initial micro structure affects hardening depth as well. The resulting difference is several tenths of a millimeter (thousandths of an inch).

Hardening Process Comparisons

In flame hardening, the depth of heating is given by the relative speed of the flame movement across the surface. The efficiency of heat transfer to the material is poor and the surrounding surfaces become heated as well. As a result, the dwell time required for austenitizing increases, the grains in the material coarsen, and their boundaries may become burnt. The surface also develops a layer of oxides. In this process, the uniformity and control of temperature are limited and inaccurate. In induction hardening, the heat generated depends on electric resistance and current. The depth of heating is given by the frequency of the power source. The lower the frequency, the shallower is the hardening depth. Even with high-frequency induction heating, however, heating takes several seconds.

In flame hardening, as well as in induction hardening, the heating stage must be followed by sufficiently rapid cooling (using a water spray or polymer solution). There is an ever-present risk that steam and a vapor blanket may form during cooling, which would retard the heat removal rate and prevent the surface from hardening. Similar to the heat-affected zone in weld joints, there is a deep transition zone beneath the hardened layer. This is the zone in which the elevated temperature, though insufficient for hardening, has caused some changes in the matrix of the parent metal. Unfortunately, these changes are typically for the worse. On the other hand, the hardening depth achieved by these methods may be up to 10 mm as opposed to the several tenths of a millimeter that laser hardening produces.

In laser treatment, the austenitizing temperature must be achieved in the surface, as with the other techniques. However, thanks to rapid heating rates of thousands of degrees per second, the resulting transitional heat-affected zone is very thin.

The maximum laser-hardened depth is approximately 2 mm (0.080 in.), depending on the heat conductivity of the material. The hardening depth can be controlled by the speed of the beam movement and by the temperature. To improve control, the process is typically monitored using a pyrometer linked to the laser source and operated by robot. By regulating the power input, this control loop can maintain a fixed surface temperature, thus providing uniform hardness and preventing local melting.

In the heating step, austenitizing temperature, that is, a temperature above Ac3, must be achieved in the surface layer of steel. Rapid heating with a laser beam is one of the factors which minimize the distortion of the part heated. As the material around the laser spot remains cold, it does not expand and the distortion does not occur. Another difference from the other surface hardening techniques lies in the heat removal mechanism which is based on self-quenching. It involves rates of several thousand degree Celsius per second. There is no need for cooling the surface using a liquid supplied from outside because the interior of the part remains cold and absorbs the heat by conduction

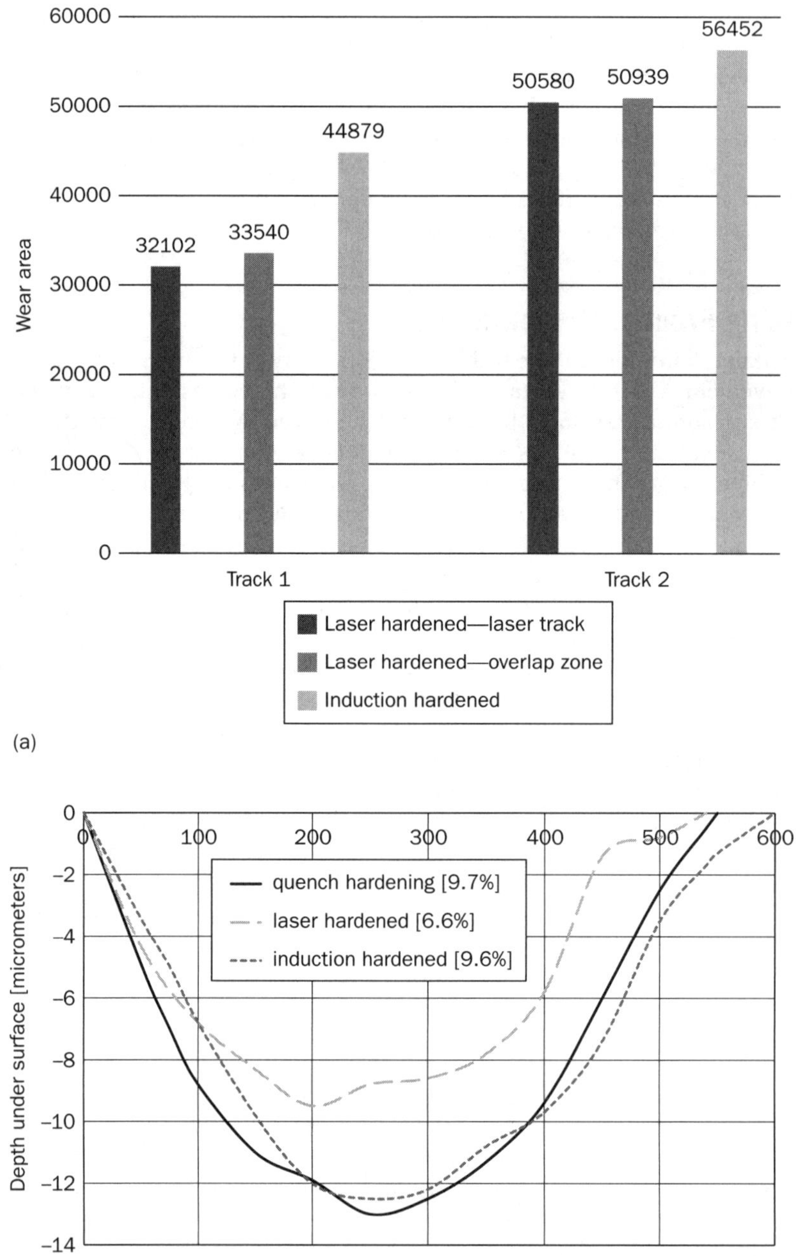

FIGURE 1.5.1 (*a*) Results of measurement of wear resistance of 42CrMo4 steel by pin-on-disc test. The columns compare the data for the centerline of a laser track, for an overlapping zone between laser tracks, and for an induction-hardened surface. (*b*) This figure compares wear resistance in 1.2311 steel after various treatments: through hardening, induction hardening, and laser hardening [1].

at a sufficient rate (i.e., cooling by convection). One can therefore assume that phase transformation begins from within the part and the surface is the last to cool down.

Laser beam hardening is much more favorable to the material in terms of crack susceptibility than other techniques. The steep thermal gradient from quenching causes severe (tensile) stresses which lead to surface crack initiation. In laser hardening, however, the cooling process begins by heat removal to the cold interior of the workpiece. The temperatures thus equalize gradually from within the part. This minimizes the resulting stresses and eliminates cracking.

The maximum hardness level achievable in a steel with 0.1% to 0.2% C is approximately 40 HRC. Surface-hardened steels with 0.4% C typically have a hardness of 58 to 60 HRC. In hypereutectoid steels, hardness may reach 65 HRC. Pearlitic cast irons may reach hardnesses above 65 HRC, particularly if their surface was melted and solidified as ledeburite. In through-hardened parts, subsequent tempering is used to relieve internal stresses. By contrast, laser hardening is not followed by any heat treatment operation, not even tempering. However, laser hardening of large surfaces creates overlapping zones between adjacent hardening tracks which experience self-tempering. The width of such zones on the surface is approximately 1 mm (depending on the hardening depth and speed). As mentioned above, the decrease in hardness also depends on the carbon content in steel.

Thanks to the finer resulting micro structure, laser hardening provides better abrasion resistance than hardening by induction. No effect of the tempered overlapping zones has been detected in the tribological characteristics of parts studied to date (Fig. 1.5.1).

Application Examples

Laser Hardening of Dies and Tooling

Automotive die materials are ordinary hypoeutectoid tool steels (most frequently 1.2311) and hypereutectoid steels (e.g., 1.2379), but the adoption of laser hardening by car manufacturers also led to creation of several specially developed steels (e.g., Toolox44). Laser hardening of rounded and shearing edges may double the number of parts formed (Fig. 1.5.2).

Laser Hardening of Gears

Gears are found in most gear boxes as well as in other torque transmission systems. They are subjected to severe cyclic loads. Their surfaces must sustain high stresses, abrasion, and alternating loads. Case hardening and nitriding are the prevailing techniques in the production of heavy-duty gears, but they share the disadvantages of part volume changes, shape distortion, surface oxidation, and, above all, a risk of crack formation. Heat treating must be followed by machining and hardening with relatively large machining allowances. Through-thickness hardened gears have somewhat lower hardness upon tempering. Consequently, the wear resistance of their surface is the poorest in comparison with surfaces treated by other above-mentioned methods.

Gears, which normally tend to exhibit numerous cracks upon induction hardening, do not crack during laser hardening. As a general rule, fine micro structure enhances the fatigue resistance of parts operating under dynamic loads. The laser process produces hardened gearwheels with long life.

FIGURE 1.5.2 Laser hardening of a forming die.

FIGURE 1.5.3 Hardening of spherical and conical surfaces.

Laser Hardening of Machine Parts

The principal advantage of lasers in machine production is that they can be used independently of the workpiece size and shape. A laser beam is easily modified for each application by the optical system and its robotic movement. Hardening of spherical, conical, or other surfaces by a laser does not pose problems, as shown in Fig. 1.5.3. It is equally feasible to harden small parts in batches of hundred thousand pieces and single products with a mass of tens of tons. It is this very field which offers a number of still unused opportunities for substituting induction hardening or carburizing with laser hardening. The most frequently used medium-carbon steel grades include C45 and the higher-alloy 42CrMo4-type steels.

Laser Hardening of Cast Iron

In a number of applications, cast iron is an irreplaceable material, owing to its affordability, favorable casting properties, and good machinability. Examples of those applications include machine tools, pulleys, and sheet forming dies for automotive industry. In terms of mechanical properties, cast iron, much like steel, offers a wide range of strength levels combined with good toughness and shock and vibration damping ability. The sliding property of cast iron is a much sought-after quality, and the wear resistance of its surface can be further enhanced by surface hardening. Surface hardening by a laser substantially improves contact fatigue resistance [2] and practically doubles the wear resistance of the material [3].

However, laser surface hardening of cast iron is not trivial, due to the material's heterogeneity and coarse-grained micro structure, especially in massive castings. Yet, it is the hardening of massive dies for automotive industry which makes interesting and sought-after applications. The normal hardness of laser-hardened cast iron with pearlite matrix, whether gray or ductile, is 800 to 900 HV (i.e., 64 to 67 HRC). However, lower hardness levels are still found in some cases. Hardening depth is normally around 1 mm.

An appropriate combination of temperature and rate of travel of the laser beam leads to an optimum micro structure of martensite and graphite (the latter provides cast iron with its good friction properties). Martensite produced by laser hardening contains notably finer laths or plates than other processes.

References

1. Fiedler L. Monitoring of Wear and Dynamic Impact Loading of Laser Surface Hardened Steels. Bachelor thesis, University of West Bohemia, 2012.
2. Zengyi T., Jiaming L. Influence of surface hardening on contact fatigue life of nodular cast iron. *Acta Metallurgica Sinica*, 1990, Vol. 3, No. 3, pp. 188–193.
3. Hwang J., Kim D., Youn J., et al. Laser surface hardening of gray cast iron used for piston ring. *Journal of Materials Engineering and Performance*, 2002, Vol. 11, p. 294 (doi:10.1361/105994902770344105).

CHAPTER 1.6

Laser Welding of Metals

LaRoux K. Gillespie

Manufacturing Consultant
Andover, Kansas

This chapter focuses on the use of lasers for hand size to ship size parts' commercial applications.[1] Laser welding is performed via a continuous beam or a pulsed beam. Pulses only a few milliseconds long weld thin materials, putting a minimum of heat into the part. Razor blade thickness parts are joined with this approach.

Pulsed welds are also used for thicker parts, but it is slow. A beam is moved slowly along a part, with each pulse overlapping the previous one. It is a slower process than continuous seam welding, but it is particularly effective on parts with high thermal conductivity—aluminum or copper for example.

Continuous laser systems in contrast are used for deep welds. Continuous lasers put more heat into the part than the pulsed laser, so they can produce welds at a much higher rate than pulsed systems do. They also are used to weld thicker parts that need deep penetration. Even with continuous laser welding, the heat-affected zones (HAZ) and the complete heating of the parts are considerably less than that obtained with arc or plasma welding, so there is less distortion. The fact that these welds can go entirely through some joints means that a single weld, from one side, provides the necessary strength in contrast to other processes that require access to both sides of the weld to get full or adequate penetration.

In all instances except one mentioned later, the parts must fit tightly together. Press fit joints provide excellent weld designs. Loose fits do not have the metal needed to fill gaps so a visible trough forms and this is less strong in many instances than a tight fit would have provided. Better fit-up requires less power, a smaller laser spot size, and results in faster welding speed.

For welding, the laser is focused to a spot whose diameter is typically 100 to 1000 μm (0.0039 to 0.039 in.). A spot size of 300 μm (0.012 in.) is a typical diameter spot size. For the micro work discussed in Chap. 3.37 the spot size can be smaller than that.

The fact that laser beams can be used in normal shop air environments makes them particularly desirable. Electron beam welding can do many of the same welds, but it

[1]Laser welding of micro parts is discussed in Chap. 3.37. That involves the smallest beam size and highest precision on the thinnest of materials and features.

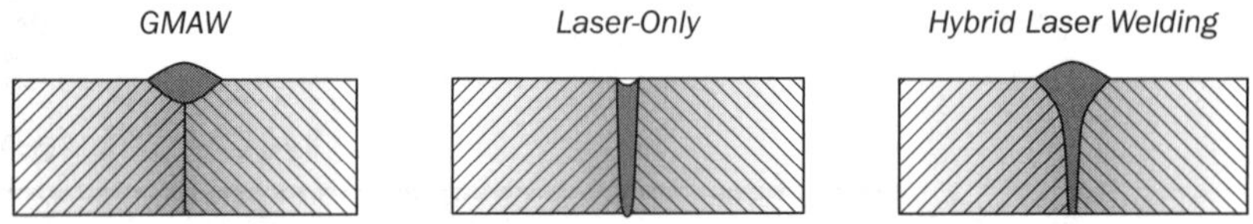

Graphic illustrating differences between GMAW, Laser, and Hybrid Laser Welding weld profiles.

FIGURE 1.6.1 Hybrid weld results compared to laser-only weld. (Courtesy ESAB.)

1. Conduction
2. Conduction/penetration
3. Penetration or keyhole

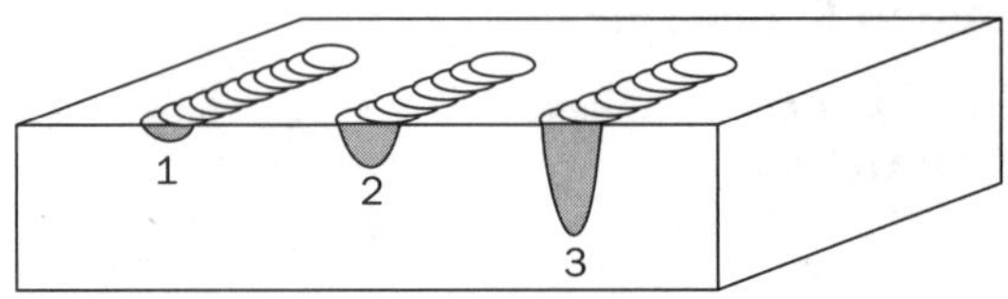

FIGURE 1.6.2 Comparison of three different laser weld approaches. (Courtesy Amada Miyachi America.)

requires a vacuum and that in turn also greatly slows production (because of entering and leaving fixed chambers and vacuum pump down times). Lasers are a highly desirable welding process because they are easily automated and the automobile companies match them with robotic machinery to reach any area on a car. With robots they can approach a weld from any angle or position, even upside down.

In addition to the variations of beam pulsing or continuous operation, another variation called laser-hybrid welding combines the laser with traditional gas metal arc welding (GMAW) by providing molten metal to fill the joint. The laser provides the intense narrow band heat, while the other process provides a wider band of heat and the fill metal (Fig. 1.6.1). The laser increases the weld speed above what traditional GMAW can do. In addition the laser provides a better weld since it reduces the potential for undercutting. MIG (metal inert gas) or MAG (metal active gas), TIG (tungsten inert gas), or plasma welding can all be used in conjunction with the laser. The laser in these instances can provide high power densities, while the arc which is broader helps heat the width that the laser may not. The combination of hybrid approaches results in faster welds.

Since filler metal is added in this process, the fit-up between parts does not have to be so closely controlled. The filler metal fills all gaps.

Lasers actually weld by two basic processes, which are described in detail here, but in the first process called heat conduction welding, the goal is to only heat the surface hot enough to cause welding. Welds only 0.02 mm (0.0008 in.) or as much as 1 mm (0.0008 to 0.039 in.) are produced in this manner. In this approach, the weld width is always greater than its depth (Fig. 1.6.2).

In keyhole (deep penetration) welding, the weld depth can be up to ten times greater than the weld width. These welds can be as much as 25.4 mm (1 in.) deep. This allows several layers to be welded together simultaneously. Laser welding can essentially weld any joint configuration. Figure 1.6.3 illustrates the three most common joints.

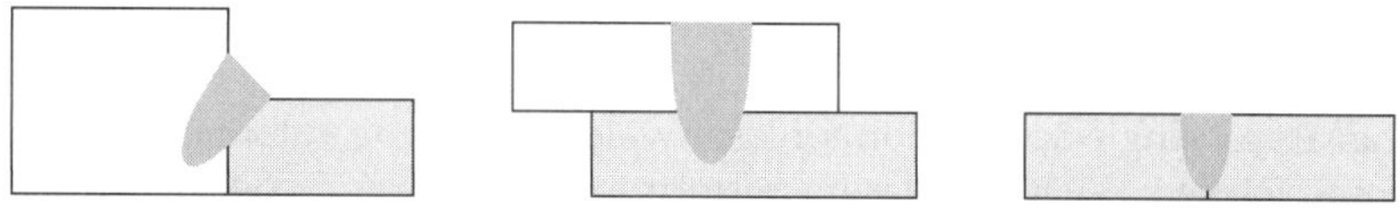

Figure 1.6.3 Common welded joint configurations. (Courtesy STI.)

Today's lasers have a power ramping feature that can prevent "last-pulse" cracking in seam welding. This had been an issue with older machines. Power ramping gradually increases the laser power at the start and reduces it at the end of the weld. The result is also a better-looking weld and no cracking at the ends.

Applications

Lasers are widely used for hermetic sealing of components to prevent damaging moisture or other atmospheres to affect internal electronics. The heat can be limited to microscopic areas that does not migrate to sensitive components nor melt thin walls (Fig. 1.6.4).

Medical implants are laser welded, and it is possible to join some ceramics to some metals using laser welding.

Lasers have not been a first choice for repair work, but they are used as a laser deposition source in two different applications. Laser cladding is discussed in Chap. 1.9 as a means to add metal back onto parts that have worn or broken. The subsequent newly added material is then re-machined to finish dimensions. A modern application is laser cladding to repair leaks and cracks in city stainless steel water reservoirs high in the air (water towers). The laser in a "welding-like" application is also used in several additive manufacturing processes as discussed in Part 2, particularly in Chaps. 2.7 and 2.12.

Figure 1.6.4 Example of micro part and wires laser welded. (Courtesy Superior Jet.) Larger laser-welded parts weigh tons.

Equipment

The largest single application for laser welding is in the automated lines in automobile manufacturers. There the lasers perform spot welds and continuous welds to join all the structural parts of the car or truck. The automotive industry uses laser welding because of its abilities, speed, and adaptability to full automation.

This application uses large systems just as the shipbuilding industry does. Laser welding is used in manufacturing airframes for massive aircraft. Desktop laser welders are used by the jewelry industry for low-volume production. Lasers are provided that are on crawler tracks to be mobile while welding. Lasers can weld at up to 4 m/min (160 in./min) and even 30 m/min (1200 in./min) although many welds are made at slower speeds of 1000 mm/min (40 in./min). Laser power, part material, and weld depth and width determine the maximum speed that can be used.

Laser equipment comes in flatbed systems, lathe-like rotary systems, tube forming and welding, and the versatile robot lines. Laser welders vary from giant machines to desktop units and even mobile floor roll around units. While most are programmable, manual systems are also available.

They are used to join wind turbines, pipelines (in place), bridge components, aircraft structural parts, earth moving equipment, and other large structures to parts that are microscopic.

At the opposite end of the scale, desktop dental units weld dental prosthetics and orthodontic pieces. Most of the body invasive products and tools use laser welding to join components.

Systems are also available that allow laser welding in a vacuum if needed. This capability provides higher aspect ratio welds, deeper penetration (up to 2×; see Fig. 1.6.5), less heat input, less weld spatter, and cleaner parts after welding. Welds made in air will have some amount of oxides on surfaces that may not be allowable for very clean applications. The oxides, if they exist, are normally removed in subsequent

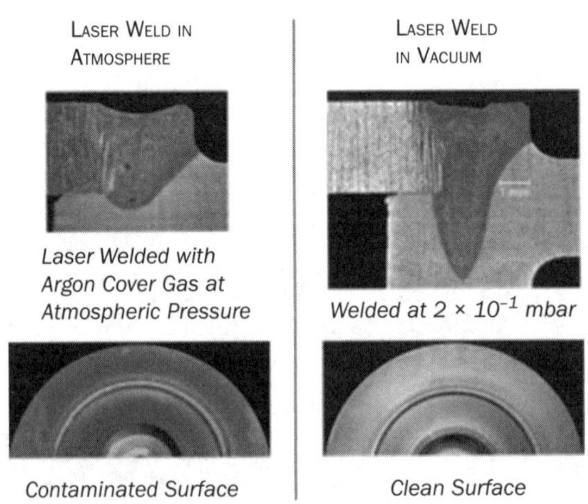

FIGURE 1.6.5 Comparisons of laser welds in vacuum versus welds in air. (Courtesy PTR-Precision Technologies Inc.)

operations when required by such processes as nonwoven abrasive disk brushing. Doing the welding in a vacuum, which has no free oxygen, can prevent that subsequent step. Cover gasses can be used in open air welding to prevent oxides but they add some complexity (Herrmann, 2011). Despite added costs and other minor issues, open air welding is faster just from a material handling factor.

Materials

Laser welding joins carbon steels, HSLA steels, stainless steel, aluminum, and titanium, although high carbon steels have a tendency to crack because of high cooling rates. Unitek Miyachi notes that the most commonly welded material is steel, and the general selection rule is to keep the carbon content under 0.12%. For stainless steels, ensure that the Cr/Ni ratio is greater than 1.7. Stainless steel alloy ANSI 303, many 400 series alloys, and high carbon steels should be avoided due to a high carbon, phosphor, and sulfur content. Generally, nickel alloys and titanium are highly weldable, with aluminum alloys and copper being case specific.

Some other common materials routinely welded include titanium, nickel, platinum, Kovar, and molybdenum. Table 1.6.1 provides insight into welded material issues.

Hybrid welding often has as its base materials structural steels, CrNi steels, ferritic and austenitic steels, duplex steels, and magnesium and aluminum materials, as well as galvanized sheet metal.

Welding of dissimilar materials can be performed but that requires close coordination between designer and manufacturer to assure that the bond meets all expectations. First the metallurgical properties of the two must be compatible and second the thermal expansion and contraction has to be nearly the same to prevent high stresses and cracking. From a metallurgical standpoint, good solid solubility is required for good weld properties. That is a result of metals that have compatible melting temperature ranges. Brittle intermetallics often form when that is not the case.

Welding titanium to aluminum alloys creates brittle intermetallic compounds, making poor quality joints.

Stainless steel to copper is similarly challenging because melted copper penetrates the grain boundaries of the steel and particles of the steel can be found surrounded by copper.

Aluminum and copper welds lead to significant cracking at weld boundaries and within the weld nuggets.

Despite these issues international research is ongoing to get past these issues, particularly for copper to steel, copper to aluminum, and aluminum to steel since there are many applications once a good weld can be made in high production. Femtosecond lasers have come close to providing the necessary quality for stainless steel to aluminum joints, but femtosecond lasers are not commonly used at the moment for welding applications.

Nitinol is joined to stainless steel in production applications daily, even with weld joints 3 mm (0.118 in.) thick.

Many of the medical devices inserted in the human body require some form of visible (detectable) marker on them. For most this means the marker has to be seen on X-ray or similar machines. The materials welded together for these applications include nickel titanium (Nitinol) to a gold marker, cobalt chromium L605/MP35N to a tantalum (Ta) marker, and stainless steel 316LVM to a Pt-Ir (platinum-iridium) marker. Lasers can join each of these biocompatible materials to each other.

Material	Comments
Aluminum 1100	Welds well; no cracking problem or transformation
Aluminum 2219	No cracks; no filler metal required
Aluminum 2024/5052/6061	Requires filler metal of 4047 Al to make hermetic, crack-free welds
Cu-Zn brasses	Out-gassing of Zn prevents good welds
Beryllium copper	Alloys containing higher percentages of alloying agents weld better due to lower reflectivity
Copper	High reflectivity may cause uneven welds; for material less than 0.254 mm (0.01 in.) thick, coating may enhance weldability
Hastelloy-X	Requires high pulse rates to prevent hot-short cracking
Molybdenum	Usually welds brittle; welds may be acceptable where high strength is not required
Inconel 625	Some tendency for porosity in deep welds
Monel	Good ductile welds; good penetration
Nickel	Must be cleaned; good ductile welds and penetration
Steel, carbon	Good welds with carbon content under 0.25%; for greater carbon content, may be brittle and crack
Steel, galvanized	Severe Zn boil-off causes porosity
Steel, 300 stainless	Welds well, except 303 and 303SE, which crack
Steel, 400 stainless	Generally welds somewhat brittle; may require pre- and post-weld heat treating
Steel, 17-4PH stainless	Needs post-weld heat treating to strengthen
Tantalum	Ductile welds; special precautions against oxidation required
Titanium	Ductile welds; special precautions against oxidation required
Tungsten	Brittle welds; requires high energy
Zirconium	Ductile welds; special precautions against oxidation required

Table 1.6.1 Laser Welding Application Guideline (Courtesy U.S. Laser.)

Quartz glass components can now be commercially laser-welded together up to 100 mm (3.9 in.) thick, providing essentially the same monolithic properties of the base material. This is not a process used by many companies, but it is a commercial process.

At the research stage lasers have welded fused silica, borosilicate, and sapphire to a range of materials that include borosilicate, fused silica, silicon, copper, aluminum, and stainless steel. Glass has been welded to aluminum in the laboratory although it is not widely reported yet.

Design Considerations

- All welded designs require some definition of the weld. Thickness, depth, location, extent, etc. are all needed, but in addition to this some definition is required for validating and controlling the weld. The details of weld specification are better defined in weld-specific handbooks such as those produced by the American Welding Society than in this focused handbook because the standards and definitions are extensive.
- For laser welding the weld gap should never be more than 10% of the thinnest material or of the weld penetration, whichever is less.
- Weld spatter or splatter is common to all welding operations. It is not blatant in laser welding, but critical designs must define what level of freedom from spatter is expected.
- Weld speed is proportional to the quantity of power supplied and on the type and thickness of the workpiece.
- Welding over plated materials can be a problem. The phosphorous in electroless nickel and any gold over 50 μm (0.002 in.) thick will have cracking or other problems. Electroplated nickel deposits do not cause the problems that electroless plating does.
- For welding aluminum to hermetically seal semiconductor packages, use type 4047 aluminum. It helps prevent buildup of shrinkage stresses.
- When dissimilar metals must be joined, there are two ways it can be done for such materials as copper to stainless steel. The concept also applies to other material combinations: (1) add a transition material between the two dissimilar metals. For copper to steel, add nickel, which is soluble with both materials. Then nickel may be plated or welded onto the steel or (2) use a piece of nickel between the two.
- Another way to weld sandwiches of two different metals together is to weld the top and bottom ones together through cutouts in the middle layer. This locks in the middle foreign layer mechanically rather than from welding. If the middle sheet is thin, holes in it can provide an opening that allows the laser to melt the top layer down into the bottom layer.
- Weld speed is proportional to the quantity of power supplied and on the type and thickness of the workpiece.
- Welding over plated materials can be a problem. The phosphorous in electroless nickel and any gold over 50 μm (0.002 in.) thick will have cracking or other problems. Electroplated nickel deposits do not cause the problems that electroless plating does.

References

Anonymous, Laser welding of dissimilar materials, *Materials World Magazine*, February 1, 2008.

Anonymous, Welding of dissimilar metals, http://www.totalmateria.com/, August 2006.

Carter, R. M., J. Chen, J. D. Shephard, R. R. Thomson, and D. P. Hand, Picosecond laser welding of similar and dissimilar materials, *Applied Optics*, vol. 53, no. 19, 2014, pp. 4233–4238 (doi: 10.1364/AO.53.004233).

Herrmann, J., The role of process gases in laser welding, *www.laser-journal.de*, no. 5, September 2011, pp. 24–27.

Huang, H., L.-M. Yang, S. Bai, and J. Liu, Femtosecond fiber laser welding of dissimilar metals, *Applied Optics*, vol. 53, 2014, pp. 6569–6578.

Mai, T. A., and A. C. Spowage, Characterization of dissimilar joints in laser welding of steel–kovar, copper–steel and copper–aluminum, *Materials Science and Engineering A*, vol. 374, 2004, pp. 224–233.

Naeem, M., R. Jessett, and K. Withers, Fiber laser welding of dissimilar materials, *Industrial Laser Solutions*, March 30, 2012.

CHAPTER 1.7

Laser Welding of Plastics

Andrew S. Geiger

Leister Technologies
Itasca, Illinois

LaRoux K. Gillespie

Manufacturing Consultant
Andover, Kansas

Laser welding of plastics is a standard commercial joining process, but there are four basic approaches and at least four other fabrication variations of these that facilitate manufacturing. The underlying process is referred to as "laser transmission welding" or "through-transmission IR" (TTIR) welding. The top layer nearest the laser has to be transparent (transparent to the laser wavelength which does not mean visually transparent to human eyes) and the bottom layer has to be a layer that absorbs the beam's energy. The beam passes the transparent plastic and is turned to heat in the absorbing plastic (Fig. 1.7.1). Both parts must be pressed together for the heat to facilitate bonding. The heat from the absorbing layer conducts to the transparent layer and both materials melt and mix creating the bond. Heated polymers thermally expand while the force on the parts to be joined induces mixing. As the laser propagates along a designated weld path, increased mixing of the polymer optimizes bonding conditions. All of these factors together create a strong bond or weld.

Contour Welding

The laser is focused to a spot most commonly using a spot optic. This spot is moved along the desired path to cause bonding. Generally, the welding cycle is complete at the conclusion of one pass. Most users agree that the ideal spot size for this kind of welding is 1 to 2 mm (0.039 to 0.079 in.), but it can be slightly larger or smaller, depending on the application.

Simultaneous

Optical elements are used to deliver laser energy simultaneously throughout a calculated path or area. Since the entire bonding path or plot is heated in the same instant,

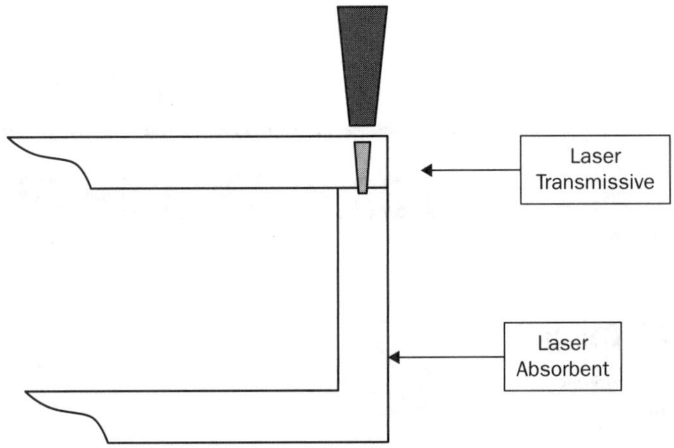

FIGURE 1.7.1 Schematic of welding plastic components with lasers. (Courtesy BASF.) The top layer A is transparent and the bottom layer B is absorbent of the energy.

cycle times can be greatly decreased from that of traversing a path like a milling cutter would do. The path typically is the same narrow path that is desired in contour welding, but the entire path is heated in the same instant. Increased laser power may be required, and in addition heat load within the welded components is generally higher than contour methods.

Quasi-Simultaneous

Piezoelectric motors position two mirrors to deflect the laser spot and guide it along the welding contour at a very high speed. This setup is sometimes referred to as a scanner or galvo optical head. The joining surface is traversed several times per second, effectively heating and plasticizing the entire welding seam at the same time. Heat loads similar to that of simultaneous welding can be expected.

Mask

A line-shaped laser curtain is passed over a stencil (mask) of a weld profile (Fig. 1.7.2). The areas not covered by the mask are irradiated as the line passes over the area. These areas hit by the beam generate heat as in the other methods and create a bond. Extremely fine structures measured in micrometers can be heated by this means and it allows seam structures of any 2D design conceivable, including curved weld lines of different widths, skipped areas, circular, square, wavy, logos, patterned lines—all resulting from a single process step. Weld lines can be as narrow as 100 μm (0.0039 in.) with this approach.

Line

While not one of the four basic approaches, the laser can distribute its energy in a straight line rather than a spot as implied in the mask approach. For some machines,

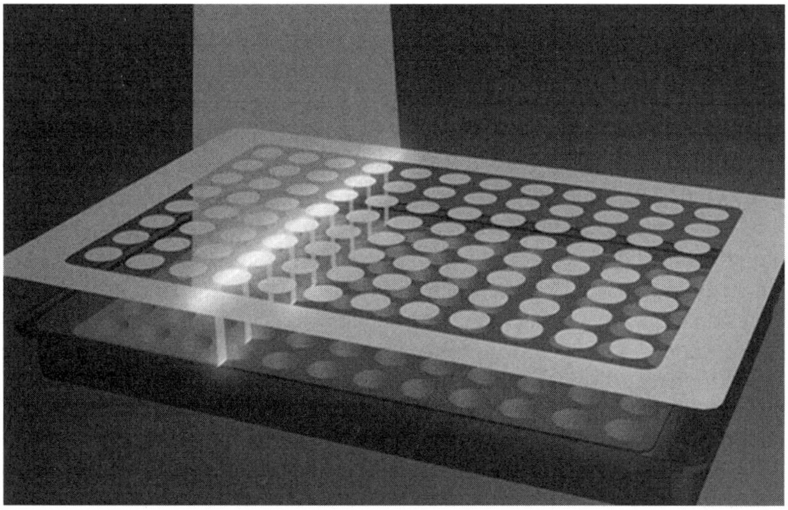

Figure 1.7.2 Mask approach to laser welding of plastics. (Courtesy Leister Technologies.)

this line provides a 1- to 2-mm (0.039- to 0.079-in.) by 30-mm (1.18-in.) weld joint in a cycle time of 1 to 2 s.

Unique Variations

While most plastic welding is performed on two-dimensional surfaces (flat surfaces), the process also works in three dimensions. In one process variation which works in two or three dimensions, a laser is focused onto a freely rotating glass sphere that pushes down on the two materials to be welded. While the sphere rolls on the component, it applies continuous pressure at the contact point and the ball is traversed along the desired weld path. The glass sphere transmits the beam through the parts similar to traditional contour welding. Since the ball is also the clamping device, no external clamping is required, and since it can be moved in three dimensions, it can track in three dimensions continuously. This contour variation is highly suitable for large automotive assembly applications requiring robotics to full-scale textile seaming operations carried out on vacuum tables.

In a similar manner, a glass roller (in some sense similar to a kitchen rolling pin, but much narrower) can be used that deliver a line-shaped laser beam equal to the width of the roller itself. This welding concept, although able to provide wider, potentially stronger seam strength, is limited to relatively linear paths. This variation is referred to as roller welding. This approach allows welding to internal or external features which are at slightly different heights like internal ribs higher or lower than the outer case.

Radial bonding refers to laser welding close-fitting cylindrical components (rod and collar for example) by deflecting the beam so that it hits radially on the outside symmetrical surface of the inner part. By using tight fits, no clamping pressure is required. This variation is suitable for different diameters, requires no rotational movement and no clamping, and can provide high throughput.

Diffractive optical elements (DOEs) can be designed to change the shape of a normal spot laser beam to produce a tailored power density for complex weld needs. A DOE can also be used as a variant of traditional simultaneous welding to produce weld contours of varying power densities and focal distances. This can be useful for sensitive part areas as well as non-planar assemblies. DOEs may be used in simultaneous, quasi-simultaneous, and contour welding processes. These special lenses must be calculated and simulated to achieve the desired task-specific optical properties in advance of their manufacturing.

Another variation denoted as "hybrid technology" welds complex 3D parts by combining the normal welding laser with special halogen lamps to apply additional preheating energy to the welding zone. That increases welding speed and reduces the internal stresses on the weld seam. One of its biggest uses is in assembling structural components for automobiles. The specific applications are automotive lighting and motor assemblies.

Weld energy is generally absorbed at a depth of 0.1 to 0.5 mm (0.0039 to 0.020 in.) within the lower part.

Applications

Laser-welded plastics are found in all commercial industries and applications. They can simplify closure interfaces, gas tight seals, and characteristically invisible seams. They can be performed on colored materials, and materials translucent or opaque to the human eye.

In addition to automotive applications like taillight bonding, the medical world also employs the process to bond parts having internal pressures of 4 bar (58 psi), welds free of particles, and parts that must meet stringent hygiene requirements for human use. Seams only 2 mm (0.079 in.) wide are used in some of the medical applications. Many applications require the welds to be tight closures that prevent moisture seepage into or loss of environment internally. Micro fluidic components are a popular application because they are sensitive to many of the manufacturing concerns aforementioned in this text. Laser welding effectively minimizes or eliminates the majority of these issues and is well on its way to being a household name in joining and sealing blood analysis and micro fluidic lab-on-a-chip products.

Ostomy bags are another laser-welded product. In this instance, it is for a more flexible product than other applications. Catheters and atomizers are other medical applications that use laser joining.

The advantages of joining with laser heating include the following: It provides noncontact delivery, flexibility in part geometry and design, and a minimum thermal load on the welded parts and subcomponents; the welded seams are of simple flat-flat geometry; no particles, flash, or expulsion are released; it is highly accurate and repeatable; and it is fast. In addition to what one might consider hard or firm plastic parts, the process can weld thin polymer films to conventional plastic parts.

Materials Selection

Almost all thermoplastic plastics and thermoplastic elastomers can be welded. The list includes ABS, PA, PC, PP, PMMA, PS, PBT, and fiberglass reinforced plastics. The welding seam strength is typically about the same as that of the base material. Transparent (to laser frequency) plastic can be welded to dark plastics, but color pigments in the plastic can affect the weld and spread the laser light in unwanted ways. Soot and special pigments are often added to weld dark-to-dark and transparent-to-transparent combinations, as well as colored plastics. Designers need to work with material suppliers or manufacturers to define the internal constituents that lasers weld well. Some of these constituents may not be biocompatible and some may not have passed required federal or FDA testing for medical-related applications.

Table 1.7.1 provides some material combinations that do and that do not work in laser welding.

Joint Design

For laser welding (plastic assemblies), lap joints, T joints, and radial joints are preferred. Butt welds are largely not compatible with plastic laser welding and generally do not

Materials combinations

Absorbing joint partner / Transparent joint partner	EVA	PE-LLD	PE-LD	PE-HD	PP	COC	COP	ABS	ASA	SAN	PMMA*	PC	PC/ABS	PS	SB	PET	PBT	PVC	PUR	POM	PA 6	PA 66	PA 11	PA 12	PPS	LCP	PES	PEI	PI	PSU	MABS	TPE	PEEK
EVA	○																																
PE-LLD		○																															
PE-LD			○	●	●						●			●		●				●	●	●											●
PE-HD			●	○	●						●									●	●	●											
PP			●	●	○						●									●	●	●										●	
COC						○																											
COP							○																										
ABS								○	●	●	●	●	●	●		●	●	●		●							●			●	●	●	
ASA								●	○	●	●	●	●				●	●												●			
SAN								●	●	○	●	●		●		●	●	●												●			●
PMMA*			●	●	●			●	●	●	○	●	●	●			●	●		●										●			●
PC								●	●	●	●	○	●	●		●	●	●				●			●		●			●		●	
PC/ABS								●	●		●	●	○																				
PS			●					●		●	●	●		○			●	●			●	●					●			●			
SB															○																		
PET			●					●		●		●				○	●	●															
PBT								●	●	●	●	●		●		●	○	●			●						●			●			
PVC								●	●	●	●	●		●		●	●	○												●			●
PUR																			○														
POM			●	●				●			●									○							●						
PA 6			●	●	●									●							○	●		●									
PA 66				●	●							●		●							●	○										●	
PA 11																							○										
PA 12																	●				●	●		○			●						
PPS										●		●								●	●	●			○		●						
LCP																										○							
PES								●				●					●			●							○			●			●
PEI																												○					
PI																													○				
PSU								●	●	●	●	●		●			●	●									●			○			●
MABS								●																							○		
TPE					●							●										●										○	
PEEK			●							●	●							●									●			●			○

*PMMA: Caution with plate material cast PMMAs are not weldable.

- ○ Excellent weldability when a laser transmission is present
- ● Good weldability when a laser transmission is present
- ● Weldability is possible but should be checked with trial runs

TABLE 1.7.1 Material Compatibility Chart (Courtesy Treffert GmbH & Co.)

create sufficient bond strength if any at all. As explained earlier in this chapter, successful bonding requires an upper laser-transparent layer and a lower laser-absorbent layer on the assembly to be welded.

Joint design and welding adequacy can be inspected by several means. Leak rate and burst pressure testing is one approach. Water bath leak testing is used and up to 6 bar pressures are commonly used in both hot and cold environments. Light microscope evaluation is sometimes useful as are transmission testing and differential scanning calorimeter (DSC) analysis of plastics. DSC thermal analysis for polymers and plastics is used for research and development, quality control, and product and process development.

Design Considerations

- Welding of transparent/transparent or white/white parts is a major restraint. There are very limited solutions for these combinations. The use of 1.7-μm wavelength to 2.0-μm wavelength lasers solves some of these issues and an IR-absorbing coating also removes some of the restraints.
- A clear polypropylene (PP) tubing can be welded to other PP components at 1.7-μm wavelength, which is not a standard laser weld frequency for plastics. At this frequency, a high-power diode laser can weld successfully with a run time less than 10 s. This is possible because of a unique carbon-hydrogen bond in the material. Using a high-power diode laser emitting at this wavelength, both parts of the tube are welded together successfully with a process time of below 10 s. That frequency is now being introduced on some machines to weld other clear materials also.
- Adding absorbers in parts destined for medical applications absorber often results in costly and time-consuming NDA-approval efforts.
- One solution for material combinations that have not been successful in the past is to work directly with material suppliers to develop a material variation that can be welded.
- Another solution to many of the near impossible to weld components is a solvent-based coating called Clearweld.[1] Clearweld is applied with a liquid dispenser onto the interface before welding. The coating absorbs light and becomes a focal point for the laser. That creates local heating and joining. Any of the basic welding processes above can be used with Clearweld and joining will only occur where the laser beam hits Clearweld. Thus if the beam is a line of curtain which traverses faster than the contour method, only the areas having the coating will be affected.
- The quality of transmission welding depends on the flatness of the mating parts, the thickness of the transparent component, and the welding pressure.
- Intimate contact is required between the top and bottom pieces in the desired weld area.
- Clamping force is required directly above and below the desired weld area.

[1]Clearweld is a registered trademark of TWI, Ltd. and the (Oval) logo is registered trademark of Gentex Corporation. Clearweld was developed by TWI in conjunction with Gentex Corporation.

- Weld area needs to be in the same plane for weld width to be consistent.
- Clean surfaces are mandatory.
- The transmission and absorptive properties of the materials are the first consideration especially when using semi-crystalline materials containing fillers (i.e., glass, mineral, etc.)
- PVC contains DEHP which is a potential carcinogen, but elastomers like TPU and TPE are good substitutes which do not contain DEHP. These substitutes often used in medical applications are optically clear and are weldable with lasers that work in the 1500 to 2000 nm frequency.
- In assemblies subject to high pressures and requiring hermetic sealing, a tongue and groove design can be incorporated to give parts linear stability in the weld area by reducing the effect of shear forces.
- For thin-walled parts, add a flange to provide reasonable clamping pressure in the weld area.
- When two materials cannot be directly welded by lasers, it may be possible to employ laser staking which involves melting slightly a projecting cylinder such that it overlaps the hole in which it projects. No mechanical staking is required and it is a process that has been used widely before lasers came to the plastic frontier (hot tool melting).
- Welding of any nature requires some definition of "how much," "where," "how," and "how it is to be verified." Designers need to work with the manufacturers to define these in satisfactory terms. Definition of acceptability of weld invisibility and visibility need to be part of that definition.
- Bonding two flat plates together requires a lot of pressure to have the plates fit tightly. A raised rib in the bottom part allows melt collapse without needing extreme clamping pressure.
- Mechanical limiting stops can be molded into one or both of the parts, but they are not normally used or recommended because they can cause internal stress after the parts cool.
- If some of the melted material flows out to form flash, add a small gap along the weld path to collect the weld flash.
- Clamping requires some area for the clamps. Allow 0.5 to 1.0 mm (0.020 to 0.038 in.) on both sides of the weld seam.
- Rough or textured surfaces scatter light, both visible and NIR wavelengths. That reduces the amount of energy that will either be absorbed or transmitted by the material. Smooth surfaces in the bond area are desired.
- Light deflects away from radius edges or ribbed surfaces.
- Butt welds are the least strong welds, so they should be avoided where possible. Use a lap joint, T joint, or a radial joint.
- Keep molding gates out of the path of the laser.
- For best results, upper layers should not be thicker than 0.3 mm (0.012 in.).
- While most parts that are laser welded can be held in the hand, parts as large as 1.4 m (55 in.) have been welded as well.

- If precise control of upper and lower positions is required, design in features (lugs or pins) that capture the second part.
- Stacked thin transparent layers can be joined from the melt up action of the bottom part. It is not a common approach but is done.

References

Anonymous, Innovations in Laser Welding of Thermoplastics: This Advanced Technology is Ready to be Commercialized, BASF Corp. Report 2002-01-2011, 2002.

Anonymous, *Laser Plastic Welding Design Guidelines Manual*, Version 4.0, LPKY Laser & Electronics, February 2016.

Gubler, U.; Vogler, D., Diffractive Beam Shaping for Enhanced Laser Polymer Welding, Proceedings of the SPIE, Volume 9351, id. 935110, 8 pp., March 2015.

Kocheny, S. A.; Zybko, J., Three Approaches in Utilizing High Power Diode Laser to Join Thermoplastics, ANTEC 2002, Annual Technical Conference, San Francisco, CA, May 5–9, 2002, Volume 1, pp. 1115–1119, 2002. Brookfield, CT: Society of Plastic Engineers.

CHAPTER 1.8

Designing for Laser Soldering

Rick Schiffer

Apollo Seiko Ltd.
Bridgman, Michigan

Mitsushige Seto

Apollo Seiko Ltd.
Japan

The Process

Laser soldering is the process of using a semiconductor diode as the heat source to provide very localized and focused laser energy to heat up a pad or component lead on a substrate in preparation for soldering. This focused energy does not disturb any of the components in close proximity nor does it reflow any previously soldered joints on a PCB (printed circuit board), substrate, or other part.

This technology is also referred to as noncontact soldering as there is no physical touching of the items being soldered such as with conventional iron tip or hot bar soldering. A laser can be utilized to reflow solder paste, solder pre-forms, or to melt cored wire solder being fed into the joint area.

A laser beam is created by a laser generator (Fig. 1.8.1) which is directed into a small diameter fiber optic cable. Typical diameters are 0.2 to 0.6 mm (0.0078 to 0.0236 in.). The laser light is then focused via an input lens and output lens which shapes the beam to the desired size. Many lens combinations can be utilized to achieve the desired beam diameter and focal distance. Typical focal distance is approximately 75 mm (2.95 in.) from output lens to substrate being soldered.

Specific laser wavelengths are considered ideal for laser soldering. Typically, wavelengths in the spectrum of 800 to 1000 nm are well suited for soldering with a focused laser beam. A wavelength of 808 nm is better suited for ceramic substrates while 980-nm wavelength is typically utilized for soldering FR-4, wires to components, magnet wire, etc. However, this is not a hard and fast rule but only a general guideline. Testing is required in order to determine the best wavelength for specific applications.

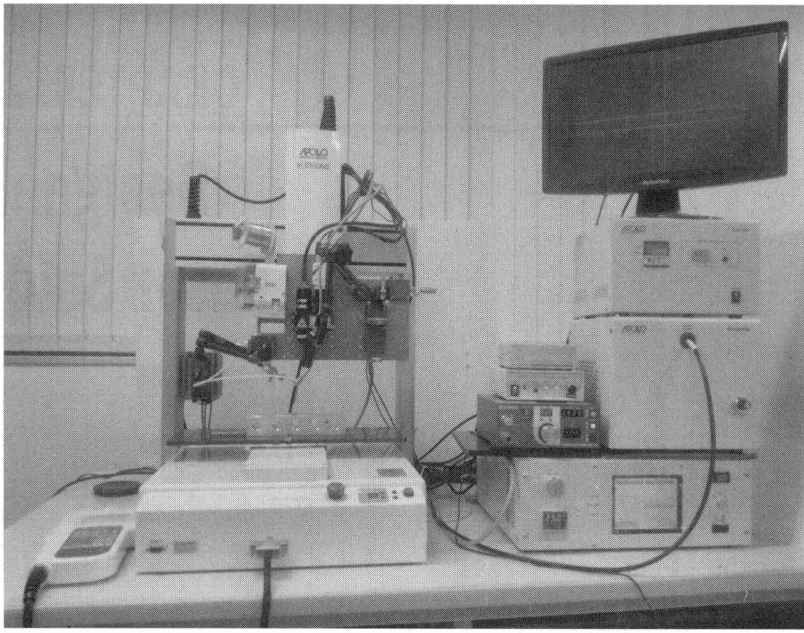

Figure 1.8.1 Tabletop laser soldering system.

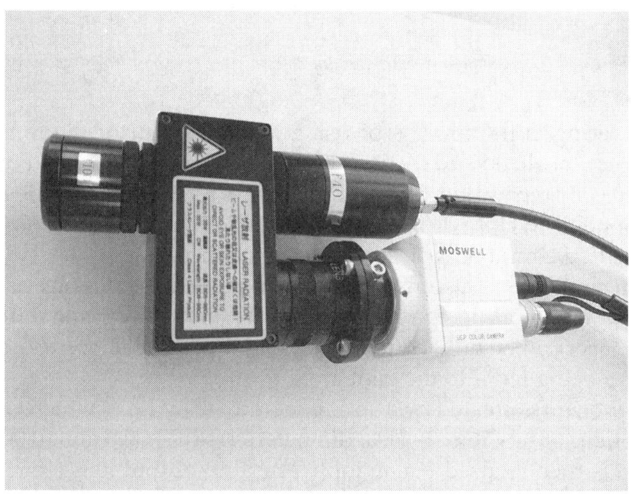

Figure 1.8.2 Camera and laser soldering lens system.

A coaxial camera is normally paired with the lensing system (Fig. 1.8.2) to allow for teaching with the actual laser beam. A pilot light (low laser power: ~500 nm) is used to create a beam to be used as a guide for teaching. A line generator is also utilized to create "crosshairs" to help determine the location of the laser during soldering (Fig. 1.8.3).

The shape of a focused laser beam is normally round. However, with the use of a "slit" plate (aperture cut to conform laser to shape of pad), the laser can be shaped to match the pad geometry to focus the laser energy on the complete solder pad. This is

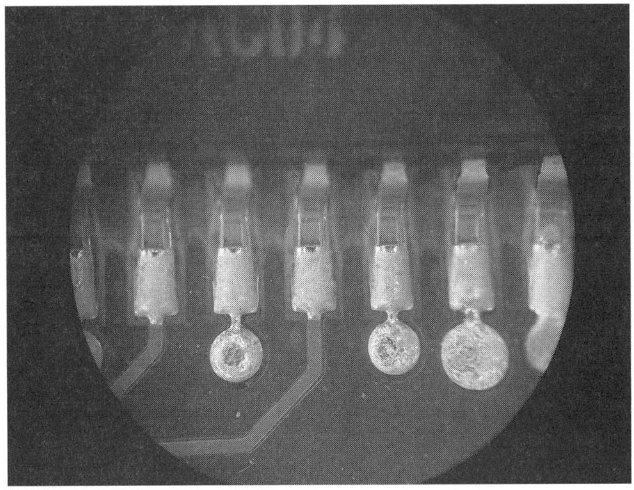

FIGURE 1.8.3 A line generator helps position laser focus.

common in micro soldering applications where very low laser power used as pad sizes can be as small as 20 × 50 μm (0.0008 to 0.0020 in.).

Laser power is normally measured in amperes or watts. Typical laser wattages utilized for soldering are 30, 50, and 80 W. However, in micro soldering applications, wattages can be as low as 1.5 to 5 W of power.

Typical Characteristics and Applications

Laser soldering is normally selected over conventional soldering for specific applications such as ceramic substrates which are normally white colored and absorb laser light easily. The laser process is also chosen for applications that are densely populated with surface mount technology (SMT) and through-hole components and there is not enough real estate to locate an iron tip and feed solder wire. Technology is providing for smaller and smaller electronic components and packages and it is just not physically possible to solder these with conventional methods. A focused laser beam must be utilized. Micro soldering applications in the medical and electronics industries such as body implants, catheters, and flexible connectors with very small gauge wires (40 AWG and above) require the use of a micro laser and slit plate. Spacing between pads can be as little as 0.1 mm (0.004 in.). A standard focused laser is too large for these types of applications and would burn the solder mask between the pads.

Surface finishes (plating) and component base metals should be considered when designing, since materials such as gold (Au), silver (Ag), and copper (Cu) are not well suited for laser soldering as they reflect the laser light, thus making heating more difficult (Fig. 1.8.4); however, it is possible to laser solder to these surfaces. Plating and base metals such as tin (Sn) and nickel (Ni) absorb the laser light well and are heated up easily. Tin plating is easily fused with solder wire and/or solder paste as these materials are also high in tin content, normally a minimum of 96.5% in most lead-free alloys.

Since most laser beams for soldering are round in shape, a round or teardrop solder pad provides for optimal thermal energy dispersion.

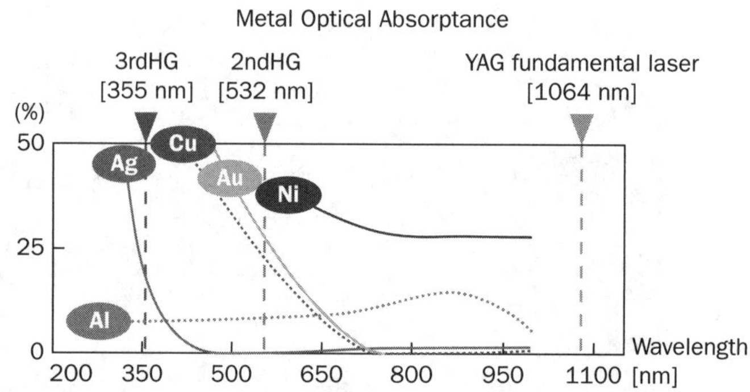

FIGURE 1.8.4 Metal optical absorptance.

Economics

Laser soldering equipment is typically more expensive than other soldering equipment. Lasers for soldering are considered as Class IV diodes by ANSI Z136. These types of systems must be guarded in order to bring the classification for safety down to a Class I laser.

Laser soldering and contact soldering (iron tip) typically have very similar cycle times. In many cases, a laser can provide higher yields than an iron tip.

Also, with a laser, there are no consumable costs as compared to conventional selective soldering methods. Over a specific period of time, using a laser will allow for a better return on investment (ROI) than an iron tip system which requires replacement of iron tips that wear out.

A major advantage of soldering with lasers is cleanliness of the process. There is no contact with the substrate or components so there is no contamination left on the PCB such as burnt flux residue that can migrate from an iron tip or copper that may leach from an iron tip as the tin plating wears away. A laser will not physically come into contact with other components on the assembly. This is important in densely populated mixed technology PCBs.

Typical cycle times of a laser solder joint range from 0.5 s up to ~5 s depending on thermal mass and copper layers of PCBs. Commercial systems range from job shop short run to high-production continuous lines.

Suitable Materials

Lasers can provide solutions for selective soldering challenges on most types of substrates. These substrates include (but are not limited to):

- FR4 PCB
- Ceramic PCB
- Wire harnesses
- Flexible printed circuits (FPC)
- Hybrid assemblies
- Insert molded connectors

Plating and base metal finishes should be a metal that is non-reflective and substrates with a lighter colored background are best suited for laser soldering.

Good candidates for laser soldering are applications that are difficult to solder by hand due to space as well as complex scenarios, for example, densely populated PCBs in which joint access is difficult. Small profile applications which may require microscopes for soldering are well suited for soldering with a laser.

Design Recommendations

Through-Hole Pad Design

When designing pads for through-hole laser soldering, use a teardrop shape where possible. The reason is that the laser light can be focused on the larger portion of the pad (to absorb the laser energy) and not directly on the terminal or through the space around the terminal. Laser light can penetrate through the open space around the terminal and contact the component body underneath the PCB. In some cases, the energy is strong enough to provide damage to the component. Figure 1.8.5 shows a teardrop pad, terminal and laser beam focal point. This is ideal for obtaining optimal results without any damage to the substrate or component.

Lap Joint Pad Design

For a "LAP" style joint (wire to solder pad) it is recommended to have pad exposure outside of the wire (on both sides) approximately 50% of the diameter of the wire (Fig. 1.8.6). Pad length is determined by the amount of stripped wire end to be soldered. The wire end should also be placed on the pad to meet IPC standards.

Connector Selection

Due to the fact that the laser energy can travel through a plated through hole, selecting a connector that has a light body color, namely white color will make the laser process much more robust (Fig. 1.8.7). The white color of a connector body will not absorb the laser light as readily as a dark-colored connector, thus reducing the risk of burning the component.

Fixturing

Proper fixturing is very important to the success of any laser soldering application. In the example above of the LAP joint application, it is important to have a fixture to locate wires repeatedly. Most robots for soldering have a repeatability of ±0.1 mm (±0.004 in.).

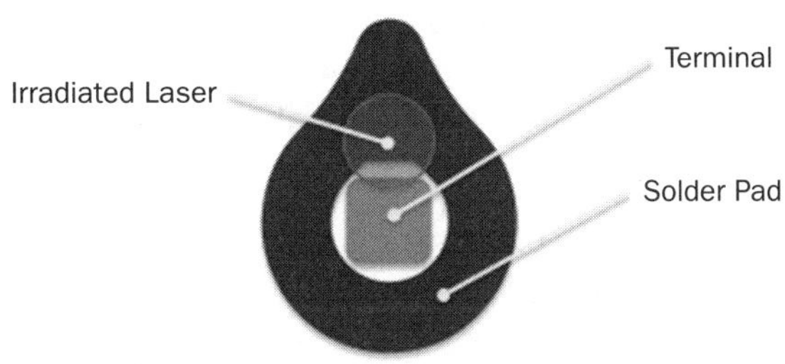

FIGURE 1.8.5 Teardrop shape pad for optimum energy dispersion.

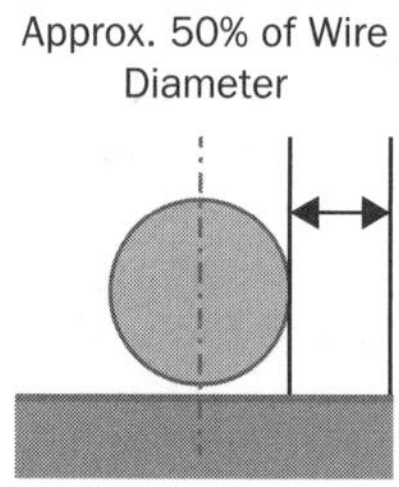

Figure 1.8.6 Good lap joint design.

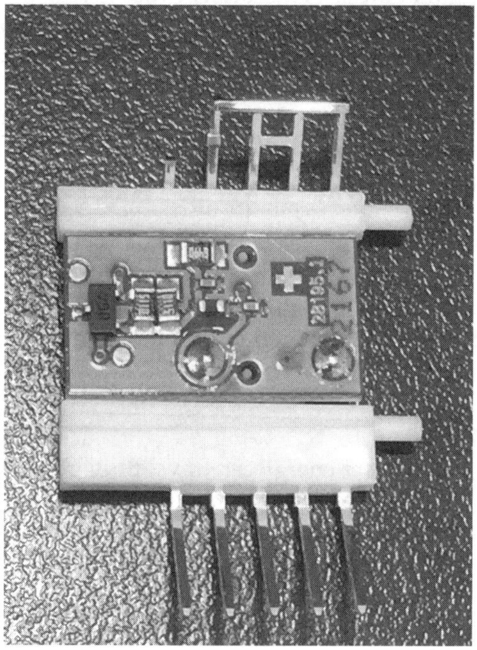

Figure 1.8.7 White connectors reduce burning risk.

If the wires do not locate consistently, the laser may not contact the wire but only the pad and thus cause burning of the substrate.

It is also important to consider the material of the fixtures being used. Aluminum or stainless steel is commonly utilized for all types of soldering fixtures. Many times, inserts of Teflon or phenolic material are used at the soldering points to minimize the heat sink to the part being soldered. The goal is to be careful not to create a large heat sink in the fixture, thus making soldering more difficult and cycle time longer. Also, the length of time that the laser irradiates on a specific pad, component, etc. is directly related to the possibility of burning.

Another method of designing fixtures to minimize heat sinking from the parts being soldered is to create cutouts in the fixture under the solder area (Fig. 1.8.8).

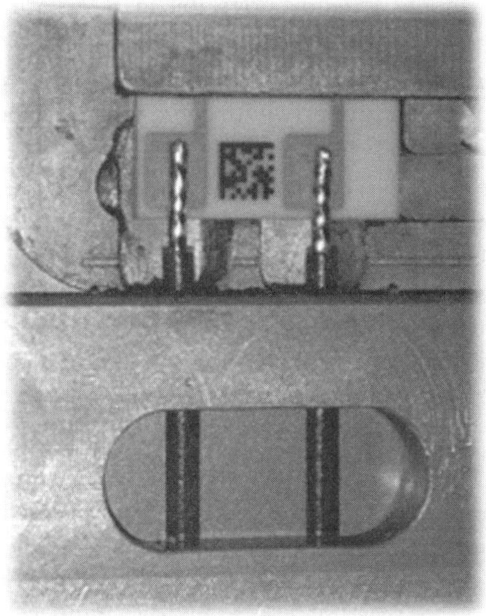

Figure 1.8.8 Fixture uses recess to position wires.

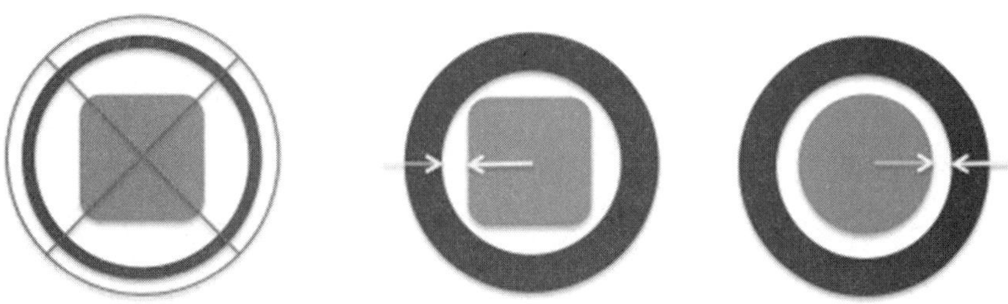

Figure 1.8.9 Gaps between outer lead and inner pad need to be greater than 0.5 mm (0.020 in.).

Lead-to-Hole Ratio

Designing the PCB pad and selecting a lead diameter is an important factor in the success of laser soldering. Guidelines to be used when designing the pads include:

- Larger width pads are better than smaller width pads as they absorb the laser energy more easily.
- Lead shape does not have much impact on laser results (round, square, or rectangular); all can be soldered with a laser.
- The gap between the outer lead edge and inner pad should be approximately 0.5 mm (0.020 in.) (Fig. 1.8.9).

- The lead should extend above the PCB surface by 0.5 to 1.0 mm (0.020 to 0.040 in.)—not taller if possible.
- Plating on the terminal and pads should be tin to create the intermetallic bond and fuse together properly with cored wire or paste (other metals can be used if required).
- Allow for a gap in relation to other components by approximately 0.5 to 1.0 mm (0.020 to 0.040 in.) from the outer edges of the solder pads (if possible to avoid reflection onto components).
- Components taller than 5 mm (0.197 in.) should be kept at a distance of 1.5 to 2.0 mm (0.060 to 0.080 in.) away from the laser solder area. The laser beam is conical in shape until it reaches its maximum focus, so it has a larger diameter above the PCB surface.

CHAPTER 1.9

Design for Laser Cladding

Wayne M. Penn

Alabama Laser
Munford, Alabama

The Laser Cladding Process

Laser cladding is a form of additive manufacturing where filler metal is fusion welded (joined together by melting the filler to a shallow molten pool) to a base metal. Generally, the purpose of the clad is for selection of an alloy that will give superior part performance over the base material. The laser provides the heat source producing the weld, but may not be the only heat source as in the case of hybrid welding (filler material preheated as in hot wire).

Laser cladding adds a performance layer to help mitigate environmental situations where the base material is weak. For example, for improved corrosion performance laser cladding with a high chrome nickel alloy may be chosen, or for improved erosion performance a laser clad of metal matrix tungsten carbide materials may be applied that improves the part wear properties. Laser cladding can also be used to improve the part toughness, for example as a wear surface on hydraulic cylinders.

The laser is focused to a sufficiently high power density for rapidly melting the surface of base material and the incoming filler material that results in a lower heat-input weld than conventional welding. The lower heat input results in a clad-weld-metal chemistry that is closer to the original filler-material chemistry that has a lower-weld metal dilution with less part distortion. Weld dilution is a mixing of the metal elements of the filler with the metal elements of the base layer. The goal of the lower heat-input weld is to preserve the chemistry purity of the clad layer, thereby preserving the protection properties of the clad material.

Laser cladding produces a layer of deposited material that is metallurgically bonded to the surface of the part. Laser cladding single layers may be as thin as 0.635 mm (0.025 in.) and as thick as 2.54 mm (0.100 in.) yet they preserve metallurgical integrity. The chemistry of the deposit and the metallurgical bond can be maintained within this range with proper laser cladding techniques and parameters. Multiple layers can be deposited for a thicker clad or for additive manufacturing by creating 3D features on the base part.

The capabilities of the technology will vary depending on the power and process of the cladding machine or facility. The results depend on several factors:

- Material handling and stages (important for large volume jobs)
- Base material preparation (cleaning and possibly pre-machining)
- The technology of the laser cladding machine and operations
    - Laser power
    - Laser cladding head
    - Laser cladding deposition technology (powder or wire)
    - Clad base material fixturing and cooling
- Post-clad heat treatment

Laser Cladding and Conventional Welding

Successful cladding facilities have a practice and working knowledge not only with lasers but also with conventional welding. Experience and depth with conventional welding lends a lot of capability and understanding for laser welding and cladding. Laser cladding is still welding; however, when optimized it can produce a lower heat-input weld.

Lasers are not the only processes that can overlay or clad surfaces. Figure 1.9.1 illustrates several approaches and among them are:

- GTAW—Gas tungsten arc welding, commonly known as TIG
- GMAW—Gas metal arc welding, commonly known as MIG
- PTAW—Plasma transfer arc welding protected tungsten electrode
- SMAW—Shielded metal arc welding, commonly known as stick welding

Conventional welding processes, however, cannot produce the high power densities achievable with the laser. One of the net results of this is that the laser with the

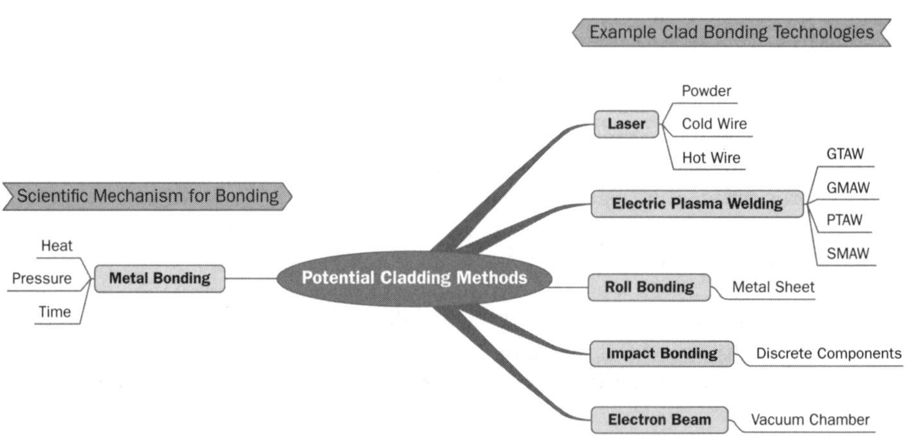

Figure 1.9.1 Cladding-bonding mechanism and example technologies.

higher power density will produce welds at a higher speed, thereby lowering the heat per unit length.

The laser heating mechanism is totally different from conventional welding. When the laser light (made up of photons) is focused onto the base material, several effects happen, including:

- Absorption—metals absorb some of the laser energy
- Reflection—metals reflect some of the laser energy
- Transmission—only for materials that have some transparency

Photons are not charged, enabling focus ability of the laser beam. Focus ability increases the power density by making the beam diameter smaller. Then the absorbed laser light or photons rapidly heat the material directly on the surface. The power density for cladding or welding has to be high enough to produce the molten pool very quickly. This fast melting of the surface or welding pool enables rapid deposition cladding for thin layers that will have good metal chemistry for the filler material.

The laser provides the power to weld the filler material to the base or previous clad pass. Laser sources used for laser cladding include, but are not limited to, direct diode lasers, fiber lasers, disc lasers, or carbon dioxide lasers. The direct diode laser, fiber laser, and disc laser operate in the near-infrared spectrum where the light absorption of metals is higher than the medium-infrared wavelength of the carbon dioxide laser. Cladding with a carbon dioxide laser beam generally works with powder where the scatter and multiple reflections of the individual powder grains generate enough absorption to melt the powder and create a molten puddle on the base material.

Laser Cladding with Powder versus Wire

The cladding filler material or weld filler material can be either powder or wire. Wire can be fed as either cold wire or hot wire. Hot wire is preheated before it enters the molten pool. Powder nozzles deliver the clad powder material to the molten pool either coaxially or laterally. If the powder flow is too high, lack of fusion or porosity may result at the base material interface.

Laser Cladding with Powder

Traditionally, the most popular method for producing a clad layer with laser technology uses powder as the filler material (see the powder option in Fig. 1.9.2).

Laser Cladding with Wire

Hot-wire laser cladding uses a hot-wire power to reduce the cooling that results when cold wire is fed into the molten pool. Using cold wire results in a slower deposition rate and increased heat input. For lower heat and higher production speeds, hot wire is a better choice.

A hybrid-wire feed process combines an external heat source for the filler wire with the laser for producing a clad weld. By preheating the incoming filler material to a plastic temperature, just below melting, then the laser only needs to add minimal heat to the filler to bring it to the melting temperature.

Additionally the laser beam absorption on the base material brings the base temperature to the melting point, creating a molten pool. The goal is to adjust the laser

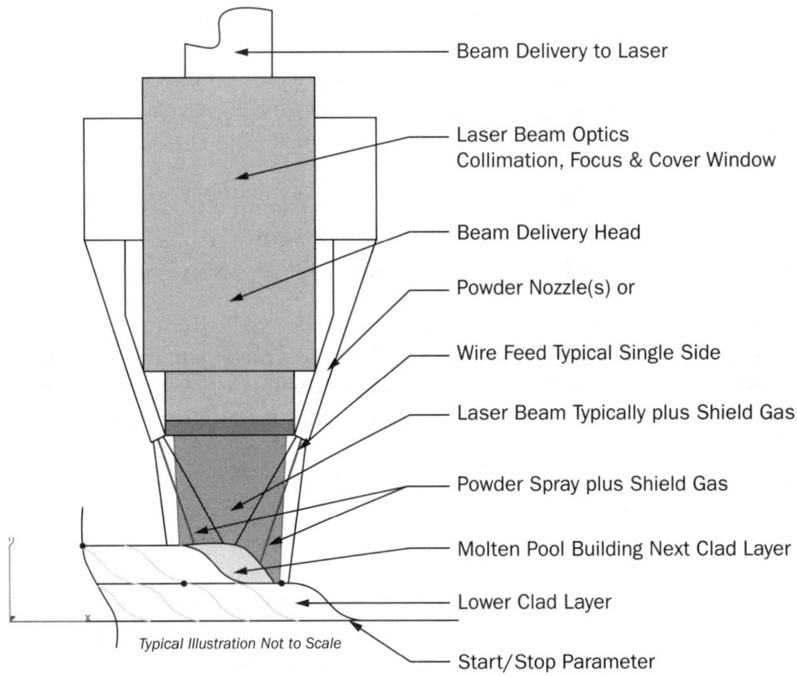

Figure 1.9.2 Cladding head illustration.

power density and process speed to produce a shallow molten-pool depth. The shallow molten pool reduces the dilution of the chemistry of the base with the chemistry of the filler clad material.

With the addition of the preheated filler, the metal flows at high velocity into the molten pool with the motion system moving the pool over the part providing for the deposited layer. Higher velocity motion (feed rate) for a given filler material rate results in a thinner clad thickness. The power input of the laser is proportional to the material speed; higher speed or feed rate needs higher laser power.

Applications and Cladding Variables

Applications for laser cladding often address two needs. The first is to improve corrosion prevention and the second is to improve wear properties. Cladding for corrosion prevention improvements may be accomplished using super alloys that are known for corrosion resistance. Improvements to the wear properties may be accomplished with cladding alloys that are harder or tougher than the base material. Application industries include:

- Power plant applications for corrosion and erosion improvements for boiler components may use super alloys to extend the life of various components.
- Oil and gas production utilizes laser cladding for corrosion and erosion protection of new components and the repair of components.
- Steel mills have many rollers that wear out—laser cladding can be used for rebuilding the surface with superior materials.

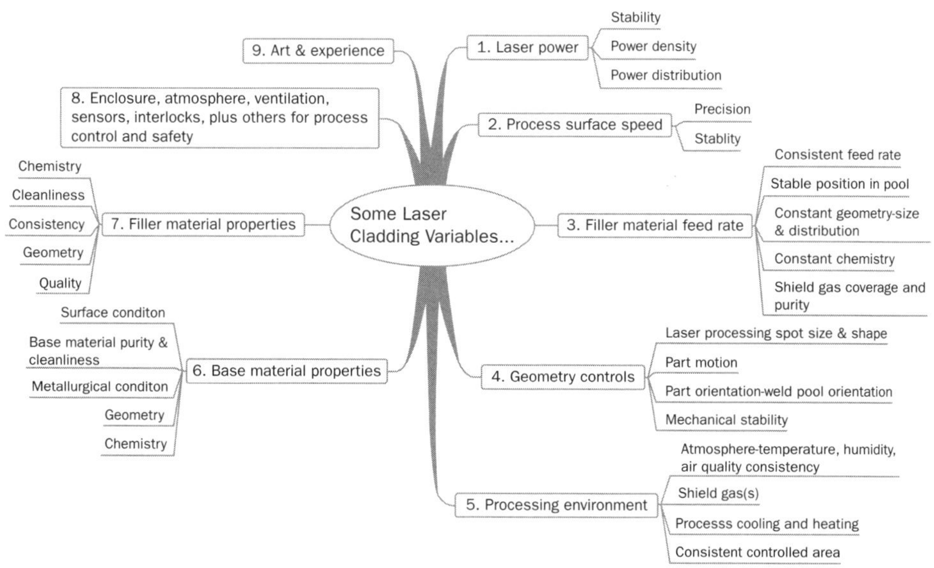

Figure 1.9.3 Some cladding variables.

- Aerospace components—with the necessary aerospace qualifications and process procedures, laser cladding has potential applications for surface rebuilds and repair work.
- Part remanufacturing and repair cover an extensive area of potential applications. Precision components and the need for solving issues with part corrosion or erosion performance are typical enablers for laser cladding.

Pre-machining and post-machining are often needed to enable and support the application. Remanufactured parts may have contaminated surfaces with debris embedded into the base layer. This contamination can produce defects in the laser clad bond and deposit. Pre-machining can be used to remove the contamination and machine down into pure base material. Then the missing material is added back to the laser clad deposit, resulting in a layer of protection for the base material with the improved properties of the clad.

After the cladding is complete, precision parts will need a post-machining operation. A lathe, milling, or grinding operation can then be used to bring the part into final dimension. The machining process needed will be driven by the machining characteristics of the clad. The clad material is often very tough and will require specialized cutting capabilities for machining the robust layer. Care must be taken on measurements and fixturing to ensure one does not cut through the clad layer. Bringing the part and clad layer into final dimensions where the desired clad layer thickness is achieved requires precision skills and techniques generally specialized in clad layer machining.

Figure 1.9.3 illustrates the many different parameters that affect laser cladding. Additionally the parameters are a function of one another, changing one may require changing a synergistic parameter also.

Laser cladding must be maintained in thermal equilibrium for consistent results. If the part heats up significantly as the process runs, it can change the metallurgical condition. Therefore either the part mass must be large enough for a constant rate of heat sinking and/or the processing time must be short enough so as not to significantly change the part temperature during the run. Cooled heat sinking may be needed for helping to maintain the temperature condition. The laser is constantly pouring in heat that must be utilized somewhere to maintain the equilibrium.

Filler Materials

The application of stainless steels and super alloys is common for laser cladding with both powder and wire. Materials that present extra challenges would include the harder alloys and materials that may contain extra ingredients especially for arc-based processes. For example, the extra ingredients may be utilized for assisting arc stabilization. For laser cladding, these ingredients are not needed and could result in erratic behavior from interaction with the shielding or metallurgy.

The base material condition, consistency, mass, and geometry are important for successful laser cladding. Generally, the base material will have to be cooled to an equilibrium temperature for consistent results from the start to the finish. The thinner the base material, the more important it is to keep it cool. For large mass parts with small clad areas, active cooling may not be necessary.

The rapid cooling of the clad-weld puddle generally makes the deposited alloy harder with a smaller grain structure. Table 1.9.1 lists some examples of clad materials and their filler forms that are used for applications on alloy steels.

Laser Cladding Production Performance

For consistent results, the parameter matrix for qualifying the procedure and the cladding processing procedure must be held constant. The requirements and details for work on code-qualified products are covered by the applicable standards.

Clad Material	Typical Laser Clad HRC	Filler Form	Example Application
Super Alloy 622	20s range	Powder or wire	Corrosion resistance
Super Alloy 625	20s range	Powder or wire	Corrosion resistance
SS 309	<20 range	Wire	Corrosion resistance
SS 316	<20 range	Powder or wire	Corrosion resistance
SS 420	50s range	Powder or wire	Corrosion and wear
Cobalt-based Alloy 6	40s–50s range	Powder or wire	Wear resistance
Maraging steel	40s range	Powder or wire	Strength and toughness
Tungsten carbide–nickel base	WC: >85, Ni: 50s range	Typically powder	Wear resistance

Table 1.9.1 Some Typical Laser Clad Materials

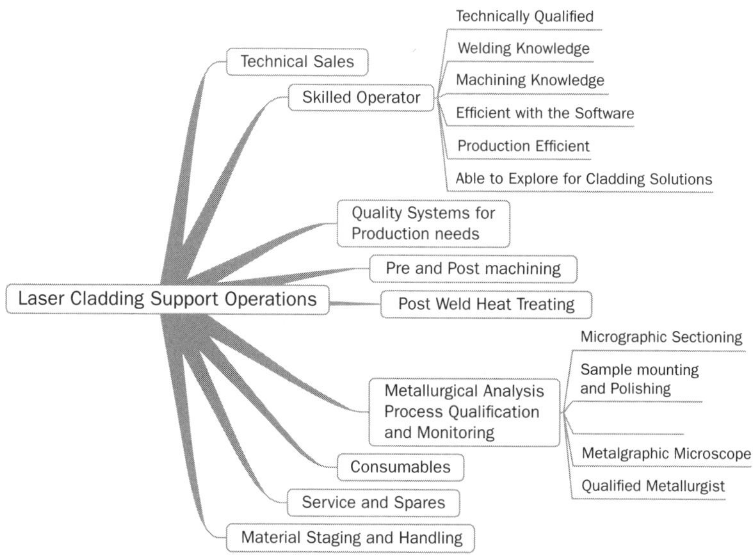

FIGURE 1.9.4 Laser cladding support operations.

Figure 1.9.4 shows the diagrams of typical elements of laser cladding support operations.

Metallurgy examination is important for finding and solving cladding issues which can include porosity, lack of fusion, cracking, and other metallurgy-related anomalies. Designers need to clearly specify the metallurgy details they need and measurement or sampling requirements. Metallurgical micrographic samples need to verify clad, bond, and heat-affected zone (HAZ) expectations. Micro hardness measurements should be specified across the clad, bond, and into the base material to document the HAZ extent and the parameters of the clad including the clad thickness.

A post-weld heat treatment (PWHT) could be needed to modify the hardness of the base material and/or clad. For applications requiring the cladded part to be bent or formed, the base material HAZ may be reduced by a PWHT. The parameters of the laser clad will affect the final size and hardness of the HAZ.

Optimized laser cladding can produce a narrow HAZ. Micro hardness testing specifications for clad products will generally contain the maximum values for this region. Lower heat input produces a smaller HAZ; however, the lower heat must be balanced out with the needs of the weld bond and metallurgical properties. Too much heat not only will affect the HAZ, it will also result in diffusion or movement of metallurgical elements across the clad and base material boundary, resulting in a different alloy than deposited. For example chromium will diffuse into the base material and iron in the base will dilute up into the clad. Heat-input effects include the weld "tie-in" bond, HAZ, and weld chemistry. Clad thickness will also be a factor in the severity of the weld dilution.

Laser cladding achieves lower dilution welds with smaller HAZs than those obtained by conventional plasma welding processes. The low-dilution laser clad enables chemical purities comparable to the filler material. The overlapping grains at

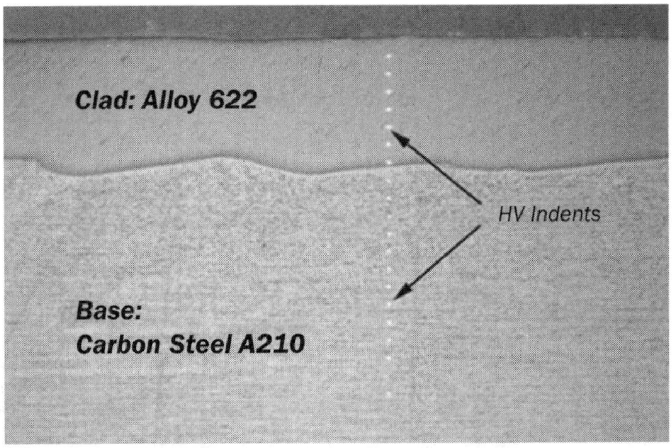

FIGURE 1.9.5 Cladding a steel part with Alloy 622. Vickers Hardness measurement points are shown as dots. In this view the clad surface is relatively flat due to the narrow view. The sides of the weld pool pass may normally show some uneven overlap (not visible here).

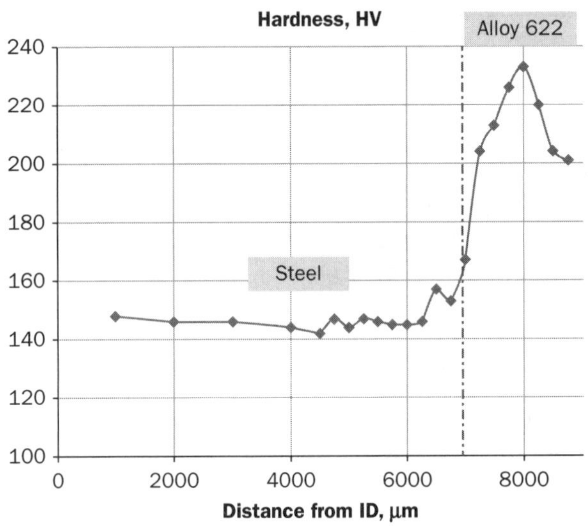

FIGURE 1.9.6 Micro hardness readings across the clad interface. Note small HAZ peak just above 6000 μm location, right below the clad layer in the base material (1000 μm = 0.039 in.); Y axis: Vickers Hardness.

the metallurgical bond may be only a few tens of micrometers across. Diffusion of the alloys will cover a wider band and should demonstrate a narrow HAZ in the base material.

Preheating and post-process heating may be utilized for heat treating and stress relieving to help achieve final material properties.

The laser clad micrographic sections in Figs. 1.9.5 and 1.9.6 show the transition from clad material on the top to the base material below the clad. Here it is evident just how defined and narrow the weld metallurgical bond line is.

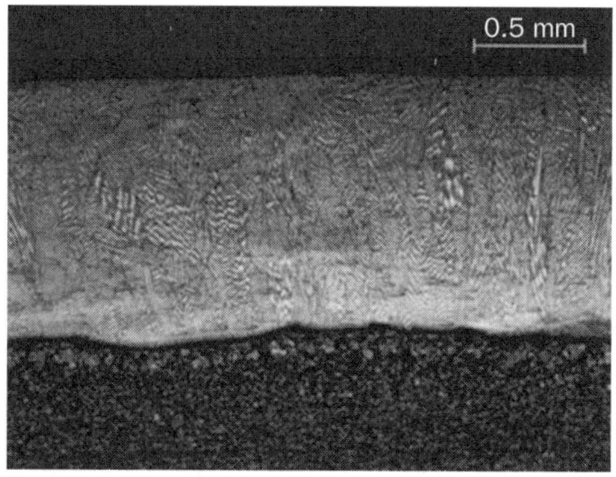

FIGURE 1.9.7 A stronger etch shows the grain structure of the clad as compared to the previous image.

Figure 1.9.7 shows a micrographic section of the grain structure of the clad material. The chemistry of this type of clad can be relatively the same down to within 254 μm (0.010 in.) of the base.

Depending on the materials and the condition of what is used, one can get various results by doing an experimental matrix and measuring the results. Results vary by all the parameters listed earlier.

CHAPTER 1.10

Laser Marking and Engraving

L. Mike Heglin

Rofin-Baasel
Devens, Massachusetts

Laser Marking

Both solid-state lasers—Nd:YAG (yttrium aluminum garnet), ND:YVO4 (vanadate), ytterbium fiber and gas—produce high-power monochromic beams that when focused through a lens produce very small spot diameters with enough power density to melt or remove metal or create an oxide layer. In nonmetals like wood and plastics, the beam creates marks by charring, browning, or melting the surface. With solid-state lasers, an optical device called a Q-switch is used to produce nanosecond pulse widths and very high peak powers at rates up to a 100,000 pulses per second which makes it the best choice for metal marking.

The laser by itself is just a light source and couldn't become a marking tool without the development of a high-speed two-axis optical galvo scanning head incorporating various optics elements (Fig. 1.10.1). To give the beam controlled positional motion, a special vector graphic driver and software program called Visual Laser Marker (VLM) controls the operation, the laser, and the motion of the mirrors in the X and Y axis to produce alpha numeric characters and complex graphics such as logos.

Marking costs are a function of many variables such as the quality required, materials marked, and real-time factory conditions including production line rates, capital, and operating costs. Lasers are often the obvious choice when traceability, quality, and repeatability are important factors in both high production and precision job shop applications. Marking up to 5000 characters per second makes lasers one of the fastest marking methods available.

Materials That Can Be Marked

Almost any material can be laser marked, including organic and inorganic materials. Hard and soft surfaces, shiny and dull, plated or bare, cast, machine, extruded, etc.

FIGURE 1.10.1 Galvo head with flat field lens to steer and focus the laser beam.

don't usually cause problems. Table 1.10.1 lists some materials that can be successfully laser marked. However, designers and manufactures need to note that some materials, especially plastic, and certain metal alloys, such as beryllium copper, give off toxic fumes, so designers need to consider the health risks associated with these materials and manufacturers should ensure the system is properly enclosed and exhausted. Lead alloys will similarly give off fumes that are hazardous.

The type and model of a laser that best meets the designer's requirements are determined by the plant manufacturing manager or process engineer, who determines the selection based on the following factors:

Material	Function of the mark (text; logos; 1D, 2D barcode)
Surface anneal or engraving	Depth of the mark, if engraved
Process rate or line speed	Permanence
Contrast for better readability	Surface roughness

There are many variations with each application, so several different marking methods may also be considered based on cycle time, mark quality, durability, and cost. Be sure that the suppliers have an on-site demonstration lab with the exact systems being recommended, so the engineer can witness their parts being processed and examine the results fully.

Selecting a Laser

Figure 1.10.2 shows the type of laser medium (solid or gaseous) and its comparative operating wavelength. Ultraviolet to near-infrared beams couple into metals and

Metals		Organics
Carbon steel	Nickel	Wood
Stainless steels	Titanium	Leather
Alloy steels	Carbide	Paper
Aluminum	Gold	Rubber
Copper	Silver	Glass
Brass	Tin	Stone
Molybdenum	Platinum	Diamond
Plastics		**Coatings**
Polyethylene	PTFE	Paint
Polystyrene	PET	Anodizing
Polycarbonate	Nylon	Inks
PVC	Epoxy	Oxides
ASA	POM	Films
ABS	PEI	Ceramics
PBT	PPS	Foils
PA	San	Zinc oxide
Acrylics	PMMA	
Ceramics		**Semiconductors**
Aluminum oxide		Silicon
Titanium carbide		Gallium arsenide
SiC		AiN
Boron nitride		Telluride
Porcelain		Sapphire
Zirconium		

TABLE 1.10.1 General List of Materials That Can Be Laser Marked

many plastics, while far-infrared (10.6-μm) lasers work best only on nonmetals but can remove coatings off metal. While the majority of lasers operate in the infrared region, certain benefits like good color contrasts are achieved in some nonmetals such as high-molecular-weight polyethylene (HMWPE) with lasers operating in the green (532-nm) or UV (355-nm) region.

Today, the main laser types used for marking include the fiber and Nd:YVO4 (vanadate) families. Fiber lasers are available from 10 to over 1000 W, but models configured for marking are from 10 to 100 W. Vanadate models are from 10 to 40 W plus these units can be modified to operate in the green (534-nm) and UV (355-nm) regions to achieve good marking contrast for some plastic materials. Most of these lasers are now air cooled, thus greatly increasing their operating life and reducing downtime and maintenance costs.

Both solid-state and CO_2 laser sources are compact and easily integrated into part handlers which can consist of conveyors, robot pick-n-place, and other

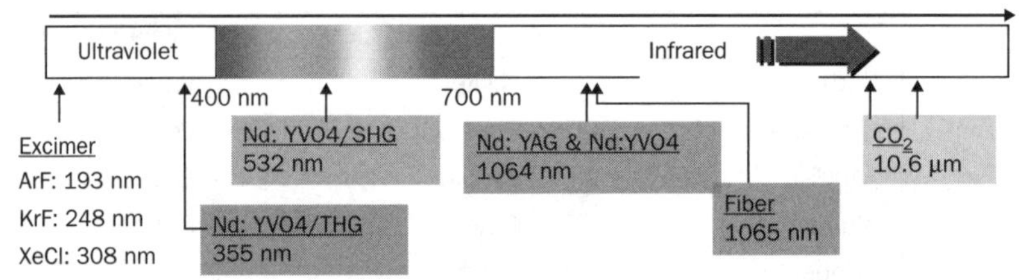

FIGURE 1.10.2 Marking laser wavelength comparison.

FIGURE 1.10.3 Tabletop full function marker. A wide range of custom stand-alone and automated workstations are also available.

machinery involving autoload handlers for high production. Manually loaded stand-alone machines are also widely available for small lot job shop production. A wide range of standard and custom workstations are available. Figure 1.10.3 shows a small tabletop manual load machine.

Marking Metals

Anneal Marking

This is the most often used surface altering method for marking metals. The process will produce a durable high contrast with no surface damage mark in metals. Studies have shown that the marks are typically only 1.25 to 5.0 μm deep (0.000050 to 0.0002 in.). A high power density and good absorption are needed to achieve a rapid thermal buildup in a matter of nanoseconds causing an oxidizing (dark) layer to form. Anneal marks can be produced on machined, cast, or ground surfaces with often negligible increase in part temperature or the generation of debris, and it is permanent in normal use. Examples are shown in Fig. 1.10.4.

Almost all metals will achieve good contrast marks with YAG, YVO4, and fiber lasers. This includes aluminum, titanium, copper, steel, stainless steel, nickel, brass, etc. When marking medical products, especially surgical tools and dental and orthopedic products made of stainless steel, titanium, and 17-4 stainless, if Product Engineering or

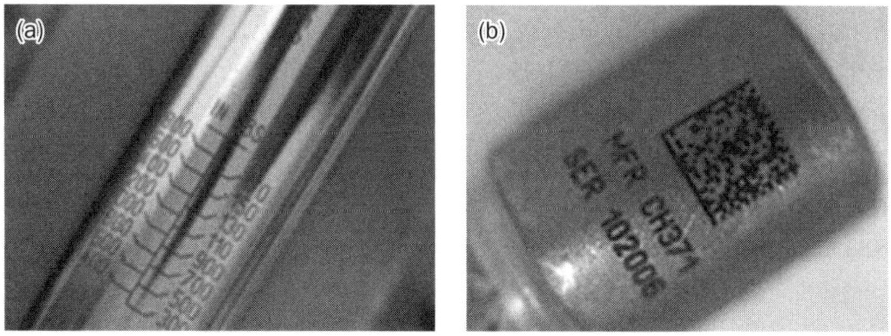

Figure 1.10.4 (*a* and *b*) Typical anneal marking of steel.

Figure 1.10.5 Two-step laser marking process used to produce both the background enhancement and also the data matrix and barcode patterns.

QC permits marking to be done following passivation, then the problem of possible fading or dulling of the mark can be eliminated. Otherwise, extra precaution is needed. These steps usually overcome most problems.

Anneal marking in cast surfaces such as aluminum, steel, and cast iron often need to have the surface pretreated with the laser, creating a uniform background surface color so that the dark anodized mark stands out, which is required in most cases to achieve the necessary contrast for barcode readability (Fig. 1.10.5).

Engraving and Etching

If the mark needs to penetrate the material to a certain depth to ensure greater permanency or to maintain readability following many coats of paint during the part lifetime, the depth often specified is up to 300 μm (0.012 in.). The depth is achieved by having the laser ablate through the surface with multiple passes until the desired depth is achieved. The as-marked surface may be rough and show remelted material along the edges. Furthermore, expulsed metal debris may reattach or deposit on surrounding surfaces and will need to be cleaned off. Typical applications include serial numbers on firearms, gears, aluminum hydraulic manifolds, and some metal identification tags. Deep engraving is often used in injection molds to create reverse raised characters in plastic and metal parts.

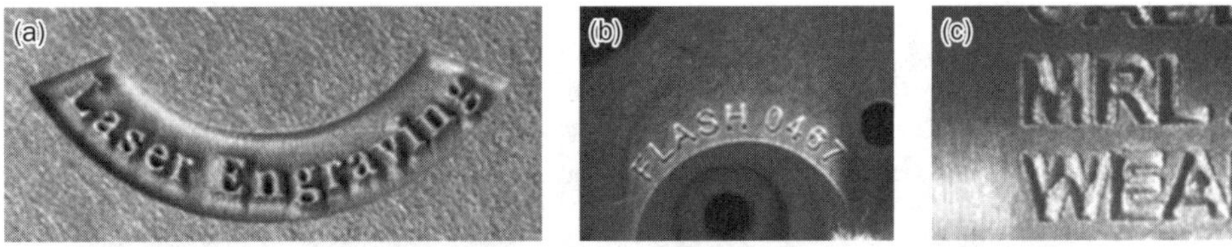

Figure 1.10.6 (*a–c*) Relief and deep engraving.

Figure 1.10.7 (*a* and *b*) Examples of high-speed (up to 5000 characters per second) CO_2 marking.

Etching involves a shallow modification of the surface to achieve minimum depth of about 25 μm (0.001 in.) or less so just some texture is achieved. This produces a more permanent mark than with just an anneal mark because should the anneal mark disappear the etch surface is still readable.

Note that many aerospace engine components have strict limits on recast and micro cracking depth and usually do not permit laser marking on many parts so pay attention to the functionality as well as the materials and process specifications of the part (Fig. 1.10.6).

Plastic, Glass, Coated, and Paper Marking

Carbon dioxide lasers produce a 10.6-μm wavelength, which is much better absorbed in organic materials such as plastic, glass, paper, and many other materials and which reflects off metals unless very high powers are used. Applications are diverse, including bowling balls, automotive window glass, date codes on bottles and paper labels, and high-speed serializing of plastic pharmaceutical devices. The characters marked in glass will appear frosted, but in plastic the laser does not produce a contrast but rather the characters are melted into the surface (Fig. 1.10.7).

In addition to the CO_2 laser, both the fiber and vanadate solid-state lasers can mark many plastics and produce surprisingly good contrast. However, with certain plastics no contrast is achieved because the material is transparent to the 1064 nm wavelength. Since some lasers can be converted to produce beams of green at 532 nm or UV at 355 nm, higher absorption and darker mark contrasts can be achieved. As an alternative, there are special proprietary laser reactive additives that most plastic compounding companies developed to provide the required contrast the customer needs.

Figure 1.10.8 Laser marking of day/night buttons.

Marking on Painted Surfaces

Another example of the versatility of the fiber and vanadate is in the manufacture of automotive and aerospace day/night control buttons. A typical day/night device is molded in clear plastic then is painted white and finally painted black or other top coat. The laser ablates the top coating but not the white layer. In daylight the white characters are bright and at night the button is back-lighted, highlighting the white against the black (Fig. 1.10.8).

CHAPTER 1.11
Laser-Assisted Forming

LaRoux K. Gillespie

Manufacturing Consultant
Andover, Kansas

Laser forming allows forming at lower forces and can provide both faster response to design changes and less costly tooling than conventional forming processes. Lasers have been used as part of hydroforming, roll forming, and conventional press work forming either as an in-line feature or as an intermediate step. It is as viable as a rapid prototyping process as it is a high-production process.

It is faster in most cases than well-established flame or torch bending for heavy plates but the basic concept is the same—heat the metal sheets or plates to soften them so they bend with less force. Flame heating has been widely used for decades in both small shops as well as larger high-production facilities, and now lasers provide a similar advantage, but at higher speeds and with much more control.

A defocused laser heats a defined spot, line, or bar on the part. The induced thermal stresses cause plastic strains which in turn cause the metal to bend or buckle. No melting is involved, just heating. With proper focus and control, the focused laser can even shorten parts as well as straighten them somewhat. Laser forming or laser-assisted forming involves three basic approaches: (1) the laser causes all the bending (laser forming), (2) the laser heats the metal to allow the conventional tools to bend the part (laser-assisted forming), and (3) micro part forming in which special approaches are required (laser-assisted micro forming).

Laser Forming

Figure 1.11.1 illustrates the three basic forms of forming. The temperature gradient approach, the most widely used, causes a steep temperature gradient across the cross section and this then causes a difference of expansion across the thickness. While heated the part bends away from the laser, but after cooling the sheet bends upward around the beam as shown in the figure. The bends are slight (1 to 2 degrees) for each pass of the laser. This is called line-scanning laser forming.

For buckling bending as shown in the figure a beam much wider than the material thickness is used and the heat is raised all the way through the thickness. The bends can be directed either upward or downward by a careful selection of prebending

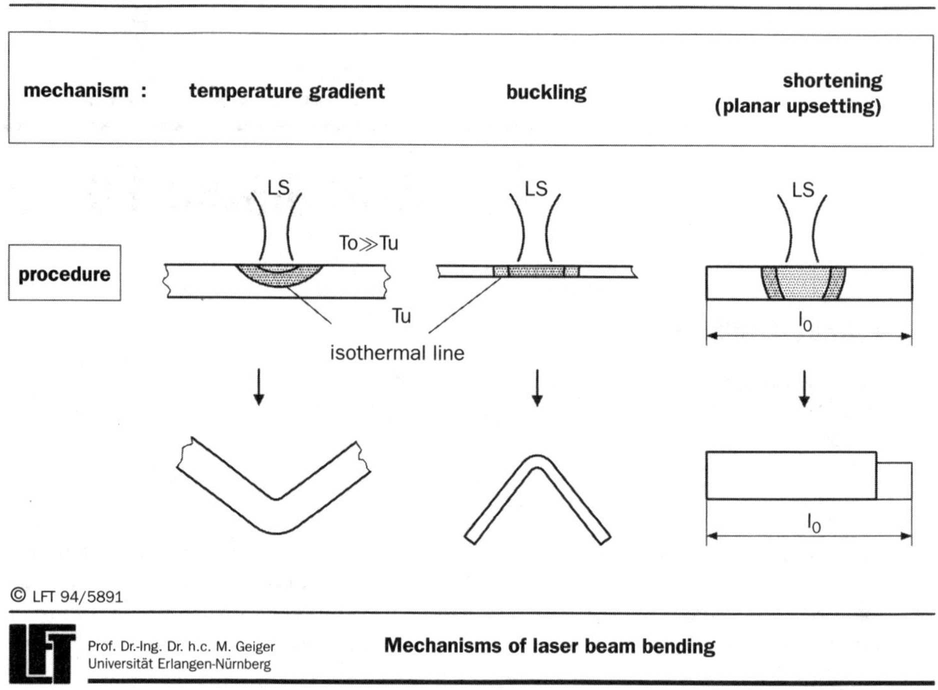

FIGURE 1.11.1 Schematic of the three main laser-forming mechanisms. (Courtesy F. Vollertsen.)

orientation of the sheet, existing stresses, and using other stresses such as a cold blast of air on the bottom of the sheet. With this approach the sheet can be bent anywhere from 1 to 15 degrees.

For some geometries, thick pieces in the instance shown, the heat results in some small shortening of the part. With experience, some shapes such as extrusions, tubing, frames, and unique shapes can bend or shorten the part out of the plane shown here.

While the focus is on forming into a desired shape, laser forming is also a technique used to correct misalignment or straightness in parts, just as in torch heating. Straightening shafts, rods, and small tubes has been a process used for decades. The laser provides much more control and repeatability than torches however.

Because only localized areas are heated, the bulk properties are generally not altered significantly. Titanium, high-strength steels, and aluminum are common materials that are laser formed. These processes can solve springback problems where only small deflections are needed. Figure 1.11.2 illustrates two laser-formed configurations.

Laser forming is not limited to 2D bending as shown in the illustrations. Saddle shapes, pockets, or depressions can also be produced. Steel plates up to 25.4 mm (1 in.) thick are being researched, so the process is not limited to thin materials. Removing the distortion from welding, which is a common issue, is another area in which laser forming can be applied.

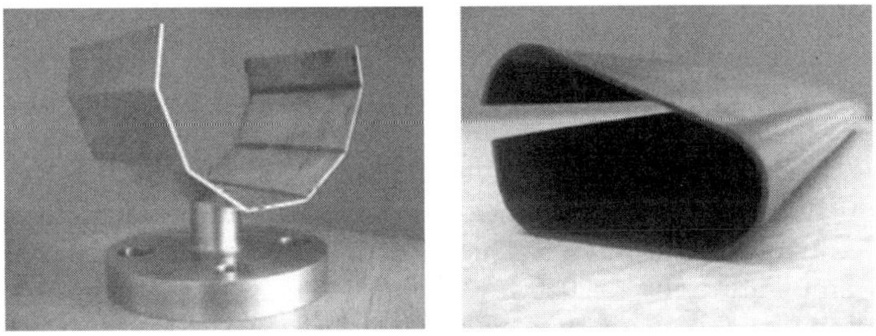

Figure 1.11.2 1.5-kW CO_2 laser-formed parts: left 250×100×2-mm-thick aluminum; right 400×200×1.5-mm-thick mild steel. (Courtesy University of Liverpool.)

Laser-Assisted Forming

In many applications the laser merely heats the part in the desired area and normal bending or forming tooling is used to bend the material at much less force. Shops having cutting and welding lasers can typically use them for forming with normal forming tools, so dedicated forming lasers may not be needed. Lasers are also now being used as an intermediate step in hydroforming to heat-treat portions of the part that have changed from austenitic to work-hardened martensite which is not possible to further work. Hydroforming requires several steps to produce the final product, so intermediate laser heating converts the material to easy to form austenite. For this chapter however the discussion is based on laser application during forming operations.

By focusing the laser directly ahead of the forming tool, the workpiece is softened to allow easier bending. It works in presses, roll forming, and even deep drawing. With the right temperature distribution and forming times, much larger bending is possible than with laser-only forming and with less time.

Studies have also shown that by heating wire while pulling on it, it is possible to draw the wire to a smaller size without using any die. For thin wires a 1-kW laser is all that is required. As in the instance of laser-only forming, a gas stream cools the wire under the laser spot.

Laser-Assisted Micro Forming

Forming micro parts in which the thickness may only be 5 to 10 grains thick and the feature sizes extremely minute (100 μm (0.004 in.) for example) is an area still requiring more work for most applications. However, laser heating during the process lowers the forces required and makes ultrathin micro parts more feasible. Micro pins, springs, screws, shafts, connectors, and gears have been fabricated on parts and features somewhat larger than these.

As an example of what can be done, laser-cut tabs (5 mm (0.2 in.) square) in untreated 30-μm-thick (0.0012 in.) aluminum foil were bent nearly 90 degrees by 200 laser passes in rapid succession.

Friction, surface roughness, grain size, structure and orientation, thermal equalization, and the necessary forces are all restraints in micro forming to a much much larger degree than they are for traditional size parts.

Lasers on micro parts have an advantage that they can cover the entire micro part or just minute spots or lines. They do not have to touch the parts so handling is minimized, they heat rapidly so time is minimized, and lasers are highly repeatable.

While the basic approach is same as described in the previous two forms of laser forming, the latest approach involves the use of transparent punch or dies. Sapphire is one first choice for soft metals since it is transparent to wavelengths greater than those in the ultraviolet range and it has high hardness and compressive strength. It also withstands temperatures up to 2050°C (3722° F).

Aluminum is soft enough to emboss with laser-assisted sapphire tools, and grooves and holes in the range of 100 to 400 μm (0.004 to 0.016 in.) have been produced. Making the sapphire punch or die, however, at micro sizes also requires great experimentation.

Another approach to micro forming is laser shock forming. It is still in its infancy, but is directly applicable to micro sheet metal forming. Spherical cups with a height of 1 mm (0.040 in.) have been formed by this process in aluminum, copper, and stainless steel. In addition to being a very rapid process, it can cause superplasticity in metallic thin films. With a laminating process, it is even possible to form MEMS thin-wire resistors over brittle molds.

Commercial companies performing micro forming by any means are reluctant to describe their propriety processes, but numerous academic reports provide insight into what can be achieved. Only a handful of companies have any production experience with truly micro product forming.

Laser-assisted manufacturing is growing at such a rate that it has its own research journal (*Laser Assisted Net Shape Engineering*) and hundreds of pages are published each year.

References

Cheng, G., Laser Induced Superplastic 3D Micro Forming, Proceedings SME Micro Manufacturing Conferences, SME, Detroit, Michigan, 2009.

Gordon D., and S. P. Edwardson, "Laser Assisted Forming for Ship Building," 2003 (available from: lasers.org.uk).

Mahdi T., K. Samm, F. von Scotti, J. Quintanilla Sanchez, A. Ostendorf, and J. Wulfserg, Laser-Assisted Micro Forming with Laser Structured Sapphire Dies, Proceedings 2nd International Conference on New Forming Technology, Bremen, Germany, September 20–21, 2007.

CHAPTER 1.12

Laser Peening

Stanley Bovid

LSP Technologies, Inc.
Dublin, Ohio

Laser Peening Process

Laser peening, or laser shock peening, enables new and corrective design possibilities by increasing material performance parameters. It is a surface-enhancement process that utilizes shockwave mechanics to generate beneficial compressive residual stresses in materials. The effects of laser peening are routinely observed with magnitudes equaling or surpassing conventional peening technologies with treatment depths extending up to ten times deeper. As a result, components processed via laser peening typically realize increased failure resistance in conditions that include fatigue, corrosion, fretting, stress-corrosion, thermal conditions, wear, and rolling contact fatigue among others.

Multiple variations of laser peening processing are used in industry for different parts, materials, and geometries. As a general guideline, conventional laser peening consists of the following steps:

1. Component is placed in position for processing via CNC device.
2. Opaque overlay is applied to the surface.
3. Transparent overlay is applied to the surface.
4. Laser energy is delivered to the surface for processing.

Schematically, the physical process of laser peening is illustrated in Fig. 1.12.1. When the laser beam first interacts with the target, a small amount of the opaque material is vaporized. The continued delivery of laser beam energy converts the vapor to a plasma that rapidly expands. While the plasma expands, it is confined by the transparent overlay generating an approximate tenfold increase in the pressure observed at the part surface. The rapid rise in pressure at the surface of the target results in generation of the shockwave. The shockwave propagates in-depth through the material and cold works the atomic structure, creating plastic strain and residual stresses.

Laser beam energy is typically delivered over circular or square areas spanning from fractions of a millimeter in each direction to several millimeters depending on

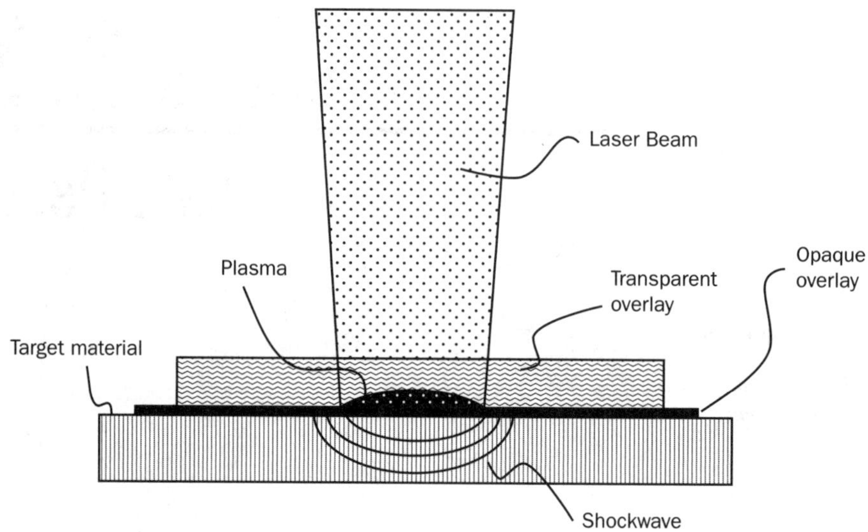

FIGURE 1.12.1 Schematic of the laser peening process showing the overlays in place and the laser beam irradiating the material surface. The image is not drawn to relative scale.

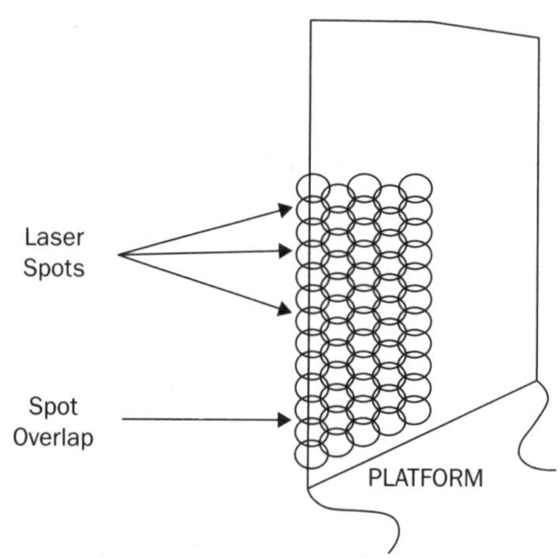

FIGURE 1.12.2 Depiction of laser peening pattern that may be applied to a blade. Placement of individual laser spots is achieved by CNC placement of part or laser beam delivery mechanism.

particular equipment capabilities. When processing areas are larger than a few square millimeters, patterns of spot applications are developed. This is illustrated schematically in Fig. 1.12.2 and is accomplished through CNC placement of the component being processed. In general, spot patterns consist of partial overlap of spots to ensure complete surface coverage. In addition, spot patterns also may be repeated to achieve a beneficial effect by applying multiple layers of coverage.

Typical Characteristics and Applications

Laser peening causes physical changes that should be evaluated on an application-specific level. Controllable process parameters can be tailored to specific applications. For example, less intense processing can be used to minimize impressions made by the plasma on the surface of the component, but that also is likely to yield a lower percentage of benefit in comparison to higher intensity processing.

Residual Stress Magnitude and Depth

Residual stresses from laser peening are dependent on material, heat treatment, and geometry. The process can be expected to yield compressive stress magnitudes on the order of a minimum of 70% of the material yield strength. Different alloys have different responses, with some materials achieving 100% of the yield strength in compression.

In-depth residual stress profiles typically exhibit a gradient from a compressive stress at the surface to a tensile stress at a location beneath the surface. This transition from compressive stress to tensile stress is often referred to as the penetration depth for laser peening. The process is capable of achieving a wide range of treatment depths depending primarily on the material strength and the component geometry. Thin geometries [typically <5 mm (<2 in.)] can develop through-thickness compressive residual stress profiles.

Residual Stress Stability

Due to the low percentage and deep gradient of cold work that is induced by laser peening, residual stresses remain stable during cyclical loading and exposure to elevated temperatures, even when approaching typical stress relief temperatures for various alloy systems. This permits surface-enhancement usage in some elevated temperature applications.

Surface Roughness Effects

During laser peening, each interaction of the laser beam with the part surface will create an impression on the surface. As most processing requires patterns of spot applications, the surface will be modified according to the pattern. The depth of the impressions will be dependent on the material strength and the processing conditions used. Typical surface impressions are on the order of 5 μm (0.0002 in.), with surface area sizing approximately equal to the laser beam diameter [typical beam diameter ranges from 2 to 5 mm (0.08 to 0.20 in.)].

Material Properties

It is not unusual to obtain minor increases in hardness and strength as a result of the induced cold work. Highly work-hardenable alloys may see significant increases. From a metallurgical standpoint, unstable phases of alloys can also be transformed by the shockwave, such as the retained austenite conversion to martensite in steel alloys.

Compensating Stresses and Deformation

Stress systems, whether residual or applied, develop an equilibrium static state. This requires that compressive residual stresses are balanced by tensile residual stresses.

Industry	Applications
Aerospace	Gas turbine engine components, blades, vanes, bearings, injectors, compressors, structural airframes, landing gear, welds
Marine	Propellers, driveshafts, welded structures, aluminum hulls
Power generation	Steam turbine blades, nuclear containment, reactors
Automotive	Bearings, crankshafts, camshafts, valves, springs, driveshafts, knuckles, wheels, boost systems
Medical	Implants, tools, prosthetics
Industrial/Manufacturing	Tools, forging dies, rolling dies
Agriculture/Heavy equipment	Engine components, drivetrains, structures, fittings, joints

Table 1.12.1 Potential Applications for Laser Peening in Various Industries

Typically, the magnitudes of tensile residual stresses are insignificant as the bulk of the material can adapt to the tensile stresses while the compressive stresses are limited to near-surface areas. However, the stress states can also lead to deformation in parts and components. A surface placed into a high magnitude compressive state will tend to deflect the surface in compression outward (curls up in an arc). High-precision applications require assessment of potential deformation from processing.

Common Applications

Laser peening can be applied to increase component performance based on two different ideologies:

1. In-service applications that do not meet desired performance characteristics.
2. New designs that can benefit from enhanced material performance.

In both cases, laser peening can be used to prevent or reduce fatigue failures, stress-corrosion cracking, fretting fatigue failures, rolling contact fatigue, foreign object damage stress concentrations, etc. In addition, laser peening can be used to increase stiffness of designs, shape parts, and allow removal of material. Several examples of these applications in different markets are listed in Table 1.12.1.

Economics

Laser peening can be considered as a treatment independent of, or in combination with, other conventional surface-enhancement methods. Modern laser peening equipment has process-throughput capability comparable to other robotically controlled surface-enhancement processes with processing rates up to 6 cm^2/s (0.93 $in.^2/s$). The throughput of the process has made mass production operations affordable for many applications. As a general guideline, the cost of laser peening is proportional to the surface area being processed. Therefore, one option to control the cost of laser peening is to restrict treatment to critical locations.

Laser peening equipment consists of both laser systems and part handling systems. Equipment costs range from several hundred thousand dollars (US) to several million based on the capability requirements. As the part handling systems require CNC control, development costs of laser peening applications require additional time and tooling compared to hand-controlled applications.

General Process Design Considerations

Design Methodology

Designing with laser peening involves evaluation of the total stress state of a component in a service environment. Benefits of the process may be evaluated through analysis of the superposition of residual and applied stress states. The design aspect of laser peening is the residual stress generated by the process. Subsequent paragraphs of this chapter include residual stress curves that may be used in approximating material performance. The most effective laser peening design is often accomplished through computerized finite element analysis but may also be completed using other methods. As a general guideline, any type of stress concentration feature, whether from design or in-service conditions, will observe a benefit from laser peening. Several examples are shown in Fig. 1.12.3.

Pattern Size and Location

Historically, laser peening has been used in the treatment of the most critical locations on components. When evaluating locations to be laser peened, accessibility to the area is an important consideration. Laser peening requires that the transparent overlays be applied to the surface and that the laser beam can be directed to the location. This

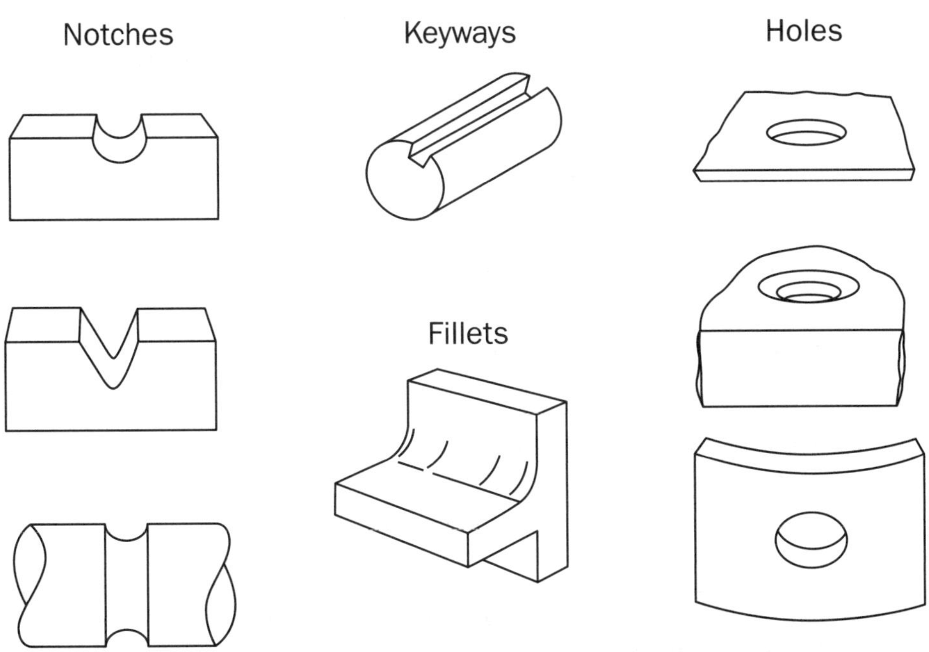

FIGURE 1.12.3 Depiction of stress concentration-generating geometries that are ideal candidates for laser peening to increase service life.

requires that complicated geometries, particularly internal geometries, needing treatment be evaluated for practicality of processing.

Intensity and Coverage

Laser peening parameters for industrial applications are most commonly specified according to power density, pulse width, and iterations of processing. The power density is a measure of the amount of energy delivered over an area in a given amount of time. Energy (E) is measured in joules, the pulse width (pw) is measured via full-width at half-max in nanoseconds, and the spot size (A) is typically measured in cm^2. In equation form, this can be shown as

$$Pd\left(\frac{\text{GW}}{\text{cm}^2}\right)=\frac{P(\text{GW})}{A(\text{cm}^2)} \quad \text{where } P(\text{GW})=\frac{E(\text{J})}{pw(\text{ns})} \tag{1.12.1}$$

Coverage of a surface may be defined as number of layers of 100% coverage, percentage of overlapped laser beam shots, or in pulse density (pulses/cm^2). Industrial laser peening practice has shown that residual compressive stresses are generally increased in both magnitude and depth with increasing coverage. The effect has been documented for more than eight layers of coverage; however, the usual coverage is two to four layers as additional layers have diminishing levels of return in benefits.

Suitable Materials

Laser peening can be effective on all classes of engineering alloys. Several experimental residual stress curves for various classes of alloy systems are provided below. These residual stress curves may be used to approximate the stress-related benefits that would be achieved on sections with thicknesses greater than 5 mm (0.2 in.).

- Aluminum alloys (Fig. 1.12.4)
- Steel alloys (Fig. 1.12.5)
- High-strength steels (Fig. 1.12.6)
- Carburized steels (Fig. 1.12.7)
- Titanium (Fig. 1.12.8)
- Nickel-based alloys (Fig. 1.12.9)

Detailed Design Considerations

Variations Based on Supplier

Commercial laser peening service providers use laser systems capable of outputting a maximum amount of energy. The maximum amount of energy that can be delivered per pulse limits the maximum spot size that can be used for processing. For most applications, this will dictate a spot size with cross-section dimensions of 2 to 5 mm (0.08 to 0.20 in.). Spot shapes are most commonly circular or rectangular. Spot shape has not been demonstrated to have measurable impact on the quality of processing. Supplier variations are also observed in the selection of overlay materials for processing.

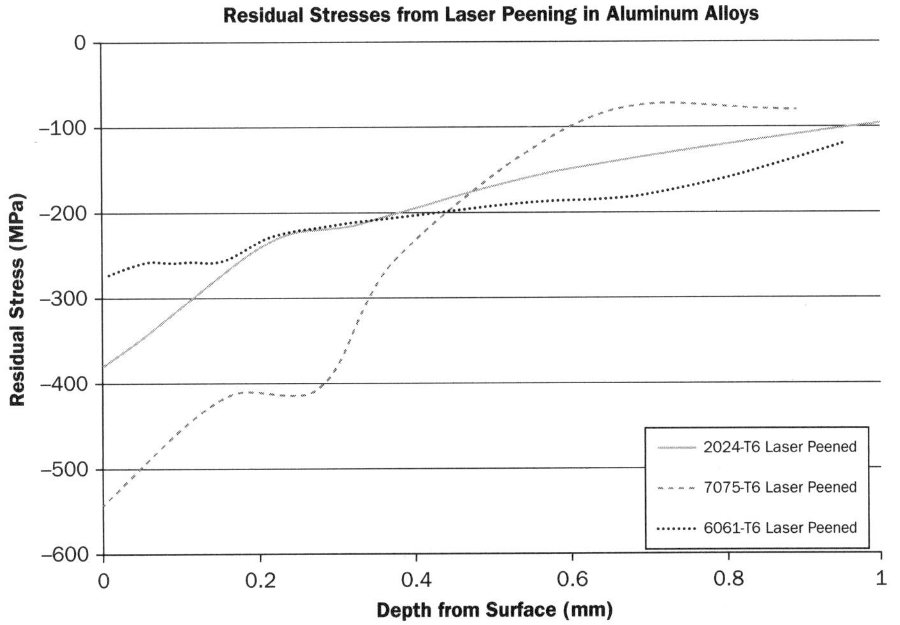

FIGURE 1.12.4 Residual stresses obtained via laser peening on three different aluminum alloys including 2024, 6061, and 7075.

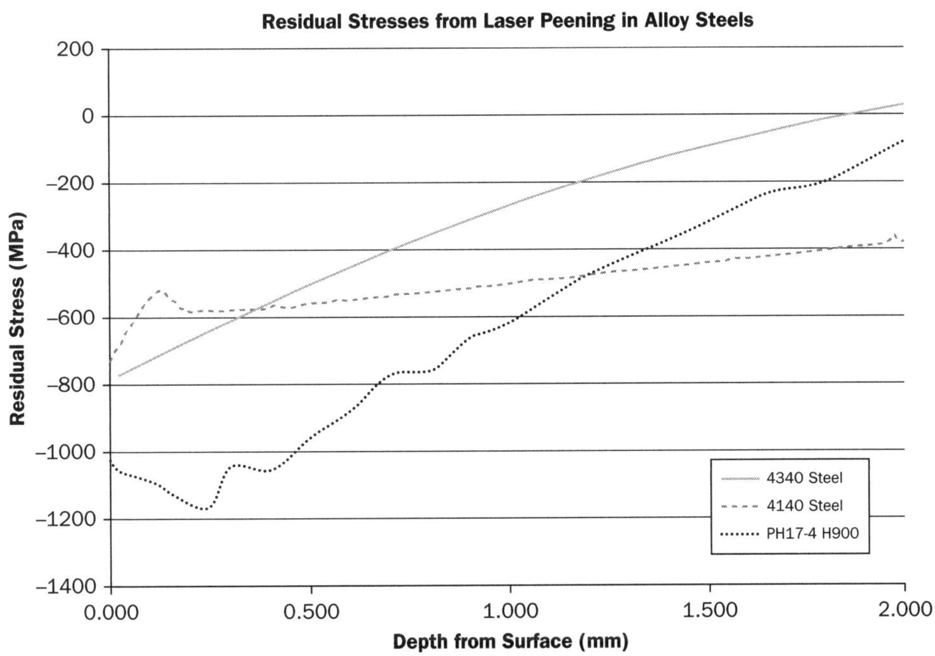

FIGURE 1.12.5 Residual stresses obtained via laser peening on alloy steels. Listed steels include SAE 4340, SAE 4140, and PH17-4 stainless steel.

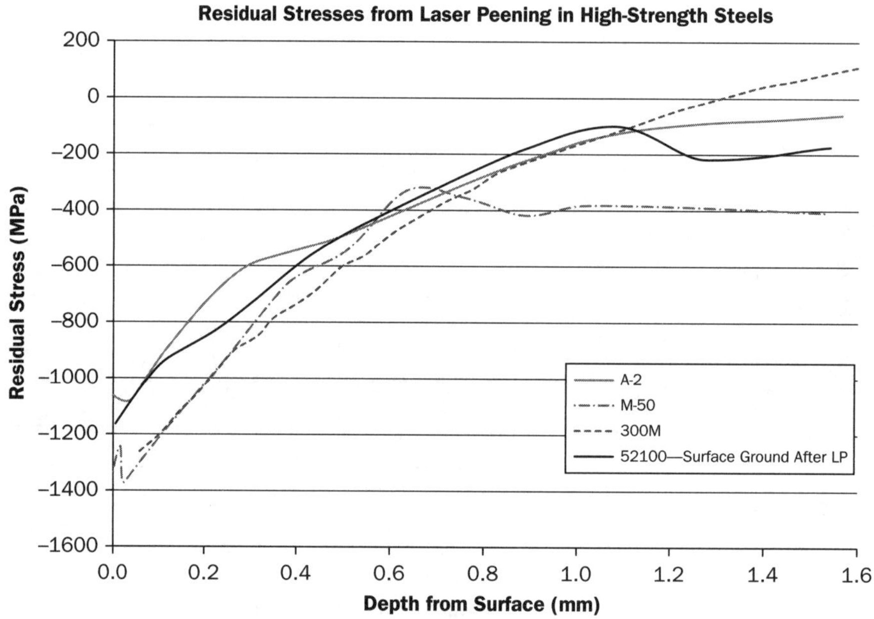

FIGURE 1.12.6 Residual stresses obtained via laser peening on high-strength steel alloys including A-2 tool steel, M-50, 300M, and SAE 52100.

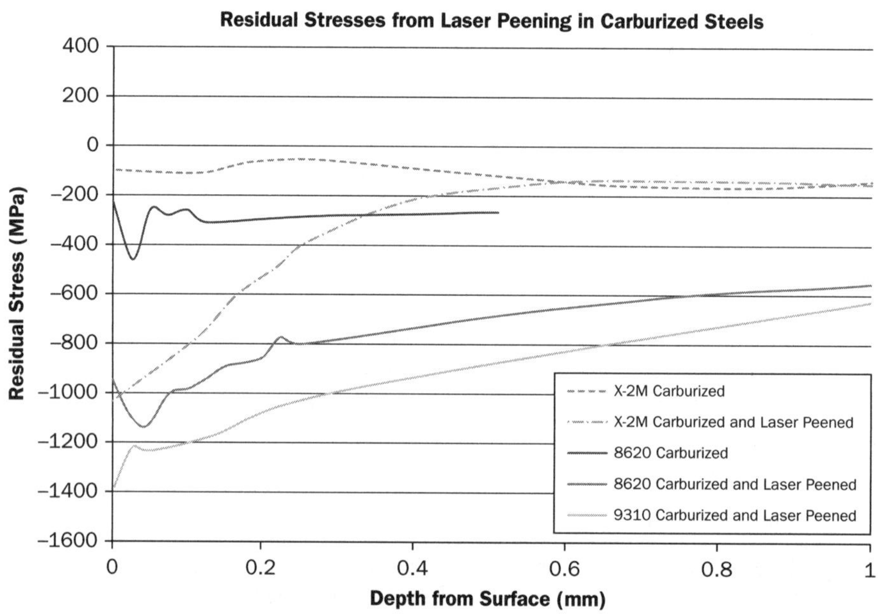

FIGURE 1.12.7 Residual stresses obtained via laser peening on three different carburizing steels including X-2M, SAE 8620, and SAE 9310.

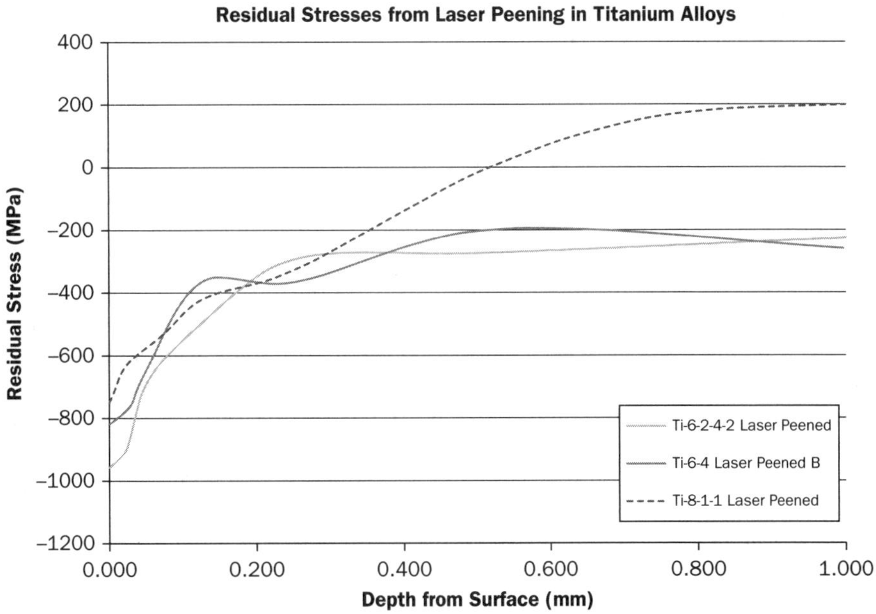

Figure 1.12.8 Residual stresses obtained via laser peening on three different titanium alloys including Ti-6Al-4V, Ti-6Al-2Sn-4Zr-2Mo, and Ti-8Al-1Mo-1V.

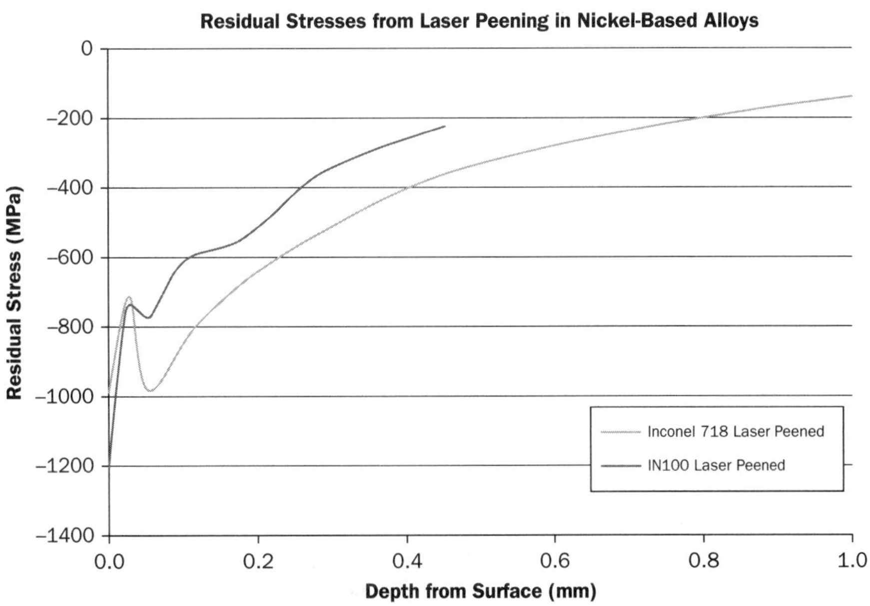

Figure 1.12.9 Residual stresses obtained via laser peening on two nickel-based alloys including Inconel 718 and IN100.

The most common opaque overlays include black poly-vinyl tape, aluminum tape, and a proprietary liquid formulation. All of the opaque overlays listed generate nominally the same shockwave and resulting residual stress profiles. Transparent overlay material is mostly standardized across the industry in the means of clean, filtered water.

Intensity and Coverage Specification

The laser peening process is fully scalable from minor treatments to heavy saturation. Selection of the proper scale for an application needs to be evaluated in terms of the failure mode being alleviated and the material geometry. For example, to prevent surface-initiated failures it will be most beneficial to obtain the highest magnitude compressive stresses on the surface possible. In contrast, to increase the damage tolerance of a component it may be best to drive compressive residual stresses as deep as possible. The control of the residual stress profile is accomplished primarily through selection of the treatment intensity. As shown in Fig. 1.12.10, the residual stresses that are developed from laser peening are highly controllable based on the laser process intensity.

Although the plastic strain induced by subsequent layers of coverage is not as influential as the processing intensity, additional coverage still plays a significant role in the development of residual stresses. This is demonstrated in Fig. 1.12.11. To achieve the deepest penetration of compressive residual stresses, more layers of coverage are necessary. For most applications, three layers of coverage provide the ideal residual stress profile.

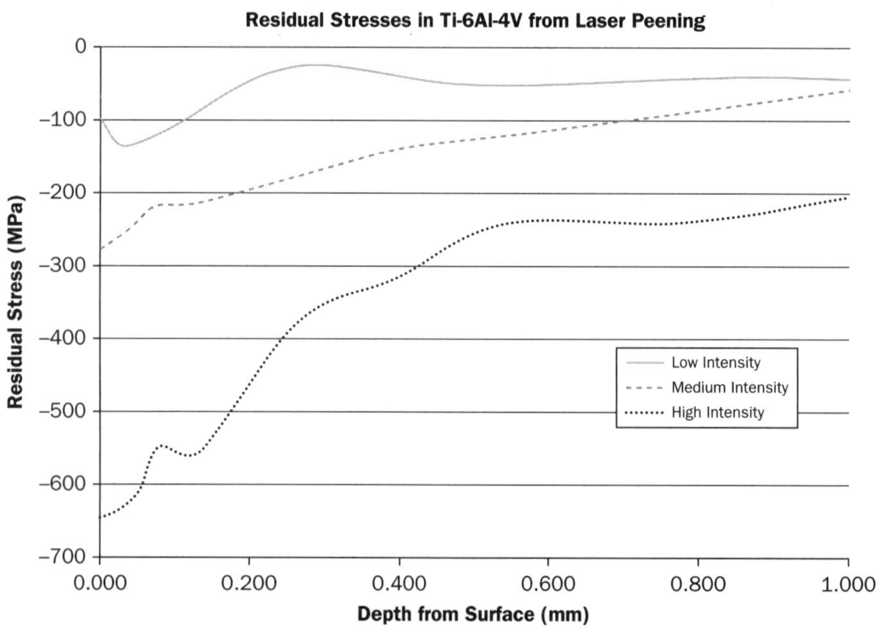

FIGURE 1.12.10 Residual stress profile in Ti-6Al-4V exhibiting the effect of different laser processing intensities on residual stress development.

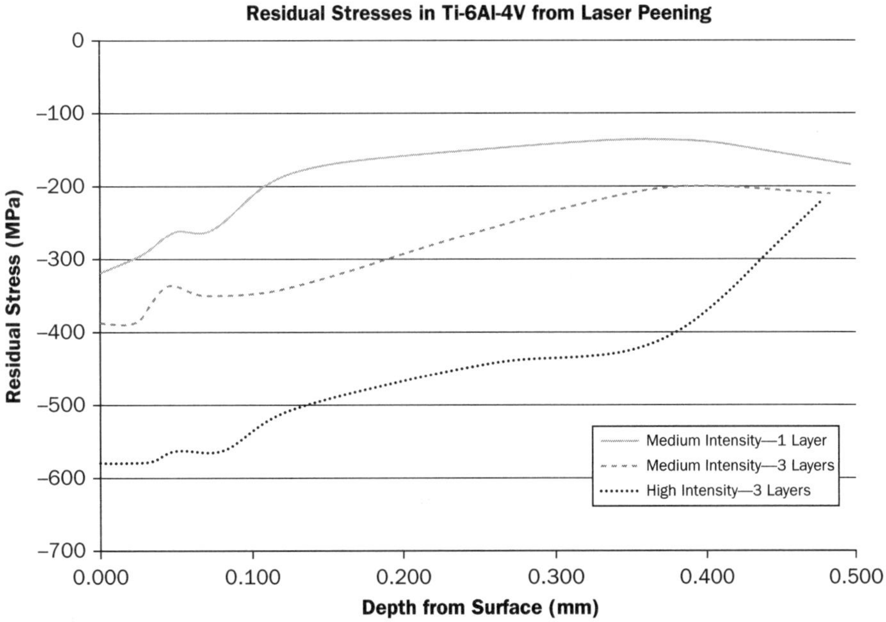

FIGURE 1.12.11 Residual stress developments in a 1-mm-thick (0.040 in.) piece of Ti-6Al-4V from laser peening with different intensities and layer combinations. It is observed that additional layers increase the depth and magnitude of residual stress.

Alloy Type	Power Density
Aluminum alloys	2–5 GW/cm^2 (0.31–0.76 GW/in.^{2})
Steel alloys (30–45 HRC)	7–10 GW/cm^2 (1.09–1.55 GW/in.^{2})
High-strength steels (55–60 HRC)	9–10 GW/cm^2 (1.40–1.55 GW/in.^{2})
Titanium alloys	7–9 GW/cm^2 (1.09–1.40 GW/in.^{2})
Magnesium alloys	3–6 GW/cm^2 (0.46–0.93 GW/in.^{2})
Nickel alloys	7–9 GW/cm^2 (1.09–1.40 GW/in.^{2})

TABLE 1.12.2 Typical Power Density Ranges to be Used in Selecting the Processing Intensity for a Given Material Type

For the selection of generic treatments for alloy groups, Table 1.12.2 provides general power density requirements. The ideal treatments are typically related to the dynamic material yield strength, known as the Hugoniot elastic limit (HEL). The limitations to processing intensity are based on the dielectric breakdown of air which occurs around 11 to 12 GW/cm^2 (1.7 to 1.9 GW/in.^{2}). Pulse-width requirements will vary depending on equipment provider availabilities but are typically 8 to 25 ns.

When specifying laser peening conditions, it is typically necessary to indicate the power density, pulse width, and layers of coverage. On occasion it is also beneficial to include the desired spot size.

Patch Size and Location on Drawings

Areas to be laser peened are typically shown on drawings as identified areas requiring laser peening and drawing notes indicating the intensity and coverage requirements. If multiple layers are to be used with a specified offset, the layers are often called out by order and offset.

Processing of Thin Sections and Shot Orders

Processing of thin sections, such as a compressor blade, requiring dual-sided treatment is best accomplished via alternating processing between the opposing faces as frequently as possible. While the residual stresses will be approximately equal on both sides of a thin section, there is a tendency for deformation to exist in applications where one side is fully processed prior to the other.

Minimum Thickness

To prevent severe deformation, and possibly fracture of materials, laser peening should be applied to sections capable of withstanding the shockwave impact. This is generally a minimum of 0.25 mm (0.010 in.) of cross section. In applications where this minimum thickness is not achievable, it may be necessary to use backing plates to support the component being processed.

Designing with laser peening can be summarized into several steps for discussions with industrial suppliers of the process:

1. Determine potential failure modes of component being evaluated.
2. Assess residual stress profile that would reduce or prevent selected failure mode(s) from occurring.
3. Select appropriate laser peening conditions and patterns to obtain residual stress profiles.
4. Complete design and evaluate laser peening performance.

PART 2

Manufacturing with Additive Processes

CHAPTER 2.1

Overview of Additive Manufacturing Processes

Carolyn Conner Seepersad

University of Texas
Austin, Texas

Additive manufacturing (AM) is making a disruptive impact on not only the types of things we make, but also the way we make them. Compelling examples of additively manufactured products are emerging almost daily. Mainstream aircraft manufacturers, such as Boeing, are utilizing AM for fabricating ductwork. As shown in Fig. 2.1.1, the capability of AM to fabricate complex parts allows aircraft manufacturers to consolidate multiple parts into a single part, thereby saving weight and assembly time and effort. In the dental industry, AM enables fabrication of braces that are customized to individual patients. The Invisalign braces illustrated in Fig. 2.1.2 are fabricated directly from additively manufactured molds, each of which is customized from digital models of an individual patient's teeth. Lack of part-specific tooling makes it economical to additively manufacture these personalized parts. In 2012 alone, according to 3D Systems CEO Avi Reichental, Invisalign fabricated more than 17.2 million unique braces using this technique [1]. Finally, AM is also being used to enhance the functionality of parts. GE is using additive manufacturing to fabricate fuel nozzles for their new LEAP engines. The nozzles are lighter and simpler because of AM's ability to make complex integrated parts; in this case, reducing the number of welds from 25 to 5 [2]. The nozzles are also more durable as a result of specialized cooling pathways.

ASTM defines AM as "the process of joining materials to make objects from 3D model data, usually layer upon layer, as opposed to subtractive manufacturing methodologies" [3]. Since parts are built additively in a layer-wise manner, it is quite distinct from conventional fabrication techniques. Conventional fabrication techniques typically either form the object volumetrically, as with injection molding or vacuum forming, or fabricate the object by selectively removing material from a precursor, as with milling, cutting, or water jetting. Tooling costs for forming processes, such as injection molding,

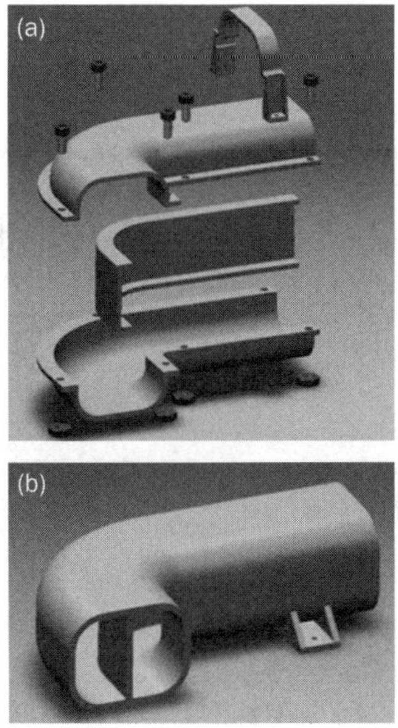

Figure 2.1.1 An example of aircraft ductwork manufactured conventionally (*a*) and additively (*b*), illustrating the merging of multiple parts into a single complex part. (Courtesy Stratasys.)

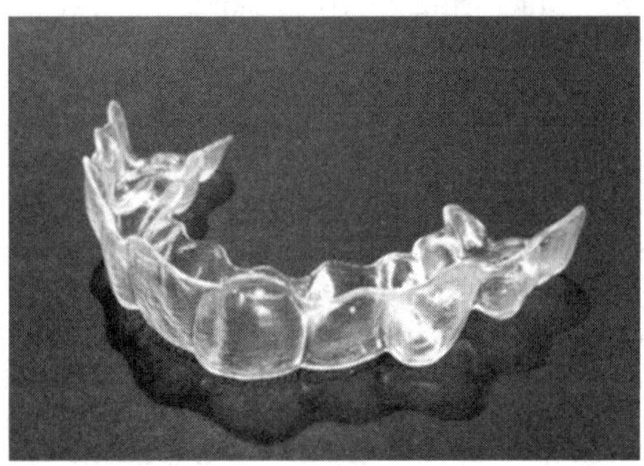

Figure 2.1.2 An example of a dental brace from Invisalign, illustrating the ability of additive manufacturing to fabricate one-of-a-kind, personalized parts. (© Smikey Io/Wikimedia Commons/CC-BY-SA-3.0/GFDL.)

can be very high, and parts are typically subjected to geometric restrictions such as draft angles. Unlike forming processes, AM requires no part-specific tooling and imposes very few restrictions on part geometry, making it much more economical for producing small quantities of parts. Advanced cutting technologies, such as laser cutting and water jetting, enable production of complex parts directly from digital models, but they limit the part complexity to 2D. CNC machining enables 3D subtractive manufacturing directly from 3D digital models, but geometric complexity is limited by tool paths and tool access routes. Unlike subtractive processes, AM builds parts layer-by-layer, making it possible to fabricate complex internal and external geometries. The disadvantage of AM is that it is newer and less well-characterized than most conventional manufacturing technologies and in some cases lacks the speed, accuracy, and precision of conventional manufacturing technologies. Also, AM can be slower and more expensive, on a per-part basis, than conventional manufacturing techniques, particularly when large quantities of identical parts are required. Its niches include small production runs, personalized or customized parts, and complex parts that cannot be manufactured in any other way.

Overview of Primary Additive Manufacturing Technologies

ASTM has categorized AM into seven technology groups: binder jetting, directed energy deposition, material extrusion, material jetting, powder bed fusion, sheet lamination, and vat photopolymerization [3]. Since each technology provides its own set of capabilities and limitations, it is important for the designer to become familiar with each technology and the design opportunities it affords.

Binder jetting is "an additive manufacturing process in which a liquid bonding agent is selectively deposited to join powder materials" [3]. Binder jetting typically involves the use of inkjet technology to selectively dispense liquid binders onto a powder bed, thereby holding the powder together into a desired cross-sectional geometry. The process is repeated in a layer-wise fashion to build a 3D object. Binder jetting was originally known as 3D printing and developed by researchers at MIT. The technology is embodied in several commercially available machines, including those from the ExOne Company and Voxeljet Technology. 3D Systems manufactures a ProJet line of binder jetting machines, which are directly descended from the ZPrinter systems commercialized by Z Corp.

From a design perspective, binder jetting's advantages include its low cost and high speed relative to other AM technologies, along with the capability of printing colors. However, parts made with the binder jetting process can be weak and suffer from poor accuracy and surface finish relative to other AM technologies (Fig. 2.1.3).

Directed energy deposition is "an additive manufacturing process in which focused thermal energy is used to fuse materials by melting as they are being deposited" [3]. Focused thermal energy could be provided by a laser, electron beam, plasma arc, or other energy source that can be used to melt materials as they are deposited onto a substrate. In one of the most widely available commercial systems, LENS (laser engineering net shaping), a focused laser creates a melt pool on the substrate, and metal powders are injected into the melt pool as it moves along the surface of the substrate. Other commercial systems include direct manufacturing (DM) from Sciaky, Inc., which uses an electron beam as the energy source and wire feed for material delivery, and Fusion Formation from Honeywell Aerospace, which uses a plasma torch as the energy source.

Directed energy deposition technologies are particularly advantageous for fabricating metal parts with highly controlled micro structures and excellent material properties. Because the powder mix can be varied dynamically, it is also possible to fabricate functionally graded structures. The technique of melting materials as they are deposited leads to poor resolution and surface finish, however, and makes the technology slow and expensive relative to other technologies for building bulk parts. One of its most common uses is for repairing or resurfacing existing, high-value components such as turbine blades (Fig. 2.1.4).

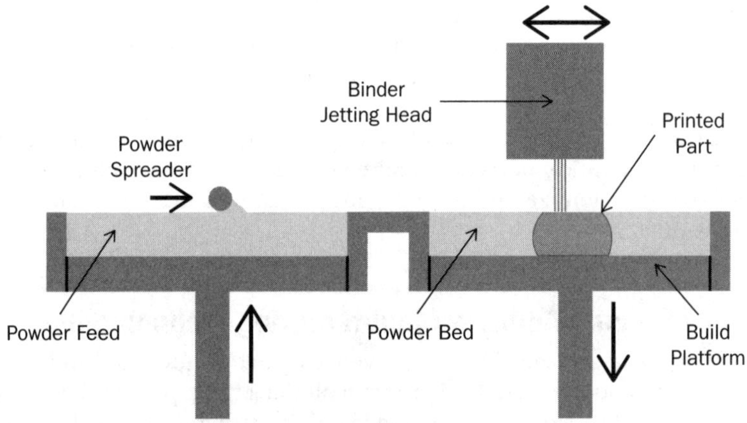

Figure 2.1.3 A binder jetting technology, similar to 3D printing.

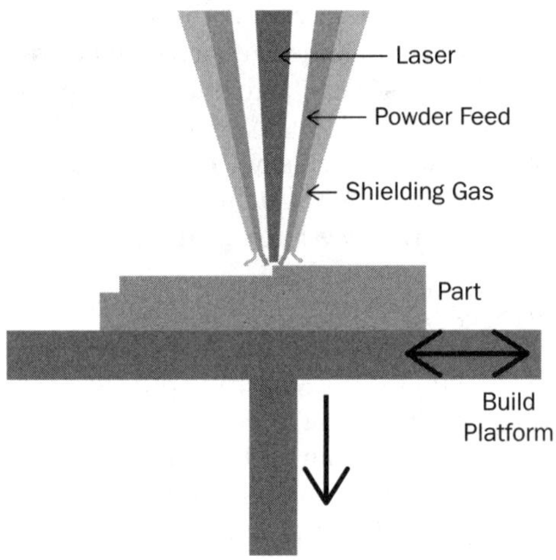

Figure 2.1.4 A directed energy deposition process similar to LENS.

Material extrusion is "an additive manufacturing process in which material is selectively dispensed through a nozzle or orifice" [3]. Materials are typically forced through a nozzle which moves relative to a platform or substrate to selectively deposit a layer of material. Then the nozzle moves up or the platform moves down to deposit the next layer. One of the most common commercially available systems is the fused deposition modeling (FDM) technology developed by Stratasys in which a spool of thermoplastic filament is extruded through a high-temperature nozzle, which melts the thermoplastic and deposits it onto the substrate. The nozzle moves in the plane as each layer is deposited, and the platform moves downward relative to the nozzle after each layer is

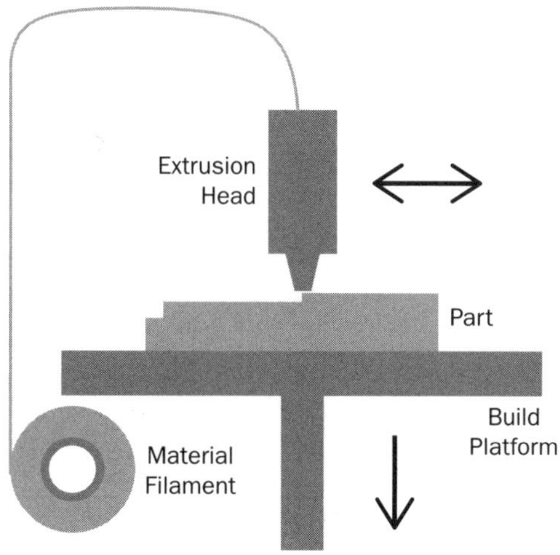

FIGURE 2.1.5 Fused deposition modeling (FDM), a material extrusion technology.

deposited. Material extrusion technology underpins a wide variety of personal systems, including those from Makerbot, Aleph Objects, Fabbster, and others.

There are more material extrusion machines in the world than any other technology. Material extrusion is so widely used because it can build parts in a wide range of materials with an office-friendly technology, and depending on the particular machine and feed stock, it can be quite inexpensive as well. The disadvantage is that the material properties can be highly anisotropic (weaker across layers) because of the deposition pattern; the surface finish is generally poor, especially relative to photopolymerization or material jetting processes; and the build process itself is relatively slow. Unlike powder bed fusion or material jetting technologies, which can support overhangs and cavities with build material, material extrusion processes require support structures, which must be removed during post-processing (Fig. 2.1.5).

Material jetting is "an additive manufacturing process in which droplets of build material are selectively deposited" [3]. Unlike the binder jetting process, which selectively deposits binders onto a powder bed, the material jetting process directly deposits all of the build material using inkjet technology. The most common materials are photopolymers or waxes. One of the most commonly available commercial versions of material jetting is the Connex technology from Stratasys, which is derived from systems commercialized by Objet. Connex machines deposit photopolymers that are cured by ultraviolet light as they are deposited. They are capable of depositing multiple materials simultaneously in various mixtures. One of those materials is used for removable support structures to support subsequent layers, especially when overhangs or cavities are present. Other commercial providers include 3D Systems, which offers a line of multi-jet modeling machines, and Solidscape, which specializes in machines that print wax patterns for casting.

Material jetting technologies are capable of fabricating with multiple materials simultaneously. For example, a prototype key fob could be fabricated with a rigid casing and elastomeric buttons in a single build. Material jetting technologies are also known for rapid build times and excellent accuracy and surface finish. The Connex

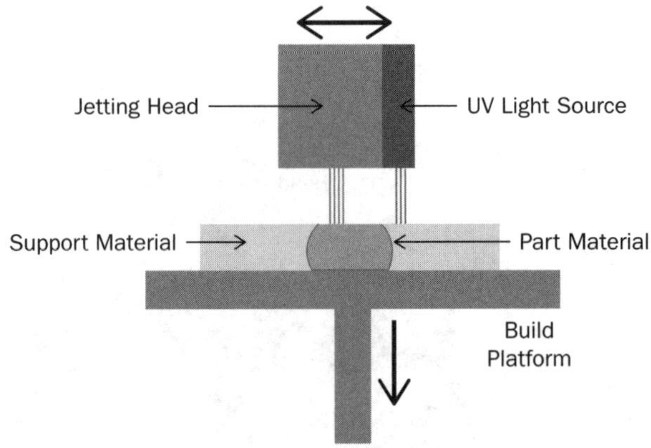

Figure 2.1.6 A material jetting technology similar to the Connex technology.

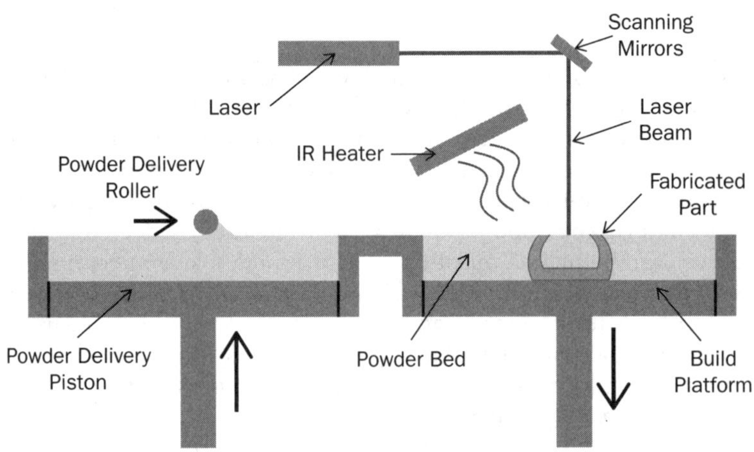

Figure 2.1.7 Selective laser sintering, a powder bed fusion technology.

technology, however, is limited to photosensitive materials whose properties can change over time (Fig. 2.1.6).

Powder bed fusion is "an additive manufacturing process in which thermal energy selectively fuses regions of a powder bed" [3]. The thermal energy is typically provided by a focused laser or electron beam, which selectively melts powder in a powder bed in a layer-wise fashion (Fig. 2.1.7). Two of the most common commercial suppliers of powder bed fusion systems are 3D Systems and EOS GmbH. EOS's EOSINT M machine is one of the most widely used metal powder bed fusion technologies, which EOS calls direct metal laser sintering (DMLS). Other commercial providers of metal machines include SLM Solutions, Renishaw, and Arcam AB, which provide electron beam melting (EBM) machines.

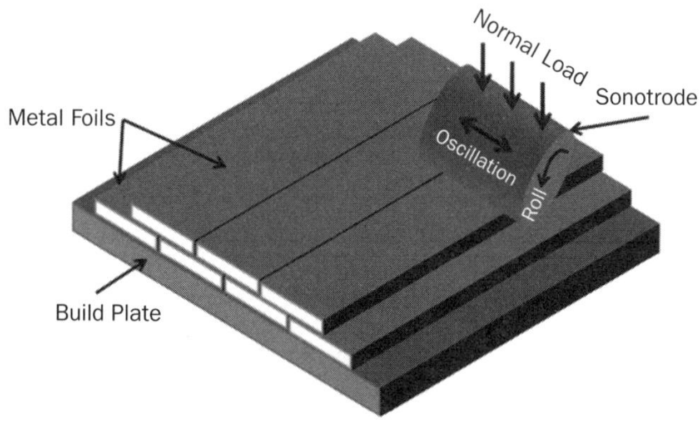

FIGURE 2.1.8 Ultrasonic additive manufacturing, a sheet lamination technology. The sonotrode oscillates at ultrasonic frequencies while applying a normal force to the metal foil and rolling along its surface.

Powder bed fusion systems are the dominant AM technology for fabricating functional end-use parts. It is possible to fabricate near fully dense parts in a wide range of materials with thermoplastics and metals being the most common. Since parts are built in a powder bed, no support structures are required, except for metal parts, which are often anchored to a build platform to prevent thermally induced warping. The lack of support structures enables fabrication of parts with channels and other complex internal features. Polymer parts require very little post-processing, although metal parts require support structure removal and thermal post-processing. Surface finish and accuracy are limited by particle size and typically much worse than liquid-based processes such as vat photopolymerization. The powder bed fusion process itself can be quite slow, with warmup and cool down times added to the fabrication cycle time, but the powder bed allows multiple parts to be nested into a single build, thereby increasing production rates relative to other AM technologies such as vat photopolymerization.

Sheet lamination is "an additive manufacturing process in which sheets of material are bonded to form an object" [3]. A very early technology called laminated object manufacturing (LOM) bonded sheets of paper together to form an object (Fig. 2.1.8). Ultrasonic AM from Fabrisonic uses ultrasonic welding to bond thin layers of metal, combined with a machining system to machine the appropriate cross sections as the layers are deposited.

Ultrasonic AM builds parts in a wide range of metals, including aluminum, Inconel, copper, and stainless steel, and it is highly accurate because it includes a high-precision machining system. The underlying ultrasonic welding process is a low-temperature fusion technique, which allows the designer to embed sensors, actuators, optical fibers, and other heat-sensitive components, and build the metallic part around them. The disadvantage is that the technology is not as widely available as other metals technologies (e.g., powder bed fusion of metals). Also, it can be difficult to fabricate large internal cavities because the ultrasonic welder, which fuses each layer, must be supported by a rigid feature beneath it.

Vat photopolymerization is "an additive manufacturing process in which liquid photopolymer in a vat is selectively cured by light-activated polymerization" [3]. Stereolithography (SLA) systems from 3D Systems use ultraviolet lasers to scan a vat of

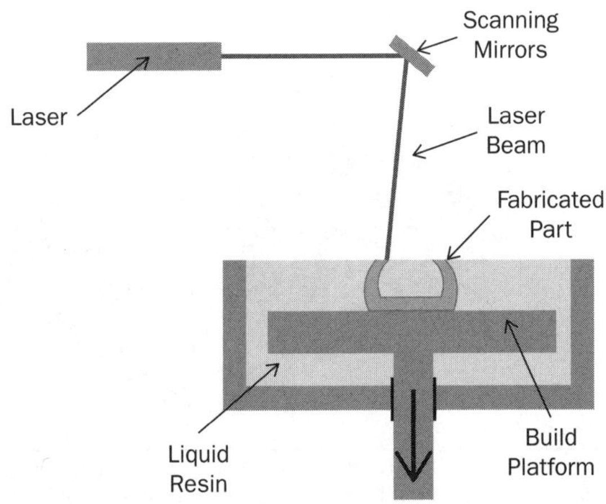

Figure 2.1.9 Stereolithography, a vat photopolymerization technology.

photopolymer resin (Fig. 2.1.9). Other systems, such as those offered by Envisiontec GmbH, cure the photopolymer with ultraviolet lamps paired with DLP technology for selectively exposing the photopolymer to ultraviolet energy.

Vat photopolymerization, in the form of SLA, was the first commercially available AM technology, introduced by 3D Systems in 1987. The systems are limited to photosensitive materials, but their accuracy and surface finish are excellent relative to other AM technologies. These capabilities make vat photopolymerization a popular choice for patterns and molds, including the Invisalign braces described earlier in this chapter.

General Design Considerations for Additive Manufacturing

Additive manufacturing has been promoted as a disruptive technology and a democratizing force for design and manufacturing worldwide. The strength of AM is NOT in replacing conventional techniques of mass production but in fabricating parts that are not suitable for mass production, either for economic or technical reasons. So, one of the first considerations of the designer is deciding whether AM is appropriate for a specific part and what types of value AM could add to the part. There are several drivers for selecting AM for production of a specific part:

- *Small lot or one-of-a-kind production.* Even if an item is suitable for fabrication via conventional forming or subtractive techniques, it may be more economical to fabricate the part with AM technologies if production quantities are small. AM does not require an investment in tooling, which can be very expensive on a per-part basis if only a few parts are produced. Many parts in the aerospace industry, for example, are produced in small quantities. Replacement parts can be fabricated in small quantities, as well, and with AM neither an inventory of parts nor an inventory of specialty tools are required—just an inventory of digital files. Some companies are also using AM for bridge manufacturing in which AM is used to build a few parts and keep production on schedule while tools are being developed for long-term mass production.

- *Material complexity.* Several AM technologies, including material jetting and directed energy deposition technologies, are capable of fabricating parts with multiple materials and functionally graded material compositions. For example, a cutting tool may require high levels of hardness and wear resistance at the tip but much more ductility and toughness at the base. The ability to deposit and process material differently at each point in a part is a unique capability of AM.
- *Geometric complexity.* Since AM technologies build parts layer-by-layer, they are uniquely suited to fabricating complex internal and external geometries, without being limited by tool access or mold removal considerations. By incorporating lattice structures into parts, for example, it is possible to remove mass while maintaining required levels of stiffness and strength. It is also possible to use internal passageways for cooling, acoustic dissipation, or other functionalities.
- *Personalization or customization.* Since AM technologies build parts directly from digital files, they are particularly suitable for personalizing or customizing those parts. Dental braces and prosthetics, for example, can be uniquely customized to the geometry of an individual patient's body. Cranial implants can be customized to individual patient's injuries. Hearing aid shells can be customized to an individual patient's ear canal.
- *Part integration.* Integrating multiple parts into a single assembly offers several benefits. By eliminating fasteners and mating surfaces, the integrated parts are often much lighter. Fewer assembly tools are required, and assembly time can be reduced significantly. Part integration is enabled by AM's capability of fabricating complex geometries, which must be subdivided into manufacturable subcomponents for conventional fabrication, and preassembled moving parts (e.g., bores and shafts), which are particularly well-suited to powder bed fusion and binder jetting processes that support parts in a powder bed during fabrication.

Once the decision has been made to use an AM technology to fabricate a part, the next step is to consider the capabilities and limitations of the chosen technology. Although popular impressions of AM technology often suggest that AM can fabricate any part, anywhere, in a matter of minutes, in reality there are some important constraints and drivers associated with AM technologies.

- *Diminished material properties.* Since AM technologies build parts layer-by-layer, material properties tend to be anisotropic with the strongest properties parallel to the build plane and the weakest properties orthogonal to the build plane or across layers. Some processes, such as material extrusion, can suffer from losses in mechanical strength as high as 25% or more in planes orthogonal to the build plane, while others, such as powder bed fusion, experience very little anisotropy. Partial density also impacts properties significantly. Partial density could be caused by air gaps in material extrusion processes or lack of complete fusion in powder bed fusion processes, for example. Whereas the impact of partial density on stiffness tends to be linearly proportional to the density, there can be a nonlinear relationship between density and strength, resulting in significant reductions in strength, for example, in the event of incomplete powder bed fusion.

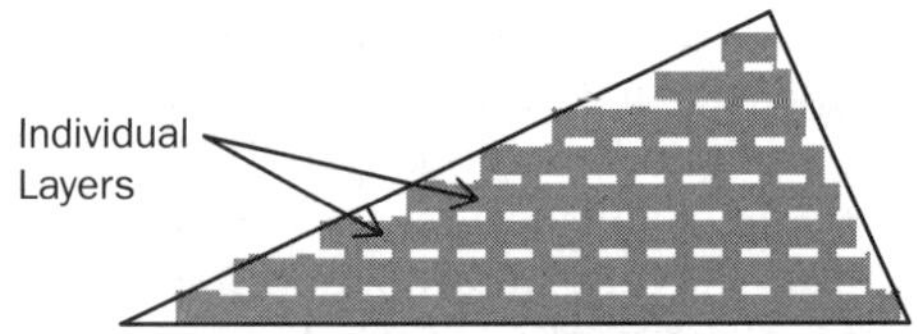

FIGURE 2.1.10 Stair-stepping in an additively manufactured part.

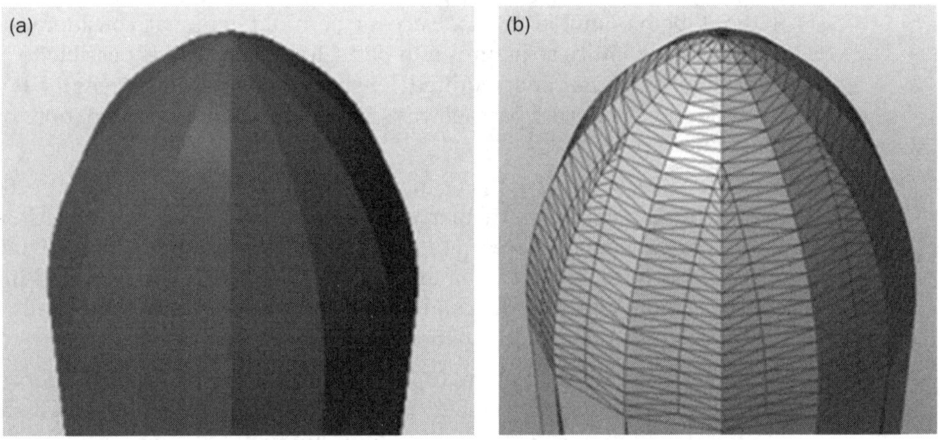

FIGURE 2.1.11 (*a* and *b*) The conversion of a CAD model to an STL model, showing the triangular representation of the surface.

- *Limited accuracy, resolution, and surface finish.* Accuracy, resolution, and surface finish capabilities vary among AM technologies, but for the most part, they cannot compete with conventional fabrication processes. For many commercially available technologies, minimum feature sizes are on the order of 0.1 to 0.6 mm (0.004 to 0.024 in.), and accuracy is typically in the range of ±0.1 to 0.2 mm (±0.004 to 0.008 in.), with some select technologies outperforming those values. Accuracy and resolution are often limited by the fidelity of the deposition tool or energy delivery mechanism (e.g., nozzle diameter, laser spot size), combined with the precision of the moving mechanical components in the machine, and the characteristics of the material (e.g., particle size in powder bed fusion). Surface finish can be influenced by some of the same characteristics (e.g., particle size in powder bed fusion and nozzle diameter in material extrusion), but it can also be worsened by a stair-stepping effect inherent in layer-based manufacturing, as shown in Fig. 2.1.10.
- *Data conversion.* The most common way to transfer part information to an AM machine is to generate an STL file from a 3D CAD model within a CAD system. An STL file represents the boundary of a part with a mesh of triangles, as shown in Fig. 2.1.11. This conversion process can introduce approximation errors, particularly for curved surfaces. Whereas planar surfaces, such as the surfaces of a cube, can be represented precisely with a triangular mesh, curved surfaces,

such as a sphere or the sides of a cylinder, are approximated. The denser the mesh of triangles, the more closely the surface is approximated; and most STL file generators allow the user to specify a desired offset error, indicating the maximum distance between the desired surface and the STL approximation. It is important for that error to be less than the resolution of the AM machine, to avoid the appearance of triangles on the surface of the part. In addition to approximation error, other problems are frequently encountered with STL files. For example, units are not embedded in STL files, so a file created in English units could be converted to metric units and rescaled unintentionally. Also, faulty STL files can be created with leaky or non-watertight surfaces, inverted normals, or triangles with poor aspect ratios. These errors can lead to faulty builds when the machine fails to fabricate the appropriate cross sections of the part. Commercial software is available for error-checking and repairing STL files.

- *Process planning.* Obtaining high-quality AM parts is a function of not only the AM technology, and the part itself, but also the way the part is processed by the AM machine. Part orientation is particularly important for desirable surface finish and material properties. The smoothest surfaces are typically obtained either parallel to or perfectly orthogonal to the build plane. Angled surfaces are typically much rougher due to the stair-stepping effect caused by layering. Also, if strength is more important along a particular axis of the part, then it is typically important to build that axis parallel to the build plane where material properties are the best. Also, if the part has cavities or voids, it is important to think about removal of either powder or support structures. If the part is fabricated with binder jetting or powder bed fusion, it is important to design access channels for removing powder from the cavities. If the part is fabricated with material extrusion, material jetting, or some other technology that requires support structures, it is likewise important to ensure access for removal of support structures, especially if those structures are not water soluble and must be manually removed. Finally, some technologies, such as powder bed fusion and binder jetting, allow multiple layers of parts to be nested within the powder bed, and doing so can greatly reduce the time and expense of fabricating the parts. In some cases, small design changes can enable highly efficient nesting of parts. For example, two identical cylinders cannot be nested, but if the sides can be tapered, then they could be nested in a manner similar to stacking plastic drinking cups, thereby making much more efficient use of the build space. Another way to make efficient use of the build space is to design parts to occupy minimum vertical height in an AM machine, since build time is often dependent on the overall height of the build and therefore the number of layers that must be fabricated.

Overall, AM is poised to make profound changes to the way we design and fabricate components and systems, especially when those systems benefit from the complexity and customization capabilities offered by AM. Designing effectively for AM requires not only new knowledge about process capabilities and part and material properties but also new design tools that can support the design freedoms afforded by AM [4]. The possibilities are endless!

References

1. On 3D Printing, "on3dprinting.com," July 11, 2013. [Online]. Available: on3dprinting.com/tag/invisalign.
2. GE, "GE Reports," June 23, 2014. [Online]. Available: http://www.gereports.com/post/80701924024/fit-to-print.
3. ASTM International, "Standard Terminology for Additive Manufacturing Technologies," ASTM Standard F2792-12a, West Conshohocken, PA, 2012.
4. C. Seepersad, "Challenges and Opportunities in Design for Additive Manufacturing," 3D Printing and Additive Manufacturing, vol. 1, no. 1, pp. 10–13, 2014.

CHAPTER 2.2
Binder Jetting

Rick Lucas

The ExOne Company
North Huntingdon, Pennsylvania

The Process and Materials

Binder jettings, as an additive manufacturing process, is similar to traditional paper printing: The paper page is replaced by a layer of powder particles, and the printhead ink is replaced by a liquid binder that joins the particles together. As the printhead rasters across the layer of powder, binder droplets (Fig. 2.2.1) are precisely jetted to selected locations that collectively form a two-dimensional (2D) slice through the three-dimensional (3D) part that is to be produced.

The part is built on a platform that is repeatedly lowered, one slice at a time, followed by spreading a new layer of powder (Fig. 2.2.2) and printing the next 2D slice until the 3D part is formed. A CAD solid model represents the 3D shape in computerized form, and is the basis for control of the printhead as it progressively forms each 2D slice.

The printed material may consist of plastic, metal, or ceramic powders, and includes particulates such as sand, starch, and plaster. Binder selection depends on the material to be printed, and ranges from water for starch and plaster to polymer-based mixtures for metals and ceramics. After the 3D shape is produced, the article may be used as-is in its bonded form, or it may be subjected to further treatments that strengthen the bonds between particles and/or eliminate the porosity between the particles.

FIGURE 2.2.1 Jets dispensing binder.

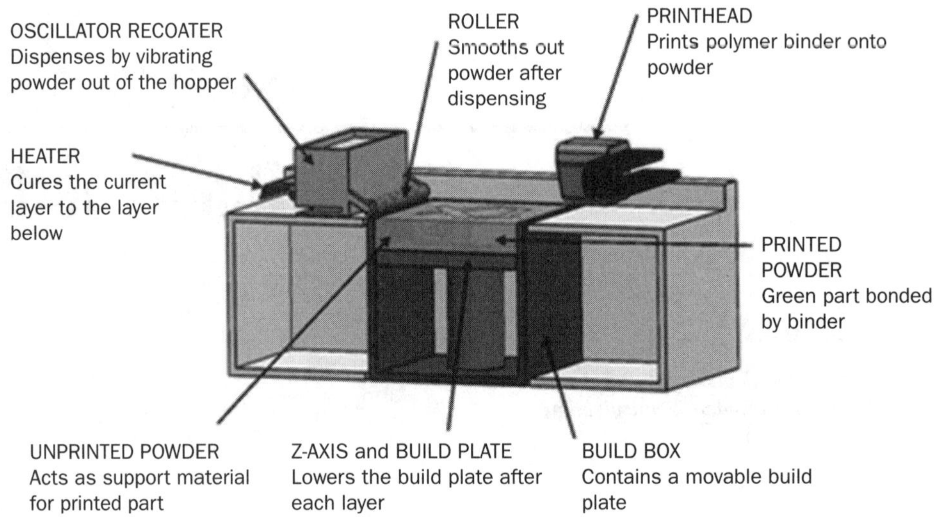

Figure 2.2.2 Binder jetting schematic.

Typical Characteristics and Applications

Low-end binder jetting machines use plaster and starch build materials with water as the binder. These binders may contain dyes, allowing for color printing. After printing, the parts are typically strengthened by infiltrating with an epoxy. Articles produced in these materials are used for prototyping, form and fit models, architectural models, and relief maps.

High-end binder jetting machines use metal and ceramic powders, which require complex polymer-based binder systems. The resulting articles can be used in one of four forms: (a) as bonded; (b) lightly sintered; (c) lightly sintered and infiltrated; (d) highly sintered. The form used is dependent on the desired product and application.

As Bonded

The resulting part consists of bonded powder whose strength is controlled by the strength of the binder. Typically surface finishes are coarse and require coatings if a smooth surface is desired. One of the more common applications for the bonded product is printed sand molds. Sand molds provide a cost-effective and time-reducing alternative to traditional methods for creating prototype and low-volume castings (Fig. 2.2.3). 3D printing sand molds and cores directly from CAD data eliminates the need for physical patterns and pattern storage, since parts can be stored digitally. Design freedom is the largest advantage of the process. Manufacturing and composite tooling is another good application for this product, but generally requires a surface coating after the printing.

Lightly Sintered

A lightly sintered metal or ceramic product would be used when porosity is acceptable or desired for the part. Typically a lightly sintered product would have porosity of

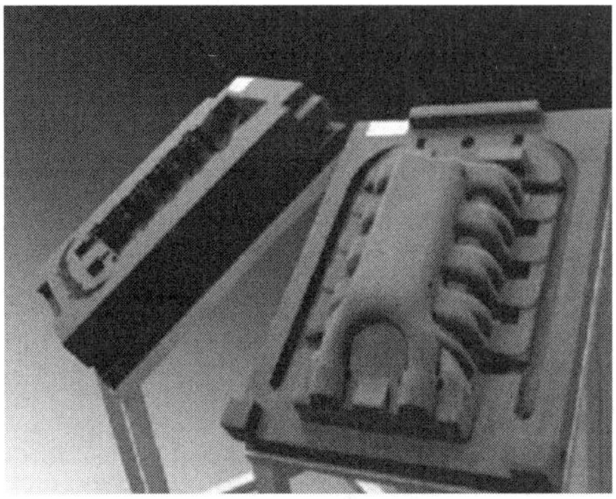

Figure 2.2.3 3D-printed sand mold.

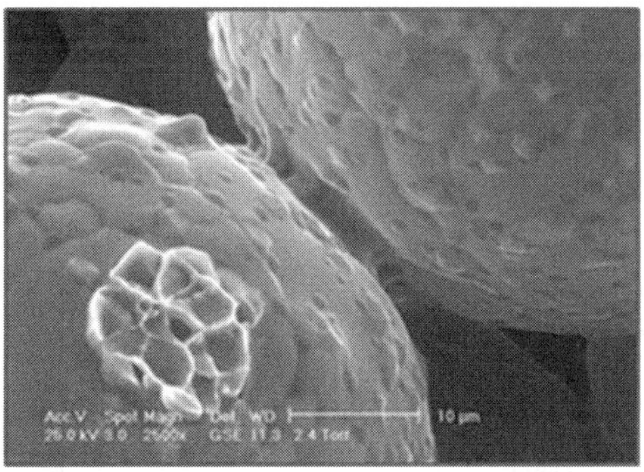

Figure 2.2.4 Metallurgical bond between two stainless steel particles after light sintering.

approximately 40%. Example applications include filters and tooling for conventional manufacturing processes.

Sintered and Infiltrated

In the infiltrated form the resulting product is a composite material. The printed part is cured and then sintered and infiltrated with another material. The infiltrant is generally used to strengthen the part. The infiltrating material can be an epoxy, adhesive, wax, or a metal having a lower melting point than the build material. Polymers and wax infiltrants are used commonly for investment castings. Metal infiltrants are used for more functional parts needing high strength (Fig. 2.2.4).

Highly Sintered

Highly sintered or monolithic materials are very similar to conventional pressed and sintered powder metallurgy parts and metal injection molded parts. The basic steps of press and sinter powder metallurgy include powder blending, die compaction, and sintering, while in metal injection molding the part is formed by injecting a mixture of metal powder and binder into a die and then sintering. In binder jetting the powder is printed instead of die compacted or injection molded. The advantage of binder jetting is

Figure 2.2.5 Examples of monolithic parts in 316 stainless steel.

that the creation of the mold for compaction or injection molding is not necessary. This provides a distinct cost advantage for prototyping or low-volume production runs. The binder jetting process also overcomes the limitation in part size in conventional powder metallurgy processes, although controlling geometry due to distortion caused by large shrinkage remains a challenge, as is true with conventional powder injection molding. Figure 2.2.5 illustrates examples of highly sintered parts produced by binder jetting.

Advantages of Binder Jetting

Various methods of 3D printing have been developed in addition to binder jetting. In contrast to joining particles by a jetted binder, these processes build parts through a melting or micro welding process using high-energy laser or electron beams to fuse each layer together. These processes require a base plate on which the part is built for stability throughout the build process. Such processes require significantly more material because support material must be used for overhangs and bridge features. The build process itself is also time-intensive.

Binder jetting is unique, in that it does not involve high heat during the build process. Other additive techniques utilize a laser or electron beam heat source that creates residual stresses in the parts. These stresses must be relieved in a secondary post-processing operation. Additionally with binder jetting, the parts are supported by the loose powder in the job box, thus eliminating the need for separate support structures and allowing multiple layers of parts. Spreading speeds significantly outperform other processes. Binder jetting has the ability to print large parts and is often more cost-effective than other additive manufacturing methods.

Economics

Binder jetting is best suited for low- to medium-level production quantities. In cases where parts are complex and a large amount of machining is required, large production quantities can be economical.

In the case of bonded products such as sand molds and cores, the process is well suited for high-volume production of over 100,000 parts annually. This is due to the fast print times of the process. Current machines can print up to 0.028 m^3/h (14 ft^3/h) resulting in less than 0.625¢/cm^3 (10¢/$in.^3$) printed mold cost.

FIGURE 2.2.6 Complex component produced by BJ.

Sintered and infiltrated as well as fully sintered materials are limited to low to medium quantities due to the slower print times. These materials require finer powders, thinner layers, and the additional post-processing. The fastest binder jetting metal printer on the market can produce 1966 cm^3/h (120 $in.^3$/h), which is one the fastest 3D printing technologies. The affordability of the process is demonstrated by the fact that, among all additive manufacturing processes, most of the small, complex metal and ceramic parts sold on the consumer market today are produced by binder jetting.

General Design Considerations

The major advantage of the binder jetting process is its ability to produce complex geometries (Fig. 2.2.6). Design considerations depend on the material being printed, the printer being used, and the application.

Print resolution is driven primarily by powder particle size, layer thickness, and the droplet size. The powders can be less than 10 μm (0.0004 in.) and as large as 500 μm (0.020 in.). Generally, the binder droplets are between 40 and 60 μm (0.0016 to 0.0024 in.) in size and can be as low as 10 μm (0.0004 in.), which gives the extremely high resolution of 1200 dpi (dots per inch).

Typical part sizes can range from 0.065 cm^2 (0.1 $in.^2$) to 516 cm^2 (80 $in.^2$) in area and from 0.25 to 97 cm (0.1 to 15 in.) in length. The larger print platforms produce parts up to 3696 l (130 ft^3) in volume.

Suitable Materials

Binder jetting has the largest material set of all the 3D printing technologies. Practically any material available in powder form is suitable for binder jetting. The powders can be angular or spherical and the powder size can vary from less than 10 μm to over 500 μm (0.0004 to 0.020 in.). Molds and tooling generally use powders above 100 μm (0.004 in.). Infiltrated materials work well with powder above 30 μm (0.0012 in.) and highly sintered monolithic materials need to be around 10 μm (0.0004 in.).

Common metal materials readily available in powder forms include iron, stainless steels, tungsten, nickel, Monel, Inconel, copper, brasses, titanium, aluminum, and magnesium. Common nonmetallic materials available include plastics, plasters, silica and ceramic sands, glasses, alumina, zirconia, silicon, and aluminum nitride powders. Printing reactive materials does require special controlled conditions.

Build Dimension (in.)	Minimum Wall Thickness (in.)	Build Dimension (mm)	Minimum Wall Thickness (mm)
0.125–3	0.04 minimum	3.175–76.2	1.01 minimum
3–6	0.06 minimum	76.2–152.4	1.52 minimum
6–8	0.08 minimum	152.4–203.2	2.03 minimum
8–12	0.125 minimum	203.2–305	3.175 minimum

Table 2.2.1 Recommended Minimum Wall Thicknesses

Detailed Design Considerations

Wall Thickness

The brittle nature of gypsum and plaster materials even after a coating of super glue requires extra design considerations. A minimum 2 mm (0.080 in.) wall thickness is needed and detail features of 500 μm (0.020 in.) are possible.

As seen in Table 2.2.1, directly printed and sintered metal and ceramic materials require increased wall thicknesses as the part size is incremented. Wall thickness adjustments are recommended in order to ensure success of part throughout the printing process. Detailed features of 254 μm (0.010 in.) are possible

Bonded sand and ceramic materials used for molds generally use course powders greater than 100 μm (0.004 in.) with thick layers 200 to 500 μm (0.008 to 0.020 in.) with a part resolution of 100 μm (0.004 in.).

Uniform Wall Thickness

Sintered materials with differences in thickness lead to distortion, internal stresses, cracking, and voids and can be avoided by using consistent wall thickness throughout. These differences can cause nonuniform shrinkage and make it difficult to maintain dimensional and tolerance control. When it is not possible to have uniform thicknesses due to the part restraints, a gradual transition is recommended. Thicknesses in the range of 1.3 to 6.3 mm (0.05 to 0.25 in.) are preferred, but exceptions in both directions are possible.

The use of webs and ribs is recommended for thin wall sections. This helps the strength and limits the distortion that could occur during sintering (Fig. 2.2.7).

Inside Edges

Sharp corners and edges should be avoided by including fillets and radii to reduce stresses at feature intersections. Sharp intersections and wall terminations should have

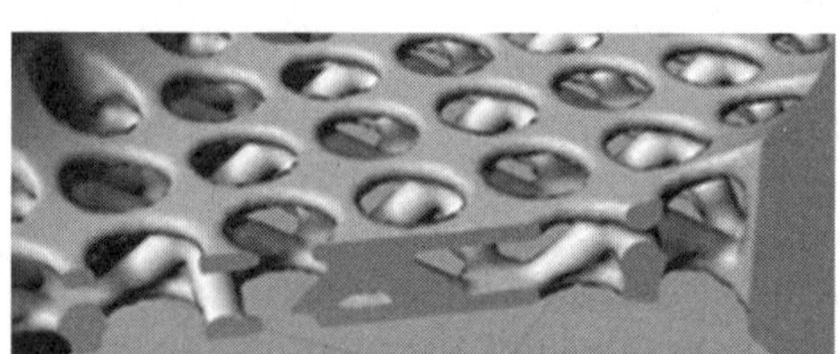

Figure 2.2.7 Use of webs and ribs for thin wall sections.

a fillet based upon the wall thickness. Variable radii are acceptable as long as the wall transitions are undeviated.

Interior Holes

As a general rule, for infiltrated materials, all through holes must have a minimum diameter of about 1 mm (0.040 in.). Holes less than this are at risk of being filled with infiltrating materials, such as loose powder, which cannot be successfully removed. Also, it is difficult to remove the supporting media or powder from the interior through holes that change direction because of its limited accessibility.

Part Connections

Connection points between features must have significant surface area interactions. Parts that have lower intersection areas have a higher risk of cracking or breakage. The lower cross section can choke the flow of the infiltrating material and lead to unbalanced infiltration.

CHAPTER 2.3

Directed Energy Deposition

LaRoux K. Gillespie

Manufacturing Consultant
Andover, Kansas

Directed energy deposition (DED), also known as laser-engineered net shape (LENS), direct metal deposition, construction laser additive direct (CLAD), or electron beam additive manufacturing (EBAM), is an additive process in which metal powder or wire is fed directly into a melt pool caused by either a laser or electron beam (Fig. 2.3.1). The head that carried the additive material and energy source is focused on a substrate and the material melts on the substrate or to any metal already there. The machine's head is programmed to produce whatever contour is required and by repeated passes and melting the top of previous layers while adding more material one can produce three-dimensional parts. While powder bed fusion (PBF) uses similar powders and heat sources, the PBF process lays a layer of powder down which is subsequently fused to whatever is under it. Directed energy deposition in contrast melts the powder or wire as it goes without any previous placement of material. Metal cladding

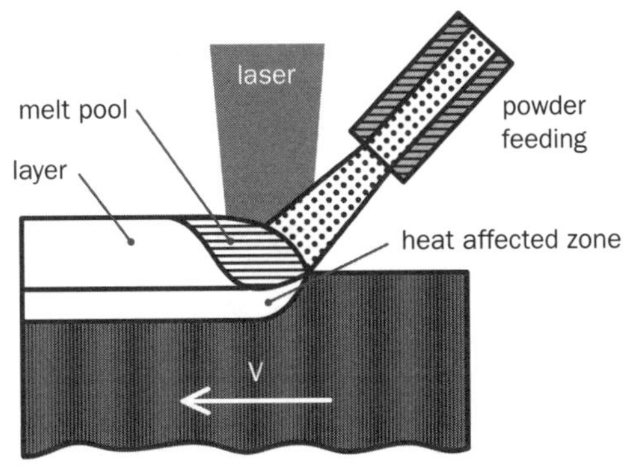

FIGURE 2.3.1 Illustration of directed energy deposition. (Courtesy European Powder Metallurgy Association, © 2015.)

(discussed in Chap. 1.9) uses the same style of equipment as DED, but the purpose of most of this process is to make three-dimensional products as opposed to just cladding an existing product.

The directed energy process is typically performed in an inert atmosphere or a vacuum in order to assure the deposited material is as pure as the starting material. A plasma arc version of the process exists using the name ion fusion formation (IFF). All of these approaches are available on commercially proven equipment. The electron beam system requires a vacuum system and uses powders only.

The process is a faster process than PBF and the chamber sizes in general allow much larger parts to be made than PBF. One of the most unique aspects of this process is that the gradient of the material within a part can be changed. Stated differently, a part may be built with different levels of purity, or different materials in different layers, or different materials slowly changing as the part is built. This ability to build in heterogeneous properties offers new design opportunities. The surface finishes produced are lower than PBF as is the precision of the components that can be made. The process creates fully dense parts, and in addition to aerospace applications it has been effective for creating implants. Like most of the other additive processes, designers provide a three-dimensional model which is then converted to computer numerical codes. The resulting code then causes the machine to lay down metal layer by layer, until the final part configuration is complete. Equipment can be provided with up to 5 axes of motion to provide a variety of configuration capabilities. Robotic arms are the most usual form of equipment moving the head.

Another advantage of the DED process is the fact that the nozzle is normally held at 90 degrees to the manufacturing surface. This means that metal can be deposited on complex contours such as spheres, cylinders, pyramids, or similar nonplanar metallic substrate. Subsequent heat treatment will often be required to provide the best material properties and machining may be needed for appropriate applications. However, some 99.999% dense parts have been fabricated without needing subsequent heat or pressure treatments.

Metals

The metals available in wire form at the time of writing include:

- Titanium and titanium alloys
- Inconel 600, 625, 718
- Nickel and copper-nickel alloys
- Stainless steels 300 series
- Aluminum alloys 1100, 2318, 2319, 3000 series, 4043, 4047, 5183, 5356, 5554, 5556
- Alloy steels
- Cobalt alloys
- 4340 steel
- Zircalloy
- Tantalum

- Tungsten
- Niobium
- Molybdenum

Powder costs are higher than wire costs of the same material. Currently available powders include:

- Steels such as 316L, 17-4PH, etc.
- Nickel and cobalt base superalloys: 625, 718, CoCr F75, etc.
- Titanium alloys: Ti6Al4V, CPTi, etc.
- Aluminum alloys: AlSi10Mg, etc.

Powder alloys in development include copper alloys, magnesium alloys, precious metals such as gold, silver, platinum, and refractory metals such as molybdenum alloys.

The EB wire version can deposit from 2.3 to 9.1 kg (5 to 20 lb) of metal per hour which is faster than powder systems. Wire systems make more efficient use of the raw metal.

Applications

Aircraft engines, air frame structures, or other large mechanical systems are a major area for this process, but it also fits well with the smaller parts, particularly for those that do not need fine details. Special tools, motor sport quick turnaround parts, and near-net turbine blades are built and finished today directly onto the machined shaft (finish machining occurs on the same setup as deposition). The process can make parts that are only a few millimeters in size to several meters if large enough chambers are provided. Figure 2.3.2 illustrates both how parts can be fixtured and positioned as well as the rather complex configurations that can be made.

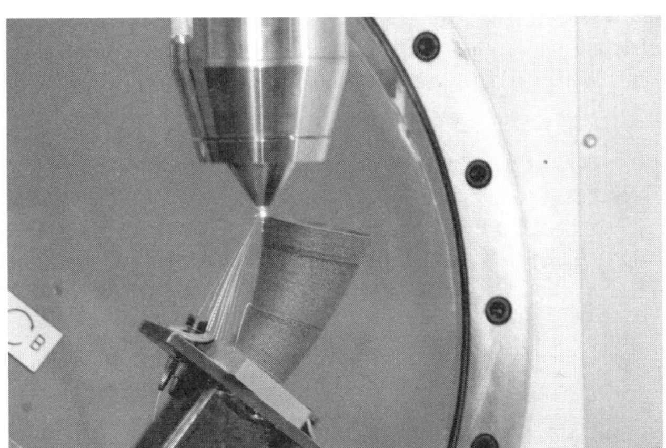

Figure 2.3.2 Directed energy deposition part in process. (Courtesy Be-AM Machines, Hebron, KY.)

Design Issues

The parts need a supporting structure that limits some geometries and can limit complexity. DED cannot make geometries having significant projection or overhang without support (i.e., the melted metal will fall or droop unless supported by something). While overhangs have been made that lean out 45 degrees, that is not the norm and the metallurgy may not be the same as a conventional vertical DED build (Nassar, 2015). Typical substrates or platforms on which parts are built include metal, metal wire, glass, and ceramics. Supports may need to be machined away after the deposition is finished, but they do not represent a significant impediment to process use.

Do not design sharp corners. Acceleration and deceleration around sharp corners can result layer height variations. Design with curves instead.

Typical accuracies vary from 0.1 to 0.25 mm (0.004 to 0.0098 in.) and details can be sized to 0.5 to 1 mm (0.020 to 0.040 in.). Surface finishes range from 60 to 100 μm (0.0024 to 0.004 in.) compared to 30 to 50 (0.0012 to 0.002 in.) with the SLM process. Layer thickness can be from 0.03 to 1 mm (0.0012 to 0.040 in.) [SLM can vary between 0.03 and 0.1 mm (0.0012 to 0.0040 in.)]. A few machines have been made that combine milling or grinding on the same platform so that parts may be finished when they come out of the chamber. This hybrid approach allows in-process machining to produce closer tolerances and better finishes as well as to machine outer surfaces without moving the part (see Chap. 2.1). Machines employing hybrid processes typically use an enclosed machine similar to a CNC milling machine design.

The European Powder Metallurgy Association notes that the mechanical properties or parts produced from the laser-fused powders are superior to the properties obtained with investment casting process and inferior or sometimes close to the conventional wrought part (EPMA, 2015).

References

Nassar, A. R. and E. W. Reutzel, Beyond Layer-by-Layer Additive Manufacturing—Voxel-wise Directed Energy Deposition, *Proceedings of SFF Symposium*, University of Texas, 2015.

EPMA, *Introduction to Additive Manufacturing Technology: A Guide for Designers and Engineers*, European Powder Metallurgy Association (EPMA), Shrewsbury, UK, 2015.

CHAPTER 2.4

Material Extrusion

LaRoux K. Gillespie

Manufacturing Consultant
Andover, Kansas

The material extrusion process involves drawing material through a nozzle that heats it, and then depositing it layer by layer (Fig. 2.4.1). Typically the nozzle moves horizontally in x and y planes and the platform upon which the first layer is placed moves up and down vertically. This is one of the most used processes since it is simple and inexpensive, and a variety of polymers are available to use. In most cases, the raw material is purchased in the form of a reel or spool of polymer filament (wire). It is also called fused deposition modeling (FDM).

The raw material is pushed through the nozzle under constant pressure and in a continuous stream. For accurate results the pressure must be kept constant and it must exit at a constant speed. The deposited layers are bonded by temperature control or by chemical agents. For most materials and machines, the falling material is hot enough to bond readily to the layer below it without additional heat.

The most common polymer used is the widely available ABS plastic. ABS has good structural properties that allow it to make relatively accurate models which can be functional. In fact, some educational models allow users to make their own mating parts and covers for their machines—no need to send off for replacement parts. The process is used for prototype models and early or low-volume production. For high-volume needs for cost efficiency, the process would be injection molded. Accuracy and speed are low when compared to other processes. The quality of the final model is limited to material nozzle diameter which for many machines is 1 mm (0.040 in.). Smaller diameter nozzles allow thinner and more tightly controlled part geometry. Tight tolerances require special controls for gravity and surface tension effects. Commonly layer thickness will vary from 0.178 to 0.356 mm (0.007 to 0.014 in.). Shrinkage can occur over time. Smoother finishes can be produced by subsequent sanding or chemical solvents.

The polymers in wide use include acrylonitrile butadiene styrene (ABS), nylon, polycarbonate (PC), polylactic acid (PLA), AB, polyether ether ketone (PEEK), and polyurethanes. Polyurethane-based materials have been used to make shape memory polymers (SMPs) that can recover large strains. For the SMPs annealing the specimens at 85°C (185°F) for 2 h improved shape recovery and consistency of mechanical test results. Some equipment can also provide ULTEM, PPSF/PPSU, PS, PET, and PUA

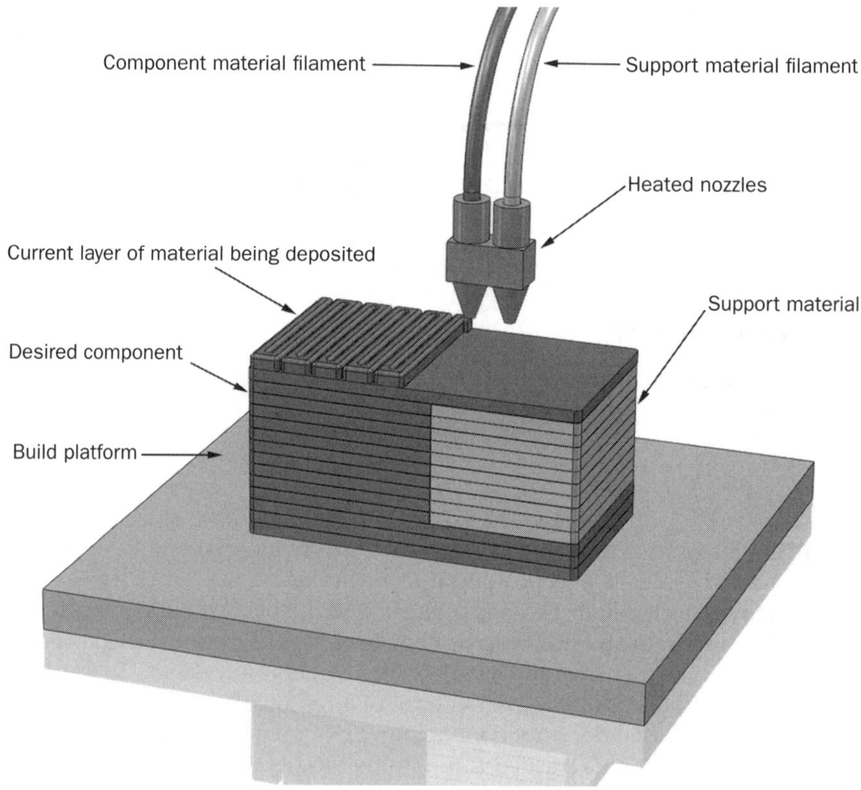

FIGURE 2.4.1 Material extrusion concept. (Courtesy Olaf Diegel.)

filaments. Extrusion of multi-material fiber-reinforced composites is in work, but not commercial at the time of this writing. A machine to extrude optically transparent glass for 3D components has also been made. Thermoplastic and thermoset polyimides, made by reactive extrusion, have been used in this process as have chopped carbon fibers in experiments, but there is no commercial capability yet. Reactive extrusion describes a process in which chemicals react within the material hopper before being extruded.

A machine with ability to make 400 × 356 × 400 mm (16 × 14 × 16 in.) parts can do so with an accuracy of ±0.127 mm (±0.005 in.) using ABS-M30, FDM® Nylon 12, and ULTEM™ 9085. Most machines are not as accurate as this. For large parts the accuracy approaches ±0.038 mm/mm (±0.0015 in./in.) of length.

While filament raw materials are the standard, syringe-based materials that use a chemical reaction to cause solidification have been made. For this approach a reservoir is filled with the desired material and heated or cooled to processing temperature. A plunger pushes the material out of the reservoir. A screw-based ejection system has also been used, but neither is currently commercial. Syringe-based systems could lead to finer resolution of finished parts [because they used nozzle diameters of 0.2 to 0.5 mm (0.008 to 0.020 in.)] as well as to stronger or harder components. Work done on this equipment design is for polystyrene (PS), polypropylene (PP) a semi-crystalline plastic, and biodegradable polycaprolactone (PCL) and PLA.

Applications

Because the process is limited today to materials that soften and melt at low temperatures, the resulting product has these same restraints. In general, the parts are used as rapid prototypes to demonstrate fit, form, and function. They are also used as demonstration models, teaching tools, for packaging studies, and can serve as mold masters but the latter is not a usual application. They are easily dissected to show interior features.

At the same time materials are being developed for a new world of applications. The University of Texas at El Paso for example has blended ABS and PC with TiO_2 to print materials for electromagnetic (EM) applications. ZnO nanorods, carbon fibers, glass fibers, tungsten (for radiation shielding), and other materials have been blended with polymers to make unique material extrusion products. For these materials screw style injection is used rather than traditional filament feed. One company specializes in making ceramic extrusions by additive material extrusion.

Studies on PEEK indicate that hot extrusion head design, extrusion temperature, and ambient temperature are key factors for producing PEEK structures that did not have warpage, delamination, or polymer degradation.

Considerations

Like all additive manufacturing processes, the material extrusion process allows designers almost unlimited freedom for shapes. Unlike machined blocks or cylinders of metal, the parts may be filled with meshes, pockets, or internal configurations of many shapes that are not possible when starting from solid blocks of material unless made into many fitting parts.

Features that hang over those below it will sag or not form correctly unless some intervening support material is provided to support the overhanging layer. Supports are readily made, but users have to make sure they can be broken free of the desired shape once completed. Since the process "writes" the layers by moving in x-y motions back and forth until a layer is complete, by having two nozzles on a machine one can print the support material while the other prints the desired part. Having multiple nozzles allows multiple materials or multiple colors to be printed in the same setup. Machines can have as many as eight nozzles although that is not common. There is no limit on the number of nozzles other than placement in the needed work area.

With some manual or robotic intervention, it is possible to embed other components within the layers. Fluid line connectors for example can be encased in the material; electrical circuits can be laid in the layers; magnets or a host of other items can become integral components with this process. Wire meshes have been overlain in semifinished parts to provide higher strength components. Ultrasonic bonding embeds them into the semifinished part.

Support structures can be made from water-soluble materials which allow easy separation of parts and supports.

Because the process makes parts by layers, the z-axis properties will vary somewhat from x-y properties, and designers need to consider the possible effects of some stratification of properties of layers. Properly made that should not be an issue, but designers need to assess that on early builds and specify expectations on designs.

Because the produced parts do not necessarily carry raw material properties fully into the finished part, tensile strength, elasticity, hardness, and other properties need to be identified for both the starting material and some form of testing described for acceptance. Cross sections of important areas may need to be analyzed in both prototype stage and in regular manufacture to assure consistency. Micro bubbles in raw filament stock may reduce expected strength of finished products.

For critical parts or to obtain useful information on new materials, the use of test coupons made in the same operation while making the parts may be advisable. These would be made with the same equipment, general process details, and material as the real component. Properly done, these test samples can reveal an exceptional amount of information about the effects of a variety of features (internal and external) as opposed to just an overall performance assessment.

Test samples can be simple coupons, flat, square, cylindrical, or they can be rigorous designs that contain real part features and include all or many of the following details:

- rectangular holes, bosses, and tubes (in multiple directions)
- round holes, bosses, and tubes (in multiple directions)
- spherical holes and bosses
- conical bosses
- L-shaped bosses
- ramps
- overhangs
- angles
- side notches
- thin walls and fine features
- freeform structures, and
- towers

A singular artifact can provide all these aspects and provide data useful for many design considerations. Moylan (2014) from whom the above listing comes provides an extensive discussion on this aspect of capturing data for design impact studies. The following references provide a broad range of additional useful information.

References

Chaung, C., T. Hughes, R. Avakian, and L. Hu, Reactive Extrusion of High Temperature Resins for Additive Manufacturing. National Aeronautics and Space Administration Glenn Research Center Report, Cleveland, Ohio, 2012.

Espalin, D., L. Herrera, M. S. Hossain, and R. Wicker, Wire Reinforced 3D Printed Parts Made by Using Material Extrusion Additive Manufacturing Process. In *CAMX 2014—Composites and Advanced Materials Expo: Combined Strength. Unsurpassed Innovation*, 2014.

Guessasma, S., W. Zhang, J. Zhu, S. Belhabib, and H. Nouri, Challenges of additive manufacturing technologies from an optimisation perspective. *International Journal for Simulation and Multidisciplinary Design Optimization*, 2015, 6:A9. (doi:http://dx.doi.org/10.1051/smdo/2016001).

Klein J., M. Stern, G. Franchin, M. Kayser, C. Inamura, S. Dave, J. C. Weaver, P. Houk, P. Colombo, M. Yang, and N. Oxman, Additive manufacturing of optically transparent glass. *3D Printing and Additive Manufacturing*, 2015, 2(3):92–105. (doi:10.1089/3dp.2015.0021).

Moylan S., J. Slotwinski, A. Cooke, K. Jurrens, M. A. Donmez, An additive manufacturing test artifact. *Journal of Research of the National Institute of Standards and Technology*, 2014, 119:429–459. (doi:10.6028/jres.119.017).

Park, Sang-in, D. W. Rosen, Seung-kyum Choi, C. E. Duty, Effective mechanical properties of lattice material fabricated by material extrusion additive manufacturing. *Additive Manufacturing*, 2014, 1–4:12–23.

Raasch, J., M. Ivey, D. Aldrich, D. S. Nobes, and C. Ayranci, Characterization of polyurethane shape memory polymer processed by material extrusion additive manufacturing. *Additive Manufacturing*, 2015, 8:132–141. (doi:10.1016/j.addma.2015.09.004).

Ramon, A. and T. Perez, Defeating Anisotropy in Material Extrusion 3D Printing via Materials Development. Paper AAI3714271, ETD Collection for University of Texas, El Paso, January 1, 2015.

Roberson, D. A., C. M. Shemelya, E. MacDonald, and R. B. Wicker, Expanding the Applicability of FDM-type Technologies Through Materials Development. In *Proceedings of SFF Symposium*, University of Texas, 2014.

Vaezi, M. and S. Yang, Extrusion-based additive manufacturing of PEEK for biomedical applications. *Virtual and Physical Prototyping*, 2015, 10(3):123–135.

Valkenaers, H., F. Vogeler, A. Voet, J. P. Kruth, Screw Extrusion Based 3D Printing, a Novel Additive Manufacturing Technology. In *Proceedings of International Conference on Competitive Manufacturing Screw Extrusion Based 3D Printing (COMA 13)*, 2013.

CHAPTER 2.5

Designing for Material Jetting Additive Processes

Christopher B. Williams

Virginia Tech
Blacksburg, Virginia

Nicholas A. Meisel

The Pennsylvania State University
University Park, Pennsylvania

Material jetting is an additive manufacturing (AM) process wherein droplets of a liquid material (e.g., waxy thermoplastics, photopolymer resins) are selectively deposited onto a substrate via a series of inkjet printheads. As the liquid material is deposited, it quickly undergoes a phase change to its solid form via exposure to ultraviolet energy (i.e., when jetting photopolymer resin) or temperature change (i.e., when jetting a waxy thermoplastic). The substrate lowers by a distance equal to one-layer thickness, and the subsequent layer is jetted directly on top of the previously formed layer. This process repeats until the entire part is fabricated in a layer-wise fashion. The presence of multiple printheads within a single print block allows the simultaneous, selective placement of separate materials, thus making material jetting one of the very few commercially available AM systems that is able to fabricate multi-material components in a single build. One of the printheads is dedicated to jetting a sacrificial material that is used to support printed features. These sacrificial materials must be removed in a post-process, typically via melting and/or washing.

Machines

Several embodiments of the material jetting AM concept have been commercialized since it was first demonstrated in 1987 as "Ballistic Particle Manufacturing," which featured a single print nozzle depositing polymer primitives [1]. Printers featuring multiple print nozzles capable of selectively jetted heated waxy thermoplastics were introduced by Solidscape and 3D Systems in 1994 and 1996, respectively [2]. Due to their reliance on waxy materials, the printed products were effective only as conceptual prototypes or as patterns for investment casting.

The use of material jetting as a platform for making functional prototypes and end-use products arrived in 2000 when Objet (now merged with Stratasys) introduced a system capable of jetting an acrylic photopolymer resin, termed "PolyJet." 3D Systems shortly followed suit with their line of InVision and ProJet platforms, which leveraged their earlier-developed "Multi-Jet Modeling" technology. These systems are capable of producing functional prototypes and products, and are the primary focus of this chapter. Process and material details are centered around the Stratasys PolyJet platform as more research literature has been published on these systems.

A schematic of the Stratasys PolyJet technology is presented in Fig. 2.5.1. The most recent embodiment of the PolyJet technology, the J750, contains eight total printheads in its print block, which enables the simultaneous deposition of six build materials and one sacrificial support material. Each printhead contains 96 jetting nozzles, resulting in an XY resolution of 600×600 dpi. As a layer of droplets is deposited [30 μm (0.0012 in.) layer thickness in standard mode; 16 μm (0.0007 in.) in high-resolution mode], two lamps (one leading and one trailing) irradiate the printed layer with UV energy to cure the deposited photopolymer. A rotating roller, which is integrated into the print block, levels the printed layer and removes excess material to ensure a flat surface for the subsequent layer (and also contributes to achieving a small layer thickness).

It is important to note that the manner in which support material is used in material jetting is unique among AM systems. Whereas other AM processes use support material only to provide scaffolding for overhanging features, material jetting relies on support material to act as a physical boundary for the build materials' liquid droplets. To prevent the jetted droplets from spreading before curing, support material is jetted around every printed part layer, and thus ensures accuracy of the printed features. When printing is complete, all of the part's surfaces are encased in the sacrificial support material. This includes the topmost surface, which is covered in cured support material in order to ensure a consistent finish across all surfaces (dubbed "matte finish"

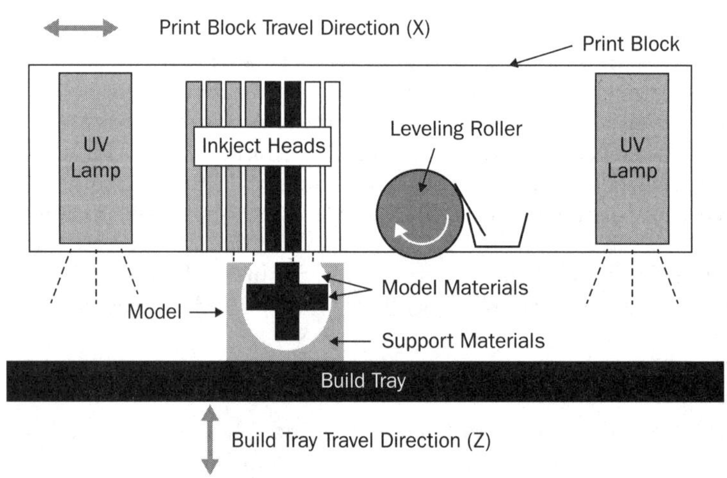

FIGURE 2.5.1 Stratasys PolyJet printhead assembly block [3].

in Stratasys systems). While this feature can be turned off to decrease print time and to preserve small features on the top surface, the topmost surface will have a different finish than other part surfaces as a result (termed "glossy finish" in Stratasys systems).

In general, material jetting AM systems provide a higher resolution and smoother surface finish than most other AM systems due to the small size of, and precise control over, jetted droplets. The presence of hundreds of jetting nozzles enables the fabrication of the entire part layer (or sections of the layer, depending on its width) in a single printing pass. This results in a much faster material deposition speed than other "1D deposition" AM processes such as extrusion and powder bed fusion [4]. In addition, as subsequent layers are deposited directly onto the previous layer, there is no separate recoating procedure, which can also increase print speed. These speed increases are balanced, however, by the processes' extremely small layer thicknesses, which result in total build times that are equivalent to other polymer AM technologies.

Still, the fundamental technology, drop-on-demand jetting, can be efficiently scaled to increase print build volume and to decrease print speed by adding additional jetting nozzles to the printhead, as commonly seen in low-cost large-scale 2D plotters. 3D Systems' fab-grade "racetrack" printer concept demonstrates this potential, claiming the capability of depositing 4 billion drops of colored photopolymers each minute [5]. In addition, as the process occurs without the presence of thermal energy, there are no physics-based limitations on print volume, as is commonly found in other AM technologies (e.g., extrusion and powder bed fusion). As such, the print volume is limited only by the length of the gantry. For example, the largest material jetting system is the Stratasys Objet1000 Plus system, which has a print volume of 1000 × 800 × 500 mm (39.3 × 31.4 × 19.6 in.).

Materials

While the use of inkjetting as a means of selectively depositing material provides high feature resolution, process scalability, and the opportunity for fabricating functionally graded materials, its broad adoption is hindered primarily by a limited material selection. Fundamentally, to be compatible with the drop-on-demand inkjetting process, a material must have a sufficiently low viscosity [0.02 to 0.04 Pa*s (0.0004 to 0.0008 lbf-s/ft^2) [6]] to pass through the inkjet nozzle [~50 μm (~0.002 in.) in diameter], and an appropriate surface tension to form into a droplet after ejection. The material must also be able to relatively quickly transition from liquid to solid once it impacts the substrate. Furthermore, the jetted material must appropriately wet the surface of the substrate, as well as the material of the previously deposited layer. These constraints pose significant limitations on the synthesis of new processable materials. The viscosity limitation is particularly troublesome, as viscosity is directly correlated to the molecular weight, and thus, mechanical properties, of a polymer. As such, the current catalog of commercially available working materials is limited to acrylate photopolymers.

Base Materials

Photopolymer-based material jetting processes are able to create parts featuring a wide variety of material colors and properties, including opaque, transparent, rigid, and flexible. Properties of Stratasys' [7] and 3D Systems' [8] representative rigid (VeroWhitePlus and VisiJet CR-WT) and flexible (TangoBlackPlus and VisiJet CF-BK) materials are presented in Table 2.5.1.

Property	VeroWhitePlus	TangoBlackPlus	VisiJet CR-WT	VisiJet CF-BK
Tensile strength (MPa)	58	1.5	56	2.2
Tensile strength (psi)	8412	218	8122	319
Elongation at break (%)	10–25	218	8.1	290
Modulus of elasticity (MPa)	2500	—	2400	0.7
Modulus of elasticity (psi)	362,595	—	348,091	102
Heat deflection temperature (°C @ 0.45 MPa)	48	—	54	—
Heat deflection temperature (°F @ 65 psi)	118	—	129	—

TABLE 2.5.1 Material Properties of Representative Material Jetting Materials

As the jettable photopolymers are viscosity-constrained, they lack the epoxy networks found in traditional resins used in vat photopolymerization. While versatile, these photopolymers in general suffer from brittleness and sensitivity to exposure to light, heat, and humidity. For example, the heat deflection temperature (HDT) of the jetted materials is ~50°C (122°F), which is roughly half of the HDT of ABS polymer, a common thermoplastic processed by extrusion AM [9]. While separate "high-temperature materials" are offered for today's material jetting systems (e.g., Stratasys' RGD525), their HDT is still limited to 75°C and requires a thermal post-treatment. Due to these limitations, material jetting is infrequently used for fabrication of end-use production parts since they are less dimensionally stable over time and under an applied load [10].

Composite Materials

The true strength of material jetting AM lies in its ability to selectively deposit multiple materials in a single build, which are then cross-linked together through UV irradiation. Current systems allow for simultaneous deposition of up to six different materials, which enable a designer to create parts featuring multiple, discrete material properties (and colors), as seen in Fig. 2.5.2*a*. For example, printing with stiff and flexible materials (e.g., VeroWhitePlus and TangoBlackPlus) would enable the direct fabrication of a stiff tool featuring a soft, compliant handle—these features are typically only possible through costly injection overmolding.

Material jetting's drop-on-demand jetting technology also enables blending of the loaded discrete materials for the fabrication of composite materials. For example, by blending stiff and flexible materials at different ratios, 13 predetermined gradient material blends—with elastic modulus ranging along the continuum of the two material extremes—can be printed. The flexibility of the intermediate gradient materials is directly proportional to the percentage of flexible material present in the printed composite. It is important to note that the jetted materials are not physically mixed; instead, these composite materials are created through combining printed droplets of the discrete materials via prearranged dithering patterns.

In principle, this capability provides a designer the power to specify material composition at a voxel-by-voxel (i.e., three-dimensional pixel) basis. This would enable fabrication of parts featuring functionally graded material properties that are tailored to

(a)

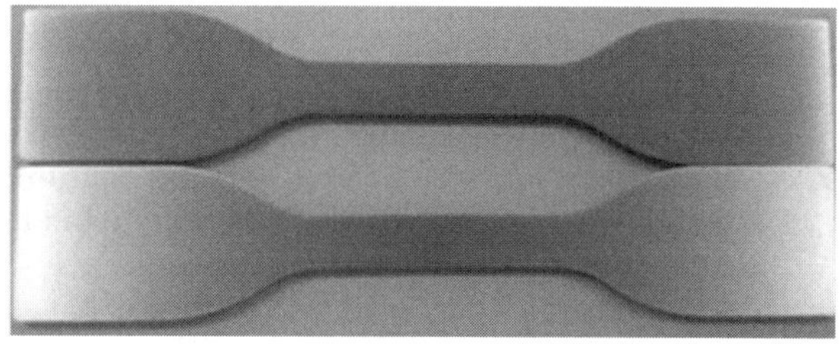

(b)

Figure 2.5.2 Examples of multi-material parts fabricated via material jetting; (*a*) cellular cube featuring flexible (black) and stiff (white) trusses; (*b*) tensile specimens featuring functionally graded elastic modulus.

specific design objectives, as seen in Fig. 2.5.2*b*. However, in current embodiments of the material jetting technology, the loaded materials can only be blended according to the manufacturer's preset patterns. In addition, a separate STL file is needed for each discrete material type in order to print multi-material parts.

Support Materials

Sacrificial support material is required for material jetting systems to support overhangs and undercuts, and also to contain the deposited droplet in order to improve print accuracy. 3D Systems' material jetting systems employ a waxy thermoplastic as a support material. This support material is melted away from the printed part in a low-temperature thermal post-process. The Stratasys systems use a photocurable, hydrophobic gel-like material as a support material. The gel-like support material is reinforced during printing with thin pillars of build material. These thin pillars act similar to rebar in concrete, in that they provide the gel material the necessary stiffness to support subsequent layers of deposited build material. As the material is not water soluble, it is removed manually via a forced jet of water in a post-process cleaning operation. This support material has been shown to have a direct impact on elements of both part quality, including minimum feature size and surface finish, and material properties. To address this issue, Stratasys has very recently introduced a water-soluble support material, though its use is not yet currently widespread.

Process Variable Impact on Part Quality

Several process variables have been shown to have an impact on material jetting part quality [11]. These include support material usage (including the effects of "glossy" and "matte" modes), orientation, and print quality setting. The orientation variable refers to the orientation of the printed part with respect to the XY, XZ, and YZ printing planes. Print quality settings within the material jetting process allow the user to specify a "high-quality" print mode or a "high-speed" print mode. These alternate modes determine the thickness of the specimen's layers [16 and 30 μm (0.0006 and 0.00012 in.) respectively].

Minimum Feature Size and Accuracy

As with other AM processes, the material jetting process has a minimum feature limit which drives the smallest object that can be manufactured. Surface finish and orientation have significant effects on this minimum feature limit [11]. In general, the "worst-case scenario" will occur when a feature is oriented in the XY plane with a matte surface finish [0.897-mm-diameter (0.035-in.) feature], while the smallest possible feature would be achieved with a feature oriented in the XZ plane with a glossy surface finish [0.373-mm-diameter (0.015-in.) feature].

In general, material jetting processes have a high degree of accuracy when manufacturing parts. Dimensional deviations are often less than 0.1 mm (0.004 in.) at most. Testing demonstrates that much smaller deviations occur in the X-direction, with larger deviations in the Y- and Z-directions [12, 13].

Surface Roughness

While support material does serve to support the printed structure, it also controls the surface finish of the final printed part. Wherever the support material comes in contact with the part, it will give the part a dull, matte appearance. Meanwhile, surfaces that do not come in contact with support material will maintain a shiny, glossy appearance. This translates directly to a lower surface roughness for the glossy surfaces over the matte surfaces. Before printing, the user has the power to assign a "Glossy" or "Matte" surface finish to the part; if "Glossy" is chosen, support material will only be used where necessary to support the structure. This has the potential to result in inconsistent surface roughness and aesthetic qualities over the surface of the part; hence, if "Glossy" mode is selected, part orientation has a dominant effect on surface quality (e.g., upward facing surfaces will be smooth, while downward facing surfaces will be rough). In addition, there will be a sharp ridge that occurs at the points on the surface where support material ends or begins. However, if "Matte" is chosen, the entire surface will be coated with support material, which will result in a more uniform surface roughness and aesthetic quality.

Stair-Stepping

Stair-stepping in AM is the phenomenon of increased surface roughness in the printer's Z-direction due to the stacking of layers during manufacture. Overall, material jetting AM parts will demonstrate minimal stair-stepping, regardless of orientation, due to the small layer thickness. This is especially true if the 16-μm (0.0006-in.) layer thickness is chosen. (It is important to note that this "high-quality" setting can only be used when printing single-material parts on Stratasys PolyJet systems.) The material jetting process is also unique, in that its Z-resolution is higher than that of the XY plane, due to the use of a roller to smooth the liquid photopolymer droplets that comprise each layer.

Process Variable Impact on Material Properties

In addition to the general part quality, certain material jetting process variables will also affect the material properties of printed specimens. This variation of material properties must be understood by designers looking to produce load-bearing structures or critical safety parts via material jetting. Of key concern are the process effects on (i) tensile properties of printed parts and (ii) fatigue properties of elastomeric and composite parts.

Tensile Properties

As with many AM process types, material jetting specimens are prone to exhibit material anisotropy depending on the orientation of the specimen during printing [14]. For material jetting, parts are strongest along the X-axis of printing (i.e., the axis in line with the print block travel, as shown in Fig. 2.5.1) and weakest when loaded along the Z-axis, where failure occurs predominantly at the interlayer boundaries [15, 16]. This holds true for both rigid and flexible materials. Parts made from rigid materials also have their lowest modulus of elasticity and highest elongation at break in the X-axis, while flexible parts have their highest modulus of elasticity and lowest elongations when loaded along the X-axis.

Along with orientation, the age of material jetted parts has been shown to affect their tensile properties. Specifically, the ultimate tensile strength of rigid parts increases as parts age (reaching a significant level after five weeks), and the elongation at break demonstrates a decreasing trend as parts age [14].

Finally, the spacing between parts in the XY plane has a statistically significant effect on their ultimate tensile strength [17]. Specimens that were printed in close proximity [20 mm (0.79 in.)] to one another had larger tensile strength (13%), larger modulus of elasticity (11%), and lower elongation at break (12%) than specimens printed far apart from each other [300+ mm (12+ in.)]. This is hypothesized to be due to over-curing that occurs when patterning UV light across parts that overlap multiple print paths.

Fatigue Properties

With the capability to manufacture elastomeric materials in addition to rigid materials, there are several existing guidelines to ensure maximum fatigue life in printed, flexible specimens [18]. Primarily, a designer should strive to reduce the strain on printed elastomeric specimens; when specimens are elongated under 20%, it will require significantly more cycles to failure than when the specimens are elongated over 20%. A fatigue life curve for Stratasys's TangoBlack material is presented in Fig. 2.5.3. As noted by the figure's large 95% confidence interval, there is significant variation in the performance of multi-material specimens; failure can occur prematurely at the interface of the disparate materials.

Designers should also take care when designing neck-like regions in elastomeric parts. These regions will shorten the fatigue life of the specimen due to stress concentrations that occur when a curved region is approximated in the XY plane via the system's pixelated droplet grid. Likewise, if a multi-material specimen is designed for fatigue loading, interfaces between rigid and elastomeric material should be designed to avoid shear loading at the interface. When these interfaces are placed under shear loading, they will decrease the overall fatigue life of the specimen. Finally, the use of a glossy surface finish has the potential to improve specimen fatigue life by an average of 37% over a matte surface finish.

Post-Processing Impact on Design Feasibility

As already discussed, the Stratasys material jetting process utilizes a hydrophobic, photopolymer support material which must be manually removed with a high-pressure water jet after printing. Unlike some AM process types, material jetted parts are only self-supporting when their faces are a few degrees from vertical. Because of this, almost

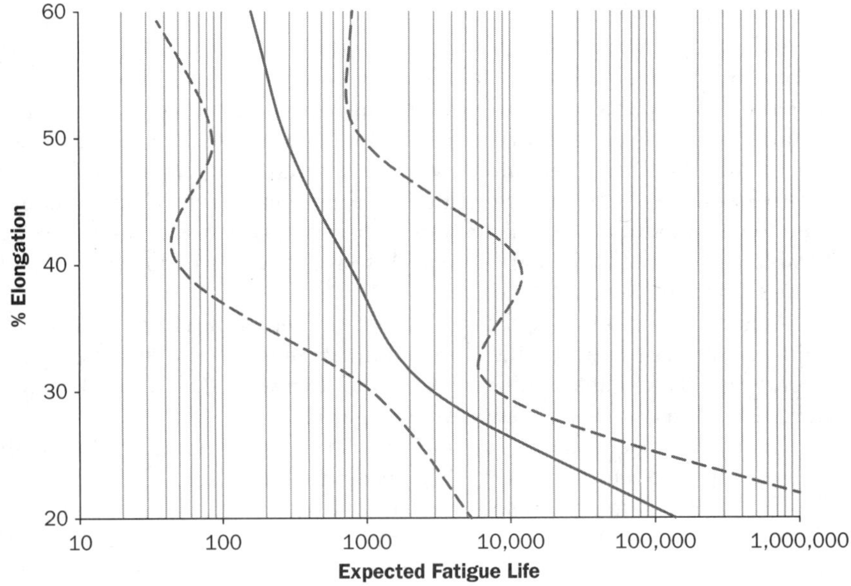

FIGURE 2.5.3 Fatigue life curve for elastomeric material jetted material with 95% confidence intervals [18].

all downward-angled faces will require support material in the final build. As such, the vast majority of material jetted parts will require post-process cleaning, which has a direct impact on the feasibility of certain design elements.

Internal Cavities

Because of the need for manual support material removal, fully enclosed, hollow cavities cannot be manufactured. If the designer still desires to create these types of features, they must understand that they will be completely filled with support material that cannot be removed.

Support Removal from Channels

Because of the manual cleaning process, care should be taken when designing channels or holes within a part. In general, it is difficult to remove the support material from channels with small diameters. The most efficient and time-effective cleaning will occur when the diameter of the channel is approximately the same as the spray from the water jet nozzle [11]. At diameters smaller than this, it will take longer to remove the support material (if it can be removed at all). At diameters larger than the water stream, the support material around the edges of the channel will not be removed with a stationary water jet, and so will require more time and effort on the part of the user to remove.

Feature Survivability

An additional unique consideration for the material jetting process is that, although the system may be able to produce fine geometric features, they may be broken while using the water jet to clean the part. To ensure that a part can survive the cleaning process, the minimum survivable feature might be larger than the minimum printed feature. This minimum survivable feature size is dependent on the material type being printed. For example, for the rigid VeroWhitePlus material, features of approximately 1 mm (0.040 in.) in diameter can be reliably cleaned, while the flexible TangoBlackPlus material requires a feature diameter of approximately 3 mm (0.118 in.) to reliably survive cleaning [11]. Note that these values are somewhat variable, depending on the care that the user shows during the cleaning process.

General Guidelines for Material Jetting

The preceding information can be distilled into a series of general design guidelines and tips for material jetting AM.

- To minimize printing time, parts should be oriented with their longest dimension in the X-direction (i.e., direction of printhead travel) and their shortest dimension in the Z-direction.
- Fully enclosed, hollow cavities cannot be manufactured because of the need to manually remove the hydrophobic support material from printed parts.
- Stratasys's support material will impart a dull, matte surface finish on any surface that it touches, which increases surface roughness, but allows for more uniform part aesthetics (i.e., the "Matte" printing mode).
- Due to their small layer thickness, material jetting systems have their best printing resolution in the Z-direction, rather than the XY plane.
- In general, material jetting processes have a high degree of accuracy; dimensional deviations are often less than 0.1 mm (0.004 in.).
- As printed channels become smaller than the diameter of the water jet, cleaning becomes more difficult and time consuming when using the hydrophobic support material common in the Stratasys PolyJet systems.
- While features <1 mm (0.040 in.) can be successfully printed using material jetting, these features are difficult to retain following the often aggressive water jet cleaning needed to remove the PolyJet support material.
- Parts are strongest in the X-axis direction and weakest in the Z-axis direction.
- Strains on elastomeric parts should ideally be below 20% elongation to maximize fatigue life.

With the speed and scalability of its two-dimensional material deposition approach, its small minimum feature size, its relatively fine surface finish qualities, and its capability to produce monolithic multi-material specimens, material jetting is well-suited among AM processes to create complex and detailed prototype parts to certify both form and fit. The capability of providing a designer point-by-point control over material properties significantly expands designer freedom in the realization of multi-functional components that are tailored to multiple design objectives. Although diverse, material jetting's currently available material selection is fairly limited for the fabrication of end-use functional products due to their brittleness, low heat deflection temperature, and sensitivity to humidity and temperature. The elastomeric-like materials do not yet feature traditional elastomeric hysteresis behavior and have limited fatigue life. The development of new materials and large-scale machines, along with further understanding of how printed material properties affect design, will soon transition material jetting as a well-suited AM process for the manufacture of end-use parts.

References

1. W. E. Masters, 1987, "Computer Automated Manufacturing Process and System," William E. Masters, assignee. Patent US 4665492 A. May 12.
2. I. Gibson, D. Rosen, and B. Stucker, 2015, *Additive Manufacturing Technologies: 3D Printing, Rapid Prototyping, and Direct Digital Manufacturing*, 2nd Edition, Springer, New York.
3. A. M. Elliott, O. S. Ivanova, C. B. Williams, and T. A. Campbell, 2013, "Inkjet Printing of Quantum Dots in Photopolymer for Use in Additive Manufacturing of Nanocomposites," *Advanced Enginerring Materials*, 15(10), pp. 903–907.
4. C. B. Williams, F. Mistree, and D. W. Rosen, 2011, "A Functional Classification Framework for the Conceptual Design of Additive Manufacturing Technologies," *Journal of Mechanical Design*, 113(12):121002-1–121002-11.
5. 3D Systems, 2014, "High Speed 3D Printing: 60 Seconds, 4 Billion Drops," http://www.3dsystems.com/blog/2014/12/high-speed-3d-printing-60-seconds-4-billion-drops; December 4.
6. B. J. de Gans, P. C. Duineveld, and U. S. Schubert, 2004, "Inkjet Printing of Polymers: State of the Art and Future Developments," *Advanced Materials*, 16(3):203–213.
7. Stratasys, 2014, "PolyJet Materials Data Sheet," http://usglobalimages.stratasys.com/Main/Files/Material_Spec_Sheets/MSS_PJ_PJMaterialsDataSheet.pdf?v=635785205440671440
8. 3D Systems, 2015, "ProJet 5500X Multi-Material 3D Printer," http://www.3dsystems.com/sites/www.3dsystems.com/files/projet-5500x-0315-usen-web.pdf
9. Stratasys, 2015, "Compare FDM Materials," http://www.stratasys.com/materials/fdm/compare-fdm-materials
10. F. Fischer, "FDM and Polyjet Printing," http://www.stratasys.com/resources/~/media/722E026DFF2F429C9DA6362BDD42ED2D.pdf
11. N. A. Meisel and C. B. Williams, 2015, "An Investigation of Key Design for Additive Manufacturing Constraints in Multi-Material 3D Printing," *ASME Journal of Mechanical Design*, 137(11):111406-1–111406-9.
12. R. Singh, 2011, "Process Capability Study of Polyjet Printing for Plastic Components," *Journal of Mechanical Science and Technology*, 25(4):1011–1015.
13. T. Brajlih, B. Valentan, J. Balic, and I. Drstvensek, 2011, "Speed and Accuracy Evaluation of Additive Manufacturing machines," *Rapid Prototyping Journal*, 17(1):64–75.
14. L. B. Bass, N. A. Meisel, and C. B. Williams, 2015, "Exploring Variability in Material Properties of Multi-Material PolyJet 3D-Printed Parts," *Rapid Prototyping Journal*, 22(5):826–834.
15. A. Cazón, P. Morer, and L. Matey, 2014, "PolyJet Technology for Product Prototyping: Tensile Strength and Surface Roughness Properties," *Proceedings of the Institution of Mechanical Engineers, Part B: Journal of Engineering Manufacture*, 228(12):1664–1675.

16. G. Kim and Y. Oh, 2008, "A Benchmark Study on Rapid Prototyping Processes and Machines: Quantitative Comparisons of Mechanical Properties, Accuracy, Roughness, Speed, and Material Cost," *Proceedings of the Institution of Mechanical Engineers, Part B: Journal of Engineering Manufacture*, 222(2):201–215.
17. M. W. Barclift and C. B. Williams, 2012, "Examining Variability in the Mechanical Properties of Parts Manufactured via Polyjet Direct 3D Printing," *Proceedings of the 23rd Annual International Solid Freeform Fabrication Symposium*, Austin, TX.
18. J. P. Moore and C. B. Williams, 2015, "Fatigue Properties of Parts Printed by PolyJet Material Jetting," *Rapid Prototyping Journal*, 21(6):675–685.

CHAPTER 2.6

Design for Powder Bed Fusion of Polymer Parts

Carolyn Conner Seepersad

University of Texas
Austin, Texas

Powder bed fusion is "an additive manufacturing process in which thermal energy selectively fuses regions of a powder bed" [1]. Thermal energy is typically provided by a focused laser beam, although a focused electron beam is used in some applications. In a selective laser sintering system, as shown in Fig. 2.6.1, a thin layer of heat-fusible powder is deposited on the part build platform. The powder is provided by advancing the powder delivery piston upward by a prescribed distance, advancing the build platform downward by a prescribed distance, and spreading the powder across the build platform with a counter-rotating roller. Other systems use blades in place of the counter-rotating roller and gravity-fed hoppers in place of powder delivery bins and pistons. The cross-sectional slice of a part is scanned by the laser which is guided

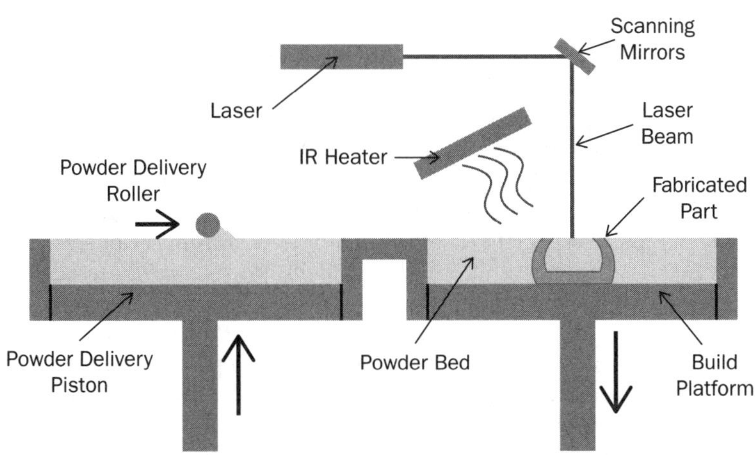

Figure 2.6.1 Selective laser sintering, a powder bed fusion technology.

by a series of galvanometers. The powder bed is typically preheated to a temperature slightly less than the melting temperature of the powder in the case of polymers. The laser elevates the temperature of the powder to its melting point in selectively scanned regions, thereby fusing the particles within the layer and into the previously scanned layer beneath it and promoting layer-to-layer fusion. After a layer is scanned, the build platform advances downward again, the powder delivery piston advances upward again, and another layer of powder is spread across the top of the build platform in preparation for another layer of scanning. Support structures are not required for polymer parts because the surrounding powder provides support. At the end of the build cycle, parts typically cool slowly in the build platform to relieve residual stresses. After removal from the machine, polymer parts are typically post-processed by sanding and bead blasting to remove excess powder.

Machines

Two of the most prominent manufacturers of polymer powder bed fusion machines are 3D Systems and EOS. The first commercially available laser sintering system was the Sinterstation 2000, introduced by the DTM corporation, which was subsequently acquired by 3D Systems. Historically, one of the most commonly used 3D Systems' laser sintering systems has been the Vanguard series with a build envelope of 370 × 320 × 445 mm. As of 2015, 3D Systems markets an sPro line of laser sintering machines with similar build envelopes, but it also offers larger production scale machines with build envelopes as large as 550 × 550 × 750 mm. Similarly, EOS offers a series of laser sintering machines with build envelopes ranging from 340 × 340 × 600 mm for their medium build volume machine to 700 × 380 × 580 mm in their modular production system. 3D Systems' machines utilize powder feed chambers and pistons, combined with counter-rotating rollers to spread powder, whereas EOS machines utilize a hopper combined with a blade.

Materials

Laser sintering machines process a wide variety of polymer powders. Thermoplastic materials that are most appropriate for laser sintering tend to have low melting temperatures, low thermal conductivities, and low melt viscosities. The most common SLS materials are polyamide-based powders, which are frequently used to make functional, end-use parts such as aircraft ductwork. Nylon 12, nylon 11, and glass-filled nylons are common materials, along with more recently introduced fire-retardant variants for aerospace applications. Elastomeric thermoplastics are also available for applications requiring rubber-like properties. Special high-temperature sintering machines can process PEEK, which is especially useful for biomedical applications. It is also possible to process metals, ceramics, and other high-melt-temperature materials using polymer laser sintering systems via indirect sintering routes. Indirect sintering involves coating the metal or ceramic powders or mixing them with a thermoplastic binder that can be melted to fuse the powders together. Parts are subsequently debinded, sintered, and infiltrated to produce a final part with properties similar to that of a fully dense ceramic or metal part. Most metal parts, however, are made via directly sintering metal powders in a direct metal laser sintering system because the resulting material properties are much closer to as-wrought or as-cast values.

The Influence of Process Variables on Part Properties

Several processing variables have a significant effect on part quality. The density of the energy delivered to the powder bed is proportional to laser power and inversely proportional to scan speed and scan spacing (the distance between parallel laser scanning paths). If the energy density is too low, incomplete sintering can occur, resulting in reduced stiffness, strength, and other mechanical properties.

Powder bed temperature is another important variable. It is typically maintained a few degrees below the melting temperature of the polymer powders, with the laser scan adding just enough energy to fuse the powders together. The elevated powder bed temperature not only reduces the required energy density from the laser scan, thereby allowing a laser with a lower power rating to scan the powder bed surface more rapidly, but also relaxes the residual stresses that build in the part as a result of layer-by-layer scanning, heating, fusing, and subsequent cooling. If the powder bed temperature is too *low*, parts insufficiently stress relax, leading to warped and curled parts. Those parts can also exhibit diminished mechanical properties due to insufficient energy density during the build. If the curling is significant enough, the parts can become caught in the powder delivery mechanism, leading to a "crashed" build. If the powder bed temperature is too *high*, oversintering in the powder bed leads to dense, solidified powder (also known as part cake) surrounding the sintered parts in the powder bed, which is difficult to separate from the sintered parts and leads to a failed build.

The condition of the powder can also have a significant effect on the quality of the sintered parts. Sintering kinetics depends strongly on the size of the powder particles. As powders are thermally cycled, the average particle size increases, making it more difficult to spread and melt the particles because larger particles require more energy for melting than smaller particles. To prevent this undesirable effect while recycling as much powder as possible, most users mix virgin and recycled powders at prescribed ratios prior to each build.

Process planning—especially the layout of parts in the build chamber—is also very important for the designer to consider. If smooth flat surfaces are required, it is important to build those surfaces either parallel to or orthogonal to the build plane. Building those surfaces at other angles results in excessive surface roughness from a phenomenon called stair-stepping, which is caused by the layer-based fabrication process as shown in Fig. 2.6.2. Also, parts that are thin, wide, and flat could be built either parallel to or orthogonal to the build plane to avoid stair-stepping; however, it is best to build them with the major axes-oriented orthogonal to the build plane to prevent warping and curling of the flat surfaces. Finally, it is important to minimize the vertical height of a part in the build chamber because the time required to build the part is typically proportional to its vertical height, and shorter builds reduce utilization of powder stocks and machine time.

Mechanical Properties of Polymer Parts

Laser-sintered polyamide parts exhibit strength and stiffness properties that are similar to injection-molded polyamide parts. Ductility, measured by percent elongation, however, is typically at least an order of magnitude less than that of injection-molded parts. Diminished ductility is often attributed to incomplete particle melting and porosity. Interlayer porosity, which can result from incomplete melting and/or

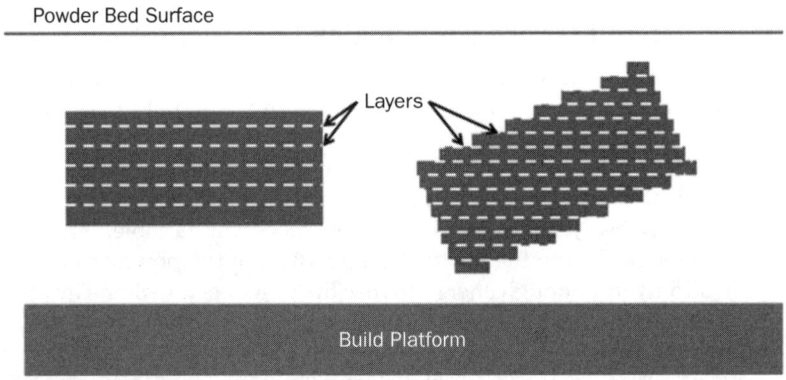

FIGURE 2.6.2 An illustration of stair-stepping as a result of part orientation. A rectangular cross section with flat surfaces built parallel and orthogonal to the build plane (left) and the same cross section with flat surfaces built at an inclined angle to the build plane (right).

inconsistent powder spreading, can cause particularly diminished ductile properties in the Z-direction, across build layers, relative to parts built with major axes parallel to the build plane, although several researchers (e.g., [2], [3]) report no significant loss of stiffness or strength in the Z-direction from production-grade builds. Bourell and coauthors [3] gathered tensile data from 3500 nylon 11 laser-sintered parts from a production-grade facility (Harvest Technologies, a manufacturer) and found tensile strength roughly equivalent to injection-molded parts with variability of approximately 10%, but they found ductility to be roughly an order of magnitude lower than injection-molded parts with much higher variability from specimen to specimen.

Dimensioning Polymer Parts

In general, polymer powder bed fusion machines can resolve features as small as approximately 0.5 mm (0.020 in.), but that dimension can differ depending on the orientation of the part and the dimensions of the surrounding part. Minimum dimensions depend on the powder particle diameter, laser spot size, and layer thickness, which are typically on the order of 50, 25, and 100 μm (0.002, 0.001, and 0.004 in.), respectively, for polymer powder bed fusion machines.

A series of metrology experiments were conducted to define the dimensional capabilities of polymer powder bed fusion more precisely [4]. Representative parts are shown in Figs. 2.6.3 through 2.6.6. Parts were fabricated on a 3D Systems' Vanguard HiS-HiQ machine at Harvest Technologies with nylon 12 material. Results may differ for different types of machines or materials.

The part in Fig. 2.6.3 was designed to determine the minimum manufacturable thickness of a feature as a function of the part orientation. As shown, the part consists of a framework surrounding walls with a variety of thicknesses, ranging from 0.2 to 3 mm (0.008 to 0.118 in.). It was built in a horizontal orientation in which the major axis of the frame was parallel to the build plane and the thin walls were orthogonal to the build plane and a vertical orientation in which the major axis of the frame was orthogonal to the build plane and the thin walls were parallel to the build plane. Flat walls as thin as 0.6 and 0.8 mm (0.024 and 0.031 in.) could be built for the vertically and horizontally

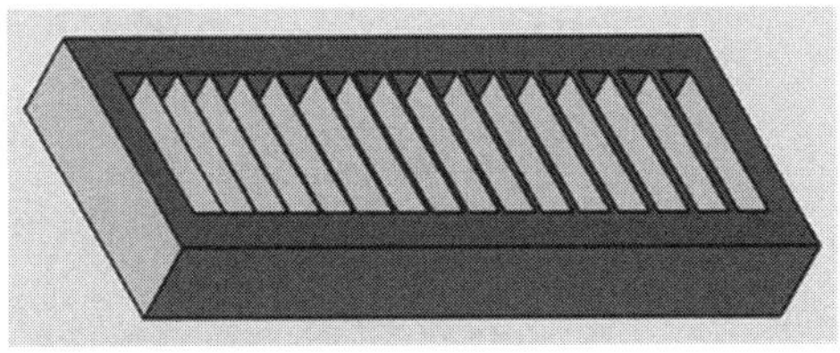

FIGURE 2.6.3 Sample part for determining minimum manufacturable wall thickness.

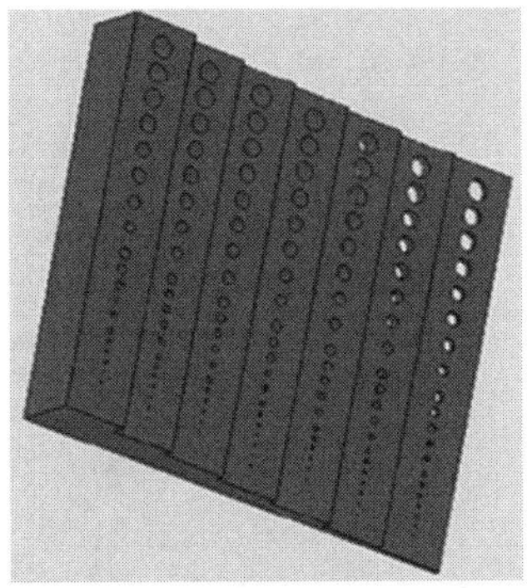

FIGURE 2.6.4 Sample part for determining minimum resolvable hole diameters as a function of part thickness.

oriented parts, respectively. Thinner walls could be built for vertically oriented frames because the thin walls were parallel to the build plane with their thicknesses determined by the layer thickness rather than the laser spot size. Even 0.2-mm (0.008-in.) walls could be resolved for vertically oriented frames, but the walls were buckled due to residual stresses in the parts from layer-wise thermal fusion.

The part in Fig. 2.6.4 was designed to determine the minimum manufacturable hole diameter in parts of various thicknesses. When investigating minimum hole diameters, it is important to consider the thickness of the surrounding part because a thicker surrounding part often leads to higher temperatures in the nearby part bed and oversintering of nearby features, thereby narrowing the diameter of the holes. As shown in Fig. 2.6.4, part thickness ranged from 0.94 to 12.7 mm (0.037 to 0.500 in.) in increments of 0.94 mm (0.037 in.); for each part thickness, hole diameters ranged from 0.125 to 4 mm (0.005 to 0.157 in.). The test part was built with the central axis of the holes oriented vertically (i.e., orthogonal to the build plane) and horizontally (i.e., parallel to the build plane). When the holes were oriented horizontally, it was possible to resolve holes as small as 1.5 mm (0.059 in.) in diameter for the thickest portion of the test part and 0.6 mm (0.024 in.) in diameter for the thinnest portion of the test part. When the holes were oriented vertically, it was possible to resolve holes as small as 2 mm (0.079 in.) in diameter for the thickest portion of the test part and 1.3 mm (0.051 in.) in diameter for the thinnest portion of the test part. To be considered resolvable, the hole needed to be well-formed with no noticeable irregularities after all reasonable attempts at powder removal.

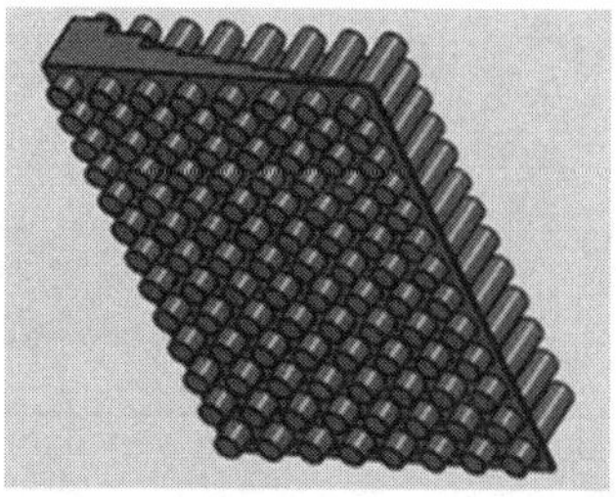

FIGURE 2.6.5 Sample part for determining minimum clearances between a bore and a shaft as a function of part thickness.

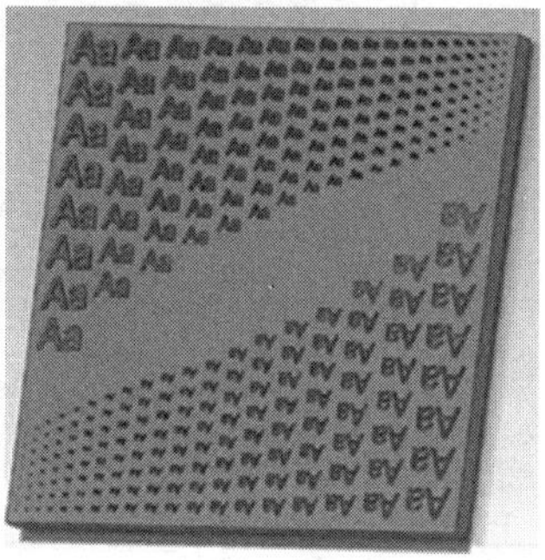

FIGURE 2.6.6 Sample part for determining minimum legible font sizes.

The part in Fig. 2.6.5 was designed to determine appropriate clearances between a bore and a rotating shaft. Each shaft was designed with a diameter of 10 mm (0.0394 in.) and a length of 40 mm (1.575 in.). The surrounding plate was designed in a series of thicknesses ranging from 25.4 to 1.27 mm (1.000 to 0.050 in.). For each step of thickness, the plate is constructed with a row of 13 bores, ranging in size from 0 to 1.8 mm (0.071 in.) of clearance around the shaft. The plate and the collection of shafts were fabricated in the configuration shown in Fig. 2.6.5, that is, pre-assembled, with the shaft axes orthogonal to the build plane. The objective of the test was to determine the tightest clearances that prevent the shafts from fusing to the bores and allow the shafts to rotate freely in the bores after they are removed from the powder bed. Those clearances were determined as a function of the thickness of the surrounding bore. For the thickest bores of 25.4 mm (1.000 in.), a uniform clearance of 0.9 mm (0.035 in.) around the shaft is required to prevent fusion of the shaft to the bore. For the thinnest bores of 1.27 mm (0.050 in.), a uniform clearance of only 0.45 mm (0.018 in.) is required to prevent fusion. Larger clearances are required for thicker bores to accommodate the oversintering that occurs with thicker parts.

The part in Fig. 2.6.6 was designed to determine the minimum legible font size as a function of build plane orientation and size and depth of the letters. Letters were either raised (upper left of figure) or recessed (bottom right of figure), and a sans serif font was used in which all portions of the letters are drawn with uniform thickness. Each column of letters represents increments of letter depths ranging from 2 to 0.25 mm (0.079 to 0.010 in.). Each row of letters represents increments of font size ranging from 36 to 1.

Letters were considered resolved if all aspects of the letters were legible with no fused or incomplete walls. For raised letters, minimum font sizes of approximately 24, 20, and 20 were required for letters on an upward facing, side facing, or downward facing surface, respectively, in the build chamber. For recessed letters, minimum font sizes of approximately 24, 14, and 12 were required for letters on an upward facing, side facing, or downward facing surface, respectively. Experiments were also conducted with serif fonts (e.g., Times New Roman) in which some portions of the letters are thinner than others. However, these fonts were not recommended because the thin portions of the letters often fail to resolve, even with large font sizes.

General Design Considerations for Polymer Powder Bed Fusion

In addition to specifying appropriate dimensions, orientations, and material properties, it is also important to consider some general design guidelines and tips:

- Sharp corners cannot be fabricated. Rounding corners and providing fillets at joints not only accommodates the rounding imposed by laser scanning and powder fusion but also prevents stress concentrations and potential splits in the part.
- Sufficient access points are needed for powder removal. Hollow, watertight solids cannot be fabricated without access points for powder removal. Similar considerations apply to bosses, which need to be fabricated with access points at the base for powder removal.
- If parts are too large for the build chamber, they can be built in multiple pieces and adhered with epoxy after the build. Parts can also be painted with off-the-shelf acrylic paints.
- Polymer powder bed fusion parts tend to be watertight as long as wall thicknesses are greater than approximately 1 mm.
- To minimize the size of the STL file, it is better to use square cut or flat surfaces, rather than rounded surfaces wherever possible, because rounded surfaces require many more triangles to be represented accurately in the STL file.

Powder bed fusion is one of the most widely used AM processes for fabricating functional, end-use parts. There are several reasons for its prominence, including favorable part properties, post-processing considerations, and build logistics. With tight process control, it is possible to additively manufacture polymer parts with strength and stiffness properties that are roughly equivalent to molded parts. Although processes that utilize photosensitive resins, such as stereolithography and material jetting, can fabricate parts with better surface finish and resolution, powder bed fusion parts do not suffer from concerns regarding degradation of part properties over time. Since powder bed fusion parts are built within a bed of powder, the powder provides supports for complex features and no support structures are required (for polymer parts). As a result, powder bed fusion can be used to build complex lattice structures and other structures without support structures and the additional post-processing steps involved in removing them and smoothing the surfaces to which they were attached. Polymer powder bed fusion parts can easily be sanded and painted and adhesion of multiple parts is straightforward, as well. Finally, build logistics favor the use of polymer powder bed fusion. Since polymer parts are arranged in the powder bed during fabrication and do not require anchoring to a build platform, multiple parts can be arranged densely in a single build. A fully dense build may require more than 24 h for completion, but a vast number of small or moderately sized parts can be fabricated in a single build, especially if they are packed efficiently.

References

1. ASTM International, "Standard Terminology for Additive Manufacturing Technologies," ASTM Standard F2792-12a, West Conshohocken, PA, 2012.
2. B. Van Hooreweder, D. Moens, R. Boonen, J.-P. Kruth and P. Sas, "On the Difference in Material Structure and Fatigue Properties of Nylon Specimens Produced by Injection Molding and Selective Laser Sintering," *Polymer Testing*, vol. 32, pp. 972–981, 2013.
3. D. Bourell, T. Watt, D. Leigh and B. Fulcher, "Performance Limitations in Polymer Laser Sintering," *Physics Procedia*, vol. 56, pp. 147–156, 2014.
4. C. Seepersad, T. Govett, K. Kim, M. Lundin and D. Pinero, "A Designer's Guide for Dimensioning and Tolerancing SLS Parts," in *Solid Freeform Fabrication Symposium*, Austin, TX, 2012.

CHAPTER 2.7

Design for Powder Bed Fusion of Metal Parts

Paul Witherell

National Institute of Standards and Technology
Gaithersburg, Maryland

Advancements in metals-focused technologies have taken 3D printing from a prototyping technology to an advanced, additive manufacturing (AM) production technology having a roadmap for even greater advancements [1]. Most of today's commercially available metal-based AM systems [2] are founded in the Powder Bed Fusion (PBF) processes. Typical applications of metal PBF include producing complex, and/or low-volume functional parts [3], particularly in aerospace, defense, medical, automotive, and even jewelry. As a production alternative, metal PBF can achieve cost savings with reduced lead times and reduced assembly part counts. Moreover, metals-based PBF supports the manufacture of complex features not possible with other manufacturing technologies.

As with polymer PBF, metal PBF includes varying processing techniques. Metal PBF processes may use low-power lasers to bind powder particles by only melting the surface of the powder particles (sintering) or high-power energy beams to fully melt and fuse the powder particles together (melting). Metal PBF most often refers to the selective laser sintering (SLS), the selective laser melting (SLM), or electron beam melting (EBM) processes. SLS is distinguished from SLM by whether or not the metal achieves a partial melt (sinter) versus a full melt. For the most part, laser-based PBF machines are capable of either, depending on exposure times and temperatures. SLS often requires secondary processing (such as infiltration), while SLM and EBM support the creation of near fully dense parts. This chapter focuses on SLM and EBM.

Electron beam melting and SLM have similar capabilities, although the beam energy from the EBM process is of a higher intensity and operates at higher temperatures than the SLM counterpart, therefore also supporting faster build rates at lower resolutions. In general, EBM processes subject parts to less thermal stresses (as platform is preheated before processing) and have faster build rates, but the tradeoff comes with general greater minimum feature sizes and greater surface roughness than SLM.

Typical PBF metals include stainless steels, tool steels, titanium alloys, nickel alloys, aluminum alloys, cobalt chrome alloys, and bronze alloys, although EBM material alternatives are a little more restricted due to different operating requirements. To control processing of materials, SLM and SLS PBF machines provide build chambers with inert environments, filling chambers with either nitrogen or argon gases. More volatile metals such as titanium alloys require argon environments. EBM PBF machines require high-vacuum environments to operate, which makes them well suited for volatile metals.

Part quality of metal PBF parts face much more uncertainty than parts created by casting or wrought processes. Many decisions that may be left to the designer, including material selection, geometry, topology optimization methods, build orientations, and support structure locations may affect the formation of micro structure, ultimately determining the mechanical properties of the part. PBF parts are susceptible to defects such as dimensional and form errors, unintended porosity, delamination of layers, and inconsistent material properties. When everything works as planned, properties have been shown to be comparable to or exceed those of wrought materials. However, unaccounted disturbances can result in lost builds, wasted materials, or faulty parts. It is important to recognize, establish, and account for correlations between the AM process parameters, material parameters, and the built part characteristics. Designers need to know the parts will be made in well-characterized and controlled processes.

Machines

Metal PBF machines are significantly more elaborate, and expensive, than their polymer counterparts. Commercialization of the technologies began to become a reality when EOS (Electro Optical Systems) GmbH introduced one of the first mainstream production machines in 2003. Although machines were produced earlier also, the field became widely viable with the 2003 machines.

Though a variety of manufacturers now exist, the basic premises and characteristics of the machines are similar (Table 2.7.1). Build volumes that are typical of metal PBF machines are on the order of 250 × 250 × 350 mm (x, y, z) (~10 × 10 × 13.5 in.), though both larger and smaller build volumes do exist. Layer thicknesses are commonly designed to be in the range of 20 to 100 μm (0.0008 to 0.0040 in.). These thicknesses will vary depending on settings, machine, and material particles sizes. For SLM, laser sources are either a ytterbium (Yb) fiber laser or a CO_2 laser (more common in sintering). Laser scan speeds may reach up to 15 m/s (49 ft/s) while electron beam speeds may exceed 8000 m/s (24,247 ft/s). Scan paths are achieved by reflecting stationary beam sources off of rotating mirrors, with beam spot sizes ranging from 0.04 to 0.5 mm (0.0016 to 0.020 in.) for SLM and 0.1 to 1.0 mm (0.004 to 0.040 in.) for electron beams.

Manufacturers continue to improve on processing techniques and capabilities as new iterations of machines are released. Manufacturers are experimenting with new ways of making parts bigger, faster, and better, including incorporating additional lasers, increased power, increased build volumes, and faster scan speeds. Machine manufacturers are also experimenting with supporting different types of build environments, for instance incorporating laminar flow, to improve process control. Improvements have also come by automating much of the pre- (including setup) and post-build processes as well, reducing potential for human error while facilitating ease of production.

	Renishaw		SLM		Concept Laser		EOS		Arcam	
	Entry	Premier	Entry	Premier	Entry	Premier	Entry	Premier	Entry	Premier
Model	**AM 250**	**RenAM 500M**	**125 HL**	**500 HL**	**M1 Cusing**	**X Line 2000R**	**M 100**	**M 400**	**Q10**	**Q20**
Build X (mm)	250	250	125	500	250	800	100	400	200	350
Build Y (mm)	250	250	125	280	250	400	100	400	200	350
Build Z (mm)	360	350	125	365	250	500	95	400	180	380
Beam	Yb fiber laser	Yb fiber laser	Yb fiber laser	Yb fiber laser	Yb fiber laser	Yb fiber laser	Yb fiber laser	Yb fiber laser	Electron beam	Electron beam
Beam power (W)	200 or 400	500	400	2 × 400 or 4 × 400	200 or 400	2 × 1000	200	1000	3000	3000
Beam resolution (μm)	70	—	60–90	80–700	50	100–500	40	50	100	180
Scan speed (m/s)	2	—	10	15	7	7	7	7	8000	8000
Build rate (cm^3/h)	5–20	—	25	50–100	2–15	10–100	5–20	5–20	55–80	55–80
Minimum layer height (μm)	20	—	20	20	20	30	20	20	50	50
Maximum layer height (μm)	100	—	75	200	80	150	80	80	200	200
Minimum feature size (μm)	300–500	300–500	140–160	160–180	300	300	300	300	600	600

Table 2.7.1 Machine Capabilities of Five Machine Manufacturers (in 2015)

Manufacturers are beginning to gradually incorporate basic monitoring into machine processes [2]. Advancements in techniques such as optical and thermal imaging are beginning to enable machines to monitor builds and catch defects early.

Materials

Material availability and processability in metal PBF machines vary depending on the machine manufacturer and the flexibility provided. Specifications in Table 2.7.1 such as minimum feature size and build rates cannot be determined until a material is selected. Part characteristics are dependent on material-process interactions. Material properties govern how the material solidifies, which determines part properties through melt pools, phase changes, and the creation of new micro structure, which is why machine manufacturers work very closely with their powder suppliers.

The characteristics of powders can very much affect the build, which is why the consistency and quality of atomized powders (commonly gas atomized) are critical. Powder characteristics such as flow rate, apparent density, morphology, and particle size are better controlled through powder consistency. Consistency allows for better tracking of properties such as particle size distribution, flowability, chemical composition, and gas content. Each of these characteristics affect finished part quality, including density, porosity, surface quality, and the grain morphology of the micro structure.

Monitoring powder usage for each layer in the PBF process is also important. Powder layer characteristics include packing density, local powder size distribution, and local powder temperature distribution. Unprocessed powders can be reused (powder recycling) in metal PBF, but the characteristics of the powder will change, including size distribution and possibly grain morphology.

Process Planning

The basic process plan (Fig. 2.7.1) consists of orienting a model, placing support structures, and setting machine parameters [4]. The decisions made at planning time will influence not only the process, but also the properties of the final part. Orientation, support-structure design, slicing, scan path, and scan speed are significant process parameters. However, what the user is able to specify will vary depending on the machine and software used.

Time and Cost Considerations

Good process planning can reduce manufacturing costs and cycle time in metal PBF far beyond conventional processes. The NIST test artifact shown in Fig. 2.7.2 is 17-mm (0.67-in.) tall, has a volume of 101,000 mm^3 (6.16 in.^{3}), and a lateral surface area of 9000 mm^2 (14 in.^{2}), takes approximately 20 h to complete in stainless steel using the machine's standard parameter set for this material [5]. This time does not include setup or any post-processing. It is important to note, however, that the build time is not dependent on the complexity of the geometric features, but rather the volume and surface area of the part. Contrast that with the time required to machine and assemble the same configuration, and the impossibility of doing so with small corner radii in the pockets geometry.

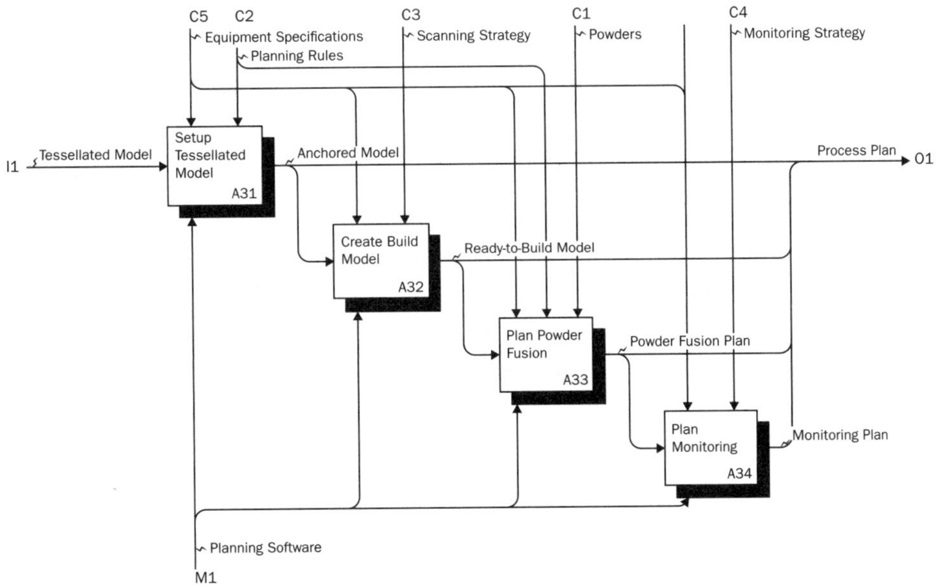

FIGURE 2.7.1 Basic decomposition of process planning (from [4]).

Scanning strategies that plan the movement of the laser and related machine components can be optimized and faster scan speeds and thicker layers can decrease the time to build beyond the cited time. Any optimization will depend on the particular machine used, the material selected, and the geometry of the part being produced. Machine manufacturers can provide recommended settings for a particular material, but some trial and error is often necessary to improve build quality.

The greatest savings, however, often come in part placement. In metal PBF, a part must be anchored to a build plate to establish a reference for part location and orientation. Some considerations that must be made when locating (x, y, and z) the part in reference to the build plate are how much material will be lost when separating the part from the substrate, and which area of the machine build plate returns the most consistent builds (often toward the center of the build plate). Orientation considerations are more complex, and can play a large role in both time and cost reduction (and in part quality).

In general, the greater the distance to the edge of the part normal to the build plate the greater the build time. By minimizing the distance between a part's z-dimension and the substrate's surface, one can reduce the build time, something that is also true for polymer PBF. Unique to metal PBF, however, is accounting for support structures. As support structures provide structural support for geometries such as overhangs or thin walls, orientation will influence the location and quantity of support structures. A function of orientation and geometry, support structures influence the cost and time of the build. The material needed for support structures can drive material costs and the processing needed to remove the structures will drive equipment costs (and time). Support structures can also decrease overall costs, as strategic placement of support

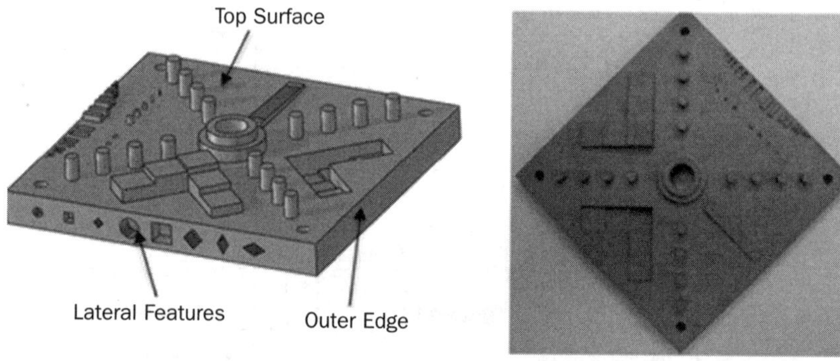

FIGURE 2.7.2 NIST test artifact (from [5]).

structures can significantly influence the chances of a successful build and acceptable quality. Support structures are generally defined by the manufacturing engineers, but designers who understand their impact can accelerate the build success in some designs.

Quality Considerations

Products need to address quality characteristics such as geometric tolerance, surface roughness, and mechanical properties. Many of the same considerations made for polymers PBF must also be made for metals. However, the implications of process decisions may drastically differ. For instance, as with polymers, layer thickness and orientation will affect the part's ability to meet geometric tolerances and to some extent mechanical performance. In metal PBF, however, the influence of these parameters is amplified. Dimensional accuracies between model and part are often specified by the machine manufacturer, and are dependent on the material. For instance, Ti64 processed in SLM machines may achieve part accuracies of ±50 μm (±0.002 in.), while EBM is closer to ±200 μm (0.0079 in.). Surface roughness follows similar patterns between the two technologies where the same Ti64 may have an Ra of 9 to 12 μm (354 to 472 μin.) with an SLM process, and 25 to 35 μm (982 to 1380 μin.) with the EBM process. In both cases the vertical surface roughness will be less than the horizontal.

As with polymers, surfaces that are non-orthogonal to the build plane are subject to stair-stepping, as layers are created with discrete steps. In metal PBF, however, layer thickness can also influence micro structure and grain formation. As builds in metal PBF are anchored, thin wall structures parallel to the recoating arm are more susceptible to damage (and thus build failure) through interference and friction forces. Because of the heating and cooling of metal, thin sheets are more subject to warping and curling when the major surface is parallel to the build area. Because of this, build orientation must be carefully considered by the manufacturer. When choosing an orientation, it is also important to note that appropriate support structure placement can relieve residual stresses.

In metal PBF, optimal placement of support structures does not directly correlate with overhang volumes and surface areas. The build plate acts as a heat sink, and any metal PBF build begins directly on the solid substrate. Unlike with sintering, the melting of metals creates a melt pool with unique geometric and temperature characteristics.

The cooling of the melt pool, or solidification, is the major source of micro structural changes, porosity formation, residual stress development, and geometric irregularities. Support structures provide thermal relief during the cooling of high melt pool temperatures. By introducing temporary material during the build, support structures can be used as a means for controlling thermal loading conditions and reducing imperfections. When properly utilized, support structures provide an in-process means of mitigating the risk of failed builds by reducing warping caused by thermal stresses and providing additional support for susceptible features.

Because the powders used in metal PBF are so susceptible to variances in processing parameters, machine manufacturers often lock in parameter settings for specific materials. However, access to these parameter sets, or "recipes," can often be obtained, allowing for greater access and control into material processing [6]. Process characteristics such as scan speed, scan pattern, and beam spot size can be manipulated to control the development of the melt pool, voids, and micro structure. Careful consideration must be paid to key process parameters to achieve desired part qualities, and further detail on many of these correlations have been identified in reference [2].

Mechanical Properties of Parts

The material properties that can be achieved through metal PBF processes are capable of meeting and even exceeding those of cast or wrought properties. SLM machine manufacturers report Ti64 tensile strengths between 1100 and 1300 MPa (160,000 and 188,600 psi), with the lower values exceeding those expected from wrought [850 MPa (123,300 psi)] and cast [758 MPa (110,000 psi)] materials. EBM machine manufacturers report yield strengths of 950 MPa (137,800 psi) for this same material. In metal PBF, the greatest challenge is getting consistent material properties throughout the part (but designers can intentionally vary properties through the part if desired and that is a major reason designers may design specifically for this process). The initial anchoring will influence how the geometry is built and in turn the anisotropic properties of the part. The initial powder state at each layer can affect the density of the part. The continuous phase changes that occur during processing cause different volumes of metal to be processed at different rates.

Powder Bed Fusion techniques can yield densities of greater than 99% (even up to 100%) depending on process parameters and materials. Nonuniform powder spreads may negatively affect these yields, however, as unintended porosity can occur as a result of voids or entrapment of unmelted powders. Porosity will affect the mechanical properties, such as fracture toughness and fatigue properties, as it increases the likelihood of micro cracking. In fact, a major consideration of AM-related cybersecurity is the potential to introduce porosity deliberately, and maliciously, into a part design so that it cannot be easily detected. For these reasons post-processing is often used, and designers should define expected material properties *after* any heat treatment.

Powder Bed Fusion processing yields anisotropic material properties; for instance the z direction may have as little as 60% (in more extreme circumstances) of the tensile strengths in the x and y directions. It is important to design and plan for anisotropy when determining build orientation, and consider the functions the part needs to perform. Different post-processing techniques are often used to improve consistency throughout part properties. Surface finishing techniques are often used to reduce surface roughness, including stair-stepping effects. Heat treatments such as hot isostatic

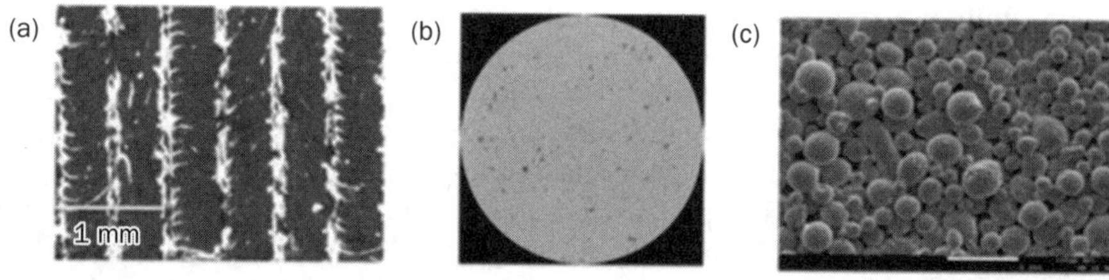

FIGURE 2.7.3 (*a*) Observing stair-stepping with optical micrograph of ramp (from [5]). (*b*) X-ray CT of cylinder cross section (from [7]). (c) SEM micrograph of stainless steel powder (from [8]).

pressing, or hipping, are used to relieve thermal stresses, improve part density, and reduce observed anisotropy in part properties.

Because geometry and micro structures can be tailored to meet specific needs, parts made with metals AM can be difficult to qualify or certify. The anisotropy of material properties can be a function of both build direction and scan paths, where scan paths are a function of geometry. This makes it difficult, if not impossible, to use directional test coupons to arrive at conclusive mechanical property evaluations. Nondestructive testing (NDT) is often a desired inspection technique, as different methods can be used in complementary manners to provide extensive insight into part properties. Computed tomography (CT) methods provide a means for evaluating both part density and validating any internal features. Scanning electron microscopy (SEM) techniques can be used to analyze micro structure, which in turn can be correlated with material properties. Ultrasonic testing techniques are also available. Regardless of what inspection technique is chosen (Fig. 2.7.3), it is important to remember that the extent and type of qualification necessary will depend on the part's intended functionality.

Supporting Infrastructure

Much attention has been given to software and related infrastructure support. Software assists decision making and parameter selection. These software resources can help in deciding on orientation and assessing the manufacturability of different features. New software will begin to account for the anisotropy of material properties and address the design of functionally graded materials (i.e., parts whose properties vary in one or more directions). Material databases such as the NIST Additive Manufacturing Materials Database are emerging. These databases support material selection for a given process through detailed material and process documentation.

Companies that develop CAD software are now integrating AM design solutions into their software portfolios. Software to guide process planning through the identification of areas with high-thermal loads (and even guide support structure placement) is becoming available. Efforts to capture material-process-geometry correlations for process modeling and the development of PBF-specific design rules within ASTM and ISO are in work (in 2016).

Modeling provides a means to monitor the processing of metals and give feedback. For example, a process model of the melt pool, which includes the pool size and its temperature distribution, can be correlated with optimal scanning speeds and laser power settings. In-process sensing is not yet readily available, but is emerging as a way to monitor and minimize variations during the AM process to ensure build completion and promote resulting product quality and production throughput. Major sensors that will emerge in the near future include dimensional measuring sensors, thermal measuring sensors, porosity measuring sensors, residual stress sensors, and chemical composition sensors such as the following:

- Cameras (thermal and optical) to detect track irregularity, layer deformation, and part accuracy
- Ultrasonic sensors to detect cracks
- Imaging systems to detect malformations due to process irregularities
- Spectral analysis to analyze chemical composition

Advancements in process planning and monitoring tools are likely to change and improve how PBF parts are manufactured in the future.

References

1. 2013, "Measurement Science Roadmap for Metal-Based Additive Manufacturing," National Institute of Standards and Technology, Department of Commerce. Prepared by Energetics Incorporated, Columbia, Maryland for the National Institute of Standards and Technology, U.S. Department of Commerce. https://www.nist.gov/sites/default/files/documents/el/isd/NISTAdd_Mfg_Report_FINAL-2.pdf
2. Mani, M., Lane, B., Donmez, A., Feng, S., Moylan, S., and Fesperman, R., 2015, "Measurement Science Needs for Real-time Control of Additive Manufacturing Powder Bed Fusion Processes." NIST Internal Report, NISTIR 8036.
3. Agarwala, M., Bourell, D., Beaman, J., Marcus, H., and Barlow, J., 1995, "Direct Selective Laser Sintering of Metals," *Rapid Prototyping Journal*, 1(1), pp. 26–36.
4. Feng, S. C., Witherell, P. W., Ameta, G., and Kim, D. B., "Fundamental Requirements for Data Representations in Laser-Based Powder Bed Fusion," *Proceedings of ASME 2015 International Manufacturing Science and Engineering Conference*, American Society of Mechanical Engineers, p. V001T002A068.
5. Moylan, S., Slotwinski, J., Cooke, A., Jurrens, K., and Donmez, M. A., 2014, "An Additive Manufacturing Test Artifact," *Journal of Research of the National Institute of Standards and Technology*, 119, pp. 429–459.
6. Moylan, S., Slotwinski, J., Cooke, A., Jurrens, K., and Donmez, A., 2013, "Lessons Learned in Establishing the NIST Metal Additive Manufacturing Laboratory," National Institute of Standards and Technology Technical Note, NISTIR 1801.
7. Slotwinski, J. A., Garboczi, E. J., and Hebenstreit, K. M., 2014, "Porosity Measurements and Analysis for Metal Additive Manufacturing Process Control," *Journal of Research of the National Institute of Standards and Technology*, 119, pp. 494–528.
8. Slotwinski, J., Garboczi, E., Stutzman, P., Ferraris, C., Watson, S., and Peltz, M., 2014, "Characterization of Metal Powders Used for Additive Manufacturing," *Journal of Research of the National Institute of Standards and Technology*, 19, pp. 460–493.

CHAPTER 2.8

Polymer Laminate Technology

LaRoux K. Gillespie

Manufacturing Consultant
Andover, Kansas

As the word *laminate* indicates, products made by polymer laminate technology (PLT) result from stacking layers of a polymer together and finding a way to fuse them into a single piece. PLT is one form of laminated object manufacture (LOM).[1] This is one of the first additive processes developed and requires the lowest investment to use.

There are two key variations of the process with others possible also. In the first a roll of polymer material (typically a polymer tape) is fed over a platform, a laser or knife traces the pattern desired for that layer, the platform is lowered, the roll is advanced to put new polymer material over the previously cut shape, a heated roller presses the fresh tape tightly against the previous shape (causing it to adhere), the laser or knife traces the shape desired for that layer, and the process is repeated until the part is complete. For the case where the laser is used, the laser cross-hatches the area which is not part of the final shape, and the cross hatching weakens the material such that after the part is finished one can shake, push, or chisel it away from the desired part. If desired the finished product is then sanded to smooth the contours and painted.

The typical material used has been a polyvinyl chloride (PVC) tape, but other polymers can be used. Solidimension of Israel uses plastic film bonded with a solvent. Another system uses simple adhesives (a polymer for ease of use, but polymers are not essential) to bond paper sheets.[2] The PVC produces a yellowish or amber-tinted transparent part. A patterned layer of adhesion prevention material has also been used by one company to allow easier part separation from waste material.

---

[1]Chapters 2.9 and 2.10 provide information for metal products. This chapter is devoted to polymer laminations.

[2]While paper is not typically a polymer, the use of polymer adhesives makes the composite part a hybrid polymer product and adds physical properties to the paper. Use of paper and some polymer paper products allows inkjet printers to add color to the layers and even to print electronics on the sheets.

Equipment has been available for $5000 per machine. Its accuracy is 0.2 mm (0.008 in.) in the x and y dimensions and 0.3 mm (0.012 in.) in the vertical direction. The minimum wall thickness via this process is about 1 mm (0.040 in.). Early machines using adhesives and paper were large enough to produce car body prototypes, but the PVC machines have a maximum part size of 200 mm (7.9 in.) or less. These machines have been tabletop units, but the process is limited by machine size, not process limits. One machine has been sold that cuts thick layers diagonally to minimize stair-stepped surfaces.

The presence of the waste material around the part provides some support to prevent sagging or movement of the desired part during build. That material, though, may inhibit separation from narrow walls or sections, and users note that care must be taken for part separation to prevent damage. Removing waste material from internal features requires thoughtful approaches before those features are covered by final layers. In general, hollow sections cannot be made unless a step is added before completing the part and that must be done carefully and involve only low forces.

Depending on the material used, the film and the adhesive material have different properties and this in turn implies that the properties of the finished part will vary in the vertical direction.

Design considerations include:

- Designing such that the longest features are in the x-y plane since the vertical dimension is limited by machine and more layers involve more time. Laser cutting is rapid.
- Design so that curved surfaces can best be cut by the laser.
- Design so that the physical properties, if important, will be strongest in the desired direction (in line or cross sheets).
- Surface roughness on "straight" walls may be as high as 3 μm (0.00012 in.). Stair-stepping will cause much larger values. Sanding improves finish.

The second approach to producing polymer products involves stacking polymer plates of the desired geometry on top of each other and bonding them by heat, adhesives, or ultrasonic means. This is a fast process that can provide many channels, holes, and features not as readily produced by the continuous roll tape method.

For one of the more popular means of making micro fluidic devices, the silicone polymer liquid polydimethylsiloxane (PDMS) is poured in a mold and cured. Curing cross-links the polymer producing an optically clear part that is then stacked onto other cured polymer slabs to make the desired final multichanneled micro fluidic part. These are used in the medical and biomedical worlds to determine the properties of the fluids being examined (blood for example).

Layers may include electronic paths and sensors as well as just channels and wells. PDMS tubing can be molded or bound into these layers as well in final assembly or during the pouring of individual layers. PDMS has a unique material advantage in that it is biocompatible, elastomeric, transparent, inexpensive, easy to mold, and copyright-free.

Hot embossing and injection molding can also make the shapes of the individual layers. Table 2.8.1 illustrates some of the materials that have been used in this approach to PLT and Table 2.8.2 provides some typical material sizes used.

In one approach a Mylar layer is sandwiched between two layers of pressure-sensitive adhesive. The adhesive layers are covered with a layer of siliconized release

Material	Fabrication Technique
Silicon	Chemical wet etch
Glass	Chemical etch, laser cutting
Polymeric films (e.g., Mylar)	Laminate laser cutting
Silicone elastomer (PDMS)	Micro molding ("soft lithography")
Photoresist, hydrogels, etc.	Photopolymerization ("micro fluidic tectonics")
Thermoplastic	Hot embossing, injection molding

Table 2.8.1 Materials and Processes Commonly Used to Fabricate Micro Fluidic and Similar Devices (Courtesy P. Yager, University of Washington)

Material	Layer Thickness
PMMA	50 μm–4.5 mm (0.002–0.177 in.)
Acetal	125 μm–3 mm (0.005–0.125 in.)
PET	12.5–250 μm (0.0005–0.010 in.)
Polycarbonate	125 and 250 μm (0.005 and 0.010 in.)
Polystyrene	6–125 μm (0.0002–0.005 in.)
Polypropylene	50 μm–1 mm (0.002–0.040 in.)
COP	50 and 175 μm (0.002 and 0.007 in.)
Silicone	125 μm–1.5 mm (0.005–0.060 in.)
Polyimide	25–125 μm (0.001–0.005 in.)
Fluoropolymers: FEP, PTFE, PVDF	25–250 μm (0.001–0.010 in.)

Table 2.8.2 Material Thicknesses Used in PLT Products (Courtesy ALine, Inc.)

material. The stack of five layers is cut with the laser, the release layers are removed, and the remaining Mylar and adhesive layer package is slid down a set of posts that align this sandwich with the previous and remaining layers. Figure 2.8.1 illustrates a final assembly.

While the tolerances and feature sizes that can be produced vary by material and process that make up the individual layers, some typical values for laser-cut layers would be:

- Channel minimum widths of 125 μm (0.005 in.) (larger are common)
- Layer thicknesses of 12.5 μm to 3 mm (0.0005 to 0.120 in.)
- Adhesive layer thicknesses of 25 to 50 μm (0.001 to 0.002 in.)
- Feature density: requires 0.5-mm (0.020-in.) spacing for 250-μm-thick (0.010-in.) channels
- Typical alignment 125 μm (0.005 in.); 50 μm (0.002 in.) possible with additional fixturing

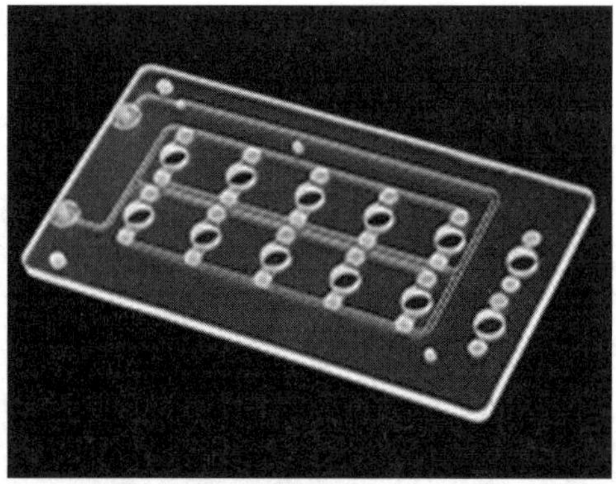

Figure 2.8.1 Top view of a polymer laminate technology product that flew on NASA's GeneSat experimental satellite. The fluidic card is constructed from layers of acrylic bonded with an optically clear bottom and a gas-permeable top. A single inlet distributes nutrients evenly across all ten sample chambers. The system telemetered data on cell growth to earth. (Courtesy ALine, Inc.)

Larger features are not an issue. Today medical devices are being made via PLT that include electronic circuits, multiple convoluted channels, attachment devices, membranes, pumps, valves, and sensors.

New materials are emerging that can provide flexible ceramics as well as a variety of electrical circuits on polymers. Partially cross-linked preceramic polymer tapes loaded with different fillers have been cast using the doctor blade method and then dried and gelled, and laminated at temperatures between 200 and 300°C. The individual tapes had thickness of 200 to 1000 μm (0.0079 to 0.039 in.) and were bonded together without additional thermoplastic-based binders.

The process is adaptable to those materials as well as many others not yet tried.

CHAPTER 2.9

Accumulative Roll Bonding

Jan Romberg

BIOTRONIK SE & Co. KG
Berlin, Germany

The Process

The shortest description of accumulative roll bonding (ARB) is "repetitive cladding" by a severe plastic deformation (SPD) process. ARB allows manufacturers to produce metal sheets with special properties, going beyond the characteristics of ordinary clad metal. Simple cladding is used to combine properties of different metals. Common examples are cooking pots of heat conductive copper or aluminum clad with a thin layer of less conductive stainless steel which does not chemically react with food. There are many further examples of cladding fabricated products like bimetal strips or coins with magnetic inlayer.

Beyond just cladding, however, the process is also used to produce ultrafine-grained (UFG) material from laboratory size to semifinished material of larger sizes.

Accumulative roll bonding is classified by the initial material combination. The initial sheets can be (1) the same material, (2) different materials, and/or (3) combined with particles or fibers. Cladding of metal sheets of the same material makes little sense but repeated cladding affects microstructure and strength of metal sheets as the repeated rolling produces a large amount of plastic deformation.

The use of SPD processes produces UFG materials and the associated mechanical properties. The SPD processes, like ARB, all occur cyclically, which presupposes the external shape of the workpiece to be identical to the original geometry at the end of the ARB cycle. The most intensively used SPD methods are high-pressure torsion (HPT), equal channel angular pressing (ECAP), and ARB. Unlike the two first methods no specialized tools are required for ARB—it is carried out with commonly used metal-forming rolling mills.

Most applications involve aluminum alloys, combinations of different aluminum alloys, and composites of aluminum and other materials such as copper, nickel, steel, and titanium. Copper-nickel, copper-niobium, and nickel-zirconium are examples of some common composite sheets. The use, effect, introduction, and distribution of particles and fibers into the metallic sheet material by ARB are also described in the literature.

Process Steps

The main process steps are (1) stack metal sheets, (2) join them by roll bonding, and then (3) cut them to length. For roll cladding a thickness reduction of 50% per rolling pass is commonly used. Thus, at negligible widening, the clad sheet has twice the length of the initial sheets. By cutting the new sheet in halves, the same external geometry as the starting sheets is produced. These sheets are stacked and bonded again. The principle is illustrated in Fig. 2.9.1, for the case of differing initial materials. By repeating the process, the plastic deformation is accumulated. In the case of 50% thickness reduction, the logarithmic deformation amounts to 0.8 per rolling pass.

During the roll bonding process, both a chemical bond by cold welding and a mechanical bond by hooking of the rough surfaces occur. Oxide layers and contamination layers covering the base metal are obstacles in the formation of a chemical bond, so the oxide layers are removed mechanically, but regrow immediately. During forming, it is essential that the area of sheet surface increases. Because the metal oxide layers are brittle compared to the underlying metal, cracks emerge and the metallic material is forced through these cracks. Bonds are created where the virgin metal of both sheets meet. The stronger the single bonds are and the more bonds are established, the stronger the bond between the two sheets. In case of weak or a low number of bonds, these bonds break in the moment when the sheets leave the space below the two rolls. This causes a threshold deformation for cold welding metal sheets by roll bonding. The minimum deformation depends on the specific material or material combination. The rolling process deforms the full thickness of the sheets. However, the deformation is not evenly distributed as shear appears mostly close to the roll-stock interface. Consequently a gradient in microstructure and mechanical features appear. Conversely, during each ARB cycle former outer surfaces that are heavily sheared become the middle sheet

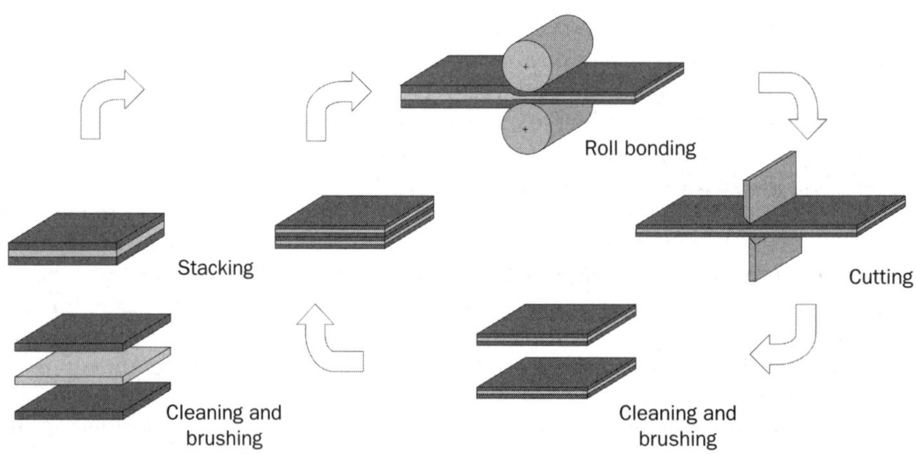

FIGURE 2.9.1 Principle of accumulative roll bonding, for example, the creation of multi-material composites.

where the shear deformation is less; after about four passes the deformation and the consequential properties are evenly distributed.

The temporary removal of contamination and oxide layer is crucial. In a normal atmosphere, oxide layers appear and grow in milliseconds but will be still thinner and therefore can be overcome during roll bonding. To obtain a clean surface and to increase roughness of these sheets at the same time, the surface is usually wire brushed. Wire brushing also strongly deforms the surface before rolling. The severely deformed zone of aluminum ranges upto 10 μm (0.0004 in.) in depth under the brushed surface.

The deformation of 50% per rolling pass requires a sufficiently ductile material, which can deform evenly in these demanding conditions. The sheets and the process (1) have to assure that the tension applied to obtain the shape change ranges between the initial yield strength and ultimate compression strength of the rolled sheets and (2) does not employ localized deformation. Localized deformation starts when hardening is no longer sufficient to compensate for the softening. In this case, local deformation fluctuations cause necking. Since the strength of a material cannot be increased infinitely, an already partially strengthened material has a reduced ability for further work hardening. Dynamic recovery or recrystallization reduces the dislocation density-based increase in strength during the deformation. The increase in strength can even be overcompensated. In particular, nanocrystalline materials quickly reach an equilibrium of strengthening and softening, which results in an unstable deformation. Intermediate heat treatment at temperatures of about 500°C (932°F) delays necking of titanium and does not result in a remarkable amount of intermetallic phases, and therefore, subsequent rolling and cladding can be carried out successfully without necking.

ARB Applications

ARB can create UFG material which has a high specific strength and can be executed without specialized equipment, in large quantities and sizes. The range of applications is enhanced by combining different material sheets. There are three basic motivations for using ARB. First, the combination of different material properties is of interest, for example, the combination of the high strength of the 5xxx series aluminum alloys with the high surface quality of the 6xxx series aluminum, used for exterior paneling of automobiles. Second, the creation of new internal interfaces by grain refinement can be used more effectively and thus a higher strength can be achieved. Third, alloying—due to a fine, laminar structure, the diffusion paths are short and numerous defects such as voids and dislocations in the material accelerate diffusion. An application of this is forming titanium and aluminum to a product shape in the metallic state first and then initiating an intermetallic phase by heat treatment to give TiAl3. This is advantageous because plastic deformation of intermetallic phases is limited while pure metals reveal comparably large deformation. Titanium aluminides are a new class of high-temperature lightweight materials. Gamma Ti-Al has low density and creep resistance at high temperatures. It is usually produced by vacuum arc remelting and processed by hot rolling or powder metallurgy. By using the ARB process so often that the layers are only nanometers thick, it is possible to make it a reactive multilayer brazing foil.

During each rolling pass of the ARB process, a new outer surface is created but the surface between the two sheets is annihilated. By covering these surfaces with particles or fibers (e.g., carbon nanotubes), these can be dispersed in the sheet material. Particles and fibers strengthen the material but make it more brittle as well. Furthermore, covering surfaces to be bonded weakens the bond of the two sheets and fibers are likely to break during the rolling process. However, it can be used to produce metals with a gradient of strength throughout and, therefore, there is one possibility to create tailored blanks with load adaptive strength.

Limitations of the Process

When cladding materials of different yield strengths, the one with the lower yield strength yields at a lower thickness reduction and, therefore, the plastic deformation takes place already somewhat outside the roll gap. In consequence, the other material undergoes less reduction in thickness and elongation during the rolling process. This effect occurs only in the first rolling cycle because during subsequent cycles, both materials are already bonded and the two metals are forced to expand synchronously during rolling and therefore keep a constant thickness ratio and equal elongation. This results in strong residual stresses. The material interaction is even more complex. Recrystallized pure titanium for example has lower yield strength than recrystallized aluminum alloy 5049 and titanium hardens more pronounced than the aluminum alloy. During subsequent cycles, it is, therefore, under tensile stress. Tensile stress favors the necking of titanium in early cycles and titanium hardening increases necking in subsequent passes.

Aluminum alloy in contrast hardens during more ARB passes and has no strong tendency for necking. Rather, in titanium/aluminum composites, aluminum functions as a matrix material and plastically flows around the necked titanium, fills occasional gaps, and ensures the cohesion of the sheet. The tensile strength increases with the number of ARB cycles. However, the elongation to failure decreases with the number of cycles and conditions, but can increase under other conditions.

Another limitation that should be considered is that the UFG microstructure that is aimed and created by ARB would not sustain during the most common welding processes. However, using friction steer welding is the preferred procedure to process ARB sheets.

Comparison of the Composite Material with Single-Material Sheet

Aluminum and titanium are metals with different deformation behaviors and, therefore, co-rolling and repetitive roll bonding require good knowledge of the process and proper equipment. It is important that the thickness reduction is constant along the width of the sheets. Consequently, using a four high rolling mill produces better results. Without a fixture the sheets tend to shift causing edge cracks and curved sheets. Therefore, a fixture that ends close to the rolling gap and a second one at the exit of the roll are recommended. The exit fixture needs to have bigger spacing but the same center as the entry fixture.

To get a fine-grained and fine laminar structure in Ti/Al sheets, there is a set of parameters to choose. The strain rate works in two respects. (1) Strain rate hardening overcomes necking in case of materials with rate-sensitive strain rate hardening. The developing neck is deformed faster and consequently increases in yield strength which stops the necking. In titanium this works better at a higher rolling speed. (2) Fast rolling increases the temperature in the rolling gap which lowers the yield strength. However, the softening due to temperature raise implies a higher danger of necking.

Effect of Ultrafine Grains	Use	Example
Enhanced diffusion and self-diffusion	Better relaxation; creation of intermetallic phases	TiAl3
Enhanced solute diffusion and solubility	Hydrogen storage and creation of alloys that don't solute otherwise	Fe-Ag, Fe-Cu, Bi-Cu alloys
Higher entropy	Higher specific heat of cooling devices	Cu
Chanced magnetic properties, e.g., lower Curie temperature and less magnetostriction	Transformer sheets	Fe-alloys
Reduced Young's modulus	Biomaterials	CaF_2, Mg, Pd
Higher hardness and strength	Lightweight constructions (e.g., airplanes)	Al

Table 2.9.1 Overview of Effects of Ultrafine-Grained Material

As strain hardening is an efficient means to overcome necking, an intermediate heat treatment can help with recovering or recrystallizing the material and therefore to enhance strain hardening during rolling. Coiling helps to keep the sheets straight and avoids it leaving the fixtures. However, as the sheets are plastic in the roll gap it can be easily destroyed by pulling. If tension is applied the forces need to be adjusted carefully.

The tensile strength of Ti/Al composites after the fourth cycle is higher than of titanium processing in the same manner. Since titanium has a higher density than aluminum, it turns out that the necked titanium lowers the specific strength while laminar titanium increases the specific strength of the aluminum matrix. ARB of aluminum without titanium does not require any intermediate heat treatment.

However, comparing the intermediate heat-treated sheets, both elongation to failure and specific strength in composite sheets are higher than the aluminum sheets. In comparison to specific strength Ti/Al composite sheets are on the same level as ultra-high-strength steel. The specific strength is a measure in lightweight construction and equals the tensile strength divided by the mass density of a certain material. Higher specific strength is inhibited by either necks in titanium layers or by intermediate heat treatment which reverses grain size reduction and strain hardening. However, by subsequent conventional rolling, grain refinement and strain hardening can be still applied on these sheets.

To put it in a nutshell, ARB provides the possibility to gain UFG material from laboratory size to semifinished material of larger sizes. The features you can get by UFG materials are summarized in Table 2.9.1.

CHAPTER 2.10

Ultrasonic Lamination Technology

Mark I. Norfolk

Fabrisonic LLC
Columbus, Ohio

The Process

Additive manufacturing (AM) is defined by ASTM (2009) as the "process of joining materials to make objects from 3D model data, usually layer upon layer, as opposed to subtractive manufacturing methodologies." In traditional machining, parts start out as large billets of material that are progressively machined, cut, and ground to a final shape. In AM, a part starts as a thin layer of material and material is selectively added only in the desired areas. By adding material layer by layer, in just the areas needed, AM can create novel features and shapes unimaginable with traditional subtractive manufacturing. AM can be used to create complex internal geometries, such as lattice structures, that are impossible with machining.

Known by several equivalent names, ultrasonic additive manufacturing (UAM) or ultrasonic sheet lamination is a metal 3D technology that uses sound to merge layers of thin metal foil stock. The process produces true metallurgical bonds with full density and works with a variety of metals such as aluminum, copper, stainless steel, and titanium. In combining additive and subtractive process capabilities, UAM can create deep slots, hollow, latticed, or honeycombed internal structures, and other complex geometries impossible with conventional subtractive manufacturing processes.

The UAM process is typically both an additive and subtractive process. UAM involves building up solid metal objects through ultrasonically welding thin strips of metal foil into a 3D shape, with periodic machining operations to create the detailed shape of the resultant objects. Figure 2.10.1 illustrates the key components of an ultrasonic welding head. It consists of two ultrasonic horns on either side of a rolling acoustically tuned tool, called the welding horn. Each transducer is made of piezoelectric elements that expand and contract with varying voltage. The vibrations of the transducer are transmitted to the rolling welding horn itself, which is an acoustically tuned device, which in turn creates an ultrasonic solid-state weld between the thin metal tape

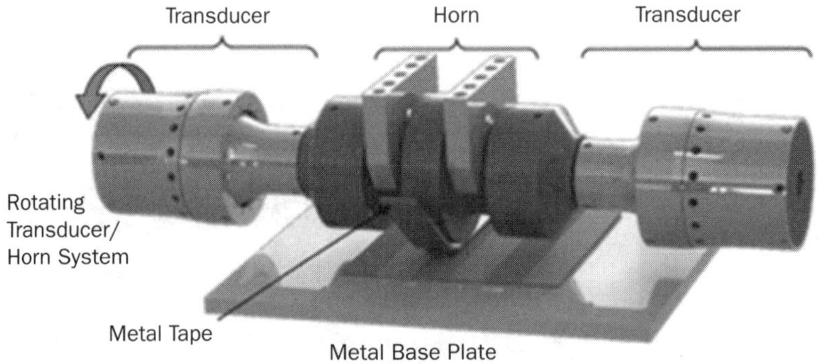

Figure 2.10.1 Ultrasonic welding of aluminum tape.

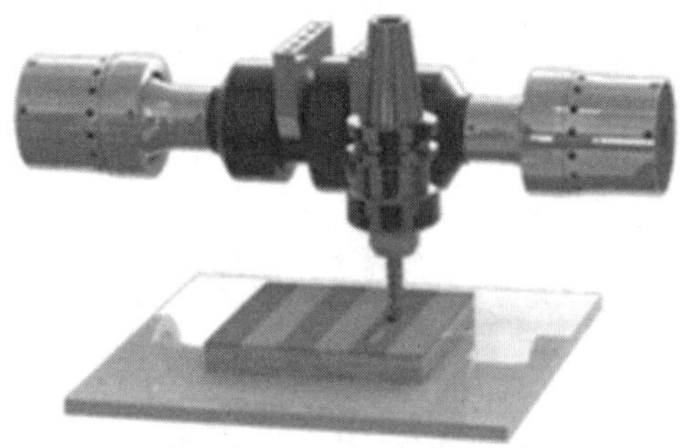

Figure 2.10.2 Periodic machining operations.

and the substrate. The continuous rolling of the vibrating horn over the plate creates a solid-state bond between the tape and the base structure.

Foils are placed side by side, and then layer on layer, in a manner similar to laying bricks. For a majority of UAM parts, the metallic feedstock is 25.4-mm-wide (1-in.) foil. By placing multiple foils side by side, a larger part can be created. Figure 2.10.2 illustrates the subtractive portion of UAM. Since the feedstock is 1 in. wide and most features are not a perfect multiple of 1 in., the additive process creates a part that is near net shape. Periodically, a computer numerically controlled (CNC) machining spindle is used to machine off the slight excess and to create a fully dimensionally accurate part. In this way internal features such as channels and voids can also be created.

One key aspect that makes UAM so unique is that the joining is accomplished in the solid state; no melting occurs during the process. In fact, temperatures at the weld have been measured in aluminums at under 94°C (200°F). This is important for two reasons. First, by avoiding melting and staying below the transformation temperatures of most metals, the inherent mechanical properties of the incoming feedstock are not reduced. For instance, precipitation-hardened alloys are not at temperature long enough to cause any meaningful precipitation reaction, thus allowing the bonding of high-strength aerospace alloys without any loss of mechanical strength, as is found in fusion-based welding technologies. The second benefit is the solid-state nature allows for welding of dissimilar metals. Typical fusion-based processes, when used on dissimilar metals, result in mechanically inferior properties as the resultant solidification structure is some type of glass. Again UAM gets nowhere close to the melting point of most alloys, allowing a true solid-state bond with superior mechanical performance.

Characteristics and Applications

Metals joined using UAM are found to be fully dense. Material properties in the x-y plane typically have similar properties to the incoming feed stock. In the z direction, tensile strength has a loss of 10% to 15% in relation to x-y due to embedded oxides that are not completely removed. As mentioned below, by combining dissimilar metals, UAM builds can be tailored to have properties unobtainable with traditional monolithic alloy systems.

Since all UAM machines are, in fact, three-axis CNC mills, internal and external features are expected to have the same fit and finish as a CNC-milled component. All types of traditional milling tools can be mounted in a UAM tool changer including drill, mills, and shaper tools. Typical accuracy is in the range of 12.7 μm (0.0005 in.) with surface finish of 0.8 μm (32 μin.) or better being easily achieved. Machines are available today that will accommodate parts from 600 × 900 mm (24 × 36 in.) up to 1800 × 1800 × 900 mm (6 × 6 × 3 ft) parts. Parts in the order of 25 mm (1 in.) and smaller can also be produced.

The ability to build fully dense metal parts in a variety of relevant engineering materials is at the core of UAM's value proposition. From this core capability, three unique opportunities arise, namely the ability to:

- Create complex 3D internal channels, which is an enabler for molds, thermal management, and chemical processing
- Embed one material in another, which is an enabler for metal matrix composites, integrated sensors, and smart systems
- Join dissimilar materials, which is an enabler for cladding and transition joints

Dissimilar Metals

The solid-state nature of the ultrasonic bonding process used in UAM permits joining of dissimilar metals without the formation of brittle intermetallics which occur in fusion processes. A wide range of material combinations have been successfully bonded using ultrasonic welding including combinations of copper, aluminum, and titanium as well as aluminum to steel. This unique ability allows for a series of useful engineering applications.

The first category of dissimilar metals is simple cladding. In a traditional cladding operation UAM is used to layer an expensive material onto a bulk substrate of a cheaper material. For instance, titanium has been welded to aluminum to act as a heat- and wear-resistant layer on the outer edges of aluminum wear components. Additionally, this technique can be used to make transition joints between two materials. Simple cladding can be extended further to create functionally graded laminates. Metal foils of alternating dissimilar metals can be welded into a layered structure. By varying the ratio of one metal to another through thickness, variations in properties through thickness can be engineered. One example is an armor produced by layering aluminum and titanium and varying the ratio through thickness to achieve specific properties.

Embedding

A second classification of dissimilar metal bonding that is uniquely suited to UAM is embedding. The low-temperature process allows embedding of all manner of wires, fibers, and sensors into a metallic substrate. Embedding reinforcing members such as boron and silicon carbide fibers or stainless-steel meshes into metal matrices is simple with UAM. Since the process can be interrupted and modified at any point, structural components can be augmented into any metal matrix to form superior, high-performance composite structures.

Additionally, sensors and actuators can be embedded using UAM into fully dense metallic structures. Numerous sensors such as thermal couples, switches, and strain sensors have been successfully integrated into UAM-built components. Electronic circuitry has also been embedded, allowing for tamper-resistant metallic devices. Since UAM bonding occurs at low temperature, electronic components can be embedded without damage to electronic traces. Electronic isolation can be accomplished by wrapping wires with a dielectric such as Kapton tape.

Complicated Geometry

By using UAM, complex parts having unique features and attributes, not possible using traditional manufacturing techniques, can be created. As with most other AM technologies, UAM can produce complex internal channels and voids that are impossible with traditional machining. As the process builds one layer at a time, the process can be interrupted at any point in z-height and the integral CNC machining can be used to remove material from any point on the part. Subsequently the process welds over these voids, incorporating them as integral channels into the structure. As machining can happen on every single layer, if needed, internal features can change size, shape, and orientation in any axis.

This particular ability has been used for creating complex 3D channels for thermal-management applications such as micro channel cooling, integrated heat pipes, and conformal cooling for molds. Additionally, the ability to create complex internal geometry has been applied to radio-frequency wave guides, light weighting, and cable runs that are completely internal to the structure.

Economics

In aluminum, build rates in excess of 492 cm^3 (30 $in.^3$) per hour have been obtained. As material yield strength goes up, the build rate lowers as the process must be run slower. In materials such as stainless steel, build rates fall to 246 cm^3 (15 $in.^3$) an hour. Another major driver of build time is feature complexity. UAM 3D prints metal near net shape and uses CNC milling to obtain both internal and external features. As feature complexity increases (fine features, undercuts, rounds), part production time increases. In an extreme case, such as micro channel heat exchangers where very small end mills must be used, the CNC portion of the build can actually dominate cycle time.

3D printing processes can be slow and expensive for general use. As such users actively attempt to minimize the amount of material that must be printed. Typically 3D printing is used to obtain a specific feature that cannot easily be created with traditional manufacturing methods (dissimilar metals, internal geometry, and embedded electronics). Thus, if a part is 76 mm (3 in.) thick but the first 3D need is not until a height of 50 mm (2 in.), the build will be started with a 50-mm-thick (2-in.) billet to which the remaining geometry will be 3D-printed. Often UAM is used to add features to pre-existing parts that have been created with more traditional manufacturing methods.

The last major characteristic that drives economics is z height. Building additional x and y dimensions is an order of magnitude cheaper than building additional z height. As such, parts are often oriented such that the thickest section lies in the x-y plane.

Materials Suitable to This Process

Ultrasonic additive manufacturing is compatible with all types of metals, assuming the metal can be purchased in a foil format. Typical foil thickness range from 127 to 381 μm (0.005 to 0.015 in.), with soft metals being able to extend slightly thicker. Most metals can be found on the open market as foil, including aluminum, copper, steel, nickel, and many others. The main exceptions are metals that work-harden at a high rate. These metals are hard to roll and thus are hard to find at commercial mills. An example of a high work-hardening material would be Ti-6Al-4V. Table 2.10.1 illustrates some proven combinations of materials for this process. Figure 2.10.3 illustrates a cross section of laminated aluminum and titanium.

	Al	Be	Cu	Ge	Au	Fe	Mg	Mo	Ni	Pd	Pt	Si	Ag	Ta	Sn	Ti	W	Zr
Al Alloys	●	●	●	●	●	●	●	●	●	●	●	●	●	●	●	●	●	●
Be Alloys		●	●			●										●		
Cu Alloys			●		●	●	●	●	●	●	●		●	●		●	●	●
Ge					●						●							
Au					●	●			●	●	●	●	●			●	●	●
Fe Alloys						●			●	●	●		●	●		●	●	●
Mg Alloys							●						●			●		
Mo Alloys								●	●		●			●		●	●	●
Ni Alloys									●	●	●			●		●	●	
Pd										●			●	●				
Pt Alloys											●	●		●		●	●	
Si													●	●				
Ag Alloys													●	●				●
Ta Alloys														●		●	●	
Sn															●			
Ti Alloys																●	●	
W Alloys																	●	
Zr Alloys																		●

● Material pair proven for ultrasonic welding

Table 2.10.1 Material Combinations Proven for Ultrasonic Joining

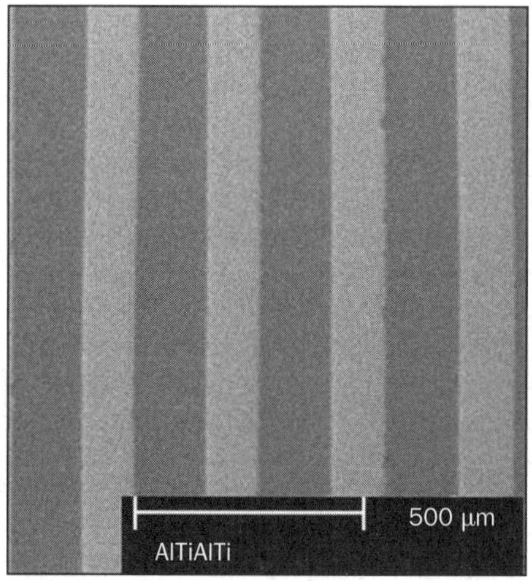

Figure 2.10.3 Aluminum- and titanium-laminated composite made by ultrasonic bonding.

Ultrasonic additive manufacturing has been used to embed all types of nonmetallics including ceramics and polymers. It must be stressed that nonmetallics are not actually being bonded at the atomic level. Instead, the UAM process is extruding metal around the nonmetallic to fully encapsulate it. In this way, strain can be shared from metal to nonmetal by interlocking surfaces, but the material will have a measurable pull-out strength.

Embedding reinforcing members such as boron and silicon carbide fibers or stainless-steel meshes into metal matrices is simple with UAM. Since the process can be interrupted and modified at any point, structural components can be augmented into any metal matrix to form superior, high-performance composite structures. These advanced materials can be stiff and lightweight, and combinations designed for high-stress, weight-critical environments.

Specific Design Recommendations

There are two main design limitations. The first limitation is best described as a height-to-width ratio. Tall, skinny structures, such as a wall, look like a simple beam. At a height-to-width ratio of about 0.7 the theoretical beam has a resonant frequency of 20 kHz. As we are inputting energy at 20 kHz, the structure is put into resonance, and it fatigues if we get to a height-to-width ratio of 0.7 to 1.0. To avoid this limitation, one typically builds tall structures wider than needed and machines them afterward to get true net shape.

The second limitation is unsupported geometry. UAM uses several thousand newtons[1] of force in order to create a bond. For structures such as overhangs, no normal force exists to push back and enable bonding. Users have found, however, that it is possible to make shallow overhangs. Additionally, UAM can build over enclosed cavities. In general, cavities of less than 15 mm (0.59 in.) are workable as the horn can weld

[1]10,000 N is a force of 2248 lb.

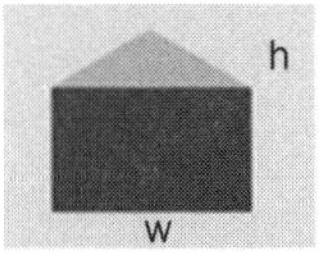

FIGURE 2.10.4 The part is rectangular with a cavity inside. Complete bonding above the cavity occurs at a height "h" above the cavity where h ~ width W.

on either side of the cavity. However, directly above the cavity there will be a region of poor bonding, even with additional layers. A rule of thumb is that the build will "heal" after a height equal to the width of the channel. This is illustrated in gray in Fig. 2.10.4. Also, welds over pockets or cavities will have a slight deformation, or bow, on the "ceiling" due to the force applied during welding.

Larger voids can be encapsulated using workarounds. For instance, in larger structures one could machine a lip around a large void and place a 3.175-mm (0.125-in.) plate on the lip. Subsequently the weld can continue over the entire structure. The 3.175-mm (0.125-in.) plate would provide support for the build to continue.

CHAPTER 2.11

Vat Photopolymerization: An Additive Process

Andrew Graves

Stratasys Direct Manufacturing
Valencia, California

The Process

Vat photopolymerization process produces plastic designs in a vat of liquid photopolymer resin using, most commonly, ultraviolet (UV) lasers to cure the resin in computer-determined patterns in a layer-by-layer method. The process involves large beds of liquid resin and begins with a 3D CAD model design in the form of an STL file. Vat polymerization can refer to multiple processes in which a photopolymer liquid is cured selectively by directed light energy; however stereolithography, the first ever 3D printing process to be patented and commercialized in 1986, remains the most popular vat photopolymerization technology today. Similar technologies include digital light processing (DLP) which projects light-emitting diodes (LEDs) onto the liquid surface—either from beneath or above the liquid platform—to form designs. The technologies available range from large industrial to small and simple desktop systems.

What made 3D printing revolutionary in 1986, and still disruptive today, is its ability to produce fully realized designs in hours without supervision or significant manual labor. Today, 3D printing can manufacture designs using the same thermoplastics or metals required to qualify as an end-use functional product. Prior to 3D printing, product development might include specialized modelers to machine scaled down prototypes, or a team could move straight to the lengthy and costly process of tool creation, relying on "first shots" to work as prototypes. Fixing design or mechanical errors involved reworking the entire tool or mold. This involved hundreds of hours of manual labor to design, build, program, and finish for a single new product, plus the cost of the tool itself, which could cost up to $70,000 or more, depending on the complexity and required tool life. 3D printing removes manual labor and cost from the equation by taking a design and physically rendering it layer by layer in a streamlined and far more automated process. Rather than moving directly into tooling, engineers can print a prototype and reprint the same part with a small design change in just one day.

Therefore, 3D printing saves engineers the costs associated with manipulating an imperfect tool. 3D printing allows for design optimization through inexpensive trial and error.

When designing for a tool, there are many inhibiting feature requirements that must be taken into account, such as draft angles, internal features without access, bosses, seams, etc. Designing for a tool has many limitations, and increasing complexity significantly increases the cost. In contrast, 3D printing provides an option to build internal zero-access features because it builds layer by layer. Designs built with 3D printing do not require release points. There is a unique freedom inherent in 3D printing technology. The engineering approach with 3D printing technology shifts into manufacturing for designing rather than designing for manufacturing. It significantly transforms the way engineers and designers approach a project and think about design.

Technology and Process Controls

Vat photopolymerization systems involve a completely enclosed bed of liquid resin to protect the resin from any outside light; due to the sensitive nature of the photopolymers used in this technology, prolonged external light from something as low level as a desk lamp can degrade the resin over time. The layer-by-layer process begins with a thin base layer of resin. The resin is cured by directed UV energy. Once one layer is cured, the build platform moves down and a fresh layer of resin is applied. The next layer is cured and the process repeats. Figure 2.11.1 demonstrates the laser energy bouncing and focusing through dynamic mirrors before curing the design into the resin. Projects begin with designs in the form of an STL file. Before a design is loaded into a system for printing, supports for the future part must be added and design orientation configured. Designs built with vat photopolymerization require build supports (Fig. 2.11.2) because unsupported designs will float in the resin and unsupported features will distort. There are many automated programs that will add supports for the designer and allow the designer to experiment with build orientation to result in the most optimum ratio between design and supports needed; fewer supports are ideal to reduce post-processing manual removal. One of the more popular programs for support creation

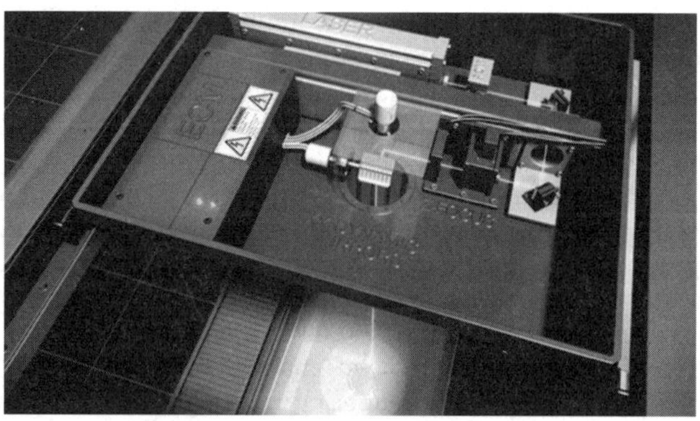

Figure 2.11.1 Schematic of vat photopolymerization operation. Laser energy is directed and focused through dynamic mirrors to cure the resin layer by layer. (Rendering by Duane Loose Design on behalf of Stratasys Direct Manufacturing.)

Figure 2.11.2 Supports anchor the design to build and support features. (Rendering by Duane Loose Design on behalf of Stratasys Direct Manufacturing.)

is Materialise Magics SG Module; however, most systems will come equipped with a support generation program.

Once the design orientation and supports have been configured, the STL file is uploaded to the system where layer thickness will be designated. In the case of stereolithography systems, solid-state lasers determine feature detail through beam width while scanning mirrors direct the laser energy onto the resin. Certain systems contain dual beam sizes. Smaller spot sizes are used for higher accuracy while less critical features—such as internal hatching—are executed with larger spot sizes for speed. The more powerful the laser, the faster the build; however, the dynamic scanning mirrors which direct the laser energy must be able to handle the speed required to scan a powerful laser, meaning there is a limit to the speed that a part can be built. UV lasers commonly used with stereolithography systems hit around 800 mW. The layer thickness of the design additionally determines the speed of the build. A middle ground between thin layers and speed ensures the design features are met without sacrificing turnaround. Layer thicknesses between 75 and 180 μm (0.003 and 0.007 in.) are ideal for the majority of designs built with any vat polymerization method. Smaller layer thicknesses around 50 μm (0.002 in.) are possible, depending on resin and application, for the majority of high-definition vat photopolymerization systems (numbers based on solid-state lasers used with SLA systems). After each layer is cured, a re-coater blade (typically involving a vacuum to suck up fresh resin) skims across the surface, depositing resin in a thin, even layer.

A part does not fully cure within the build chamber during vat photopolymerization. After a design is complete, the part may need to be drained of all uncured resin, which may be trapped in internal areas. Parts that require internal draining, such as hollow parts or patterns for investment casting, are usually placed within a centrifuge to drain out excess resin, and holes may be added to parts for the purpose of removing resin. After the design is fully drained, supports are removed, and the clean design is placed in a UV oven to complete curing.

The formula of the resin chosen for use with vat photopolymerization designs determines ultimate feature size and can have an influence on the formation of bubbles and final part accuracy. While bubbles can form in the resin, as it is a liquid substance, there are resin formulas which are less prone to bubble formation. Simply removing bubbles prior to beginning a build will eliminate formation within the design.

Materials used with vat photopolymerization are not intended for strenuous use. Tensile strength for vat photopolymerization ranges from 36 to 53 MPa (5221 to 7687 psi), with flexural modulus around 2000 MPa (290,000 psi). Impact resistance is fairly low for these polymers, as they aren't intended for long-term durable testing. The shelf life of a part built with vat photopolymerization is about five years, as long it is kept in a cool dark place during that time due to its susceptibility to light.

There are a few sources for vat photopolymerization resins. Some systems will only allow resins approved by the manufacturer to be used on their platforms. Many resins used in vat photopolymerization share very similar properties with slight variations. Table 2.11.1 shows examples of available materials from two sources with their respective mechanical properties.

Vat Photopolymerization: Systems

Vat photopolymerization systems available today range from industrial to desktop. These systems include stereolithography, DLP, and thin film photopolymerization. They share the basic technology principles of curing photopolymer material while utilizing different process parameters to achieve their designs.

Geometries and Tolerances

Vat photopolymerization offers near limitless design freedom. Because vat photopolymerization requires excess resin be removed during post-processing, geometries must factor an exit route for the resin. Often, simply a small cut that is resealed during post-processing is enough for a geometry. Certain highly complex internal features which cannot be accessed through the separation and bonding of a design are better suited to a process such as laser sintering, which does not require build supports. Vat photopolymerization is well suited to bonding through the application of epoxy resin and re-curing through handheld UV lights. For large parts, it remains economical to 3D-print designs in sections, drain and post-cure the design, and then bond the sections together.

Minimum feature size is a function of both layer thickness and resin composition. Typical layer thicknesses—or resolution—for a vat photopolymerization part range between 127 (0.005 in.) and 500 μm (0.020 in.). Layer resolution determines stair-stepping visibility. Thinner layer thicknesses are achievable, depending on geometry. Micro high-definition printing can achieve layer lines as small as 50 μm (0.002 in.) on the Z and 100 μm (0.004 in.) on the X/Y axes. Minimum feature size is determined by how thin a feature can be on the X/Y plane and therefore how small a feature is—such as a nose on a sculpture, for instance—before it becomes indistinguishable. The minimum feature size for micro SLA printing hits around 100 μm (0.004 in.) and is determined by the material, geometries, as well as laser diameter and mirror accuracy.

Tolerances for vat photopolymerization parts are ±380 μm (±0.015 in.) or ±0.0015 in. per measured inch, whichever is greater. High definition using a smaller spot (tighter laser point) can achieve tolerances at ±250 μm (0.010 in. for the first inch) or ±0.001 in. per measured inch, whichever is greater, while micro printing can encompass ±125 μm (±0.005 in.) or ±0.001 in. per measured inch, whichever is greater. Figuring out the tolerance of a part relies largely on layer resolution. Thinner layers can have a positive effect on meeting tolerances; however, resin shrinkage and beam width will be the final determinant in meeting tolerances. When a particular design requires significant strict tolerances, post-processing can provide one solution by applying or sanding away material where needed. Most vat photopolymerization resins have surfaces that are easily sanded by 220 to 800 grit sandpaper.

Applications

3D printing has opened doors to what many term mass customization, or the ability to produce dozens of small parts with slight variances in one build. 3D printing is catalyzing many paradigm shifts, especially within the realm of design freedom but also in the way engineers think and plan for production manufacturing. Aerospace companies are implementing 3D printing processes to build thousands of parts for aircraft with high-performance thermoplastics for end-use performance. Medical, energy, and consumer industries are following the trend set by the aerospace field.

Vat photopolymerization is not used for production right off the machine. Rather, it's a bridge application in many cases. Its most popular use has been in the creation of investment cast patterns. Vat photopolymerization can build mostly hollow parts with robust outer shells. These mostly hollow parts are filled with a hatch print to reinforce the design and are used as investment cast patterns. In investment casting, hollow lightweight patterns can result in better burnout; therefore, hollow vat photopolymerization investment cast patterns are desirable.

Many companies that produce 3D printed parts for customers as well as vat photopolymerization systems offer their own build methods for these patterns along with materials. Key factors that determine a good resin for investment cast pattern purposes include low and stable viscosity, high green strength, and low water absorption. A vat photopolymerization part is defined as being in its green state before any post-curing processes take place. A resin with high green strength therefore has better properties before being cured, such as higher tensile strength. Higher strength during building translates to higher accuracy and tighter tolerances and can minimize the final mass of the design. Vat photopolymerization resins are susceptible to humidity, which causes them to distort in hot and humid temperatures; therefore, it is important to have a resin with low water absorption, especially for traveling patterns. Low and stable viscosity ensures consistent drainage results in patterns. After a design is finished building, it must be drained of all excess resin. The build platform raises up and resin drains away from a part. Therefore it is important that the resin has low viscosity to ensure resin can escape the part. Uncured resin can be stored within the build chamber of a system for up to about three years, depending on resin composition and whether the system is kept in low light away from heat and humidity. Therefore, the ideal resin for an investment cast pattern involves strength, resistance to humidity, and higher tolerances than standard resins.

Traditionally, creating a pattern for investment casting begins with a tool or mold for a lost wax pattern. Creating a lost wax mold requires conventional manufacturing, typically through the machining and bonding of many parts. While there are technologies for 3D printing wax, 3D printed wax patterns are limited to very small designs and excel in jewelry and dental applications. For large applications in aerospace, energy, and similar fields, lost wax tools are required. Lost wax tooling is used when production of a metal part exceeds 100 to 200 parts. Tools involve large investments of both time and money. A typical lost wax tool can take up to six months to create. Once the tool is received from the tool maker or machine shop, it is then injected with wax. The wax is used in a slurry casting process to create the metal part. If a detail is wrong

Property	ASTM Test	DSM Somos GP Plus 14122	VisiJet SL Flex	VisiJet SL e-Stone	VisiJet SL Jewel	VisiJet SL Tough	DSM Somos NeXt	VisiJet SL Black
Tensile modulus	D 638	2510 MPa 364 ksi	1620 MPa	1630 MPa	1910 MPa	1890 MPa	2430 MPa 352 ksi	2150 MPa
Tensile strength	D 638	37 MPa 5.4 ksi	38 MPa	38 MPa	40 MPa	41 MPa	42.2 MPa 6.1 ksi	45 MPa
Elongation at break	D 638	7.5%	16%	17%	12%	18%	9%	5%
Flexural strength	D 790	67.3 MPa 9.8 ksi	57 MPa	112 MPa	61 MPa	62 MPa	69.3 MPa 10.1 ksi	76 MPa
Flexural modulus	D 790	2200 MPa 319 ksi	1420 MPa	3080 MPa	1824 MPa	1850 MPa	2470 MPa 358 ksi	2350 MPa
Izod impact strength (notched)	D 256	0.41	22 J/m	26 J/m	45 J/m	44 J/m	50 J/m 0.94 ft-lb/in.	47 J/m
Water absorption	D 570-98	0.40%					0.40%	

*Data adapted from 3D Systems and DSM Somos material datasheets, available at www.3dsystems.com and www.dsm.com. Chart data adapted from both sources to reflect mechanical property variability in popular resin brands available to the stereolithography vat photopolymerization technology. Materials are registered trademarks of their respective owners.

Table 2.11.1 Properties of Photopolymerization Resins*

or a design change is required, it takes months to receive updated patterns. Therefore, a quick and accurate method for complex low-volume parts is desired.

Vat photopolymerization investment cast patterns provide many key benefits: the ability to quickly execute a large, complex pattern for investment casting; the ability to produce one-off or low-volume production of patterns for investment casting; and the ability to customize multiple patterns with slightly varying details for investment casting (Fig. 2.11.3). Vat photopolymerization investment cast parts are most commonly used in the production of jet engine components and medical in-body hip acetabular implants. The majority of Stratasys Direct Manufacturing vat photopolymerization parts are used for investment casting, with over 10,000 patterns for knee and hip implants built over the past decade, many used in-body. Another key benefit of a vat photopolymerization pattern for investment casting over a lost wax pattern involves the composition of vat photopolymer resins. Vat photopolymer resins are thermosetting plastics; therefore, the resins are incapable of melting like wax during burnout; in many ways, these resins are vaporized during burnout which makes them ideal. Ash depends largely on foundry practices for burnout, but typically lands at similar levels as wax patterns.

While vat photopolymerization is not a functional plastic for rugged performance testing, it is frequently used in performance testing for certain low-stress applications when the engineer wishes to determine how an injection mold plastic may perform after years of wear and tear. By increasing the thickness of the vat photopolymerization

DSM Somos ProtoCast 19122 (Casting Material)	VisiJet SL Impact	VisiJet SL Clear	DSM Somos Water-Clear Ultra 10122	Somos Proto-Therm 12110	VisiJet SL HiTemp	DSM Somos NanoTool	DSM Somos PerFORM
2013–2187 MPa 292.0–317.2 ksi	2626 MPa	2560 MPa	2880 MPa 418 ksi		3390 MPa	11,000–11,400 MPa 1590–1650 ksi	10,500 MPa 1520 ksi
44.5–45.3 MPa (at break) 6.5–6.6 ksi	48 MPa	52 MPa	56 MPa (at break) 8.1 ksi	70.2 MPa 10.2 ksi	66 MPa	61.7–78 MPa 8.9–11.3 ksi	68 MPa 9.9 ksi
6%	14%	6%	7.5%	4.0%	6%	0.7–1.0%	1.1%
73–76 MPa 10.6–11.0 ksi	74 MPa	83 MPa	84 MPa 12.2 ksi	109 MPa 15.8 ksi	112 MPa	79–121 MPa 11.5–17.5 ksi	120 MPa 17.4 ksi
2135–2265 MPa 309.7–328.5 ksi	2390 MPa	2330 MPa	2490 MPa 361 ksi	3320 MPa 482 ksi	3080 MPa	10,200–10,800 MPa 1480–1570 ksi	10,000 MPa 1450 ksi
0.17–0.29 J/cm 0.32–0.54 ft-lb/in.	65 J/m	46 J/m	25 J/m 0.47 ft-lb/in.	0.12 J/cm 0.22 ft-lb/in.	26 J/m	0.12–0.15 J/cm 0.23–0.29 ft-lb/in.	17 J/m 0.32 ft-lb/in.
0.70%			1.1%	0.37%		0.23%	0.2%

FIGURE 2.11.3 By using a cellular additive process called QuickCast investment casting, patterns can be made quickly and with a minimum of resin. Rather than filling in a part to create a fully solid pattern, techniques such as QuickCast build upon a cellular-like pattern to create a mostly hollow pattern which is in some ways self-supporting through the cellular reinforcements.

part to support functional testing, engineers seek out fracture points in the vat photopolymerization design. Those fracture points are then hypothesized as weak points in the design and corrected before moving into injection molding. Therefore, vat photopolymerization provides answers to future problems that a true performance thermoplastic may experience after five to ten years' worth of wear and tear. In this way, engineers can successfully lengthen the life of their final product.

Vat photopolymerization is commonly used in the production of large models, specifically for the entertainment industry. Large hollow parts with internal scaffolding as reinforcement ideally support models while reducing weight. ID-Light, a trademark of Stratasys Direct Manufacturing, is one such technology which produces life-sized vat photopolymerization parts at 1/12th the weight of their foam counterparts.

The final most common application of vat photopolymerization parts is their use as form-and-fit prototypes or models as well as master patterns for the creation of a variety of casting processes apart from investment casting. Sand casting and urethane casting are just two examples of areas where a highly detailed and inexpensive vat photopolymerization master pattern is used. Special vat photopolymerization resins provide easily sanded surfaces which lend themselves to the urethane casting process. Urethane casting involves a silicone soft tool or mold. Liquid silicone, catalyzed by platinum is poured around a secured vat photopolymerization master pattern. Once the silicone has cured, the pattern is opened and the master pattern removed. Preparing a vat photopolymerization master pattern for use in such processes as casting—be it urethane or other—requires handiwork. Layer lines, or build lines, become starkly noticeable when transferred to the sensitive silicone mold; painted models additionally require hand sanding to remove layer lines which are noticeable once color is added. Therefore vat photopolymerization parts, as with most additive manufacturing technologies, require hand sanding or bead blasting and primer to smooth out build lines.

Starting a Project

To begin planning for a vat photopolymerization project, consider three factors: interior features, dimensional tolerance requirements, and minimum feature size requirements. Often the best use of vat photopolymerization lies not in its inherent abilities but in its partnership with traditional processes which result in the most revolutionary new products. Figure 2.11.4 illustrates the fact that many different parts can be made in the same cycle, saving hours of time required by one at a time operations.

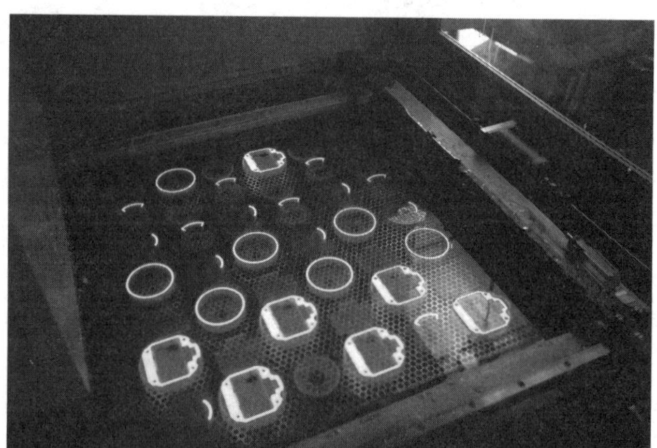

FIGURE 2.11.4 Multiple parts being made at the same time.

CHAPTER 2.12

Hybrid Additive Process

LaRoux K. Gillespie

Manufacturing Consultant
Andover, Kansas

The Process

The term *hybrid additive process* has no official definition, but for this chapter it includes any two processes employed in a single operation to produce a part that is in part made by an additive process.

As a result it encompasses:

- Two different additive processes used together to make a product
- Two variations of an additive process used together to make a product
- An additive and a subtractive process used together in one setting
- An additive and an assembly operation together in one setting

The impact of hybrid processes is that they can in many instances produce a complete part or a complete assembly in one operation. They can make the basic product with an additive process and machine it to final dimensions or other requirements on the same bed. They can produce a part requiring strong mechanical strength while incorporating colors and textures as well as electrical paths. They can impart varying mechanical properties throughout the product. And they can encapsulate an electronic or biological module inside an additive base. These approaches also allow the possibility for time-varying properties and uses.

This chapter provides design-related details for three hybrid approaches:

- Composite-based additive processes
- Additive processes that build electrical and mechanical properties in a single operation
- Additive processes that also incorporate subtractive steps

Multiple Additive Processes on a Common Platform

Additive processes allow the "printing" of structures, wires, batteries, coatings, colors, and a wide variety of materials. Makers today employ multiple nozzles or heads to add different colors, different properties, and different materials in the same operation. Properties that vary in a precise manner throughout the cross section are useful for some products. Coaxial nozzles today allow one material to be printed around another and multiple printheads or distribution ports deposit different metals and polymers routinely. While most of the current combinations involve variations of the same process, it is possible to design equipment and controls that mix processes on the same platform.

Products employing integral mechanical and electronic functions will typically employ more than one process, but not necessarily. Three common approaches are in use to add electronic paths, but other approaches integral to additive processes are also used. The three most in use or in consideration are:

1. Filament-based with inkjet-applied conductive ink
2. Filament-based with syringe-applied conductive gels
3. Filament-based with embedded copper conductors

The first is straightforward to understand. The second employs aerosol jet spraying conductive liquids loaded with particles. Typically these materials are put on the outside of a structural shape already produced (i.e., a coating). The third extrudes conductive, 3D printable inks. The common materials beyond copper include silver pastes, silver-based gels, and nanoparticle graphene substrates.

Yet another approach is to physically embed copper wires bonding them ultrasonically. The copper wires add structural integrity as well as allow higher currents to flow. In this approach a robotic arm or other mechanism can put a copper wire down and bond it as it goes.

A flexible stretchable antenna has been printed with conductive ink at Lawrence Livermore National Laboratory showing the potential for not only fixed circuits but flexible ones via additive processes. Korean researchers are also making flexible nanographene-based circuits. Other researchers are investigating 3D printing of quantum dot LEDs.

Additive Plus Subtractive

At least four manufacturers provide equipment that uses conventional additive processes and then machines away unwanted material while in the same setup. This approach provides the best of both processes and can result in more precision products as well as fine-tuning internal features. While the combination of additive and subtractive processes typically is used on metal parts, the potential is equally viable on polymer, ceramic, and composites as well. Engineers have developed cellular material not unlike balsa wood that is lightweight, yet strong because of its cellular structure and the fact that it can be bioprinted in an additive process. With a potential for turbine blades, even this biological material can be machined in a hybrid process. Add/subtract machining also has been used for wood and foam materials.

The subtractive steps can machine dimensions to more precise and more repeatable values and they can smooth the outer or inner surfaces to lower roughness.

The polishing or smoothing adds in reducing stress risers, and even surface cracks, as well as providing smoother operations in some issues and much more aesthetic appearance. Machining is generally much faster than additive processes for metals at least and the marriage of the two allows for faster total part manufacture in almost all cases. Hybrid equipment today includes 3-, 4-, and 5-axes machining centers with the additive process. They can hold the same tolerance as any machining center which includes machining 3D shapes to tolerances as fine as ±0.0003 in. for almost any material and some features can hold to even less milled variation. Grinding flat surfaces can better those tolerances.

Holes can be drilled quickly or re-drilled or reamed to provide more size control and better surface finish. The harder materials slow the cutting process, but cutting technology continues to advance production rates even for the hard metals.

While most of today's equipment uses milling machines for doing the subtracting portion of work, laser cladding is often performed on lathes. Powder metal operations could also be performed that way, but controlling the powder in the process would be more challenging. At least one manufacturer produces a machine that allows additive deposition followed immediately with machining away unwanted material. Most machines use two functions on the same bed, but this machine does both in the same cycle. That is a particularly attractive opportunity for those products that might need two different materials or two different densities on the same part.

CNC machining adds one extra issue not present in pure additive processes. It generates burrs that typically have to be removed. But that same CNC capability also provides a solution for most materials—use brushes (or nonwoven abrasive wheels) in the spindle after machining to remove the burrs. The result is a finished part coming off the machine.

The hybrid add/subtract process does not have to be limited to any one additive or subtractive process. Milling and grinding are equally possible on any of the subtractive platforms as are polishing and other finishing steps. Chips and dust are two aspects that machine makers have to address since they can become nonuniform particles in the final product.

One manufacturer produces add/subtract equipment that uses ultrasonic sound waves to weld layers of metal foil into near-net-shape parts which are finished to dimensions with CNC milling. As described in Chap. 2.10, this approach, ultrasonic sheet lamination, also called ultrasonic additive manufacturing (UAM) typically involves laminating metal strips and machining as needed during the process to generate cavities or external configuration. Thus it is inherently a hybrid process. It is used on aluminum, copper, stainless steel, and titanium to make full density metallurgical bonds. Like most other additive processes, it can make hollow, latticed, honeycombed, and micro channels. In addition, this process can weld dissimilar metals for unique properties.

The polishing of metal surface via lasers can be one means of improving the finish of layer-based metal parts, but the step heights (layer thicknesses) are generally much higher (i.e., rougher) than what laser polishing begins with (see Chap. 1.3). Despite that it is an approach that can fit in a hybrid machine.

Additive Plus Assembly Process on a Common Platform

Assembly implies several aspects that go beyond physically attaching one part to another. Additive processes can extrude metal or ceramic whiskers or strings to strengthen a product, but with whiskers that respond to magnetic or electrical fields the whiskers can be positioned in any orientation or position at will, before any subsequent step or layer occurs.

Assembly includes bonding of parts. While additive manufacturing by its nature tries to reduce part count by building many functions in a single part, because of size or materials or function some additive processes will require bonding, welding, or fusing together. Two-part adhesives can be mixed and painted onto surfaces while at least one of the products is still there. With multi-process system, or multi-headed systems, one material can be adhesively glued by finishing one part, laying down the glue, then making the mating part on top of the glue. While not ideal it still may be the least expensive and strongest approach for some materials. The PolyJet process uses cyanoacrylate glues when parts must be joined. Other glues also work with materials from that process.

Miscellaneous Adaptations

The processes described in this handbook are focused on the materials and processes used by conventional manufacturing facilities. The growth of additive processes, however, has resulted in a similar growth for biological products, both living organs and functioning tissues and structural items having a biological base but not intended to function long term (biological scaffolding and bandages for example). Biological laser printing, for example, uses a laser beam to move cells. Typically, the cells are within a micro bead or micro capsule. This process can rapidly deposit cells onto a variety of substrates. The biological adaptations include laser-based micro patterning of cells, cell assembly, and other laser-engineered micro environmental operations. Some of these same approaches are conceptually also viable for more industrial approaches.

While most additive manufacturing relies on building layer by layer, lasers are being used to build parts by writing or building 3D parts within a liquid resin (hydrogel) made of macromolecules. In other words, molecules are joined to a specially picked molecule in the hydrogel. The laser breaks bonds at the irradiated point, creating highly reactive intermediates that locally attach to the hydrogel very quickly. This point-by-point approach can build scaffolds or structures with a resolution of 4 μm (0.00016 in.). These connections can be made in 3D rather than just the 2D common today with industrial additive processes.

A variation of the 3D build process (continuous liquid interface production—called CLIP) uses tunable photochemistry instead of layer-by-layer approaches. A beam of light is projected through an oxygen-permeable window into a liquid resin. Light and oxygen team up to control the solidification of the CAD-written structure. The parts simply grow horizontally and vertically in a pool of liquid. Parts are completed in minutes, some 25 to 100 times faster than with other processes, and can have features below 25 μm (0.00098 in.). Larger features are also possible.

A new additive process using a liquid-metal alloy of gallium and indium reacts with room air oxygen to produce a "skin" that allows the liquid-metal structures to keep the shape that the printhead or process produced. These materials allow for very rapid printing of liquid-metal wires or electronic connections. The process allows both in-plane printing or vertical printing.

Aerosol jet deposition is a high-resolution layer deposition print technology for flat or 3D surfaces that allows printing features down to less than 10 μm (0.00039 in.) in size and layers ranging from tenths of nanometers to several micrometers. It involves low temperature and polymers as well as biomaterials. It is in use for printing conductive material on surfaces, even gentle curves.

Spark plasma sintering involves lower sintering temperatures and faster heating rates than other processes. It relies on mechanical pressure and electric currents to produce dense materials. The benefit is that these conditions inhibit grain growth and allow better control of phrase transformations as well as reduced energy consumption.

Marking is a process often overlooked particularly when the goal is to provide a complete part in one operation. Many products must be marked with manufacturer's code, date of manufacture, or other details. Lasers can quickly provide the necessary marking either in alphanumeric or coded form, but some plastics and some metals do not react well to the laser used by the additive process. In fact, most plastics cannot be marked well without special additives in the plastic. Laser additives enhance marking results.

Product marking can be incorporated as an as-built feature in a specific layer or it can be performed at the end of the cycle if appropriate additives or colors are in the resin or on the surface or a laser with appropriate beam conditions.

Electroforming over Stereolithography

Electroforming is an additive process based on plating onto a substrate that is discarded. While not performed in a single machine, electroforming over stereolithography mandrels is a new approach to making electroformed parts. Chapter 3.32 briefly describes this approach.

Design Considerations

Designers will need to work with manufacturing staffs to make considerations about how much additive and subtractive elements may be needed (milling away in the middle of the additive build followed by final machining, i.e., will multiple rounds of both processes be needed to complete a part).

Laser cladding, which is similar in effect to powder bed fusion, uses the clad metal to enhance wear and to repair (build layers back to original size). The resulting buildup is then machined to final part dimensions and surface requirements. That is a design approach that can be used with other additive processes as well.

Most of the additive processes allow limited overhang, because the material starts to sag if too much mass is added to the furthest portion of the overhand. That is not such a major issue for hybrid add/machine operations since extra material in one or two areas can be quickly machined away. It can be machined away for both internal and external features if properly considered.

The fact that the additive/subtractive machines support both additive and subtractive steps in any order and on any material means that the designer is free to design parts with almost any number of projections, cutouts, and geometries as can be envisioned. Some of these machines, if not all, provide programming that truly integrates the two processes in a manner that allows programmers to make the best choices and know they are optimal before producing the first piece.

Hybrid parts do not have to start with powder. They can start with a solid block of whatever materials compatible with the added material (metal or plastic). That can allow a previous part with existing functions be added to for more functions.

Designers still have to assess metallurgy requirements for the finished product. That may also require defining the starting powder or other base additive material.

The combinations of additive and other elements make them attractive for low-run production. In addition, the parts made by these hybrid processes in many instances can be the needed parts for mold patterns (interior features may be a limit however). Molded parts would be less costly for long runs.

PART 3

Manufacturing Micro Parts and Micro Features

CHAPTER 3.1

Micro Manufacturing: An Overview

LaRoux K. Gillespie

Manufacturing Consultant
Andover, Kansas

The manufacture of micro size parts, those parts or features that fall roughly at the 1-mm (0.040-in.)[1] or smaller dimensions, is a major new frontier for manufacturing. Micro manufacturing allows designers to create much lighter weight components, more efficiently fill minute volumes of space, and at least at the nano dimensions to take advantage of a world of new material properties. For some, the use of micro parts offers material cost savings. For others, micro assemblies or components face less thermal distortion and mechanical vibration just because of their minute size. For some products such as the MEMS-like processes, micro sizes mean more parts can be made on a single substrate, resulting in cutting costs by one-half or one-third.

While this section is devoted to features and parts of micro size, some of the processes literally can make dimensions less than 1 μm (39 μin.), which is 1000 nm. Today, features are being made at the 10 to 100 nm dimensions. One nanometer is one-billionth of a meter. That is a bit below what most designers will normally desire, so this section is for the manufacture of slightly larger features—think 1 to 1000 μm (0.000039 to 0.039 in.). The typical hair on your head is 76 μm (0.003 in.), so micro manufacturing includes the ability to make features or parts that are at least one-tenth and as much as 1/76 of the diameter of a single hair.

Definition of "Micro Manufacturing"

Some industrialists use the term "micro fabrication" to mean things built on silicon wafers, as if there are no other micro part types or materials or processes. There are. In fact we have identified over 50 manufacturing processes for micro parts, and those using silicon wafers for their basis only represent a handful of processes out of these.

[1]Some companies will define micro as any dimension from 10 μm to 10 mm (0.0004 to 0.400 in.). Whatever dimensions are used, most features can be defined in some scale of micrometers.

Micro manufacturing employs over 50 processes and much more when one looks at all the subprocesses that might be used to produce a part. For the most part, however, the processes described in this section are stand-alone processes, capable of making parts without the aid of other processes beyond cleaning, inspection, and packaging.

Applications of Micro Manufacturing

One of the reasons for micro manufacturing is the impact minute dimensions has on rotary speed. Large turbines, for example, fly apart at high revolutions per minute because of centrifugal force and vibration. In contrast, micro turbines can rotate at 500,000 rpm and the goal is to achieve 1,000,000 rpm. The components are so small that inertial forces have less impact if material properties are uniform throughout. The ratio of surface area to part volume can change dramatically. Micro channels provide lab-on-a-chip diagnostic capabilities for blood, serums, and other liquid analysis. These, in turn, allow micro manufacturing to provide quicker analysis of biological issues, immediate detection, and analysis.

Implants in the veins and arteries keep arteries open. Invasive medical tools must be minute enough to travel through the arteries and smaller vessels to repair the heart and leaking areas, implant sensors, and diagnose issues. The smaller the tools the more the areas that can be reached and the less stressful on the body it becomes.

Miniature probes for atomic force microscopes, electronics for micro cube satellite experiments and diagnostics, unique physics experiments, fuel injector ports, parts with impossible internal features are all common products now made by micro manufacturing processes.

Micro versus Conventional Manufacturing

The challenges of micro manufacturing are multifold:

1. To see the features clearly, high magnification is required. As an example, at least one electronic device is inspected at 400× magnification to ensure that it is burr free. High magnification requires more time to position and inspect. Working under microscopes limits active viewing of what is occurring under the scope.

2. Microscopic cutting tools deflect proportionally more than their larger counterparts because the more conventional tools can have larger shanks than small feature clearances allow. As an explanatory approach to illustrate the impact, assume that in end milling the manufacturing engineer needs to machine a 0.004-in. slot as opposed to a 0.040-in.-wide slot. The depth will remain the same and similar tools are used. Further assume that the cutting force will be based on a smaller chip size that causes a lateral cutting force of 1/10th that of the 1-mm slot. Equation (3.1.1) describes the amount of deflection at the end of the tool based on force (P), material's modulus of elasticity E, tool length (L), and polar moment of inertia (I). For this example, we leave the length of the tool the same (Fig. 3.1.1).

$$\Delta = \frac{PL^3}{3EI} \tag{3.1.1}$$

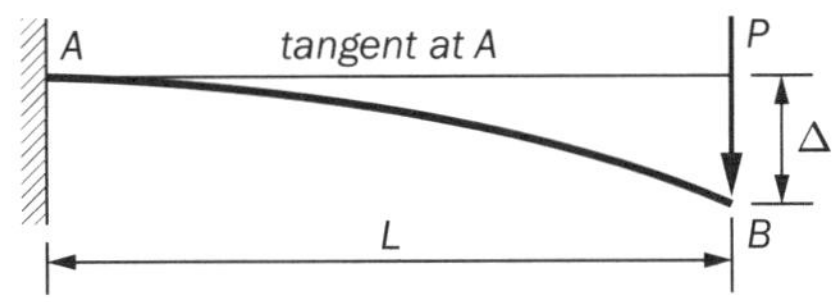

FIGURE 3.1.1 Cantilever beam deflection from a vertical load.

The moment of inertia for a solid cylinder is $\pi D^4/32$. Inserting the latter formula into the first and simplifying gives Eq. (3.1.2).

$$\Delta = \frac{(32PL^3)}{(3\pi ED^4)} = \frac{(32L^3)}{(3\pi E)} \times \frac{P}{D^4} \tag{3.1.2}$$

For this example everything but the force and diameter stayed the same. Substituting 1/10 P and 1/10 D for the small tool, the deflection to be expected for the tool under these conditions is proportional to $(0.1)/(0.1^4)$ or the deflection is 1000 times larger than that of the big tool. While the length of the tool actually used will have to be much smaller to accommodate the deflection, and somehow the cutting forces must be reduced even more, the point should be clear from this simple calculation that micro tools will bend and break much more easily than their larger counterparts.

At the minute sizes it is not just deflection that is the issue, but breakage is a great concern, because there are only a few grains of tool material making up the tool. A single minute flaw cannot withstand the cutting forces which a larger tool can with hundreds of more grains and thickness to absorb some of the stresses. Similarly, a single flaw in the workpiece which may only have 4 to 10 grains in its cross section makes it more susceptible to breaking. Broken tools and destroyed parts are much more common unless tremendous care is taken for the process.

3. The cutting edges of micro tools cannot be sharpened as sharp as needed for cutting minute bits of material. A lathe tool for example may have a 50 µm (0.002 in.) radius on the cutting edge while it removes a chip that is 0.25 mm (0.010 in.) thick. A tool 10 times smaller can only have a cutting edge radius of one-half the larger one and that causes rubbing more than cutting when taking a very thin chip. Large chips cause large cutting forces which break the tool. There are a few ways to sharpen a truly micro metal cutter, and none do it as sharp as truly needed for fine cutting. In addition looking down a cutting edge at high magnification the observer will see a number of small serrations along the edge, because the sharpening process cannot make a perfect edge at that scale. Making micro tools results in more relative run out of the tool from perfect symmetry and that adds to tools' stresses and imperfect cuts.
4. The burrs produced by micro cutting require removal without destroying other features and tolerances, and yet there are a few tools or processes to remove these irritations without affecting other aspects of the parts. The ideal case

would be to use small ants to saw off or chew off burrs. That is the relative size of tools required for minimal part impact when burrs are present.

5. Material properties are generally not as uniform for workpieces that are only a few grains thick. It is not just grain size variations, but hardness, small alloy distribution changes, and differences in strain hardening before and after cutting. Normal size stock has much more ability to average out localized differences so they are not detected or at least are not a major change in properties affecting cutting. In a similar vein, the crystal orientation of some materials greatly affects the machining properties. While that can be true for more conventional cutting, it can make a major difference in the micro world.

   Material forming faces many of the same issues as micro cutting. Grains and metallurgy factors do not scale down in size like the parts do. All of the assumed conventional force and movement models fail to account for the extra forces required in micro part manufacture. While the basic issues are the same, the mechanics and formulas behind them are different in scale.

6. Surface effects which are normally inconsequential at larger sizes become stumbling blocks to micro manufacturing. Electrostatics, van der Waals forces, and surface tension hold on to parts so distinct forces are required just to pick up and move micro parts.

7. Micro parts have both micro size dimensions and micro tolerances. The tooling used with them often also has micro tolerances. For accurate holding of parts, small cavities having tolerances of only ± 2.5 µm (± 0.0001 in.) are used to capture the parts. Removing micro parts from tight fitting cavities requires defeating all the forces just mentioned plus normal friction forces when rubbing against walls. Micro molded parts have to be ejected from the mold while still warm without leaving undesirable ejection marks. Tight parts in tight spaces are a challenge requiring great insight and innovation to overcome.

8. Micro molded parts require the liquid material to flow through micro channels and to do so without freezing to walls before the shot is fully in the mold. Driving liquids through micro channels is like pushing a tennis ball with the air blown through a straw. Capillary actions and related fluid flow restraints require careful control to accomplish that. Smaller channels require much more analysis and consideration than conventional feature sizes. High molding fluid pressures limit wall thickness of the resulting parts, and it limits the thickness that the molds can have without distortion. Distortion of microscopic features as they cool is a bigger problem than conventional molding. Again temperature control, shot size, shot pressures, and material uniformity become much larger issues.

9. Handling and assembly are more challenging because there are much finer fits between parts, because the surface forces, van der Waals forces, and electrostatics inhibit assembly—they fail to release the parts when tweezers or other tools let go. Assembly needs to be done under high magnification and the vision systems often interfere with the tools used to assemble.

Micro tools indent micro parts. Minute particles are released when parts rub against each other and those particles can be damaging to performance and life.

Micro Machines and Machines for Micro Work

There are some futurists who believe that micro parts should be better made on micro machines. A micro lathe has been built that fits within a 1-in.[2] cube. A micro milling machine and a micro forming machine as well as a micro assembly robot have been built and each fits within a 1-in. cube. That is not to say they are current production machines, but if machines can be made at that scale then entire factories could be transported to space, placed in environments too small to allow normal machines. Micro machines use roughly 1000 times less energy than their larger versions and some produce 1000 times less pollution than their big brothers.

The processes discussed in the micro section with the exception of one or two have commercial machines that support micro manufacturing. These one or two are proven processes, but no manufacturer has taken on the challenge to successfully market equipment (i.e., the process is proven and designers can count on it, but the simple machine required would have to be built for that application).

Processes

The 52 processes described in this section are unique, in that they concentrate only on the micro aspects of the process. None of the more conventional size aspects are discussed here.

Micro biological processes are progressing at the same rate as the micro fabrication processes, but we only concentrate on those processes which have some industrial relationship to metal, ceramic, glass, and polymer products. Many of the micro products are used in the biological realm, but this handbook is solely devoted to the more industrial applications of the processes—think no for biological processes, but yes to tools for biological processes.

Materials

Every chapter describes a process and the common materials that the process can fabricate. While a section on new materials is warranted, it is beyond the scope of this handbook. There are so many new and enhanced materials that there is simply not enough room to fit into this handbook.

At the same time readers will see processes that are more tuned to foils rather than flat stock in some processes. Some processes will machine the "unmachinable" materials such as glass and hard ceramics. Thin films are a part of designing for micro manufacturing. Shape memory alloys are used in micro assemblies for several reasons, and a host of new polymers are described in various chapters. The designer will see what materials are appropriate for each process.

Research

Every process described here has an industrial application. At the same time extensive research is being performed to make the processes more minute, lighter, stronger, more precise, faster, and less expensive. There will be advances and possibly even more advanced processes in the future. We are in an exciting age for each of the processes

[2]16,387 mm^3.

described in this handbook. We are beyond the forefront of the technology. We are in a semi-mature age of the technology for each of these. There are more advances to be made, but these are all practical processes that are being used industrially.

One aspect is that we are truly at the forefront in micro manufacturing. We still have a lot to learn about how we should be designing to take most advantage of these process capabilities. Many of the processes are straightforward, but origami for instance requires its own handbook of how to fold and make hinges to take advantage of all the self-release possibilities, 2D to 3D unfolding to make strong light-weight products. These design manuals have been recently developed for specific applications. Similarly there are only a few electronic novel approaches that can take advantage of the more mechanical or structural processes. More will be unveiled in the next few years to match the more mature processes.

So once again the message to all designers is that each of these chapters provides the elements they need for designing to know that the future can be produced by one or several of these processes. Design considerations are listed. The materials applicable are described. The limitations are generally defined.

Seeing and Measuring

Some form of microscopic view is required for all micro manufacturing. Scanning electron microscopes (SEM) are used in some of the more advanced facilities and their 10,000× images of features provide tremendous insight into the quality of the smallest features and parts. Some facilities need to visually see what is going on while the process is under way and CCD cameras displaying on big screens are needed. Measuring dimensions can be the biggest challenge for some parts, particularly those having hidden recesses or very soft pliable surfaces that will not stay flat. Snow white parts under scopes cause snow-blindness-like eye fatigue and shiny surface reflect the needed light into glare on some parts.

Testing and Acceptance

One chapter points out that for many of these very thin section parts and processes it will be necessary to establish new testing standards. The features require such low test forces that existing equipment may not be suitable. For thin webs or meshes of materials, none of the standard formulas and published data may be adequate. If there are only four grains of material in the cross section, would a designer believe standard published data for large parts was representative? Not likely, so the designer needs to specify what the requirements are for product testing to assure conformance to intent. Similarly, with the exception of the MEMS devices, there appears to be little data on the life cycle performance of micro parts and features.

People

Designers are not responsible for the people who work on micro products. Yet, it is significant to note that every human interaction on micro parts is more challenging than on more conventional size products. Some designs can assist that human interaction interface by making it easier through design—easier to pick up, inspect, mark, measure, make, etc. All of that reduces costs and at the same time it reduces ergonomic issues.

Facilities

Most micro manufacturing facilities do not differ in appearance much from that of other shops. Some use clean rooms or semi-clean rooms. All have cleaning facilities, desks, tools, machines, and material movement. Some use dark rooms to reduce eye fatigue from lights and scopes. All have relatively clean work areas. If you drop some of these parts they are either destroyed, contaminated, or impossible to see on the floor and under desks. Some have chemicals or solutions that are harmful to the product. When a designer has the opportunity, touring some potential fabrication probable facilities may reveal some precautions that belong on the drawing, precautions that disallow contact with undesirable elements.

Services

The processes described in this handbook are available for almost any company. They can purchase the equipment or they can subcontract that portion of work. While not widely publicized, there are service providers for almost every process described here. Equipment providers can provide this information and an in-depth search on the Internet will also, but it is not always clear which is the best choice facility without visiting.

Software

The one aspect perhaps most lacking in micro manufacturing is software that accurately predicts effects that can be used in conjunction with normal design software. Numerous academic software packages have been developed, but are not commercially available. It requires extensive effort to find them when the designer knows what he/she wants. But, in general, the software is missing along with critical aspects of the materials' data for micro features. Data exists for the raw part material, but not for the effect that the thin walls, new processes, and time has on your part design.

Design

This section provides a wealth of information about over 50 manufacturing processes used for micro parts and micro features. Without an understanding of the basic process capabilities and limitations, it is not possible to design features that can be made economically. While each of the issues presented above limit normal manufacturing at these very small scales, there are several processes with advanced capabilities for the micro parts. While old ways may not be as capable as the new ways are, the new processes and innovations described in this handbook overcome many of the former restraints. Some of the restraining forces can be used to facilitate assembly and handling and even tooling approaches. But it does take a different mind-set than "the way we have always done it, the way we have always designed it."

As the parts get smaller, designers may have to design thicker walls to provide adequate strength; they may have to change the materials to a higher strength or more durable material. Ceramics may be desirable rather than metals. Conductive ceramics may answer some needs. More than anything, however, many of the micro assemblies will need to incorporate a different design paradigm. Compression may be required rather than tension. Meshes may provide better properties than solid material. Parts may be designed to deflect deterministically rather than designing to not deflect. Innovation is the rule. Processes proven for specific designs are essential.

All the design for assembly issues remain the same for micro assembly, but the extra restraints require careful innovation to make the best use of assembly. This is one aspect in which the designer can make a huge impact on cost and consistency—design for micro assembly.

CHAPTER 3.2

Micro Mechanical Drilling

Jack Burley

KAISER Precision Tooling, Inc.
Hoffman Estates, Illinois

Introduction

Micro machining, by definition, is cutting where the volume of chips produced with each tool and tool path is very small. The workpiece may or may not be small in size, meaning that the physical proportions of a larger part may have details that require ultrasmall profiles achieved only by micro machining. This means that micro machining is not limited in scope to only miniature parts (Fig. 3.2.1). Micro holes are common in various industries including automotive, aerospace, printed circuit boards, semiconductor, electronics, medical device manufacturing, and more mundane products such as filters. Requirements for improved fuel economy principally found in fuel injectors and fuel nozzles have highlighted the need for higher accuracy holes that are smaller in diameter as well as have a specific shape and surface finish. Mechanical drilling of these components yields the best accuracy and with the right process of cutting tools and machine tool, the highest productivity.

Defining the Limits

Micro drilling can produce holes as small as 0.03 mm (0.0012 in.). "Micro" normally is considered as large as 2 mm (0.078 in.) for drilling purposes. While drills have been made and used as small as 12.7 μm (0.0005 in.), they are not widely used because of breakage and cost. The drills are produced to exceptionally close tolerances and repeatability. Most are made from ultrafine grain carbide for drilling metals, but high-speed steel is also used for softer materials. Geometrically, micro drills have the same point angle, margins, and helix angle as normal larger drills. Materials of all types can be drilled with micro tools—aluminum, brass, stainless steel, titanium, precious metals (gold and silver, for example), plastics of all compositions including Delrin,

Figure 3.2.1 This titanium part used in the medical industry has 400 holes, with diameter of 0.15 mm (0.006 in.). The hole is drilled to a depth of 0.80 mm (5× diameter) and takes approximately 3 s to complete each hole. (Courtesy Sphinx Werkzeuge AG.)

nylon, composites, and even challenging polyether ether ketone (PEEK). The first applications for micro drills were for brass watch parts; printed circuits became the next large user of these tools.

Drilling by definition requires a point geometry that is capable of producing a hole from solid by a complex action of cutting and chiseling of the material, and micro tools face the same challenges of physical forces. Micro drills, however, are exceptionally delicate and must be used very carefully. The equipment used must be sensitive enough to make minute precision motions smoothly. Operators must be trained in handling delicate tools, observing microscopic wear patterns, watching chips come off the drills (often at 30× magnification, i.e., while looking through a microscope), and listening to minute variations in cutting sounds. While most micro drilling is performed with computer numerically controlled machines, for the very small drills operators have to pay close attention to the action. For drills at the high end of the sizes mentioned, such close observation is not normally required.

Micro holes can be drilled to a diameter repeatability as fine as ±2.54 μm (±0.0001 in.), to depths 30 times the drill diameter (even longer with micro gun drills) and positional accuracies of ±12.7 μm (0.0005 in.) or better. The level of accuracy depends mostly on the depth that the hole must be drilled to and how the drill starts to make the hole. Micro drills have a very weak resistance to bending. This means that if the tool must enter the workpiece without a properly located pilot hole to guide from, the drilled hole will not be the right position due to "walking" of the drill. Drills will always follow the path of least resistance, so if the workpiece material is not pure (i.e., has grain sizes that vary significantly or has varying hardness throughout), or if the hole is located nearby another feature, the drill will try to move, causing the hole to be out of true alignment at the bottom. This is vividly shown with some long high-speed steel drills which can actually exit the part at almost 90 degrees from the direction they start. Carbide tools will not have such bending abilities; they will break before such excessive deviation.

Characteristics of Good Micro Drills

Micro drills have very tight specifications in order to perform effectively. The grade of carbide must be an ultrafine grain with nano or submicrometer grain size to assure high abrasion resistance and good toughness. The drills themselves can be quite long due to the drilling depth and for this reason, the carbide grade and grain structure must have very high yield strength. The concentricity between the shank, drill point, and margins of the tool must be made to nearly perfect tolerances [often within 2.5 μm (0.0001 in.)]. The point geometry must have very sharp cutting edges with primary and secondary relief margins for positive cutting and edge life. Drills have the unique task of not only cutting away the material, but also carrying the chips through the flutes of the tool and out of the hole. To do this, the drill must have excellent surface finish in the flutes, often 0.2 μm (8 μin.), and a helix angle designed for the specific workpiece material. The addition of tool coatings is highly recommended for applications that require drilling abrasive and heat-resistant materials such as titanium. These coatings, usually less than 2 μm in thickness to keep cutting edge sharpness, vary in chemical compositions and application techniques depending on the targeted usage. The primary advantage of the coating is to increase tool life, but it also aids in chip evacuation by producing a lubricating effect against the friction that results from hot chips sliding over the surface of the tool.

There are literally hundreds of combinations of drill sizes, shanks, designs, and coatings available for micro drilling. Figure 3.2.2 shows one common design for very small holes. In addition to twist drills, spade drills having a flat side with no spiral are used in these sizes, micro gun drills that are 0.5 mm (0.0197 in.) in diameter or bigger are used for very long holes—up to 50 mm (2 in.), and diamond-coated drills for ceramics and glass are made as small as 25 μm (0.001 in.). The diamond-coated drills grind their way through the material rather than shear as a drill does.

Starting the Hole

Micro drills require a starting hole, especially for deep hole drilling. The process serves two important functions:

1. It provides a precise position for the hole to be produced.
2. The small pilot hole allows the center point of the drill to bite into the metal easier.

Spot drills and center drills are short, stubby, and stiff drills; this stiffness minimizes drill deflection and provides an accurate starting position for the micro drill. The point

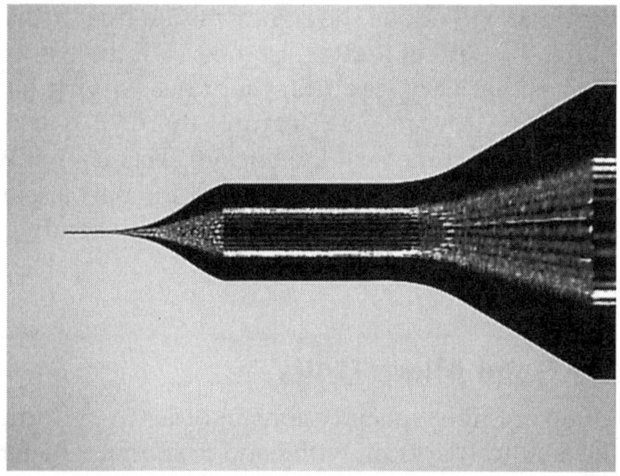

FIGURE 3.2.2 Typical geometry of micro twist drills. A 0.03-mm-diameter (0.012-in.) micro drill. The shank diameter on this drill is 3 mm (0.0125 in.); it then necks down to 1 mm (0.040 in.) and then down to the 0.03 mm drill. The cutting portion is 0.67 mm long (0.026 in.), while the overall drill tool is 38 mm (1.5 in.) long. (Courtesy Sphinx Werkzeuge AG.)

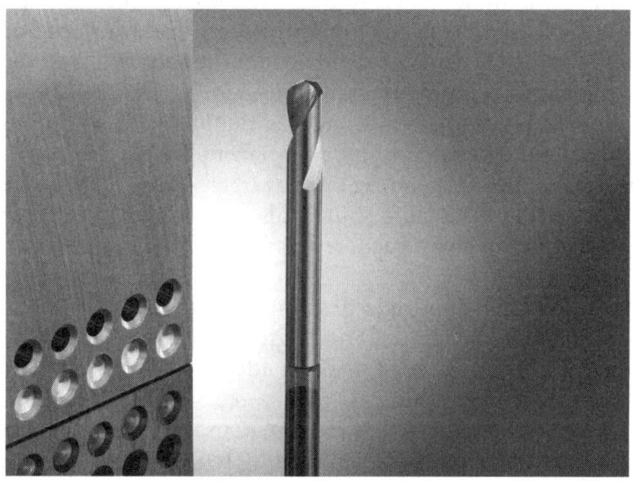

FIGURE 3.2.3 Spot drilling provides a chamfer which guides the micro drill. All these holes have been spot-drilled but only half have a subsequent micro drilled hole in them. (Courtesy Sphinx Werkzeuge AG.)

angle on the starting drill is the same as the point angle on the micro drill. Once this shallow start has been put on the workpiece, the micro drill is guided into the cone-shaped entrance by the sides of the micro drill (Fig. 3.2.3). For deep holes, the spot drill often is slightly larger than the micro drill and the micro drill slides down into this starting hole before it begins cutting (Fig. 3.2.4). The micro drill enters the hole at slow revolutions per minute (RPM) and then the speed is safely increased to operational speeds. The top of the hole down to one or two diameters deep then is about 5 μm (0.0002 in.)

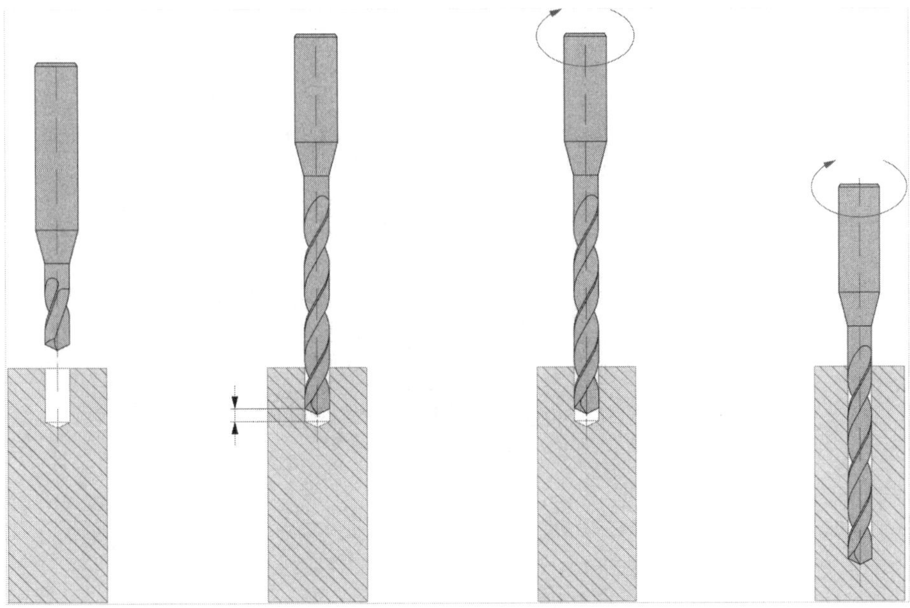

FIGURE 3.2.4 Spot-drilled hole is deep enough to guide the diameter of the micro drill safely and accurately down to long depths. (Courtesy Sphinx Werkzeuge AG.)

larger than the lower part of the hole that the micro drill made. For a 127-μm-diameter (0.005-in.) drill, this implies the top 127 to 254 μm (0.005 to 0.010 in.) depth would be as much as 5 μm (0.0002 in.) larger than the lower parts of the hole. This spot-drilling operation becomes very important when drilling into materials that are not pre-finished, such as hot-rolled steel or forgings with rough, uneven surface texture. A wisely chosen center-drilling or spot-drilling operation can also chamfer the micro hole entrance, preventing burr formation which would otherwise require a subsequent deburring operation (Fig. 3.2.5).

Operating Parameters

Spindle speed and feed rate through the material are critical. The spindle speed in RPM that micro drills must perform best at consists of two major factors:

1. The workpiece material
2. The diameter of the hole to be drilled

Table 3.2.1 defines the cutting velocity and federate normally recommended to machine some common materials. The velocity shown is the surface feet per minute (SFPM) the outer edge of the drill is traveling. Table 3.2.2 converts the SFPM values to the spindle RPM required based on the drill diameter. As seen there, drilling a 127-μm-diameter (0.005-in.) hole at 30.5 m/min (100 SFPM) requires the spindle to turn at 76,400 RPM. This is far above the conventional drilling equipment speed.

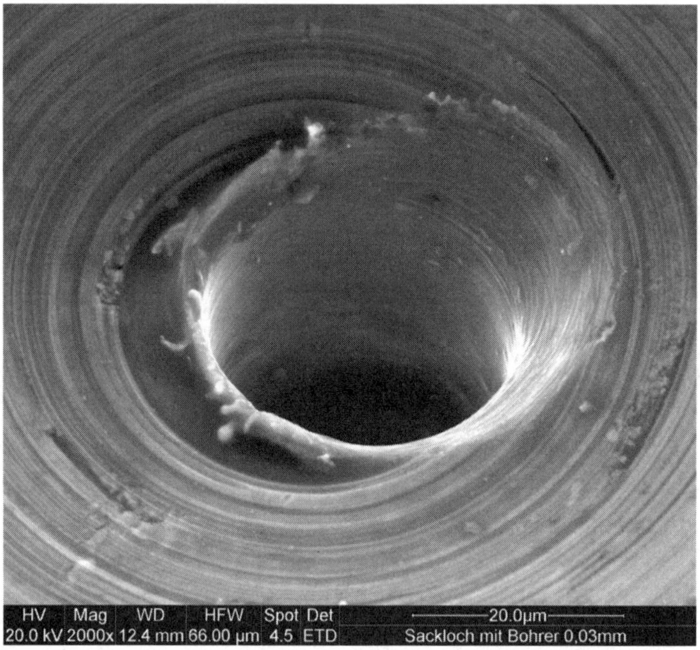

FIGURE 3.2.5 Magnification at 2000× of a 30 μm (0.0012 in.) hole drilled with a carbide drill in aluminum into the bottom of a blind counterbore. This hole is produced without spot drilling and its location is offset from where it should be as can be seen by the lack of concentricity to the surrounding circular marks. A large burr also occurred because the hole was not countersunk before micro drilling. Removing burrs from micro holes can be an expensive and slow task. (Courtesy Sphinx Werkzeuge AG.)

The basic formula to calculate speed as a function of RPM is:

$$\text{RPM} = (\text{SFPM}) \times 97/(\text{drill diameter}) \text{ (when drill diameter is in mm)} \quad (3.2.1)$$

$$\text{RPM} = (\text{SFPM}) \times 3.82/(\text{drill diameter}) \text{ (when drill diameter is in inches)} \quad (3.2.2)$$

Cutting tool manufacturers list the recommended speeds in SFPM for their tools based on experience and how the tools are designed for each specific work material. As the diameter becomes smaller, the relationship to maintain surface footage requires the RPM to increase. Most machine tool spindles are not capable of maintaining the required speeds for tools with a diameter below 0.75 mm (0.030 in.) at 300 SFM. In these applications, electric or air spindles can be attached to conventional machines to drill the holes (Fig. 3.2.6). Micro drilling machines, however, are available that are designed to specifically cut at these high speeds.

Feed rates for micro drills are always very light, in the order of 1.25 to 2.5 μm (0.00005 to 0.0001 in.) per cutting edge which for drilling is 2.5 to 5.0 μm/rev (0.0001 to 0.0002 in./rev). By multiplying the feed rate per revolution by the calculated RPM, the actual penetration or drilling rate can be defined.

For example, a drill with diameter 152 μm (0.006 in.) in titanium could run at 10 SFPM and 2.54 μm/rev (0.0001 in./rev) feed. The feed rate into the workpiece then is

Material	Drill Diameter (mm)	Carbide Drills		HSS Drills	
		Speed (SFPM)	Feed (IPR)	Speed (SFPM)	Feed (IPR)
Mild, low-carbon steel (1018, 1020, A36)	0.05–0.30	5–20	0.00005–0.00010	2.5–5.0	0.00005–0.00010
	0.31–0.50	5–35	0.00010–0.00015	5–10	0.00010–0.00015
High-carbon alloy steels (4140, 4340, 8620)	0.05–0.30	5–15	0.00005–0.00010	2.5–5.0	0.00005–0.00010
	0.31–0.50	5–25	0.00010–0.00015	5–10	0.00010–0.00015
Stainless steel, austenitic (300 series, 17-4ph)	0.05–0.30	5–10	0.00005–0.00010	2.5–5.0	0.00005–0.00010
	0.31–0.50	5–15	0.00010–0.00015	5–10	0.00010–0.00015
Stainless steel, martensitic (400 series)	0.05–0.30	5–10	0.00005–0.00010	2.5–5.0	0.00005–0.00010
	0.31–0.50	5–15	0.00010–0.00015	5–10	0.00010–0.00015
Gray cast iron	0.05–0.30	5–25	0.00005–0.00015	5–15	0.00005–0.00020
	0.31–0.50	20–50	0.00015–0.00030	5–20	0.00020–0.00030
Ductile iron	0.05–0.30	5–20	0.00005–0.00010	2.5–5.0	0.00005–0.00010
	0.31–0.50	5–35	0.00010–0.00015	5–10	0.00010–0.00015
Aluminum	0.05–0.30	10–50	0.00005–0.00015	5–25	0.00010–0.00015
	0.31–0.50	25–85	0.00015–0.00030	15–45	0.00015–0.00030
High-temp alloys (titanium, Inconel, Monel)	0.05–0.30	5–20	0.00005–0.00010	2.5–5.0	0.00005–0.00010
	0.31–0.50	5–25	0.00010–0.00025	5–10	0.00010–0.00015
Brass	0.05–0.30			5–15	0.00005–0.00010
	0.31–0.50			15–35	0.00010–0.00015

*These are starting recommendations, but actual cutting conditions depend on hole depth, workpiece homogeneity, drill geometry used, drill micro grain consistency, and drilling equipment. Many polymer materials can be cut at velocities and feeds 5 to 10 times those shown for metals.

Table 3.2.1 Recommended Cutting Conditions for Conventional Micro Drills (Courtesy Sphinx Werkzeuge AG)*

Drill diameter	SFPM							
	10	20	30	40	50	100	200	300
0.125 mm (0.005 in.)	7640	15,280	22,920	30,560	38,200	76,400		
0.25 mm (0.010 in.)	3820	7640	11,460	15,280	19,100	38,200	76,400	
0.375 mm (0.015 in.)	2547	5093	7640	10,187	12,733	25,467	50,933	76,400
0.50 mm (0.020 in.)	1910	3820	5730	7640	9550	19,100	38,200	57,300
0.75 mm (0.030 in.)	1274	2547	3820	5094	6367	12,734	25,467	38,200
1 mm (0.040 in.)	955	1910	2865	3820	4775	9550	19,100	28,650

Table 3.2.2 Spindle Speed (in RPM) for Various Surface Feet per Minute and Drill Diameters

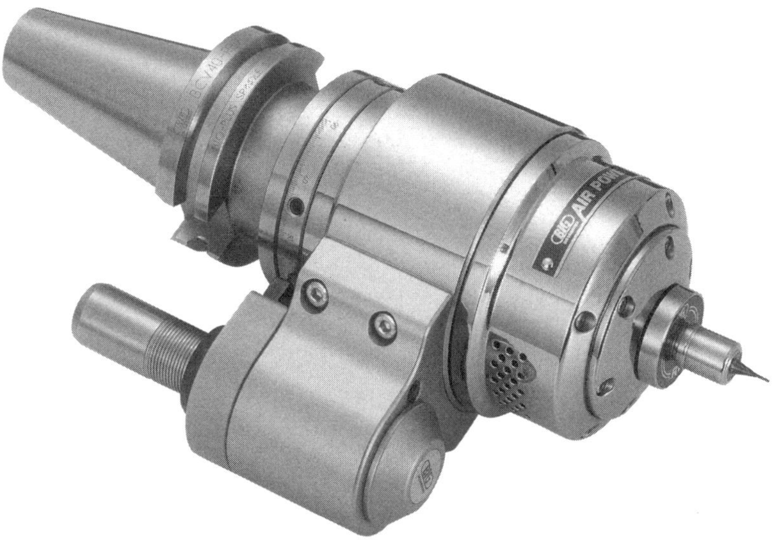

FIGURE 3.2.6 Air-driven high-speed spindle that fits in a standard machine tool machine spindle. Very low run-out is maintained even at spindle speeds up to 120,000 RPM with a device like this. (Courtesy BIG Daishowa Seiki, Ltd.)

calculated by using Eq. (3.2.1) or (3.2.2) and then multiplying that by the feed rate per revolution just mentioned, giving

$$\text{Feed rate} = \text{RPM} \times \text{IPR} \tag{3.2.3}$$

Resulting in a feed rate of 6366 × 0.0001 = 0.637 in./min (16.2 mm/min)

If the hole depth is 0.75 mm (0.030 in.), the drilling time is about 3 s per hole, not including spot drilling and any pecking cycle of the drill to remove chips. Deep holes will normally require the drill to be retracted several times to let chips be thrown out of the flutes. The drill will break unless chips are cleared from deep holes.

Machine Tool Requirements

Micro machining requires that the machine tools have very good sensitivity and fine resolution in the feed axis. Machines with very precise spindles capable of high speed and low dynamic run-out accuracy (i.e., top of spindle and bottom of spindle must rotate with no measureable offset, lack of concentricity) are essential to productivity. The spindle must also be in alignment with the direction of feed, and free of spindle or machine vibrations. In short the smaller the drill and the longer it is, the more sensitive and free of all vibrations and abrupt actions the machine must be.

While spindles are available that turn at 400,000 RPM, many are not designed for the delicate motions that micro drilling requires. In addition, the drills themselves are held in the spindles by precision collets or other tool holders. These must be accurate and each type of holder has its own limits.

CHAPTER 3.3

Micro Milling

LaRoux K. Gillespie

Manufacturing Consultant
Andover, Kansas

The manufacture of numerous micro parts requiring small channels and sculpturing is traditionally done by a milling machine. The process is well known, and can accommodate hundreds or thousands of parts or simply ones and twos. It is the way parts were traditionally made. Parts are still made that way and many micro features can be made that way, but it is a larger challenge than milling more conventional size features.

While there is no official definition of what size micro milling refers to, clearly any feature that requires a tool of less than 0.5 mm (0.020 in.) diameter falls within the micro milling field. For some companies anything under 1 mm (0.040 in.) is micro milling and for others it is anything under 3 mm (0.118 in.).

When designs need narrow slots, minute sculpturing, several alternatives exist:

- Micro laser "milling"
- Micro electrical discharge machining (plunge or WEDM)
- Micro electrical chemical machining
- Micro additive manufacture
- Ion beam machining
- Micro water jet machining
- Photochemical etching
- Micro injection molding
- Micro hot embossing
- Roll-to-roll embossing
- Micro electroforming
- LIGA

Chapters on each of these processes are found in this handbook, which describe their capabilities. Some are only useful for features that are a few micrometers deep or wide. Others apply to a very limited selection of materials (only polymers, for example). Some are very slow. Some require very expensive equipment and must be subcontracted out. For most applications, micro milling is the most straightforward and most economical process for small quantities of parts (hundreds, for example). Several manufacturers make machines specifically for micro machining. In a few minutes a feature can be made with little advance preparation, so micro milling is convenient as well as straightforward.

When features become too challenging for milling, micro EDM is the next best choice with respect to cost, ease of use, and ability to meet many differing design needs. When milling outer configurations, laser cutting for micro features can be an excellent choice as well since it is faster than milling and EDM.

Basic Limitations

Because there is a limit to how sharp micro cutters can be made, there is a limit to the size of chip that can be cut. Cutting forces are much higher in comparison to conventional size tools and because the tools are so tiny they break easily. Tool breakage is the biggest problem in micro machining.

Because the cutters are so small the machinist cannot tell how well the tool is performing. Hard spots in material cause immediate tool breakage; chips clogging flutes increase the forces and break the tool.

Materials Machined

Micro milling cutters can machine metals, polymers, and ceramics, but steels harder than about 50 Rockwell C (HRC) should be machined with micro EDM. Micro cutters break too easily with hard steels. For best machining it is critical that the materials being machined have a uniform grain structure and not have hard spots in them. Suddenly running into hard spots breaks the cutters. The process has successfully been used on mild steels, stainless steels, titanium, aluminum, brass, copper, and many polymers. In general, polymers are the best choice since cutting forces are much lower for them and tool breakage and deflection are thus less of a problem.

One recommended cutting condition for machining steels having a hardness of 40 to 52 HRC using a 0.3-mm-diameter (0.012-in.) tool is to cut with a spindle speed of 40,000 rpm, feeding at 2.5 μm (98 μin.) per tooth [200 mm/min (7.9 in./min) for a 2 flute tool], but cut only 5 μm (0.0002 in.) deep and only 3 μm (0.00012 in.) radially.

Glass (crown glass and soda glass) can be machined with micro ball end mill cutters being fed at rates roughly 0.1 to 1.0 mm/min (0.004 to 0.040 in.), but there are no known production applications for this and few individuals have done it (Foy, 2008). Graphite is yet another material successfully milled.

Grain size plays a role in successful end milling. Reducing grain size in aluminum can improve surface roughness from 0.5 to almost 0.1 μm (20 to 4 μin.) Ra. Each material appears to experience different affects from changing grain size.

Material smearing can be a problem with some materials and cutting conditions. In almost all materials, burrs are produced which can be difficult to remove without affecting part properties on complex geometries.

Studies of tool life when milling hardened steel (55 to 62 HRC) showed they varied from 0.5 to almost 5 min depending on feeds and speeds used. Because tools wear, the optimum cutting conditions in this material allowed only 20 mm/min (0.79 in./min) feed, a feed rate of 1 μm/tooth (40 μin./tooth) and 15 μm (0.0006 in.) depth of cut with a 0.5-mm-diameter (0.20-in.) tungsten carbide tool. End of tool life was defined as a reduction on tool diameter of 30 μm (0.0012 in.). Total tool breakage did not occur for some 2000 to 3000 mm (79 to 118 in.) of cut for some conditions, but effective tool life is a function of breakage, tooth chipping, cutting edge rounding, burr size, wear of end mill corners (which results in narrower channel thickness at the bottom of channels), dimensional cutter change (implying the part dimensions are changing also with time), surface finish, and workpiece smearing. The issue here is that many factors can limit tool life to much less than that where the tool breaks. In this instance life ranged only from 10 to 100 mm (0.039 to 0.390 in.), a difference of 30 to 200 to one less than breakage life!

Brass can be machined at 100 mm/min (3.9 in./min) with excellent tool life. Most plastics can also be cut at rates equivalent to brass or better. Cutting feeds are a function of tool diameter, tool material, part material, and cutting speeds and depths.

Cutters

Table 3.3.1 provides a representative overview of the smallest commercial end mill cutters. They come in several shapes even in those minute sizes (Fig. 3.3.1). The ones shown have a square bottom end, but similar ones having a small radius are available as are ball end shapes. Smaller tools have been made for research applications via ion beam machining. That approach is almost required once a cutting diameter below 25 μm (0.001 in.) is required (Gietzelt, 2012). The smallest known end mill made by a commercial tool maker is 5 μm (0.0002 in.).

As seen in the tabled data, the typical depth to diameter ratio for these cutters is either 1.5 or 3.0. This greatly limits using them for cutting deep slots. As the diameters get larger, the ability to cut deeper also increases. The advantage that end mill style tools have is that they can profile shapes that other cutters cannot.

Slotting cutters, similar to ultrathin table saw blades, have been made down to 15 μm (0.0006 in.) wide and the potential is that they can be made thinner. They have been able to cut trenches or slots 15 μm (0.0006 in.) wide by 150 μm (0.0060 in.) deep in PMMA, an aspect ratio of ten (Gietzelt, 2012).

Micro engraving cutters can produce v-shaped grooves and engraving is a milling process, not unlike end milling except it does not produce rectangular shape slots. The cutter geometries are also different.

Micro milling has been performed on diamond turning lathes which make high-precision mirrors and lenses. That application generally uses single point diamonds mounted on the bottom of a shank. One manufacturer makes flat bottom, tapered, radial, and other special diamond points to do this cutting. These applications, in general, while described as micro milling, are for larger features than those discussed in this chapter and it is a very specialized application (Scheiding, 2009).

Commercial cutters may not all arrive for use with the same quality. In one study (Aramcharoen, 2009) roughly "60% of micro tools from any given batch were unacceptable due to chipping, cracking, incorrect geometry (rake angle, etc.) or having a larger than specified cutting edge radius." This will influence a manufacturer's ability to produce the desired part quality. Another study found "deviation from the nominal tool

Cutter Diameter	LOC		Shank Diameter	OAL	Uncoated		AlTiN Coated		Amorphous Diamond	
D_1+0.0005 in./ −0.0005 in.	L_2+0.010 in./ −0.000 in.		D_2 (in.)	L_1 (in.)	2FL	4FL	2FL	4FL	2FL	4FL
0.001	**0.001**	1.5x	1/8	1-1/2	Yes					
0.001	**0.003**	3x	1/8	1-1/2	Yes					
0.002	**0.003**	1.5x	1/8	1-1/2	Yes					
0.002	**0.006**	3x	1/8	1-1/2	Yes					
0.003	**0.004**	1.5x	1/8	1-1/2	Yes					
0.003	**0.009**	3x	1/8	1-1/2	Yes					
0.004 (0.1 mm)	**0.006**	1.5x	1/8	1-1/2	Yes					
0.004 (0.1 mm)	**0.012**	3x	1/8	1-1/2	Yes					
0.005	**0.007**	1.5x	1/8	1-1/2	Yes	Yes				
0.005	**0.015**	3x	1/8	1-1/2	Yes	Yes				
0.006	**0.009**	1.5x	1/8	1-1/2	Yes	Yes				
0.006	**0.018**	3x	1/8	1-1/2	Yes	Yes				
0.007	**0.010**	1.5x	1/8	1-1/2	Yes	Yes				
0.007	**0.021**	3x	1/8	1-1/2	Yes	Yes				
0.008 (0.2 mm)	**0.012**	1.5x	1/8	1-1/2	Yes	Yes		Yes		
0.008 (0.2 mm)	**0.024**	3x	1/8	1-1/2	Yes	Yes		Yes		
0.009	**0.013**	1.5x	1/8	1-1/2	Yes	Yes		Yes		
0.009	**0.027**	3x	1/8	1-1/2	Yes	Yes		Yes		
0.010	**0.015**	1.5x	1/8	1-1/2	Yes	Yes	Yes	Yes		Yes
0.010	**0.030**	3x	1/8	1-1/2	Yes	Yes	Yes	Yes	Yes	Yes
0.011	**0.016**	1.5x	1/8	1-1/2	Yes	Yes		Yes		
0.011	**0.033**	3x	1/8	1-1/2	Yes	Yes		Yes		Yes

0.012 (0.3 mm)	**0.018**	1.5x	1/8	1-1/2	Yes	Yes		Yes		
0.012 (0.3 mm)	**0.036**	3x	1/8	1-1/2	Yes	Yes	Yes	Yes	Yes	Yes
0.013	**0.019**	1.5x	1/8	1-1/2	Yes	Yes		Yes		
0.013	**0.039**	3x	1/8	1-1/2	Yes	Yes		Yes		Yes
0.014	**0.021**	1.5x	1/8	1-1/2	Yes	Yes		Yes		
0.014	**0.042**	3x	1/8	1-1/2	Yes	Yes		Yes		Yes
0.015 (1/64)	**0.022**	1.5x	1/8	1-1/2	13915	14015	13915-C3	14015-C3	13915-C4	14015-C4
0.015 (1/64)	**0.045**	3x	1/8	1-1/2	72015	73015	72015-C3	73015-C3	72015-C4	73015-C4
0.016 (.4 mm)	**0.024**	1.5x	1/8	1-1/2	13916	14016		14016-C3		
0.016 (.4 mm)	**0.048**	3x	1/8	1-1/2	72016	73016		73016-C3		73016-C4
0.017	**0.026**	1.5x	1/8	1-1/2	13917	14017		14017-C3		
0.017	**0.051**	3x	1/8	1-1/2	72017	73017		73017-C3		73017-C4
0.018	**0.027**	1.5x	1/8	1-1/2	13918	14018		14018-C3		
0.018	**0.054**	3x	1/8	1-1/2	72018	73018	72018-C3	73018-C3		73018-C4
0.019	**0.029**	1.5x	1/8	1-1/2	13919	14019		14019-C3		
0.019	**0.057**	3x	1/8	1-1/2	72019	73019		73019-C3		73019-C4
0.020 (0.5 mm)	**0.030**	1.5x	1/8	1-1/2	13920	14020	13920-C3	14020-C3		14020-C4
0.020 (0.5 mm)	**0.060**	3x	1/8	1-1/2	72020	73020	72020-C3	73020-C3	72020-C4	73020-C4
0.021	**0.031**	1.5x	1/8	1-1/2	13921	14021		14021-C3		
0.021	**0.063**	3x	1/8	1-1/2	72021	73021		73021-C3		73021-C4
0.022	**0.033**	1.5x	1/8	1-1/2	13922	14022		14022-C3		
0.022	**0.066**	3x	1/8	1-1/2	72022	73022		73022-C3		73022-C4
0.023	**0.035**	1.5x	1/8	1-1/2	13923	14023		14023-C3		
0.023	**0.069**	3x	1/8	1-1/2	72023	73023		73023-C3		73023-C4

TABLE 3.3.1 Commercial End Milling Micro Cutters (Courtesy Harvey Tool)

Cutter Diameter	LOC		Shank Diameter	OAL	Uncoated		AlTiN Coated		Amorphous Diamond	
D_1+0.0005 in./ −0.0005 in.	L_2+0.010 in./ −0.000 in.		D_2 (in.)	L_1 (in.)	2FL	4FL	2FL	4FL	2FL	4FL
0.024 (0.6 mm)	**0.036**	1.5x	1/8	1-1/2	13924	14024		14024-C3		
0.024 (0.6 mm)	**0.072**	3x	1/8	1-1/2	72024	73024		73024-C3		73024-C4
0.025	**0.037**	1.5x	1/8	1-1/2	13925	14025	13925-C3	14025-C3		14025-C4
0.025	**0.075**	3x	1/8	1-1/2	72025	73025	72025-C3	73025-C3	72025-C4	73025-C4
0.026	**0.039**	1.5x	1/8	1-1/2	13926	14026		14026-C3		
0.026	**0.078**	3x	1/8	1-1/2	72026	73026		73026-C3		73026-C4
0.027	**0.041**	1.5x	1/8	1-1/2	13927	14027		14027-C3		
0.027	**0.081**	3x	1/8	1-1/2	72027	73027		73027-C3		73027-C4
0.028 (0.7 mm)	**0.042**	1.5x	1/8	1-1/2	13928	14028		14028-C3		
0.028 (0.7 mm)	**0.084**	3x	1/8	1-1/2	72028	73028		73028-C3		73028-C4
0.029	**0.043**	1.5x	1/8	1-1/2	13929	14029		14029-C3		
0.029	**0.087**	3x	1/8	1-1/2	72029	73029		73029-C3		73029-C4
0.030	**0.045**	1.5x	1/8	1-1/2	13930	14030	13930-C3	14030-C3		14030-C4
0.030	**0.090**	3x	1/8	1-1/2	72030	73030	72030-C3	73030-C3	72030-C4	73030-C4
0.031 (1/32)	**0.046**	1.5x	1/8	1-1/2	13931	14031	13931-C3	14031-C3	13931-C4	14031-C4
0.031 (1/32)	**0.093**	3x	1/8	1-1/2	72031	73031	72031-C3	73031-C3	72031-C4	73031-C4
0.032	**0.048**	1.5x	1/8	1-1/2	13932	14032		14032-C3		
0.032	**0.096**	3x	1/8	1-1/2	72032	73032		73032-C3		
0.033	**0.049**	1.5x	1/8	1-1/2	13933	14033		14033-C3		
0.033	**0.099**	3x	1/8	1-1/2	72033	73033		73033-C3		
0.034	**0.051**	1.5x	1/8	1-1/2	13934	14034		14034-C3		
0.034	**0.102**	3x	1/8	1-1/2	72034	73034		73034-C3		
0.035 (0.9 mm)	**0.052**	1.5x	1/8	1-1/2	13935	14035	13935-C3	14035-C3		14035-C4
0.035 (0.9 mm)	**0.105**	3x	1/8	1-1/2	72035	73035	72035-C3	73035-C3	72035-C4	73035-C4
0.036	**0.054**	1.5x	1/8	1-1/2	13936	14036		14036-C3		
0.036	**0.108**	3x	1/8	1-1/2	72036	73036		73036-C3		

0.037	**0.055**	1.5x	1/8	1-1/2	13937	14037		14037-C3		
0.037	**0.111**	3x	1/8	1-1/2	72037	73037		73037-C3		
0.038	**0.057**	1.5x	1/8	1-1/2	13938	14038		14038-C3		
0.038	**0.114**	3x	1/8	1-1/2	72038	73038	72038-C3	73038-C3		
0.039 (1 mm)	**0.058**	1.5x	1/8	1-1/2	13939	14039		14039-C3		
0.039 (1 mm)	**0.117**	3x	1/8	1-1/2	72039	73039	72039-C3	73039-C3		
0.040	**0.060**	1.5x	1/8	1-1/2	13940	14040	13940-C3	14040-C3		14040-C4
0.040	**0.120**	3x	1/8	1-1/2	72040	73040	72040-C3	73040-C3	72040-C4	73040-C4
0.041	**0.123**	3x	1/8	1-1/2	72041	73041		73041-C3		
0.042	**0.126**	3x	1/8	1-1/2	72042	73042		73042-C3		
0.043 (1.1 mm)	**0.129**	3x	1/8	1-1/2	72043	73043		73043-C3		
0.044	**0.132**	3x	1/8	1-1/2	72044	73044		73044-C3		
0.045	**0.067**	1.5x	1/8	1-1/2	13945	14045	13945-C3	14045-C3		14045-C4
0.045	**0.135**	3x	1/8	1-1/2	72045	73045	72045-C3	73045-C3	72045-C4	73045-C4
0.046	**0.138**	3x	1/8	1-1/2	72046	73046		73046-C3		
0.047 (3/64)	**0.070**	1.5x	1/8	1-1/2	13947	14047	13947-C3	14047-C3		14047-C4
0.047 (3/64)	**0.141**	3x	1/8	1-1/2	72047	73047	72047-C3	73047-C3	72047-C4	73047-C4
0.048	**0.144**	3x	1/8	1-1/2	72048	73048	72048-C3	73048-C3		
0.049	**0.147**	3x	1/8	1-1/2	72049	73049	72049-C3	73049-C3		
0.050	**0.075**	1.5x	1/8	1-1/2	13950	14050	13950-C3	14050-C3		14050-C4
0.050	**0.150**	3x	1/8	1-1/2	72050	73050	72050-C3	73050-C3	72050-C4	73050-C4
0.051 (1.3 mm)	**0.153**	3x	1/8	1-1/2	72051	73051		73051-C3		
0.052	**0.156**	3x	1/8	1-1/2	72052	73052	72052-C3	73052-C3		
0.053	**0.159**	3x	1/8	1-1/2	72053	73053		73053-C3		
0.054	**0.162**	3x	1/8	1-1/2	72054	73054		73054-C3		

Table 3.3.1 Commercial End Milling Micro Cutters (Courtesy Harvey Tool) (*Continued*)

Cutter Diameter	LOC		Shank Diameter	OAL	Uncoated		AlTiN Coated		Amorphous Diamond	
D_1+0.0005 in./ −0.0005 in.	L_2+0.010 in./ −0.000 in.		D_2 (in.)	L_1 (in.)	2FL	4FL	2FL	4FL	2FL	4FL
0.055 (1.4 mm)	**0.082**	1.5x	1/8	1-1/2	13955	14055	13955-C3	14055-C3		14055-C4
0.055 (1.4 mm)	**0.165**	3x	1/8	1-1/2	72055	73055	72055-C3	73055-C3	72055-C4	73055-C4
0.056	**0.168**	3x	1/8	1-1/2	72056	73056		73056-C3		
0.057	**0.171**	3x	1/8	1-1/2	72057	73057		73057-C3		
0.058	**0.174**	3x	1/8	1-1/2	72058	73058		73058-C3		
0.059 (1.5 mm)	**0.177**	3x	1/8	1-1/2	72059	73059		73059-C3		
0.060	**0.090**	1.5x	1/8	1-1/2	13960	14060	13960-C3	14060-C3		14060-C4
0.060	**0.180**	3x	1/8	1-1/2	72060	73060	72060-C3	73060-C3	72060-C4	73060-C4
0.062 (1/16)	**0.093**	1.5x	1/8	1-1/2	13962	14062	13962-C3	14062-C3	13962-C4	14062-C4
0.062 (1/16)	**0.186**	3x	1/8	1-1/2	72062	73062	72062-C3	73062-C3	72062-C4	73062-C4
0.065	**0.097**	1.5x	1/8	1-1/2	13965	14065	13965-C3	14065-C3		
0.065	**0.195**	3x	1/8	1-1/2	72065	73065	72065-C3	73065-C3	72065-C4	73065-C4
0.070	**0.105**	1.5x	1/8	1-1/2	13970	14070	13970-C3	14070-C3		
0.070	**0.210**	3x	1/8	1-1/2	72070	73070	72070-C3	73070-C3	72070-C4	73070-C4
0.075	**0.112**	1.5x	1/8	1-1/2	13975	14075	13975-C3	14075-C3		
0.075	**0.225**	3x	1/8	1-1/2	72075	73075	72075-C3	73075-C3		73075-C4
0.078 (5/64)	**0.117**	1.5x	1/8	1-1/2	13978	14078	13978-C3	14078-C3		14078-C4
0.078 (5/64)	**0.234**	3x	1/8	1-1/2	72078	73078	72078-C3	73078-C3	72078-C4	73078-C4
0.080	**0.120**	1.5x	1/8	1-1/2	13980	14080	13980-C3	14080-C3		
0.080	**0.240**	3x	1/8	1-1/2	72080	73080	72080-C3	73080-C3	72080-C4	73080-C4
0.085	**0.127**	1.5x	1/8	1-1/2	13985	14085	13985-C3	14085-C3		
0.085	**0.255**	3x	1/8	1-1/2	72085	73085	72085-C3	73085-C3		73085-C4
0.090	**0.135**	1.5x	1/8	1-1/2	13990	14090	13990-C3	14090-C3		

0.090	**0.270**	3x	1/8	1-1/2	72090	73090	72090-C3	73090-C3	72090-C4	73090-C4
0.093 (3/32)	**0.139**	1.5x	1/8	1-1/2	13993	14093	13993-C3	14093-C3	13993-C4	14093-C4
0.093 (3/32)	**0.279**	3x	1/8	1-1/2	72093	73093	72093-C3	73093-C3	72093-C4	73093-C4
0.095	**0.142**	1.5x	1/8	1-1/2	13995	14095	13995-C3	14095-C3		
0.095	**0.285**	3x	1/8	1-1/2	72095	73095	72095-C3	73095-C3		73095-C4
0.100	**0.150**	1.5x	1/8	1-1/2	13999	14099	13999-C3	14099-C3		
0.100	**0.300**	3x	1/8	1-1/2	72100	73100	72100-C3	73100-C3	72100-C4	73100-C4
0.105	**0.158**	1.5x	1/8	1-1/2	50200	50300	50200-C3	50300-C3		
0.105	**0.315**	3x	1/8	1-1/2	72105	73105	72105-C3	73105-C3		73105-C4
0.109 (7/64)	**0.164**	1.5x	1/8	1-1/2	50201	50301		50301-C3		
0.109 (7/64)	**0.327**	3x	1/8	1-1/2	72109	73109		73109-C3		
0.110	**0.165**	1.5x	1/8	1-1/2	50202	50302	50202-C3	50302-C3		
0.110	**0.330**	3x	1/8	1-1/2	72110	73110	72110-C3	73110-C3		73110-C4
0.115	**0.173**	1.5x	1/8	1-1/2	50203	50303	50203-C3	50303-C3		
0.115	**0.345**	3x	1/8	1-1/2	72115	73115	72115-C3	73115-C3		73115-C4
0.118 (3 mm)	**0.177**	1.5x	1/8	1-1/2	50204	50304	50204-C3	50304-C3		
0.118 (3 mm)	**0.354**	3x	1/8	1-1/2	72118	73118	72118-C3	73118-C3		73118-C4
0.120	**0.180**	1.5x	1/8	1-1/2	50205	50305	50205-C3	50305-C3		
0.120	**0.360**	3x	1/8	1-1/2	72120	73120	72120-C3	73120-C3		73120-C4

TABLE 3.3.1 Commercial End Milling Micro Cutters (Courtesy Harvey Tool) (*Continued*)

Cutter Diameter	LOC		Shank Diameter	OAL	Uncoated		AlTiN Coated		Amorphous Diamond	
D_1+0.000 in./ −0.002 in.	L_2+0.030 in./ −0.000 in.		D_2 (in.)	L_1 (in.)	2FL	4FL	2FL	4FL	2FL	4FL
0.125 (1/8)	**0.187**	1.5x	1/8	1-1/2	50208	50308	50208-C3	50308-C3		50308-C4
0.125 (1/8)	**0.375**	3x	1/8	1-1/2	72125	73125	72125-C3	73125-C3	72125-C4	73125-C4
0.140 (9/64)	**0.220**	1.5x	3/16	2		50309		50309-C3		
0.140 (9/64)	**0.562**	4x	3/16	2	72140	73140		73140-C3		
0.156 (5/32)	**0.281**	1.5x	3/16	2	50210	50310		50310-C3		50310-C4
0.156 (5/32)	**0.562**	3x	3/16	2	72156	73156	72156-C3	73156-C3	72156-C4	73156-C4
0.172 (11/64)	**0.625**	3x	3/16	2		73172		73172-C3		
0.187 (3/16)	**0.312**	1.5x	3/16	2	50212	50312	50212-C3	50312-C3		50312-C4
0.187 (3/16)	**0.625**	3x	3/16	2	72187	73187	72187-C3	73187-C3	72187-C4	73187-C4
0.203 (13/64)	**0.625**	3x	1/4	2-1/2		73190		73190-C3		
0.218 (7/32)	**0.330**	1.5x	1/4	2-1/2		50314		50314-C3		
0.218 (7/32)	**0.625**	3x	1/4	2-1/2		73193		73193-C3		
0.234 (15/64)	**0.750**	3x	1/4	2-1/2		73195		73195-C3		
0.250 (1/4)	**0.375**	1.5x	1/4	2-1/2	50216	50316	50216-C3	50316-C3		50316-C4
0.250 (1/4)	**0.750**	3x	1/4	2-1/2	72199	73199	72199-C3	73199-C3	72199-C4	73199-C4
0.312 (5/16)	**0.470**	1.5x	5/16	2-1/2		50320		50320-C3		
0.312 (5/16)	**0.812**	3x	5/16	2-1/2		15220		15220-C3		15220-C4
0.375 (3/8)	**0.570**	1.5x	3/8	2-1/2		50324		50324-C3		
0.375 (3/8)	**1.000**	3x	3/8	2-1/2		15224		15224-C3		15224-C4
0.500 (1/2)	**1.000**	2x	1/2	3		15232		15232-C3		15232-C4

LOC, length of cutting flutes; OAL, tool overall length. Presence of tool number implies tools available with those dimensions; Yes indicates tools can be obtained to those dimensions with those coatings; a blank indicates that coating not available for that size tool. 0.0005 in. = 12.7 μm; 0.010 in. = 0.25 mm.

TABLE 3.3.1 Commercial End Milling Micro Cutters (Courtesy Harvey Tool) (*Continued*)

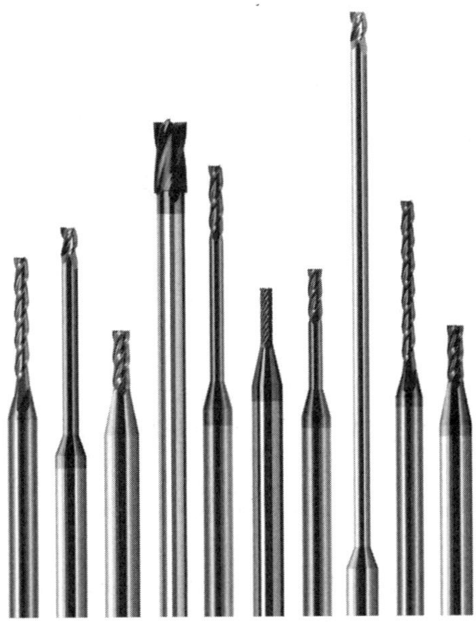

FIGURE 3.3.1 Commercial end milling tool configurations. (Courtesy Harvey Tool.)

diameter varied between 19.5 μm undersized (−0.0008 in.) and 27.4 μm oversized (+0.0011 in.) in magnitude (3.9% to 5.5% in percentage terms) over the range of tools analyzed" (Saedon, 2011).

Coatings

Conventional commercial cutters such as those shown in Fig. 3.3.1 can be coated with aluminum titanium nitride (AlTiN) or amorphous diamond if they are large enough (Table 3.3.1). These coatings provide wear resistance and extend life.

Applications

Micro milling is used for products in aerospace, medical, electronic, optics, jewelry, and mechanical watches. Micro fluidic channels 152 μm (0.0060 in.) wide and 178 μm (0.0070 in.) deep are a common application. They can be milled with resulting surface finishes of 0.127 μm (5 μin.) Ra in metal. Lab-on-chips, molds for micro features, micro engraving, micro EDM electrodes as well as micro electrodes or contacts for electronics, sub millimeter turbine wheels are other examples of applications.

A number of applications have machined a soft steel or brass mold for making fluidic components for the assay of blood or chemicals. A polymer is poured into the metal mold to become an embossing tool for other polymers that are the basis of the fluidic component. In essentially every case, after milling the part has to be sanded or otherwise finished to remove burrs. Burrs on flat surfaces may be easily removed but for complex geometries on micro parts deburring itself is a challenge (see Chap. 3.34).

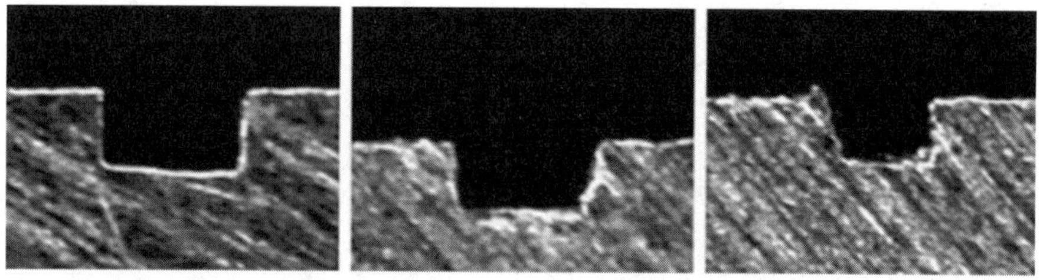

Figure 3.3.2 Grooves milled in aluminum with 0.2 mm (0.0078 in.) end mill (Monroy-Vásquez et al., 2013).

Figure 3.3.2 illustrates one common aspect of micro cutters—the side walls are often not perpendicular to the bottom of a slot.

Machines

Machines designed for micro milling can position within ±1 μm (±0.000040 in.). Conventional milling machines are not capable of this precision nor in general do they have a capability to perform micro milling. As described in this chapter, micro milling refers to using a tool which cuts like a milling cutter. The equipment for micro milling, however, can also perform micro drilling and micro grinding.

Design Issues

- Do not design sharp internal corners in pockets—use generous curves at least three times the cutter radius that will be used.
- Keep the depth of pocket or channel as shallow as possible (cutters have very short flutes).
- Define allowable radius at bottom of cut—tools will develop a radius there quickly.
- Milling produces burrs that must be removed. EDM may be a better choice.
- Specify part materials with uniform grain sizes and no hard spots. Super fine grain material may be needed.
- If surface contamination by chemicals is an issue, specify requirements (coolants are used in machining).
- Micro milling cutters deflect sideways while cutting—specifying wall perpendicularity may be required.
- When a deep slot is required, use a slotting cutter or micro EDM rather than an end mill style cutter. These can be made to 0.05 mm (0.002 in.) widths.
- When very close accuracy on dimensions are required, recognize that new cutting tools, and tools after use, vary significantly in size.
- Do not expect to use end milling on hard materials—it is just not generally effective.
- Tool life in steels will be measured in only minutes or seconds.

A number of references are given below that provide more detailed insight into capabilities and limitations.

References

Aramcharoen, A., and P. T. Mativenga, Size effect and tool geometry in micromilling of tool steel, *Precision Engineering*, 2009, 33(4), 402–407.

Foy, K., *Ductile-Regime Machining of Glass Using the Micromilling Process*, All Theses, Paper 359, Clemson University, 2008.

Gafford, J. B., R. M. Panas, M. A. Cullinan, and M. L. Culpepper, Design principles and practices for rapid prototyping of meso- and micro-scale flexures via micromilling, *Proceedings—ASPE 2010 Annual Meeting*, vol. 50, pp. 411–414.

Gietzelt, T., and L. Eichhorn, Mechanical micromachining by drilling, milling and slotting, In *Micromachining Techniques for Fabrication of Micro and Nano Structures*, Dr. Mojtaba Kahrizi (Ed.). InTech. ISBN: 978-953-307-906-6. Available from: http://www.intechopen.com/books/micromachining-techniques-for-fabrication-of-microand-nano-structures/mechanical-micromachining-by-drilling-milling-and-slotting

Guckenberger, D. J., T. E. de Groot, A. M. D. Wan, D. J. Beebe, and E. W. K. Young, Micromilling: a method for ultra-rapid prototyping of plastic microfluidic devices, *Lab Chip*, 2015, 15, 2364–2378 (doi: 10.1039/C5LC00234F).

Monroy-Vázquez, K. P., A. Attanasio, E. Ceretti, H. R. Siller, N. J. Hendrichs-Troeglen, and C. Giardini, Evaluation of superficial and dimensional quality features in metallic micro-channels manufactured by micro-end-milling, *Materials*, 2013, 6, 1434–1451 (doi:10.3390/ma6041434).

Saedon, J. B., *Micromilling of Hardened (62 HRC) AISI D2 Cold Work Tool Steel*, PhD Thesis, University of Birmingham (England), 2011.

Scheiding, S., R. Eberhardt, A. Gebhardt, and A. Tünnermann, Micro lens array milling on large wafer, *Optik & Photonik*, December 2009, 4, 41–45.

Vella, P., J. C. Borg, A. K. Grech, D. Spiteri, and J. Randich, Design for micro milling guidelines, *Proceedings of the International Conference on Engineering Design*, ICED'07, August 2007.

CHAPTER 3.4

Designing for the Swiss Screw Machine

Kenneth J. Mandile

Swissturn/USA, Inc.
Oxford, Massachusetts

Introduction

Computer numerical control (CNC) Swiss-type screw machines offer more complexity, higher precision, and faster set-up times for many cylindrical shapes being cut from bar stock.

Screw machines are a class of automatic lathes used for mid- to high-volume machining of metal and plastic components. There are a variety of screw machines, but the Swiss-type CNC is often the best choice for machining micro components 20 mm (0.787 in.) diameter down to 0.25 mm (0.010 in.), complex parts up to 32 mm (1.25 in.) diameter, parts with very high-precision requirements [0.01 mm (0.0004 in.) and less], and parts with very high length-to-diameter ratios (Fig. 3.4.1).

When fed by an automatic bar loader, CNC Swiss are often used for "lights out" operation. In an automatic bar loader, several bars are queued. When the bar being machined is completely machined into finished parts, the loader automatically removes the remnant (bar end) and loads another bar into the machine.

Despite the name, a CNC Swiss screw machine is rarely the best choice for making screws. Swiss refers to the design form of the machine, not the country where the machine was made. Swiss screw machines were first invented for manufacturing very small, high-precision watch parts.

Process

A Swiss-type machine has a sliding headstock (often referred to as the "main spindle") that holds the material in a collet and advances it through a guide bushing. The bushing keeps the raw material from flexing away from the cutting tool. It is this feature of a sliding headstock in combination with a supporting bushing that differentiates Swiss screw machines from other types of screw machines. Guide bushings are usually made

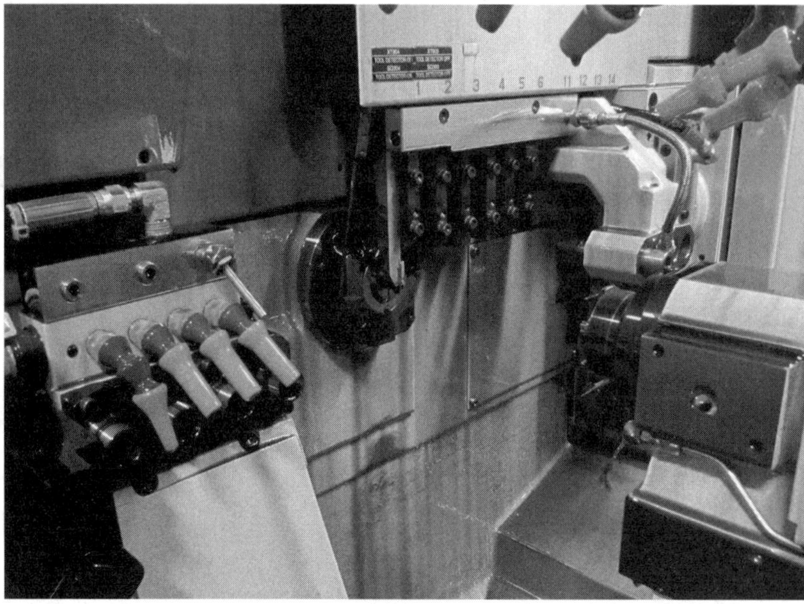

Figure 3.4.1 The work area of a CNC Swiss screw machine. In this instance 6 tool holders are located above the spindle. On the right is a secondary spindle (called a sub-spindle) used to hold the front end of the part so that the rear of the part can be finish machined, all in a single operation.

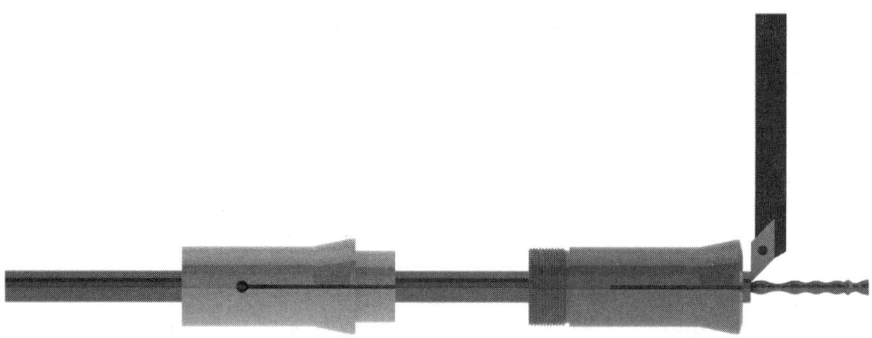

Figure 3.4.2 The collet (left) holds the bar stock to rotate it and to advance it through the guide bushing (right) and into the cutting tool (far right).

with carbide pads that the material slides through, but special applications sometimes require the use of steel (Meehanite) or plastic bushings (Fig. 3.4.2).

The cutting tools on a Swiss-type machine are positioned close to the face of the bushing, anywhere from 0.127 to 1.27 mm (0.005 to 0.05 in.) depending on the diameter of the stock. The headstock rotates the stock and advances it through the bushing. The turning tools move up and down radially, creating the form of the part through "single-point" turning. In addition to conventional turning, most CNC Swiss have live tools for milling, cross drilling, and cross tapping parts.

The headstock can be stopped and indexed to produce multiple cross drilled holes or milled features. CNC controls equipped with polar interpolation give users additional options for machining very complex shapes from round stock.

When the machining is finished on the bar, a cut-off tool and a sub-spindle advance. When the machined part is ready to be parted from the front of the bar stock, the sub-spindle closes on it. The sub-spindle is synchronized with the main spindle so that it maintains the same rotational speed and linear movement as the main spindle.

After being cut off from the bar, the part may be machined on the cut-off side using the sub-spindle. Sub-spindle or "back working" features may include turning, drilling, tapping, boring, milling, threading, etc. After any required machining is finished on the sub-spindle side, the part is ejected, usually into a parts catcher or chute.

Machining is usually done under flood coolant, either conventional cutting oil, a synthetic cutting oil, or a vegetable base oil.

Because of the gap between the sliding headstock collet and the carbide guide bushing, there is always a short material remnant at the end of each bar that cannot be machined. These are ejected from the machine before a new bar is loaded.

Characteristics

Swiss screw machines are usually the first choice for parts with a high length-to-diameter ratio.

Because cutting is done with a single-point tool (versus a form tool), Swiss screw machines also are very good at making very small parts. Turned diameters as small as 0.25 mm (0.010 in.) are possible. Depending on the material type, these small diameters can be turned over long lengths in relation to the diameter. This is the main advantage that Swiss-type screw machines have over conventional CNC lathes and over other types of screw machines.

Swiss machines are best known for their precision, a legacy of their origin in the watch industry. Using centerless ground raw material, high-precision carbide bushings, and a knowledgeable machinist, tolerances of 0.005 mm (0.0002 in.) are achievable. Often though, it may be more economical to machine parts to a larger size and then to centerless grind them to the finish diameter and tolerance (Figs. 3.4.3 and 3.4.4).

Economics

Because programming and setting up CNC Swiss is relatively complex, Swiss machines are rarely used for runs less than 100 pieces. They are most economical in mid-volume production runs of 500 to 100,000 pieces. Automatic loaders that allow for lights out operation often allow Swiss screw machine companies to be competitive for high-volume work into the hundreds of thousands and millions.

Swiss machines are usually not competitive on very simple parts, parts with tolerances larger than ±0.13 mm (0.005 in.), and parts that are over 32 mm (1.25 in.) diameter.

Materials

Almost any type of metal or plastic available in bar form can be machined on a Swiss-type screw machine. In the United States, most metal bar stock is available in 3.7 m (12 ft) lengths, though some specialty metals and plastics come in shorter lengths.

Figure 3.4.3 Example of slotted threaded pin machined on a CNC Swiss screw machine.

Figure 3.4.4 Variety of CNC Swiss screw machine parts in small vial.

Cycle time and productivity are heavily influenced by material type, stock diameter, depth of cut, and finish requirements. Free machining leaded brass (CDA 360), aluminum, free-machining stainless steels (303, 17-4, 416), and leaded carbon steels are among those most preferred by machinists.

Nickel alloys, 316 and 304 stainless steel, and unleaded carbon steels dramatically increase the cost of machining due to excessive tool wear, slow surface speed rates, and chip buildup. These materials are not uncommon, but often demand significantly higher machine hour rates.

Plastics are commonly machined on Swiss screw machines. Acetyl (Delrin®), PEEK®, nylon, PFE, and PTFE (Teflon®) are readily available in rod form for use on screw machines.

There are several issues to keep in mind when looking to manufacture plastic components on screw machines. Many parts are designed for molding, but if quantities are low, Swiss screw machining can be very competitive due to the cost of tooling for molding operations. Many plastic grades are only available as molding pellets, so a material change may be required to make a machined part. Small diameter plastic rods [under 6.35 mm (0.25 in.)] are sometimes too flimsy to handle in a bar loader, so sometimes a larger diameter than necessary will be used to increase rigidity.

Plastic rods tend to have nonuniform diameters that will not properly feed through the guide bushing. For this reason, it is usually recommended to centerless grind the outside diameter of the raw stock to a uniform diameter.

Plastic bars are often not straight enough for the screw machine bar loader. Before specifying any plastic, it is advisable to consult with a Swiss screw machine job shop.

Design Issues

- *Turned diameters.* Long-turned diameters are usually ideal for Swiss screw machines. Typical modern Swiss CNC screw machines have a headstock stroke (the distance the bar stock can be moved in a single motion to advance the bar through the guide bushing) of 89 to 203 mm (3.5 to 8 in.). If longer parts are needed, the machine can be programmed to drop the headstock back in order to advance more of the bar stock. Though this will often leave a mark where the cutting action was interrupted.
- *External threads.* There are several threading methods commonly used on CNC Swiss. Most common is probably single-point threading. Threading inserts allow for excellent profile control. Single-point threading allows for special thread pitches and forms. There is a limitation to the length that can be single-point threaded. Because multiple passes are required, the material will have to retract into the bushing to start a new pass. If the thread length is longer than the land (bearing area) on the bushing, then the material will fall out of the bushing and will be unsupported on the next pass of the thread tool. There is no hard rule about single-point threading, but 16 mm (0.625 in.) length is a good starting point as a maximum length. Smaller diameters are usually limited more than larger diameters when single-point threading. Threads as small as a 0.30 UNM (0.012 in.) can be single pointed. The success of threading parts this small is dependent on the material being threaded, the length of the thread, and the geometry of the part. The flexibility of CNC controls also allows for special pitch threads. Very fine pitch threads on larger diameters are often used for applications requiring fine adjustment.

Another commonly used threading method is roll forming. The part is turned to the pitch diameter of the desired thread and then a form die is passed over the turned diameter to force material to the shape of the thread form. The die usually consists of a set of three rolls that contain the form of the thread.

Conventional cutting dies can also be used, but because of the speeds being used, they are often not a good choice.

Thread whirling is a thread-cutting process that has become popular for many challenging external threading jobs. It involves a process with a whirling head rotating at high speed around the bar stock. The inside of the whirling head has three thread form inserts. As the head whirls around the bar stock, the material is fed forward at the pitch of the thread. Because the thread cutting is done close to the bushing, longer, more precise, and very deep cuts can be made. This operation has become very popular for cutting titanium bone screws as small as 1 mm (.040 in.) in diameter.

External thread callouts should specify whether the length dimension is to the last full thread or to a shoulder. If it is to a shoulder, it is recommended that an undercut to the root diameter of the thread of at least 1.5 times the pitch in length be allowed.

- *Internal threads.* Internal threads can also be single pointed, but they are more commonly tapped. One issue that is frequently seen as a problem for screw machinists is a requirement for a full thread to the bottom of a hole. Tapped holes require clearance at the bottom of the hole due to the lead at the front of the tap. A plug-style tap has a tip chamfer equal to four to six thread pitches, so if you ask for a 10-32 tapped hole, then you should allow for at least 4.76 mm (3/16 in.) beyond the requested thread depth (6 × 1/32). If it is absolutely necessary, a bottoming tap or a thread mill can be used, but the machinist will still need at least two thread pitches for clearance.

Cutting a thread in a blind hole with insufficient clearance can also be complicated by chips packing into the bottom of the hole. For many materials, a form tap may be preferable. This type of tap forms the thread profile rather than cutting it. The tapped hole diameter is drilled to the pitch diameter of the thread, which is somewhat larger than a pre-tap hole for a cut tap. Cut taps are drilled to the minor diameter of the thread.

Tap sizes down to 0.30 UNM (0.012 in.) are available and can be used successfully on CNC Swiss, but taps of this size are not practical for larger production runs. In designing medium- to high-production parts, tap sizes should be limited to 0.80 UNM (0.031 in.) and above.

- *Corners.* Sharp corners leave open the possibility of damaging the edge of the part. Chamfers and radii should be proportional to the size of the part. Very small parts may only have room for a 50-mm (0.002-in.) chamfer, while a 25 mm (1 in.) diameter may look better with a 0.38- to 0.76-mm (0.015- to 0.030-in.) chamfer. Because of the nature of Swiss screw machines, if a radius is preferred over a chamfer, this will usually not affect the cost or machinability.

Both external and internal threads should have a front chamfer that reaches to the root diameter of the thread.

- *Internal diameters.* When drilling on a Swiss-type screw machine, the drill tip can be positioned at the face of the guide bushing. This allows for more accurately sized holes and minimizes concentricity issues. The depth of holes is very dependent on the hole size, the material being drilled, and the tolerance required. Holes as small as 0.20 mm (0.008 in.) can be drilled, but the depth would be limited to one times the diameter of the drill. The depth-to-diameter ratio quickly increases as the hole diameter goes up. Holes above 6.35 mm (0.25 in.) diameter can be drilled to depths 8 times the diameter.

The use of through coolant drills has greatly increased the capability to drill deep holes. Coolant is pumped through the drill at pressures between 3.45 and 13.8 MPa (500 and 2000 psi). The high-pressure coolant provides lubricity and cooling at the tip and it forces the chip down the web of the drill and out of the hole. This allows for drilling at higher speeds and feeds with few, if any, pecking (withdrawing the drill to allow for coolant to enter the hole).

Very close tolerance holes or holes with tapers, undercuts, or other features can be machined with boring tools. It is possible to bore to tolerances of 5 μm (0.0002 in.) if necessary. Utilizing boring bars, wall thicknesses as small as 50 μm (0.002 in.) can be bored on a CNC Swiss. Boring also offers advantages in surface finish and concentricity. As with drilling, high-pressure coolant allows for deeper bores, especially in holes that would normally not receive enough coolant to cool the tool tip or to flush out the chips.

CHAPTER 3.5

Designing for Turning Micro Parts

Robin C. Whitmore

Medical Micro Machining, Inc.
Colfax, Washington

Any part smaller than approximately 3.175 mm (1/8 in.) in diameter is micro by many definitions. Machining down to 0.076 mm (0.003 in.) is clearly micro by any definition and this chapter will focus on dimensions of 1 mm (0.040 in.) or smaller as "micro."

Micro-size turned parts can be manufactured by several processes, each of which has distinct capabilities and limitations. Table 3.5.1 provides some basic insight into these processes.

Turning on conventional computer numerical controlled (CNC) lathes can produce diameters down to 76 μm (0.003 in.) and length-to-diameter ratios (aspect ratios) of 5 and even beyond. They are fast and can easily produce almost any configuration. Parts do not have to be round, but they have to have a longitudinal axis. For the miniature configurations, where miniature is defined as less than 1 mm (0.040 in.), CNC screw machines can be a better choice because of their rapidity and ability to make complex geometries. They can make very long length-to-diameter ratios accurately, and since they are highly automated, they produce parts at rates up to 10 a minute and in almost any material (i.e., one part every 6 s and they can cut faster than that on many parts). They have produced parts down to 0.25 mm (0.010 in.) diameter and aspect ratios of 10 or greater. Several manufacturers have produced micro lathes over the past 50 years. These are manual or CNC machines today capable of making turned parts as small as 76 μm (0.003 in.) in stainless steel. The larger of these micro lathes have high-production capacities while the smaller of them are capable of extremely minute diameters but have a production rate more typically of only one or two parts per hour. Despite this low-production rate, their ability to produce minuteness makes them an important tool for many applications. Wire EDM and WEDG because they involve cutting via sparks rather than physical contact can produce much higher length-to-diameter ratios in the smallest sizes. Separate chapters in this book describe the capabilities of screw

Process	Advantage	Disadvantage	Applications	Typical Smallest Diameter
CNC turning conventional lathes	Wide selection of machines	Limited on micro sizes	Miniature parts but larger than most micro parts	0.5 mm (0.020 in.)
CNC screw machines	Very fast, lowest possible cost for large runs	More expensive than small lathes	Commercial screws, pins, complex external geometries	0.25 mm (0.010 in.)
Micro lathes	Can make very small pins		Micro to miniature turned parts, some milling possible	25 μm (0.001 in.) Aspect ratios ~5
Micro research lathes	Make very minute parts	Slow, limited materials, tools difficult to make	One of a kind parts that are very micro	3 × 35 μm (0.0001 × 0.0014 in.) 7.5 × 94 μm (0.0003 × 0.0037 in.)
WEDM	Noncontact, easy to make shafts, outstanding aspect ratios possible, material hardness not an issue	Slower than lathes, rough surface finish	Micro hole electrodes, thin shafts	6 μm (0.0002 in.) 80 × 15,000 μm (0.0031 x 0.590 in.)
WEDG	Better surface finishes than WEDM	Slower than lathes	Micro electrodes; thin shafts	50 × 1500 μm (0.002 × 0.059 in.)

Table 3.5.1 Comparison of Some Common Micro Machining Processes and Equipment

machines, WEDM, and WEDG. One commercial hybrid machine combines the ability to turn, mill, EDM, WEDM, and WEDG on a single machine.

Micro Lathes

Several commercial micro lathes are in production use in a variety of companies. Microlution's CNC MR-4 lathe has a footprint of only 1016 × 1054 mm (40 × 60 in.); in addition to spindle rotation, the tool slides can move in X, Y, and Z directions to provide extreme flexibility for machining complex shapes. This system also uses a camera and sensors to simplify tool setup, and several tools can be used in a manner like Swiss screw machines, although the stock does not feed like Swiss machines do. With a reported accuracy of ±1 µm (±0.000040 in.), conventional CNC controls and programming, and several tools available in a single setup, Microlution's MR-4 lathe fills the majority of needs for small and micro applications.

While lathes are the typical tools for turning, that is not a requirement. Renishaw decades ago turned small parts in the spindle of a vertical mill using fixed tools on the mill bed to cut the stock. That approach offered the same potential as a lathe and could be used for conventional milling on other parts. The stock goes in the chuck or collet and the X, Y, Z movement of the bed moves the tools. Thus, some micro mills are capable of turning micro parts.

There are at least four categories of lathes that manufacture micro products:

- Manual watch maker style lathes
- CNC lathes for micro parts
- Automatic screw machines (ASM)
- Research lathes that make exceptionally minute parts

The manual lathes are very capable of making parts down to 0.125 to 0.250 mm (0.005 to 0.010 in.) and drilling holes as small as 76 µm (0.003 in.) diameter. They are slow and require great skill to operate. But they do make micro parts and are relatively inexpensive for sites that only need a few simple parts. The early manual lathes were initially called instrument lathes designed to machine watch and similar minute components. The smallest of the micro lathes can be held in your hand.

The majority of successful companies in the micro world use advanced CNC lathes designed specifically for micro machining. Some mill as well as turn. Several commercial micro lathes are in production use in a variety of companies. Microlution's CNC MR-4 lathe has a working volume of 300 × 125 × 150 mm (11.81 × 4.92 × 5.91 in.) and a spindle speed of 15,000 rpm and rotating tools of 30,000 rpm.

The Blum sensing systems used on the Microlution equipment allows quickly putting all of the tools on center in both the X and Y axis as well as locating Z very accurately for the linear dimension. This combination of equipment produces a part that tapers from 25.4 to 78.7 µm over its 0.356 mm length (0.001 to 0.0031 over its 0.014 in. length). A 25.4 µm (0.001 in.) diameter is probably the smallest diameter that can be made by these machines, but an absolute limit has not been published. The equipment also allows cutting a rectangular milled section that is 0.078 to 0.160 mm by 0.229 mm (0.0031 to 0.0063 by 0.009 in.) long in a material that is somewhat brittle and is slightly magnetic.

One might define pin turning abilities for very capable commercial micro lathes as 0.2 to 0.02 mm (0.0078 to 0.00078 in.). The commercial lathes are bench top size typically and one employs a vertical spindle while most lathes are horizontal machines.

Automatic screw machines are able to provide more complex geometries and cut outs than most lathes for micro parts. In addition, they are typically much faster and can run millions of part quickly (see Chap. 3.4). Screw machines typically, however, are only able to run parts down to about half a millimeter (0.020 in.). They have made turned parts 0.66 mm (0.026 in.) diameter by 0.41 mm (0.016 in.) long with a 0.41 mm (0.016 in.) hole in 304 stainless steel. Although these machines are more difficult to get set up, users are still able to accurately position the tools and maintain exceptional tolerance.

Picking up and handling micro parts like the ones mentioned here is a science in itself and more difficult when the parts are round which all lathe parts are (see Chap. 3.51). Vacuum suction is one common way for handling these parts.

Micro Lathe Capabilities for Micro- and Nano-Size Products—Research-Level Capabilities

The minute research lathe mentioned in this section is not typical of the micro lathes. It is in fact the world's smallest working lathe. The smallest CNC micro lathe now at Japan's Ministry of Economy, Trade and Industry and built by them and Olympus Optical Co. is 32 mm long, 25 mm wide, and 30.5 mm high (slightly bigger than 1 cubic inch). The lathe weighs 100 g (3.53 oz), turns at 10,000 rpm, and has unique PZT-micro sliders that provide high-resolution motion control. It has made brass needles 0.05 mm diameter and 0.6 mm long (0.002 × 0.024 in.) and screws that are 0.05 mm (0.002 in.) in diameter. The axial accuracy for the screw was measured at 1.5 μm (59 μin.) and a roundness of 2.5 μm (98 μin.).

The smallest pin-like shape known to have been machined strictly by mechanical cutting is a 3-μm-diameter pin 35 μm long (0.0001 × 0.0014 in.). Japanese investigators have spent over a decade refining the lathes, the tools, and the process to produce such parts. It is the current ultimate minuteness of capability on micro lathes.

With the same specialized lathe, brass needles 50 μm in diameter and 600 μm (0.002 × 0.024 in.) long, an aspect ratio of 12, were machined. Using diamond cutting tools, surface finishes of 60 nm (2.4 μin.) Ra were produced.

Cutting Tool Challenges

The challenge in making minute parts with conventional machining approaches is that the cutting tools have to be exceedingly sharp, far sharper than commercial approaches produce. The workpiece material has to be exceptionally uniform and free of any hard spots or issues that will cause changes in cutting forces. The cutting process has to gently lead into the cut and similarly come out of the cut without sudden motions. And the machine tool has to be capable of uniform movements of a few millionths of an inch, while commercial machines may move in 1.3 μm (50 millionth of an inch) motions. Producing the cutting tools is a challenge, moving the cutting tool is a challenge, and finding the best cutting conditions is a challenge.

As an example, when turning a 0.25-mm-diameter (0.010-in.) pin, the material being peeled off has to be very thin, typically less than 25 μm (0.001 in.). The cutting edge of the tool has to be at least as small as 1/10th of that which is 2.5 μm (0.0001 in.). In fact

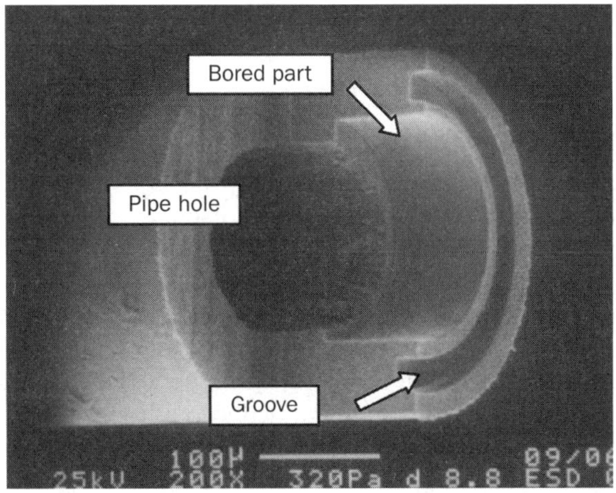

Figure 3.5.1 Micro turned part with 19 μm (0.0007 in.) groove and 100 μm (0.004 in.) hole (from Egashira, Iwata, Nomura, and Furakawa, 2011).

for some cutting, the edge radius might have to be 1 μm (0.000040 in.). A typical commercial turning tool has a radius of 25 to 50 μm (0.001 to 0.002 in.), which is 10 to 20 times too large to cut these minute chips.

When parts require features other than pin shapes, the challenge is multiplied by the number of different tools and steps required. Figure 3.5.1 illustrates some of the Japanese capabilities for turning, boring, facing off, and grooving the end of a small tube. This particular part is about 214 μm (0.0084 in.) O.D. with a groove that is only 19 μm (0.00075 in.) wide by about the same depth. Again this is one of the smallest such configurations ever produced with conventional turning.

Pin shapes are the simplest geometry to produce, even though it is a major challenge at these minute sizes. Producing shapes involving different configurations along the diameter becomes a major restraint because the tools that must have rake angles and clearance angles and must be smaller in width than the feature being cut become almost impossible to make.

WEDM and WEDG are able to make a wide variety of shapes in these small sizes, so they become the typical process of choice in these micro and nano sizes to make both tools and the parts.

While micro lathes can turn a variety of shapes, because the stock sticks out for the full length of the part, the cutting tool forces bend the part when the tools reach the free end of the part. Swiss screw machines advance the stock out of the spindle as the part is being machined, while the micro lathes described here advance the tools along the full length of the part. This makes for a more uniform diameter down the length of long parts. The micro lathes described here are, however, capable of turning smaller diameters than traditional Swiss machines. Chapter 3.4 provides more discussion on screw machines for micro parts.

Micro threads are most easily produced on a lathe as part of the turning operation. Handling both the tapping or cutting tool and positioning the tool are readily done on the lathe and not so readily on other machines or with other tooling. Do it on the lathe and minimize the additional handling of doing it any other way. The very minute 0000 to 120 thread can be made on one of the precision lathes even in stainless steel.

Micro and Nano Turned Materials

Because hard metals generate large cutting forces that both bend the part while it is being machined and cause the part to break itself as well as break and wear the cutting tool, most micro parts are made of soft materials—copper, brass, aluminum, and some precious metals like gold and silver. Several plastic materials can also be machined but they are not commonly produced that way since it is typically more economical to mold them.

303 stainless steel is a better choice for commercial micro machining than 304 or 316. It is easier to machine, results in much better tool life, and an overall reduction in cost per piece. If the part is to be welded then 304L is a better choice and if the piece is going to be implanted in the body then 316L is the stainless of choice. If surface finish or tensile strength is a consideration then 17-4 PH, a heat-treatable stainless, may be the best solution. Titanium is another common choice for turned micro products. To reiterate, the material that is chosen for the design will be the catalyst for the machining process.

Small precision lathes can make 0.41 mm (0.016 in.) diameter and 1 mm (0.040 in.) long with a 0.25 mm (0.010 in.) through drilled through hole. This leaves a wall thickness of 76 μm (0.003 in.) in each side. While that may not be an absolute minimum for wall thickness for practical machining, it certainly is near the limit.

Some materials require a unique way to machine. Turning the resorbable materials can be done very effectively using liquid nitrogen. Craniomaxillofacial bone screws couldn't be machined using standard methods so the liquid nitrogen concept was created to make the material machinable without contamination. That same process has been used to machine Teflon parts that are 1.16 mm ±12.7 μm (0.0455 ±0.0005 in.) diameter by 0.71 mm ±51 μm (0.028 ± 0.002 in.) long with a 0.343 mm ± 12.7 μm (0.0135 ±0.0005 in.) hole. A major advantage for this approach is that in these materials at low temperatures no burr is produced (all the ductility is removed by the cryogenic temperatures). At conventional temperatures, the burr can be very big and exceptionally difficult to remove. On micro parts, the deburring process can be a major restraint to successful production (see Chap. 3.34 for further discussion on deburring micro parts). Just handling the parts for the deburring process can be a significant issue, so preventing them by cryogenic machining is exceptionally beneficial in addition to the fact that the cryogenics do not add contaminants to the parts.

Being able to watch the machining in real time gives the manufacturers a much better ability to know what may be needed to better produce micro parts. Specifically being able to watch how the chips come off the part provides the information needed to determine the best cutting conditions. That is especially critical in how a drill is performing. The Microlution machine (and others) has a camera that can be used for multiple uses and is very helpful in the setup stage of a project. With crosshairs on the screen, it can be used for some unique setups that make using the Blum sensor difficult if not impossible. The more advantages the producer has, the easier it is to get a job up and running efficiently.

References

1. Brinksmeier, E., and O. Riemer, Metal cutting in microstructures, In W. Menz and S. Dimov, *Multi-material Micro Manufacture*, 2005, pp. 19–25.
2. *Masuzawa, T.*, State of the art of micromachining, *Annals of the CIRP*, Vol. 49, No. 2, 2000, pp. 473–488.
3. Furukawa, T., Y. Nomura, K. Harada, and K. Egashira, Fabrication of micropins using micro turning tools, *Key Engineering Materials*, Vols. 523–524, 2012, pp. 293–298.
4. Egashira, K., M. Iwata, Y. Nomura, and F. Furakawa, Boring and face grooving using micro turning tools, *CIRP Annals*, Vol. 60., No. 1, 2011, pp. 81–84.

CHAPTER 3.6

Design Considerations for Laser Micro Machining

Ronald D. Schaeffer

PhotoMachining, Inc.
Pelham, New Hampshire

Laser micro machining is defined as a process where the on-target feature sizes are less than 1 mm (0.040 in.), and usually much less, and also the material thickness is less than 1 mm, and usually a lot less. In order to determine if a product is a candidate for laser micro machining, there are a few basic things that need to be considered. The most important of these considerations are discussed below.

Laser Details

First, laser light is generally of a specific "color" or *wavelength* in the electromagnetic spectrum. "Laser light" as we know it goes from about 10-μm wavelength in the infrared (IR) to about 157-nm wavelength in the ultraviolet (UV), a range which also spans the "visible" portion of the spectrum. Different colors of light react differently with materials. IR light tends to impart heat or thermal energy into the material as a first-order effect, whereas UV light can in many cases, especially in polymers with pi bonding electrons, break chemical bonds as a first-order effect. In addition, the wavelength of the light influences the minimum feature size that can be attained on target, with UV photons—because of their shorter wavelength—giving the smallest attainable features. UV lasers are best utilized when part quality, usually meaning minimal heat-affected zone (HAZ), and feature resolution concerns outweigh cost concerns.

The next important parameter is *pulse length*. This is the total temporal profile of the laser pulse including any rise time or tail. In micro machining applications, pulse lengths of microseconds to femtoseconds are used. As a general rule, shorter pulses will give cleaner results on target, but cost rises as the pulse length gets shorter. Also, a "square" temporal profile would be ideal, but this is seldom seen in real-life lasers unless post-laser shaping optics (reflective, refractive, or diffractive) are used. Closely coupled with the pulse length is the *energy per pulse*. For micro machining applications this varies from a few joules to about 50 μJ on the low-energy side, with most lasers averaging a few hundred microjoules to a few millijoules per pulse. Energy per pulse divided by pulse duration gives a number called *peak power* (J/s = W) and this number is critical to clean and efficient laser processing. The higher the peak power (within a small focus area), the better the material processing will be. So, as a general rule, processing quality gets better with increasing peak power, meaning with increasing energy or decreasing pulse length.

Repetition rate is the number of times a laser is fired in 1 s. Lasers can pulse from a single shot to millions of pulses per second [megahertz (MHz)]. On the low side, excimer lasers usually operate at a few hundred or maybe up to a few thousand pulses per second. Other lasers operate in the MHz range, but these very high repetition rate lasers are frequently not useful in machining applications because the beam cannot be moved fast enough. Whereas today, a normal operating range for a majority of micro machining lasers is in the range of 20 to 500 kHz (excluding excimer lasers that operate up to about 2 kHz maximum), and the rapid emergence of new, high-power MHz lasers and beam deflecting technologies means that faster processing will be possible assuming the optics and controls can match the laser speed. On the one hand, having a high repetition rate seems like a way to increase processing speed, but there are two main issues. First, as mentioned above, at the highest available repetition rates the optics and controls are not available to move the beam quickly or accurately enough. Also, as the pulse repetition rate increases, the energy per pulse usually decreases. Repetition rate multiplied by energy per pulse is equal to the output *power* of the laser (*not* the peak power, which is described above) and many micro machining lasers are below 10 W total power. While there is no upper limit on power used for micro machining applications, a few hundred watts is considered powerful. However, one must be careful when looking at a laser with a particular output power, because this number can be met by very high energy per pulse and low repetition rate and also by low pulse energy coupled with very high repetition rate.

As a final laser parameter, we must discuss spatial beam shape. This is the actual beam profile at a specific point in time. For most lasers the beam shape is Gaussian, or very close to it. There are some lasers that do not have round, Gaussian output beams and these use imaging rather than focusing techniques, but the vast majority of micro machining lasers have a round, Gaussian beam. Round beams are convenient to work with, but the sloped sides of the beam frequently present a problem when doing very fine work as the "wings" are below the ablation threshold of the material and impart heat. Table 3.6.1 summarizes some typical laser parameters.

Laser output is directed through the optical system and onto the target. There are a few considerations that affect the beam shape and quality on target. First is the *focal length* of the imaging or focusing lens. The shorter the focus, the smaller the achievable spot—but, in the case of galvanometers, the available processing field will also be smaller. Also, shorter focal length lenses have a smaller *depth of field* (DOF). The DOF is the tolerance in the position of the focus—as you get further from the focus of a lens the beam diverges and the spot size grows, and therefore the energy density in the beam decreases. These effects can have a big impact on the on-target features and resulting target beam energy density.

Multiple manufacturers produce lasers that can be used for micro machining. Figure 3.6.1 illustrates one designed for micro work.

Wavelength	10 μm (IR) to 157 nm (VUV)
Pulse length	Microseconds to femtoseconds
Energy/pulse	25 μJ to several hundred millijoules
Repetition rate	Single shot to MHz
Output power	A few watts to a few hundred watts
Spatial beam shape	Mostly Gaussian or close to it—used in focusing mode Some lasers high-order multi-mode—used in imaging mode

TABLE 3.6.1 Typical Laser Micro Machining Parameters

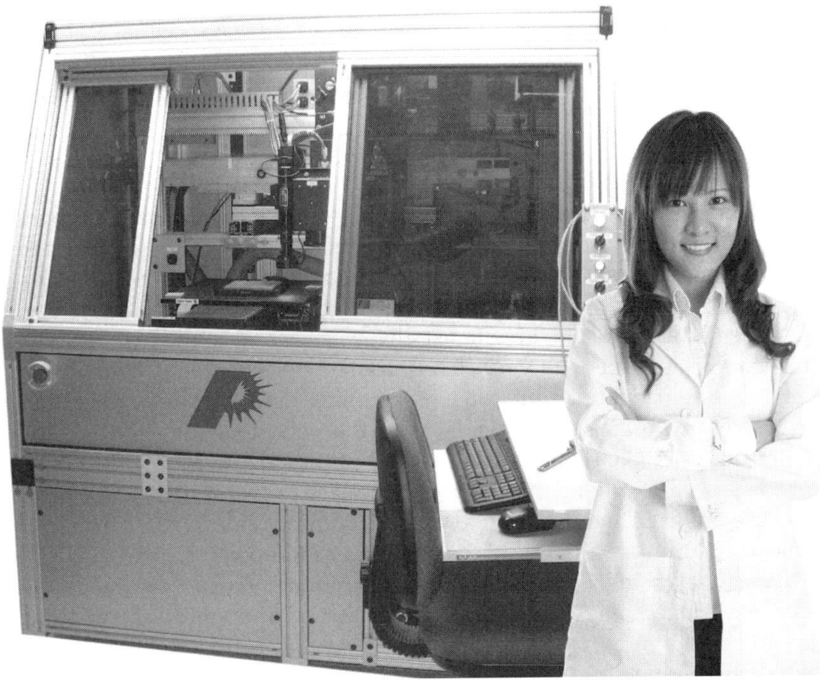

FIGURE 3.6.1 Typical micro laser system.

Product Considerations

Kerf width is one of the key issues of precision laser micro machining. This is essentially the minimum *spot size* of the laser on target. If the spot size on target is 20 μm (0.00078 in.), then this is the minimum width of cut or diameter of hole that can be made. If a larger cut or feature is required, then multiple passing is used, but this increases processing time. Other things affecting the kerf width are taper and aspect ratio. For a laser-processed feature, there will always be an inherent *taper* and it will be such that the entrance side is bigger than the exit side, unless very special optics are used to mitigate

Focal length of lens	Shorter lenses give smaller spots, but have a smaller depth of field
Aspect ratio	10:1 is pretty standard, up to 100:1 in some cases—subject to taper considerations
Taper	5–10 degrees; can be made larger by multipassing; smaller requires prism optics. Because of taper, material thickness must be very uniform to avoid variation of kerf width on target.
Heat-affected zone	Decreases with decreasing wavelength and pulse length, energy per pulse, and repetition rate
Co-axial assist gas	Can be used with fixed beam delivery systems to minimize thermal effects and carry away debris
Galvanometers versus fixed beam	Galvos are faster, but less accurate, than fixed beam
Spot size	UV lasers give smaller spot sizes than IR lasers

Table 3.6.2 General Design Considerations

this effect. Taper is affected by energy density, aspect ratio, material interaction, focal length of lens, and other things, but it is generally between a few degrees on the low side to about 10 degrees. More taper is easier to obtain than "straight walls" as this can be programmed as long as the main feature is larger than the beam size. Because of this taper, hole diameters are almost always specified at the exit, or the smaller side. This may be an issue with very closely packed holes that may actually overlap on the entrance side if the holes are too closely spaced and/or too large in diameter. *Aspect ratio* is the material thickness related to the hole diameter—for example, drilling a 20-µm-diameter (0.00078-in.) hole into a 60-µm-thick (0.0024-in.) piece of material gives a 3:1 aspect ratio. Higher aspect ratio holes are harder to drill, may have more taper (because multiple lines are used to increase entrance side kerf), and take longer to drill. Table 3.6.2 summarizes some of the above considerations.

When designing a product for laser micro machining or for investigating whether laser micro machining is appropriate for a particular application, the above discussion should be kept in mind. When dealing with a contractor, the first order of business is usually the signing of some type of mutual nondisclosure agreement.

Laser Software Considerations

Many different product design files types can be used including dxf, dwg, Gerber, HP Plot, and Solid Works. Most fabricators have translators, but the industry general standard is a dxf file, which can be generated by almost any CAD program. Two-dimensional files are the easiest to work with and usually only require x/y and possible z-motion for feature depth control. Three-dimensional files are harder to work with and require a multi-axis coordinated laser system or perhaps robotics. Some processes, like tube and stent cutting and drilling, also require the use of a rotary axis.

Laser cut areas are usually on one layer while *fiducials* are on another. Fiducials are used to align the laser-processed areas with known locations on the part. In principle, features can be used, but in many cases actual alignment fiducials are used. The best in alignment accuracy can be obtained when these fiducials are precise and crisp. CAD files should clearly state the units (English or metric) in the legend as well as the material, the material thickness, and all other information necessary to review the document completely and make informed decisions on the correct way to move forward. Another critical piece of information is the tolerance, which should be realistic and possible to achieve.

Frequently, the material choice is limited, but if there is some leeway in the material choice, it is pretty easy to identify whether a particular material will be good for laser processing. There already exists a library of information in the public domain, but the way to tell if a particular material is a good candidate for any laser is to look up the IR or UV absorption tables and see if the material has a high absorption at the laser wavelength of interest. If so, the processing will probably be possible, subject to all other considerations, but if there is little or no absorption the chance of success is small to nil. Note that here again, UV and USP [ultrashort pulse—femtosecond (fs) and picosecond (ps)] lasers have higher absorption in many more materials than IR lasers. Table 3.6.3 gives some information on commonly seen materials.

Metals	Metals process well around the 1-μm wavelength (YAG, disc, and fiber lasers). Back reflections from especially refractive metals can be problematic. Requires fairly high on-target fluence. Melt can be a problem with long-pulse and/or long-wavelength lasers.
Stainless steel	Used extensively in medical instruments and implants
Copper	Used in micro electronic applications
Molybdenum	Responds very well to laser processing with good edge quality
Titanium	Used in medical implants and other applications requiring high strength and low weight
Nickel	Another implant material, especially Nitinol®
Hard materials	Require relatively high fluence on target. Chip out on back side especially can be a problem.
Alumina ceramic	A very popular material for electronic applications
Zirconia ceramic	Used extensively as a coating for aircraft engines
Glass	The "holy grail" of laser processing is to process glass quickly and cleanly—for micro fluidics, next-generation printed circuit boards, displays, etc. Usually requires UV and/or USP lasers to process cleanly.
Sapphire	Very *hot* application area, especially for USP lasers, smart phone displays, notebook displays, etc.
Silicon	Still extensively used in solar cells and semiconductor applications
Polymers	Require much lower fluence than the materials discussed above
Teflon®	A saturated polymer that is very hard to laser-process due to poor absorption even in the UV. Requires very short wavelength (<200 nm) or USP laser.
Kapton® (Polyimide)	Probably the most popular material used in material processing—good mechanical, electrical, and thermal properties
Mylar®	Another very widely used material in many different application areas

TABLE 3.6.3 Common Materials Encountered in Laser Micro Machining Applications

Set up	From $100 to a few hundred dollars depending on the laser and complexity of set up. May be minimized or waived for recurring business or large lot sizes.
NRE—programming	Usually billed on an hourly basis—can be minimized if files are received in workable fashion
NRE—tooling	Associated with specific part dimensions and dedicated to a particular application/customer
Piece cost	Depends on laser time, number of parts, handling, shipping, packaging, and QC requirements
Laser time–lab time	Billed as "best effort" on an hourly basis. When a customer is present, the customer is working with a trained engineer or scientist. Varies from about $100 to $600 per hour depending on laser and level of operator.

Table 3.6.4 Cost Considerations

After assuring the technical issues are fit for laser micro machining, there is the issue of cost. In general, laser methods of micro machining should be used when cut or drill quality demands outweigh cost demands. This is especially true as the wavelength of the laser or the pulse length of the laser gets shorter because not only do the photons cost more, but the material removal rate is slower due to shallower penetration depth. When using a laser contract manufacturer, the costs include recurring set-up costs, nonrecurring engineering (NRE), programming, tooling, and piece cost. Set-up costs are usually incurred every time a job is run, even if it has been done previously. In some cases, if a job is recurring and if it is high enough in value, these set-up costs might be minimized or eliminated. NRE is charged only one time assuming the task does not change. If there is a change of design or material, these charges may be incurred again. NRE includes programming and tooling expenses. The per piece cost depends on a variety of factors such as the laser to be used, the time it takes on the laser, the number of parts per lot, the material, and also additional requirements like part handling, QC, packing, and shipping. Sometimes it is not possible to accurately quote a job if it is not clear that the specifications can be met, or if it is an unfamiliar material or if a design just cannot be made without a detailed investigation. In this case, straight "lab time" is usually billed on an hourly, best-effort basis. It is much more efficient if the customer can be present for this type of investigation as it saves a lot of back and forth time and mitigates risk. Table 3.6.4 summarizes some of the costs involved in laser micro machining.

Examples of Micro Machining

Figures 3.6.2 to 3.6.5 and Table 3.6.5 present just a few of the application areas for laser micro machining. "Soft" materials like polyimide and PET as well as "hard" materials like diamond, ceramics, and glasses can be effectively machined with features down to the micrometer level.

Laser micro machining is a very well-established technology. On the other hand, new laser sources, optic designs, and controls are almost monthly opening up even

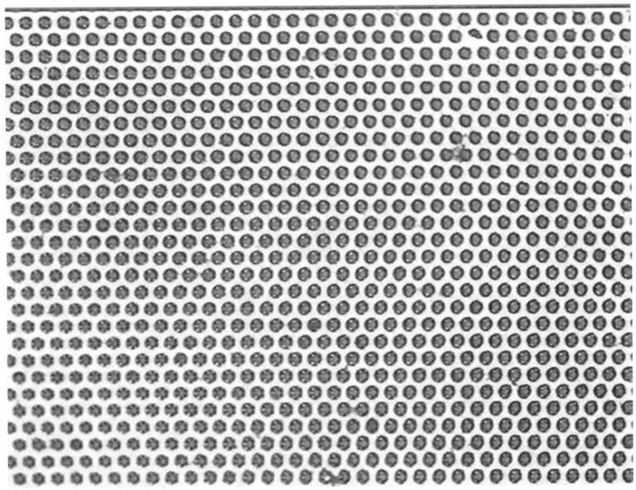

Figure 3.6.2 Biofilter with 0.001-in. (0.025-mm) holes in 0.002-in.-thick material on 0.002-in. centers.

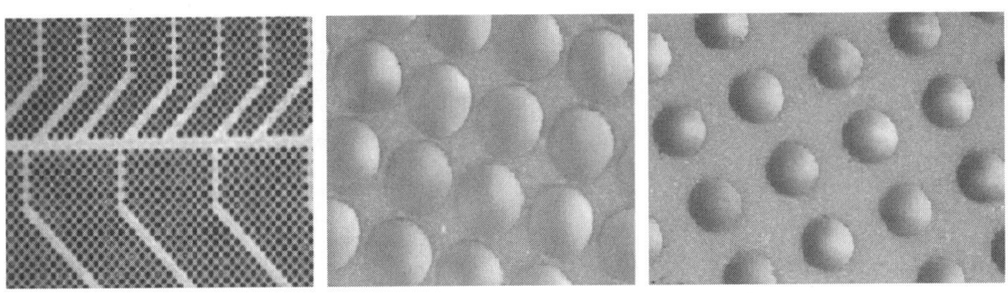

(a) (b) (c)

Figure 3.6.3 Wafer probe card with 0.100-mm holes on 0.200-mm spacing in 0.200-mm-thick material.

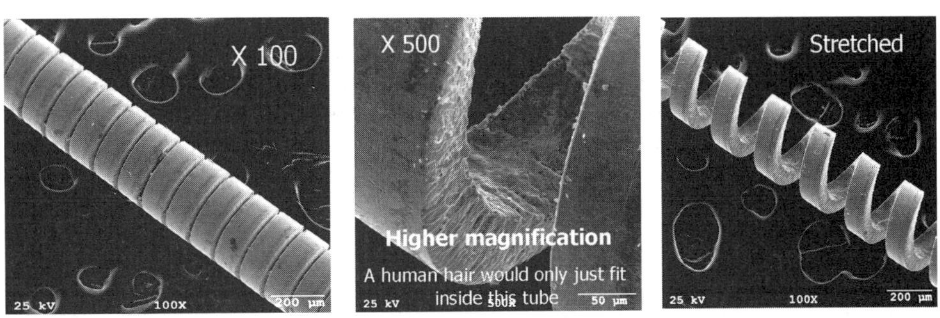

(a) (b) (c)

Figure 3.6.4 Hypo tube cut with IR fiber laser into a “spiral.”

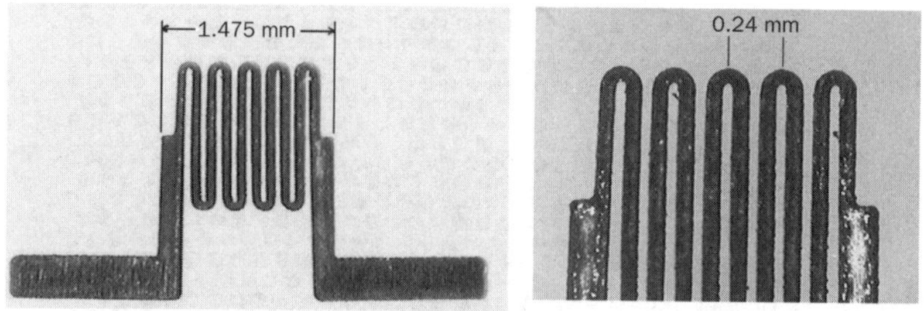

Figure 3.6.5 Serpentine pattern in 0.050-mm-thick (0.002-in.) metal.

Figure/Feature	Material	Comment
3.6.2—Biofilter with 0.001-in. (0.025-mm) holes in 0.002-in.-thick material on 0.002 in. centers	Polyimide	248-nm excimer laser drilling with mask allows thousands of holes to be produced in one image
3.6.3—Wafer probe card with 0.100-mm (0.004-in.) holes on 0.200-mm (0.0078-in.) spacing in 0.200-mm-thick material	Alumina ceramic	Note size difference in entrance (b) and exit (c) holes because of taper
3.6.4—Hypo tube cut with IR fiber laser into a "spiral"	Stainless steel	As indicated the interior diameter would just encase a human hair. Post-processing like electropolishing is generally used with IR lasers.
3.6.5—Serpentine pattern in 0.050-mm-thick (0.002-in.) metal	Molybdenum	Cut using a nanosecond, 355-nm laser using NO post-processing

Table 3.6.5 Examples of Micro Laser Machined Features

smaller applications, faster cutting, smoother surfaces, and a wider variety of workpiece materials. These advances also open and broaden the use of lasers into new fields. Their use in welding, joining, marking, 3D manufacturing, and interferometry is described in other chapters.

CHAPTER 3.7

Micro Electrical Discharge Machining

LaRoux K. Gillespie

Manufacturing Consultant
Andover, Kansas

Electrical discharge machining (EDM) provides unique opportunities for micro parts since it does not create any physical forces on the part to remove material. It can remove only millionths of a millimeter or millionths of an inch of material or it can remove millimeters and thousandths of an inch at one time. It can cut, slice, grind, polish, drill, "mill," pocket, sculpture, pattern, and roughen surfaces. It works on cylindrical parts as well as prismatic ones, thin or thick, hard or soft materials. Since it does not involve normal cutting tools it does not create burrs (which can be difficult to remove on micro parts). It can however produce recast material at edges, but modern machines have been designed using conditions that make them almost perfectly free of any projecting edge material. While normally limited to metals, it can do some machining of glass or other materials that have some electrical conductivity. It is not a process for polymers, nor for most ceramics, although some ceramics can have some conductivity.

There are four variations of the process for micro machining: conventional sinker electrode, wire electrical discharge machining (WEDM), wire electrical discharge grinding (WEDG), and electrochemical discharge machining (ECDM). As a subset of conventional electrode-based EDM, "hole poppers'" sole design is to make holes, and to do so quickly and automatically.

Solid Electrode EDM

Solid electrodes are machined to a mirror image of the features needed in the finished part (Fig. 3.7.1). In most instances this would be a mirror image of a pocket or cavity, a notch, or a hole. Holes or cavities can be of any shape and through the part or merely sunken into the part. Square holes, *Allen*-head hexagonal cavities, star shapes, tapers, and many others are common. This process is also called sinker EDM or plunge

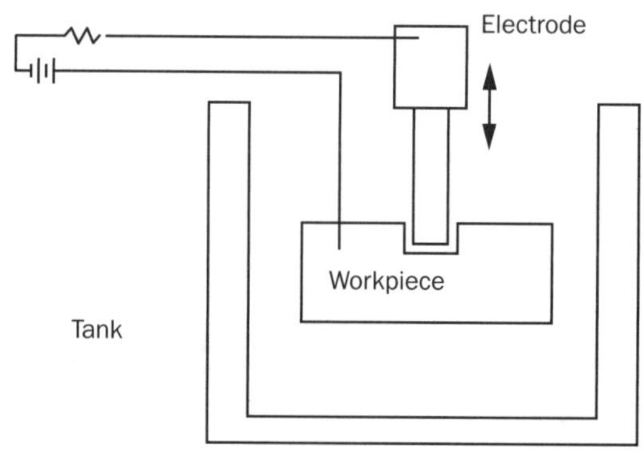

FIGURE 3.7.1 Sinker electrode machining.

EDMing. The process is the same for large parts or micro parts although the power supplied will be different.

Rather than using a mirror image tool, simple cavities can also be cut using a solid cylinder, for example, like an end mill would be used to cut the cavity. This type of tool is simply traversed back and forth over the cavity area in a manner that causes all the tool wear to be on the bottom of the cylinder so that wear does not affect the vertical shape of the cavity. This approach requires a machine that can move the electrode in x, y, and z axis, as well as rotate the electrode. This micro EDM milling can create cones, pockets, slots, and arbitrary 3D shapes. It is used to machine catheters and surgical and measurement instruments among other applications.

What is not common are EDM features that are hidden from view in the part. In other words if the feature cannot be seen by looking down on it, it is not a good choice for EDM (Fig. 3.7.2). This however is not an absolute limitation since it is possible on some machines to have a horizontal motion added to what is normally simply vertical feeding electrode. This is not standard on simple sinker machines, but many of today's machine do have lateral motion that can create these undercuts.

Wire Electrical Discharge Machining

Wire EDM uses a moving wire to cut features. The wire is fed off a reel of microscopic wire past the part and back onto a take-up reel (Fig. 3.7.3). The EDM sparks roughen and change the dimensions of the wire so it has to be constantly fed to provide a fresh surface for cutting. The wire never touches the part; it can have a spark gap of 5 µm (0.0002 in.) between it and the feature to be machined. This size gap would cut 10 µm (0.0004 in.) larger than solid electrode machines that have a similar gap in their operation. The wire can cut slots of any size by moving the wire down like a coping saw does then feeding the wire sideways and then back up. In most instances, the wire is actually held vertically and the table with the part is moved in an x-y plane. The head of the wire feed mechanism can be rotated so the wire is fed at an angle to produce tapers (if the machine is bought with that capability). The fact that the table can move in any x-y combination means the parts can have intricate "Z" shaped, circle, or other cut outs. Each can be as small as the wire plus twice the spark gap (the cutting spark

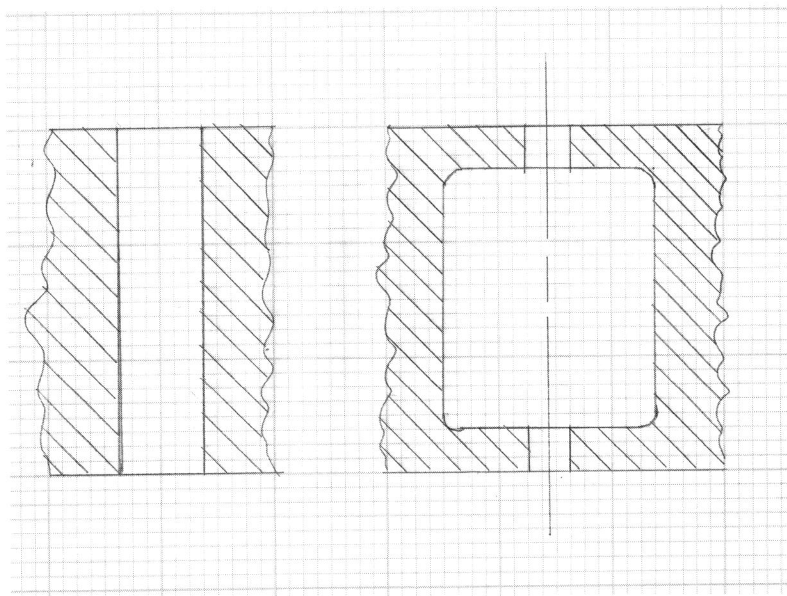

FIGURE 3.7.2 The pocket shown in the figure is not a good candidate for EDM but the vertical shaft hole is. The pocket cannot be cut with wire EDM. The pocket can be cut on some plunge machines, but not on machines having only a straight plunge capability.

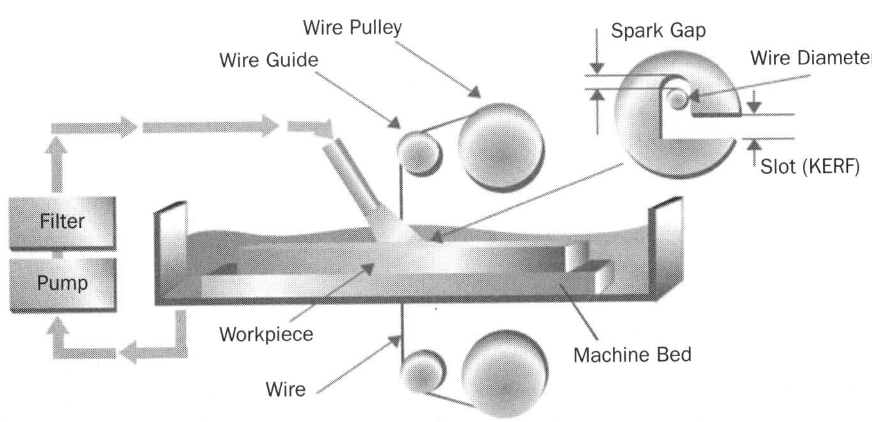

FIGURE 3.7.3 Wire EDM schematic. (Courtesy American Wire EDM.)

shoots out 360 degrees around the wire). For simple shapes the wire machines provide a less expensive tooling approach; solid electrodes are expensive to make and have to be made before the part is machined. None of this is required for wire machines. Wire machines cut on the wire plane. Because the wire is held straight the internal pocket shown in Fig. 3.7.2 is not a candidate for wire, but the wire can cut a through hole of any straight-sided shape.

While WEDM slots are limited to the diameter of the wire plus overcut, the process can make solid electrode tools that produce slots only 25 to 50 μm (0.001 to 0.002 in.) wide.

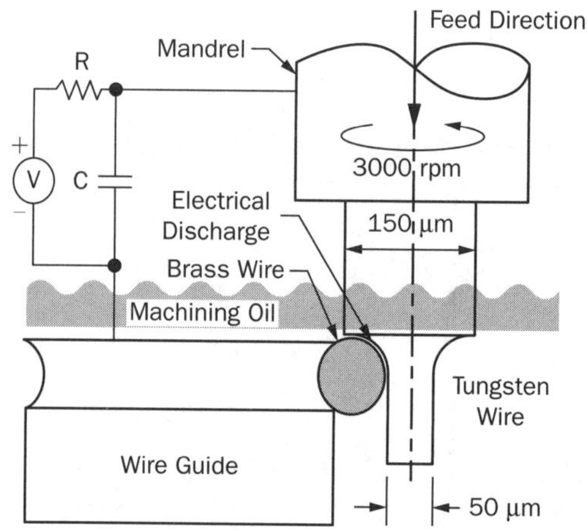

FIGURE 3.7.4 Wire electro-discharge grinding. (After Morgan, 2004.)

Wire Electrical Discharge Grinding

Wire electrical discharge grinding (WEDG) is essentially a wire machine in which the part can be rotated while the cutting action occurs so that a polishing action or grinding function occurs (Fig. 3.7.4). Sparks from the moving brass wire erode away material from the rotating workpiece. The resulting part is some form of a cylindrical shaft. Despite its grinding name, there is no physical contact with the workpiece.

The wire used for this process is typically a 100-µm-diameter (0.004-in.) brass wire and the spark gap can be as small as 2 µm (79 µin.). For this process the part is immersed in a dielectric oil. If the bottom of the workpiece end is free, the setup can machine it as well as provide a contour on the cylindrical portion of the part.

The uniqueness of the process allows it to produce very long length-to-diameter ratio (high aspect ratio) tools or parts. As an example, making a tungsten rod 50 µm (0.002 in.) in diameter by 1500 µm (0.059 in.) long is very simple and easy to replicate. No other process can accommodate this aspect ratio for such a small diameter (30:1). The process can make even smaller diameters.

With no special effort 50 µm (0.002 in.) tungsten rods 500 µm (0.020 in.) long will have surface finishes as smooth as 0.14 µm (5.5 µin.) Ra and straightness of 3.53 µm (0.00014 in.). Because of machine errors (rather than process errors) longer shafts will have more out of straightness.

Electrochemical Discharge Machining

By adding chemicals to EDM, users can etch glass substrates. The electrodes have to be attached very near the glass target site and the etching is slow.

As one example of electrochemical discharge machining (ECDM), razorblade tool electrodes can be used to etch channel-like features on glass substrates. Glass can be machined by using conventional EDM as well, but it is slow since glass has very small conductivity compared to metals. EDM has been used for decades to cut glass, but it is not a common process, nor a recommended one. In addition to EDM, ball end mills have also been used to machine shallow depression in glass. ECDM is described in Chap. 3.10.

Materials Machined

All variations of EDM can machine any hardness of metal. Stainless steel which is relatively soft is a long-time material application. Tungsten carbide, which can have hardness of 62HRC, is an excellent candidate for cutting. Titanium, hardened steel, conductive ceramic, shape memory alloys, and PCD can be EDMed. Steel, copper, aluminum, Hastelloy, Kovar, and Inconel are other materials that can be EDMed.

Polycrystalline diamond (PCD) tools, which are often used for either machining or grinding, are good candidates for WEDG machining since the cobalt binder conducts electricity. While most readers might not think of grinding wheels as something that needs forming, PCD tools often are used for form grinding. WEDG can both form the initial shape and dress the wheel after some use.

Equipment

Micro EDM equipment is generally significantly smaller than that used for larger parts. Several equipment designs have been used for desktop units, although most are standalone machines that are the size of a refrigerator. Some are portable so they can be moved around the shop to meet a variety of schedule changes or business growth.

Commercial machines are available for each of the variations described here, except for ECDM. Many allow both WEDM and WEDG in the same machine. For plunge (solid electrode) EDMing a typical plunge rate on micro features might be 400 μm/min (0.016 in./min) for a 0.6-mm-diameter (0.024-in.) hollow electrode. For a WEDM machine, it might be as low as 2 μm/s and up to 10 μm/s (79 μin./s to 0.00039 in./s).

Applications

As mentioned above, dressing PCD tools is a major application for WEDG since even the smallest grinding wheel can be formed without any physical contact. Micro pipes/nozzles and ink-jet nozzles are two applications in wide use. Because it is not a contact process, EDM of any variation can machine structures leaving very thin and uniform webs or walls. Spinneret nozzles for fiber production as well as some fuel nozzles are good EDM applications, although some laser micro machining can accomplish the same thing quicker in some designs. EDM machines are used to develop medical parts, micro fluidics circuits, surgical needles, implants, precision surgical instruments, micro probes for CMM-like measurement, and intra-ocular lenses. WEDG is used to make most of the micro diameter high aspect ratio EDM drilling electrodes. In essence, the process makes its own electrodes.

For micro machining, EDM is one of the first choices that should be considered.

Figures 3.7.5 through 3.7.12 illustrate some of the micro features produced by micro EDM on production parts.

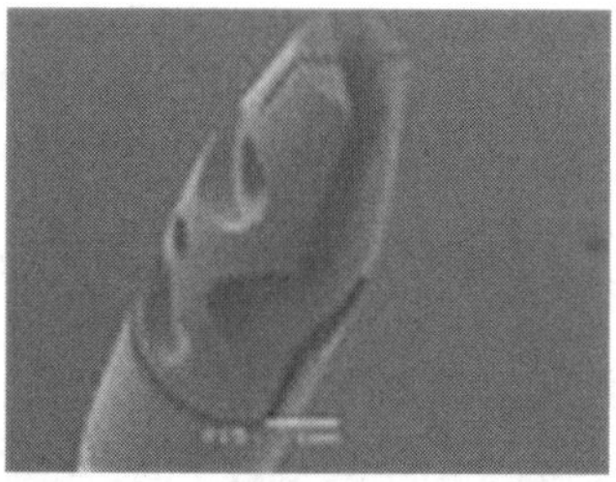

Figure 3.7.5 Slots produced by micro wire EDM. (Reproduced from *Journal of Materials Processing Technology*, vol. 210, p. 133, 2010, with permission from Elsevier.)

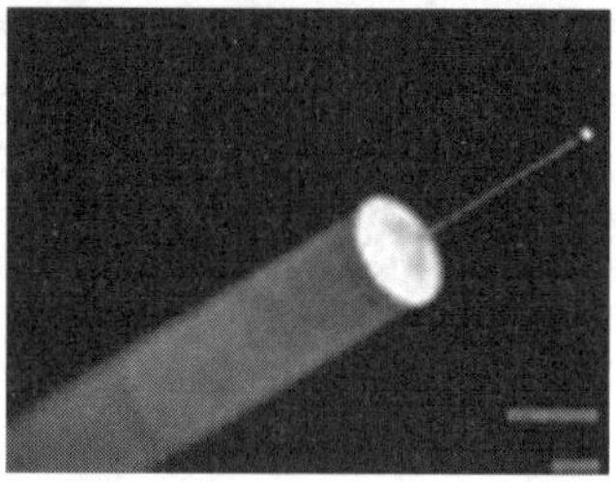

Figure 3.7.6 Micro probe produced by micro EDG. (Reproduced from *Journal of Materials Processing Technology*, vol. 210, p. 133, 2010, with permission from Elsevier.)

Figure 3.7.7 0.4-mm-pitch (0.016-in.) micro electrode array, 5 × 5 mm (0.2 × 0.2 in.) with 144 electrodes of 5 mm height (0.2 in.). Material: doped silicon. (Courtesy Viteris Technologies.)

Figure 3.7.8 0.4-mm-pitch (0.016-in.) micro electrode array, 5 × 5 mm (0.2 × 0.2 in.) with 144 electrodes of 5 mm height (0.2 in.). Cross section of electrodes is varied along length to increase flexibility. Material: doped silicon. (Courtesy Viteris Technologies.)

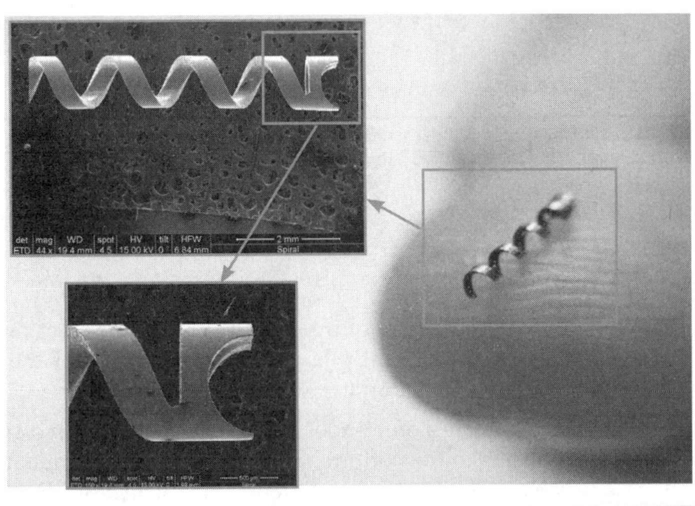

Figure 3.7.9 1.0-mm-pitch (0.04-in.) helical micro swimmer machined with rotary axis. Material: 1-mm-diameter nitinol tubing. (Courtesy Professor Jake Abbott, University of Utah.)

FIGURE 3.7.10 50-μm 3D island left in EDM pocket. (Courtesy Kamlakar Rajurkar, University of Nebraska & Zuyuan Yu, Dalian University of Technology.)

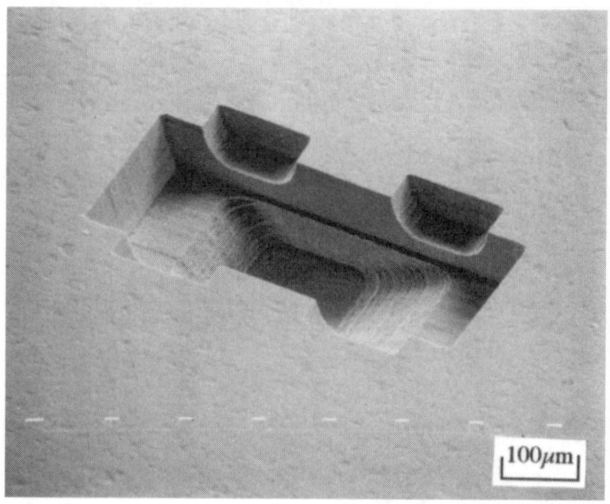

FIGURE 3.7.11 Microscopic car cavity made by EDM. (Courtesy Kamlakar Rajurkar, University of Nebraska & Zuyuan Yu, Dalian University of Technology.)

Design Considerations

- If a change in diameter is required using WEDG, the fillet radius between the two diameters will be slightly larger than the radius of the wire that cuts it [typically 50 μm+ (0.002 in.)] (see Fig. 3.7.4).
- Some tool wear occurs on all tool surfaces. Starting with a solid cylinder tool when the tool gets to the bottom of the desired hole it is making, tool wear will have generated a conical-shaped tool which will also produce a conical-shaped hole. The amount of wear and taper can be controlled and a round hole can be achieved by various means.

FIGURE 3.7.12 Microscopic pyramid produced by micro EDM. (Courtesy Kamlakar Rajurkar, University of Nebraska & Zuyuan Yu, Dalian University of Technology.)

Characteristic	WEDM	Plunge EDM	WEDG
Geometric accuracy (μm)	±1 (39 μin.)	±1 (39 μin.)*	±1 (39 μin.)
Maximum positioning accuracy (μm)	1 (39 μin.)	1 (39 μin.)	1 (39 μin.)**
Finest surface finish Ra (μm)	0.08 (3 μin.)	0.2 (8 μin.)	0.06 (2 μin.)
Minimum channel widths (μm)	50 (0.0002 in.)	20 (0.0008 in.)	120 (0.0047 in.)
Maximum aspect ratios for hole		100	
Maximum aspect ratios for shafts			30+
Smallest internal radius (μm)	20 (0.0008 in.)	10 (0.0004 in.)	n/a
Smallest electrode made (μm)	20 (0.0008 in.)	10 (0.0004 in.)	5
Smallest electrode used (μm)		5 (0.0002 in.) [commercial applications 20 μm (0.0008 in.)]	

*The best commercial tolerance claimed for size and location is ±0.0002 in.
**One manufacturer of micro EDM machines claims 0.1 μm (4 μin.) stepping capability.

TABLE 3.7.1 Capabilities of Micro EDM Variations (After Fleischer et al., 2004)

- Holes made by wire EDM can be made perfectly round as long as they are through holes.
- A heat-affected zone (HAZ) results from the sparks. It is measureable and may be a concern for thin parts, even though it is very thin on any micro EDMed surface.
- Designers need to consider the stresses and fatigue environment for EDMed parts. Fatigue life should be better for micro EDM than for conventional EDM

simply because the spark energy is less, resulting in thinner HAZ, but little data exists for the thin conditions on micro parts.

- Surface finishes are rougher for any EDM than for polished or lustrous surfaces.
- Part surfaces will have carbon on them and other elements found in the dielectric.

A review of published data indicates that the capabilities shown in Table 3.7.1 are characteristic of the best or smallest that these processes can do on micro level work.

References

Bamberg E., and S. Heamawatanachai, Orbital electrode actuation to improve efficiency of drilling micro holes by micro-EDM, *Journal of Materials Processing Technology*, 2009, 209(4), pp. 1826–1834.

Dave, H., V. Mathai, M. Mayanak, H. Raval, and K. Desai, Studies on Tool Wear Characteristics in Micro Electro Discharge Slotting Process (μ-EDS), *Proceedings of the 5th International & 26th All India Manufacturing Technology, Design and Research Conference (AIMTDR 2014)*, December 2014, pp. 295-1–295-7.

Fleischer, J., J. Schmidt, M. Knoll, S. Haupt, C. Müller, R. Förster, and A. Gehringer, Mikrobearitung durch abtragen, *Werkstattstecknik*, 2004, 94(4), pp. 1070–1110 (also online).

Hung, J.-C., J.-K. Lin, B.-H. Yan, H.-S. Liu, and P.-H. Ho, Using a helical micro-tool in micro-EDM combined with ultrasonic vibration for micro-hole machining, *Journal of Micromechanics and Microengineering*, 2006, 16(12).

Liu, Q., Q. Zhang, G. Zhu, K. Wang, J. Zhang, and C. Dong, Effect of electrode size on the performances of micro-EDM, *Materials and Manufacturing Processes*, 2016, 31(4), pp. 391–396.

Morgan, C. J., Micro electro-discharge machining: techniques and procedures for micro fabrication, University of Kentucky Master's Theses, Paper 327, 2004.

Peng, Z., Z. Wang, Y. Dong, and H. Chen, Development of a reversible machining method for fabrication of microstructures by using micro-EDM, *Journal of Materials Processing Technology*, 2010, 210(1), pp. 129–136 (doi:10.1016/j.matprotec.2009.08.002).

Qian, J., J. Wang, E. Ferraris, and D. Reynaerts, A holistic approach for zero-defect micro-EDM milling, *IEEE International Symposium on Assembly and Manufacturing (ISAM)*, Xi'an, 2013, pp. 226–229 (doi:10.1109/ISAM.2013.6643533).

Rakwal D., and E. Bamberg, Slicing, cleaning and kerf analysis of germanium wafers machined by wire electrical discharge machining, *Journal of Materials Processing Technology*, 2009, 209(8), pp. 3740–3751.

Rakwal D., S. Heamawatanachai, P. Tathireddy, F. Solzbacher, and E. Bamberg, Fabrication of high aspect ratio silicon micro-electrode arrays using micro-wire electrical discharge machining (micro-WEDM), *Microsystem Technologies*, 2009, 5(5), pp. 789–797.

Zhang, L., H. Tong, and Y. Li, Precision machining of micro tool electrodes in micro EDM for drilling array micro holes, *Precision Engineering*, January 2015, 39, pp. 100–106.

CHAPTER 3.8

Precision Electrochemical Micro Machining

Don Risko

DGR Consulting
Jamestown, Pennsylvania

The Process and Capabilities

Precision electrochemical machining (PECM) for micro parts can produce feature geometries (holes, slots, cut outs) in metal surfaces in the dimensional range of 10 to 100 μm (0.0004 to 0.004 in.). Careful selection of parameters and the use of precision cathodes and fixtures make machining on internal and external surfaces (and both simultaneously) possible to ±1 to ±5 μm (±40 to 200 μin.) tolerance. Surface finish is typically very good 0.025 to 0.2 μm (1 to 8 μin. Ra) depending on material type. The process is controllable and repeatable and is very useful for high-precision, high-volume production due to the short cycle times and long tooling lifespan. Since no cutting forces are employed, the edges of machined surfaces are free of burrs.

PECM or micro ECM is a process to remove material while holding micrometer range dimensional tolerances. PECM uses the same principles as traditional electrochemical machining but requires a greater degree of tooling and process control precision. The process is typically used to machine precision features in metals that require intricately shaped geometry and smooth surfaces. Advantages include the ability to machine metals in the hardened state; machine with very low force; generate little heat; and machining multiple parts simultaneously without cycle time penalty.

Process Principles

Electrochemical micro machining is a metal removal process based on Faraday's law of electrolytic dissolution of metals. The principal components of a PECM system are the workpiece (anode), the tool (cathode), which are both conductive metals, and an aqueous electrolyte. The electrolyte is a conductive solution typically consisting of a salt such as sodium nitrate ($NaNO_3$) dissolved in water. Other electrolytes of varying

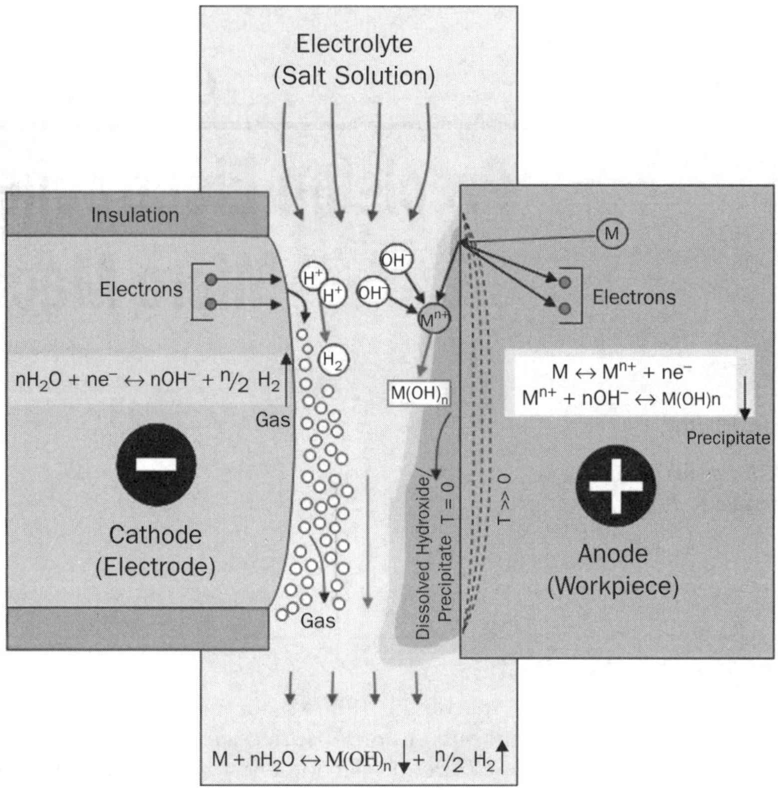

Figure 3.8.1 Electrochemical machining reaction schematic.

concentrations have also been used but sodium nitrate is the most common. In an electrochemical cell, the electrolyte is flushed between the anode and the cathode gap, thereby completing an electrical circuit. Decomposition of the salt in the water permits the breaking of the weak ionic bonds holding the sodium ions (Na^+) and the nitrate ions (NO_{3^-}) together. Once dissolved, these ions are free to move about the solution, carrying charge and permitting current flow (electron transfer) when an electrical potential or voltage is applied between the anode and cathode electrodes. The electric charge is carried by the positive ions moving toward the cathode, or negative pole, and the negative ions moving in the opposite direction toward the anode, or positive pole.

Application of the voltage, and the resultant electrical current flow through the electrolyte, initiates chemical reactions that occur at the surfaces of the electrodes as shown in Fig. 3.8.1. At the anode or workpiece surface, electrons are removed from the metal (M) surface which causes the oxidation reaction of the metal anode according to the reaction:

$$M \rightarrow M^{n+} + ne^- \tag{3.8.1}$$

where the variable n is the valence of the particular metal comprising the anode. At the cathode surface, electrons must be accepted causing a reduction reaction to occur.

However, in an ECM circuit, this reaction involves the reduction of water, not the metal surface of the cathode according to the reaction:

$$nH_2O + ne^- \rightarrow n/_2\, H_2 + nOH^- \tag{3.8.2}$$

This typically yields hydrogen gas but not always depending on the overall solution pH. As the hydroxyl ions (OH^-) migrate toward the positive pole (anode) in the electrode gap, a reaction with the dissolved metallic ions (M^{n+}) occurs according to the reaction:

$$M^{n+} + nOH^- \rightarrow M(OH)_n \tag{3.8.3}$$

which produces a metal hydroxide that typically precipitates out of solution as an insoluble solid, depending on the solution pH. The net chemical reaction is

$$M + nH_2O \rightarrow M(OH)_n\,(\downarrow) + n/_2\, H_2\,(\uparrow) \tag{3.8.4}$$

Because the metal hydroxide that is produced in the electrode gap is typically not soluble in the aqueous solution and is electrically neutral, it behaves like a resistive or insulating material.

The electrochemical machining reaction is governed by Faraday's law, which indicates that the mass of material removed (W_M) is equal to

$$W_M = M \times I \times t/n \times \mathbf{F} \tag{3.8.5}$$

where

M = atomic weight of anodic material
I = current (A)
t = duration or run time (sec)
n = valence of anodic metal
F = Faraday's constant (96,500 C)

In the Faraday's equation, M, n, and F are constants (M and n are material-dependent constants) and can be combined to a single constant k_M. The material constant can be calculated for alloy compositions as well. Therefore, this equation can be simplified to

$$W_M = k_M \times I \times t \tag{3.8.6}$$

The theoretical amount of material removed from the anode surface is proportional to the current applied and the duration. Ohm's law dictates that the potential or voltage (E) is equal to the current (I) times the resistance or impedance (R) of an electrical circuit.

$$E = I \times R \text{ or } I = E/R \tag{3.8.7}$$

Putting Eq. (3.8.7) into (3.8.6) results in

$$W_M = k_M \times E \times t/R \tag{3.8.8}$$

This equation demonstrates the relationship of material removal with respect to the electrical resistance across the electrode gap. For an electrochemical circuit the conductivity, which is the inverse of impedance or resistivity, is not constant. More important than the "bulk" electrolyte conductivity is the "local" conductivity of the solution in the electrode gap. The conductivity is dependent on several factors which themselves can be functions of time. The electrolyte conductivity in the gap is dependent on parameters such as the concentration of salt in the solution, local hydroxide concentration in the electrolyte, bulk and local (gap) temperature, electrolyte flow rate, and velocity profile.

Even though most metals and alloys will electrochemically machine at a wide range of voltage potentials and current densities, adequate results such as defined edges and good surface finishes can only be achieved at a much narrower band of parameters.

Electrolyte Type and Concentration

In any ECM operation, proper electrolyte selection is an essential step. The electrolyte selected directly affects the machining characteristics such as surface finish and feature geometry. Electrolyte type, concentration, pH, and temperature must be chosen for the particular anodic metal or alloy that is being machined.

Sodium nitrate is the typical electrolyte used in PECM because of its propensity to passivate the anode surface. While passivation prevents unintentional machining of some areas, passivation of the intended machining area is undesirable and is controlled by correct choice of machining parameters which vary by material type. For instance, under the exact same process conditions some materials will not machine at all due to passivation that occurs at the anode surface. Once a surface is passivated, current flow is prevented and the electrochemical reaction is terminated.

In PECM applications, where the electrode gaps are small and the processing voltages tend to be low, electrolyte concentration and temperature are critical to controlling surface passivation or selective machining of alloy components.

PECM System

PECM systems consist of a machine structure that incorporates a precision machining axis (or multiple axes); tooling (consisting of a cathode, anode fixture, and electrolyte flow components); an electrolyte supply; precision DC power supply; and control with a user interface.

Electrochemical Tooling

The tooling is the most important part of a PECM system. A fixture must hold the cathode and the workpiece within tens of micrometers (a few thousandths of an inch) of one another, and they additionally must be centered with respect to one another within only a few micrometers (less than 0.001 in.). Electrical contact must be made to both electrodes as well for machining to occur. The fixture also must introduce electrolyte into the electrode gap. A typical PECM tool is shown in Fig. 3.8.2. The flow velocity and patterns are very important to the end machining result and a repeatable flow pattern is necessary to achieve consistent machining results.

The cathode must be precisely micro machined using micro milling, grinding, micro EDM, laser, and even micro ECM. The machined cathode may be insulated except for an exposed cathodic surface that is nearly the mirror image of the target machining geometry. However, certain cathode dimensions must be undersized to account for the electrical field expansion across the gap. The dimensions of the cathode and its position in the fixture will determine the initial electrode gap. This parameter is critical because the gap, and the cathode dimensions, will determine the achievable micro ECM feature

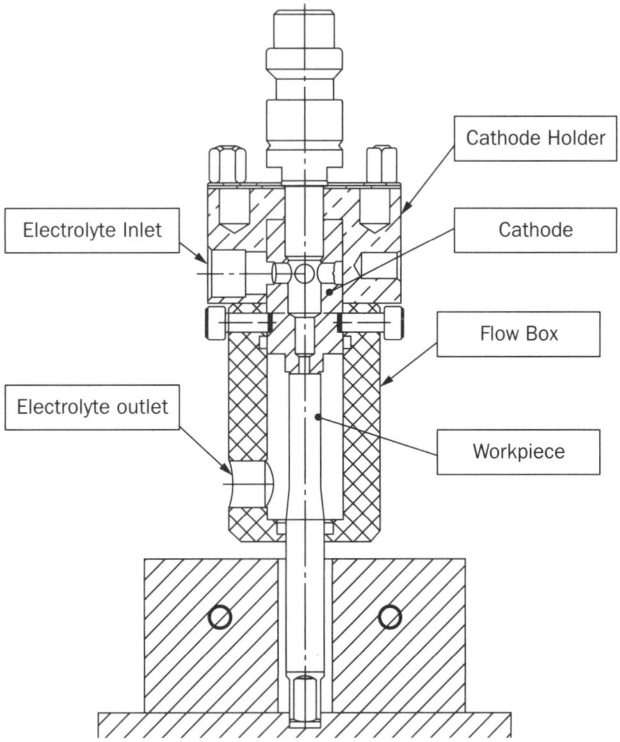

FIGURE 3.8.2 PECM fixture diagram.

geometry. The larger the gap between the electrodes, the higher the voltage required to achieve the proper current density for micro ECM. Higher voltage potentials result in relatively wider field profiles which predictably result in wider machined features. During production involving repeated cycling, processing with higher potentials can also increase the heat added to the system requiring greater temperature control. Higher temperatures increase the activity of the electrolyte which, in turn, increases the solution conductivity. This can result in an increased current density which yields deeper and wider dimensions. Also, the by-products of the ECM process (metal hydroxides, gas bubbles, and temperature) build up during current flow between cathode and anode. (This is particularly true in frontal gap machining.)

Since the machining gap is typically in the 10 to 25 μm (0.0004 to 0.0010 in.) range, and the total path length can be 100 times larger, there is only a small volume of electrolyte in any given cross-sectional area at any particular time. Therefore, the electrolyte near the flow outlet can have a considerably lower conductivity compared to the cleaner incoming electrolyte pumped from the bulk electrolyte supply. Since the current will follow the path of least resistance, preferential electrochemical machining will occur nearer to the flow inlet, and little or no machining may occur at the flow outlet. In fact, depending on the flow rates used, there may be hydroxide accumulation at the flow outlet region of the anode workpiece, which may totally insulate that area of the part from any ECM processing. This can be controlled by a combination of current pulsing, electrolyte flow rate, and cathode oscillation.

Cathode Oscillation

A precision machining axis oscillating the cathode allows the gap to close down to typically 10 μm (0.0004 in.) and reciprocated to 100 to 1000 μm (0.004 to 0.040 in.). This allows a small gap to be established for precision electrochemical anode dissolution at relatively high current when a DC pulse is applied as the cathode is in close proximity to the anode (workpiece). Then as the cathode retracts, the larger gap allows sufficient flushing to reestablish the desired electrochemical machining conditions.

Electrolyte Flow

The electrolyte "flow box" is a directed flow passage chamber that provides one or more inlet and outlet passages for electrolyte flow to the cathode-anode gap. The function of the flow box is critical to the PECM metal removal process. Electrolyte velocity, pressure, and flow uniformity determine material removal rates during the ECM process.

Setting up the proper flow patterns and selecting the flow rate that gives the optimal flow velocity are integral to the success of any PECM application. While the amount of material removed from the anode surface may be controlled by the reaction rate (proportional to the current and time), the material removal is also bound by the transfer rate of insulating hydroxides and hydrogen bubbles away from the electrode boundaries. If too large a concentration gradient is allowed to develop across the flow path, variations in depth, definition, and surface finish will be evident on the workpiece.

Power Supply

A pulse power supply is necessary for electrochemical machining to micrometer tolerances. The power supply settings that must be optimized for the system include current (control) set point or voltage (control) set point, pulse ON time; pulse OFF time; pulse ratio; and cycle time. From Faraday's law, the pulse ON time and OFF time will determine the definition and uniformity of the micro ECM features.

Pulse parameter selection is critical to processing results. If the pulse ON time is too long or the pulse OFF time (relaxation time) is too short, the conductivity of the electrolyte near the flow outlet may be different enough to result in uneven feature machining along the flow path. Also, selection of the pulse parameters permits optimization of the machined geometry and surface finish.

Proper selection and optimization of pulse parameters and phase relationship with the oscillation cycle will provide the best possible surface finish and geometry.

PECM Equipment

The electrochemical machining process requires a machine designed specifically for PECM because of the salt-based electrolyte used. All areas exposed to electrolyte must be fabricated of materials that will not be attacked by salt electrolyte, such as stainless steel, plastic, or other similar materials. All joints need to be sealed and bellows used on moving actuators. A typical PECM system consists of the following elements: machine

Figure 3.8.3 PECM machine.

structure; electrolyte supply (including filter, and chiller with heat exchanger); pulse DC power supply; and integrated control (see Fig. 3.8.3).

Process Capabilities

The PECM metal removal rate is dependent on DC current flow. PECM machines are capable of removing around 1.5 cc per minute per 1000 A (0.091 in.^{3}/min per 1000 A); however, with pulsed current the average current will be limited by the pulse ON time ratio. The surface finish and edge geometry will be dependent on material and electrolyte flow parameters. Maximum machining depth will be determined by the maximum stroke length of the machine, an effective electrolyte flow path length, and cathode design.

Typical PECM capability:
Feed rate: 1 to 0.4 mm/min (0.004 to 0.016 in./min)
Metal removal rate: 1.3 cc/min (0.08 cu in./min)
Current density: 1 A/mm^2 (650 A/in.^{2})
Tolerance capability: 10 to 20 μm (0.0004 to 0.0008 in.)
Maximum surface area: depends on machine type

Some Typical Examples of PECM Parts

Example 1: Rotary Shaver Head

Thin, preformed stainless steel rotary shaver head blanks (Fig. 3.8.4) are electrochemically machined with very smooth features to allow hair to be cut close to the skin. Details of the operation are shown in Table 3.8.1. Tooling consists of a cathode with thin

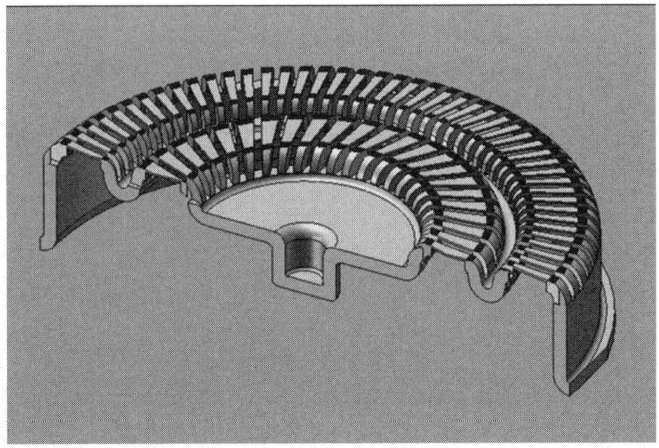

FIGURE 3.8.4 Rotary shaver head.

Workpiece Name	Specifications	Machining Data	Comments
Rotary shaver head	0.28-mm-wide (0.011-in.) slots with specified edge radius; 0.1 mm (0.004 in.) thick; material: chrome steel; 180 slots (typical)	35 s machining time per cycle; one part per cycle	See Fig. 3.8.4
High-precision gears	50 mm (1.96 in.) diameter; pre-machined hardened steel	20 min per cycle; four parts per cycle; surface finish 0.05 μm Ra (2 μin. Ra)	See Fig. 3.8.5
Diesel valve plate	25.5 mm (1 in.) diameter; 0.30 mm (0.012 in.) depth (typical)	Four parts per cycle; 7 min per cycle; surface finish 0.1 μm Ra (4 μin. Ra)	See Fig. 3.8.6
Prototypes	Array of 50-μm-diameter (0.002-in.) holes in 100-μm-thick 304 stainless steel	6 V pulses of 60 ns ON, 1 ns OFF were used	
Electrode	10-μm-diameter (0.0004 in.) tapered needle that is 4573 μm (0.180 in.) long (aspect ratio 450)	Electrode was μECMed to be used as a tool for μECM of deep holes	ECM produces no measureable side forces so very thin and long tools can be fabricated with good finishes
Prototypes	50 μm (0.002 in.) wide × 80 μm (0.0032 in.) deep slots in 304 stainless steel	6 V pulses of 60 ns ON, 1 ns OFF were used	

TABLE 3.8.1 Examples of Micro ECM Applications

FIGURE 3.8.5 High-precision gears.

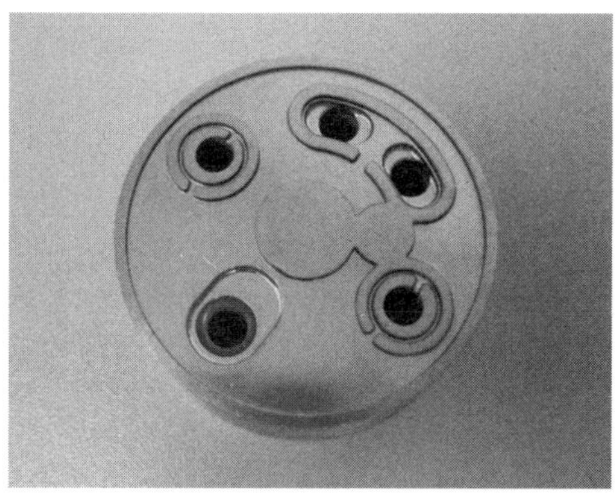

FIGURE 3.8.6 Diesel valve plate.

fin type structures for each slot; a precision locating fixture for the blank; and a flow chamber to control electrolyte flow.

Example 2: High-Precision Gears

Small-precision gears (Fig. 3.8.5) can be machined to final shape and surface finish in hardened metal. Tooling consists of four individual cathodes each surrounded by a flow chamber to control electrolyte flow, and a precision locating nest for four blanks.

Example 3: Diesel Valve Plates

Flow circuit features in valve plates for diesel injection systems are PECM-machined into pre-hardened blanks (Fig. 3.8.6). Tooling consists of four individual cathodes each surrounded by a flow chamber to control electrolyte flow; and a precision locating nest for four blanks.

Summary

PECM can produce feature geometries in metal and alloy surfaces in the dimensional range of 10 to 100 microns. Careful selection of parameters and the use of precision cathodes and fixtures make machining on internal and external surfaces (and both simultaneously), possible to ± 1 to ± 5 micron tolerance. Surface finish is typically very good (1-8 μ in R_a) depending on material type. The process is controllable and repeatable and is very useful for high precision, high volume production, due to the short cycle times and long tooling lifespan.

CHAPTER 3.9

Electrochemical Micro Deburring

Don Risko

DGR Consulting
Jamestown, Pennsylvania

The Process

Electrochemical deburring is a targeted metal dissolution process that uses stationary tooling to focus a deburring current on only those areas where burrs are to be removed. Since the deburr tooling known as a cathode never touches the workpiece, there is virtually no tool wear in the process. Typical deburring times are extremely fast (10 to 40 s for most applications as per Fig. 3.9.1). Depending on production requirements and the workpiece size, a multiple-part fixture can be used to obtain high production rates.

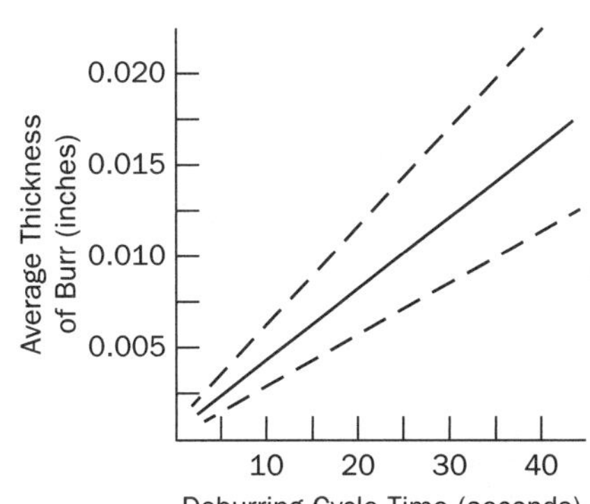

FIGURE 3.9.1 Relationship between time and material removal for typical edge-finishing application.

Process Principles

Electrochemical deburring is a metal removal process based on Faraday's law of electrolytic dissolution of metals where the theoretical amount of material removed from the anode (workpiece) surface is proportional to the current applied and the duration. A DC voltage is applied to the cathode (a shaped conductive tool) and anode while an electrolyte (saltwater solution) is flushed between the anode and the cathode gap [1 to 4 mm (0.040 to 0.160 in.)], thereby completing an electrical circuit.

Application of the voltage (typically 7 to 20 V), and the resultant electrical current flow [typically 0.5 to 1 A/mm^2 (325 to 645 A/in.^{2})] through the electrolyte (typically a 15% sodium nitrate and water solution), initiates chemical reactions that occur at the surfaces of the electrodes. At the anode or workpiece surface, electrons are removed from the metal surface which causes metal removal. The metal ions quickly form neutral metal hydroxides that are filtered from the recirculating electrolyte stream. Quick removal of metal hydroxides, small gas bubbles (from the hydrolysis of water), and heat from the gap by the electrolyte flow permits high machining rates.

Tooling—Cathode and Anode Fixtures

Cathode and anode alignment is important to achieving desired deburring and edge radius results. Figure 3.9.2 depicts metal removal results based on cathode alignment to the workpiece edge. The fixture must provide conditions for a consistent and uniform

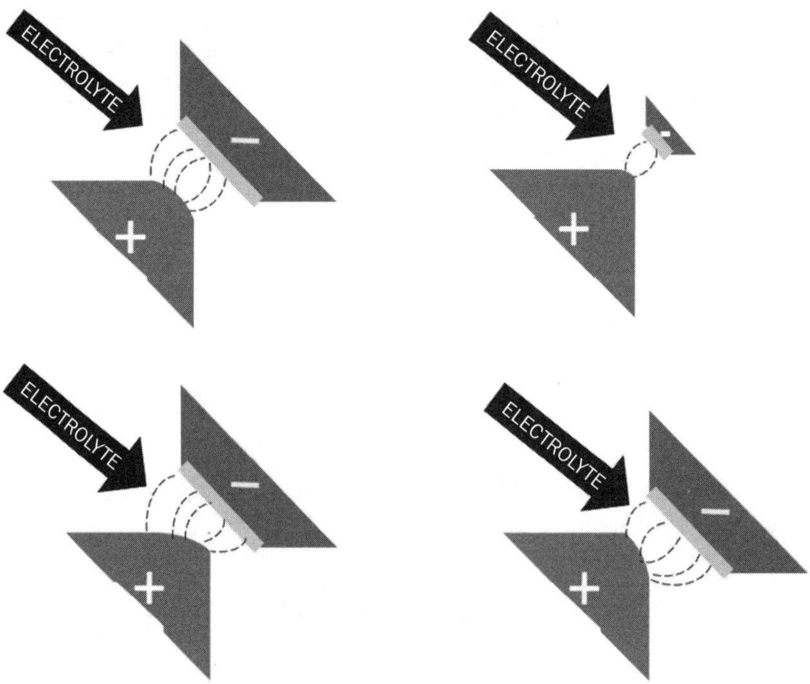

FIGURE 3.9.2 Edge radius form based on cathode alignment.

machining gap between the cathode and the workpiece. A machining gap that is not uniform will result in uneven depth and width in the workpiece edge. Since the process is driven by current flow between the electrodes, if one cathode region is closer than another, then the current will take the path of least resistance. If the tooling is too far off-center, then damage to the cathode can occur if a burr contacts the cathode.

The following are the typical characteristics of the ECD components.

Anode (Workpiece)

Materials that are electrochemically deburred include most conductive metals. They electrolytically "dissolve" at approximately the same rate, regardless of hardness or tensile strength. The rate of dissolution is influenced by chemical and electrical properties, such as atomic weight and valence, rather than by mechanical properties. Anode contacts are made from materials that are less susceptible to electrolytic dissolution.

Cathode (Tool)

Brass, stainless steel, or copper tungsten is used for electrodes. Tooling tends to be robust and cathode tools do not wear or dissolve in the process which eliminates the need for costly frequent tool replacement encountered with other processes.

Fixtures

Fiberglass-reinforced epoxy or plastic materials are used for the tool base and other structural parts of the fixture because these materials do not absorb moisture from the electrolyte and are good insulators. The workpiece positioning surfaces are usually made of titanium, plastic, or ceramic material depending on the application.

Electrolyte

Electrolytes used in the electrolytic process are typically aqueous salt solutions, containing 10% to 20% concentration of either sodium chloride (NaCl) or sodium nitrate ($NaNO_3$). Typical temperature of the solution normally ranges from 15°C to 40°C (59°F to 105°F), with a pH ranging from 7 to 8.5. Various additives have been used with these two basic electrolytes but they are not required in most applications.

Process Capabilities

As with all specialized deburring techniques, there are certain part configurations or requirements that make a particular method most advantageous. Some potential electrochemical deburring applications are:

- Intersecting holes or ports
- Inaccessible or hard-to-reach areas
- Complicated shapes
- Areas requiring consistent radiusing
- Areas that cannot be abraded
- Materials that may work harden when machined conventionally
- Parts processed on automated lines where work flow cannot be interrupted
- Areas where large radii are required

Edge radii can be produced repeatedly with cycle times of 10 to 40 s.

The process requires that all grease, oil, hanging burrs, and loose chips be removed from the workpiece prior to the operation.

Equipment

Electrochemical deburring equipment is generally built to suit specific machining tasks. A complete installation consists of the following components.

Machine	Machine base (which may include a feed unit and work enclosure), fixture platen, and machine control
Power supply	DC power source with features suited to the application, i.e., constant voltage, constant current, pulse, or variable voltage
Electrolyte system	Supply circuit, parameter regulation instruments, and metal hydroxide removal circuit
Tooling	Cathode, flow chamber, fixture, and anode contacts

The equipment design for the electrolytic process is largely influenced by the workpiece to be processed. The dimensions of the workpiece, the production rate, and quantity of material removal will determine whether a single station, dual station, or a special fully automatic machine is required. Machines may also be equipped with indexing tables, locating devices, or other special configurations for workpiece manipulation. Some fully automated systems will also have integrated parts washing systems. Semi-enclosed machines have an upper platen that holds anode contacts, part clamping points, and electrolyte sealing surfaces (see Fig. 3.9.3). Small machines the size of a desk with self-contained electrolyte supply and power unit and utilize a removable fixture consisting of a lower platen that holds workpiece(s); upper platen that holds cathode(s) and quick-connect electrolyte and power connections.

ECD Examples

ECD Example 1: Aluminum Manifold

Multiple internal intersections are simultaneously deburred in 30 s. ECD cathodes are pneumatically inserted such that the exposed cathodes are adjacent to the burrs at the internal intersection. The process is then initiated (see Fig. 3.9.4).

ECD Example 2: Gear-Edge Deburring

Small burrs are electrochemically removed from a stainless steel gear in seconds with one cathode. Six parts are deburred simultaneously in 25 s. Figure 3.9.5 shows before and after ECD.

ECD Example 3: Air Bag Housing

Multiple inflation holes in air bag canisters are deburred and radiused in 45 s (see Fig. 3.9.6).

Removing burrs from micro parts and micro features is difficult for many reasons, but chief among these is the inability of many processes to attack narrow slots and

FIGURE 3.9.3 Four-station modular electrochemical deburring system.

FIGURE 3.9.4 Fixtured manifold with cathodes.

almost-invisible micro holes and reach internal features. Removing burrs without affecting adjacent surfaces is yet another constraint on many processes.

ECD can quickly and effectively address each of these issues. Holes as small as 0.005 in. have been deburred. Many parts could be done in a single 5- to 8-s cycle.

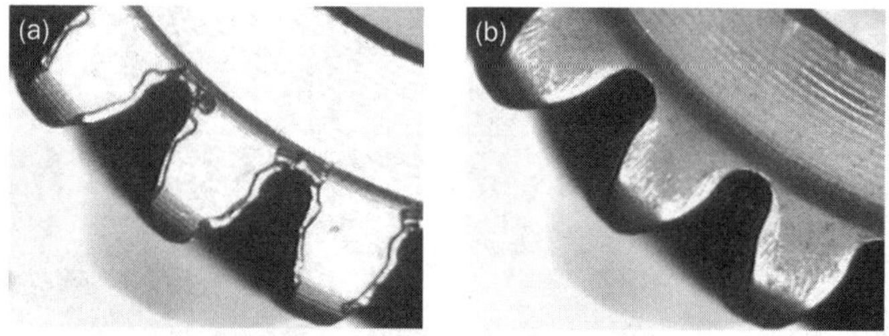

FIGURE 3.9.5 Gear deburred electrochemically.

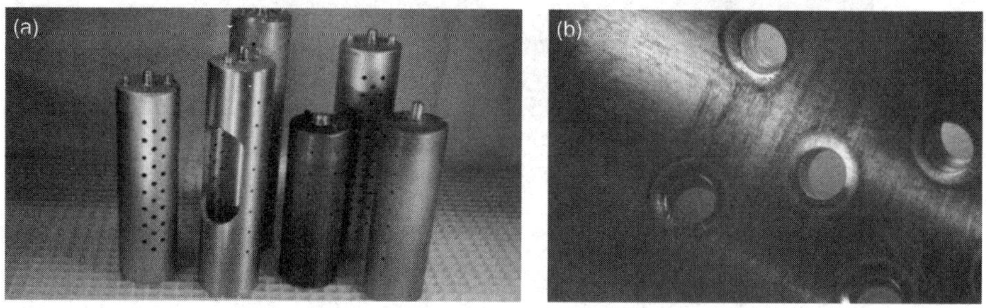

FIGURE 3.9.6 ECD of inflation holes in air bag canisters.

Feature	Material	Comment
0.5-mm-dia (0.020-in.) hole	Brass	76.2-μm (0.003-in.) edge radii required after deburring
10-tooth gear, 6.35 mm (0.250 in.) dia	17-4 Ph stainless steel	76.2-μm (0.003-in.) edge break max. 303 stainless is a better workpiece material because it does not surface etch like 17-4.
0.5-mm (0.020-in.) holes	High-pressure turbine blades of U-700 R80 and R120 high nickel super alloy	Edge radii were 127–254 μm (0.005–0.010 in.)
1.25-mm-dia (0.050-in.) holes	Hastelloy X	700 holes per blade
1 × 20 mm (0.040 × 0.80 in.) slots	Molybdenum	0.05-mm (0.002-in.) edge radius
1.25-mm-dia (0.050-in.) pins	CPM-S90-V	Deburr with 0.05-mm (0.002-in.) radius
0.8-mm (0.032-in.) rib	Kovar	Deburr with min edge radius

TABLE 3.9.1 Examples of Micro Size Features That Have Been Deburred by Micro ECD

Table 3.9.1 illustrates some of the micro size features that have been deburred successfully. These are not limits, but they do represent some production applications. Deburring smaller features might require some more specialized machines and tooling. Conceptually there is not a lower limit to size that can be accommodated as long as good electrical contact to the workpiece can be assured.

Summary

The electrolytic process can be characterized as a high-rate metal dissolution process used to remove material from designated areas on the workpiece. The material to be removed can be an unwanted burr, excess edge material, or improved surface roughness. Components requiring an edge radius from 0.5 to 2.5 mm (0.020 to 0.098 in.) are excellent applications for the electrolytic process, and for micro parts edge radii less than 0.003 in. have been produced in production. Surface finishes can be as fine as 8 μin. The tooling is relatively simple and machines can accommodate a wide range of part sizes and materials. System components and parameters are typically chosen from factors that influence the process, including the electrochemical properties of both electrodes (the cathode tool and the anode workpiece), electrolyte type (concentration, pH, and temperature), aspects of component geometry, and process voltage and current.

CHAPTER 3.10

Electrochemical Discharge Machining

Apurbba Kumar Sharma

*Indian Institute of Technology
Roorkee, India*

Chandra Shekhar Jawalkar

*PEC University of Technology
Chandigarh, India*

Introduction

Electrochemical discharge machining (ECDM) is a nonconventional, electrical-spark-based machining process. The process is especially useful for machining hard and electrically nonconducting materials. Material removal in the process is basically through thermal erosion; however, the process has capabilities for micro machining. It was first used for drilling glass. The process is identified by different names such as electrochemical spark machining, discharge machining of nonconductors, and electrochemical anode machining.

Materials with high resistance to corrosion such as different grades of glass, quartz, granite etc. are suitable to be machined by this process. Machining of other materials such as purified Al_2O_3 alumina, Kevlar fibers, and glass composites has also been achieved with ease. It is possible to machine even the conducting materials using ECDM, but, usually, these materials tend to react with the electrolyte used in the process and the resulting compound may get attached to the tool (i.e., get micro welded, when the gap is a few nanometers); hence such applications are of little use.

The preferred applications of ECDM include machining of micro holes and micro channels, which are used in various applications such as in lab-on-chip, surgical appliances, micro filters, micro reactors, and three-dimensioned micro structures.

Working Principle of ECDM

The basic requirement of electrochemical discharge machining is to form an electrolytic cell. A schematic diagram illustrating the principles of the ECDM process is shown in Fig. 3.10.1. In this process, the work material is immersed in a base electrolyte such as NaOH or KOH. A potential difference, either DC or pulsed, is applied between a pair of electrodes, namely, the tool electrode (normally cathode) and the counter electrode (usually anode). Reverse polarity is also possible, but not preferred owing to a few disadvantages over the direct polarity. The tool-electrode tip is dipped a few millimeters in the electrolyte while the counter electrode (larger in size about a factor of hundred) is placed a few centimeters away from the tool-electrode. Electrolysis starts at a low voltage (lower than the critical voltage, typically between 20 and 30 V). As the potential difference between the electrodes is increased, gases are released at both the electrodes and a gas film starts forming on the tool electrode. Hydrogen bubbles are formed at the tool-electrode and oxygen bubbles evolve at the counter electrode. As the process continues, the gas bubbles coalesce and form an electrically insulating gas film. The gas film, however, breaks at the critical voltage (around 29 V) and discharge starts occurring, leading to sparking between the electrodes, that is, the tool and the electrolyte (which acts as another electrode). The sparking breaks the gas layer through the shortest thickness and forms a tiny discharge channel. The thermal energy released in the discharge causes the material in the vicinity to get removed through melting and evaporation. The thermal erosion that occurs leaves behind a small crater on the immediate near work surface. Consequently, the locations of the sparks keep changing in order to follow the minimal channel path. A gap, usually below 1 mm (0.039 in.), is maintained between the cathode and the workpiece using some tool-feeding mechanisms to enable formation of an effective gas layer. The craters formed in ECDM process are almost same as those formed in the EDM process, along with some recast effects, which are mainly due to the sparking action.

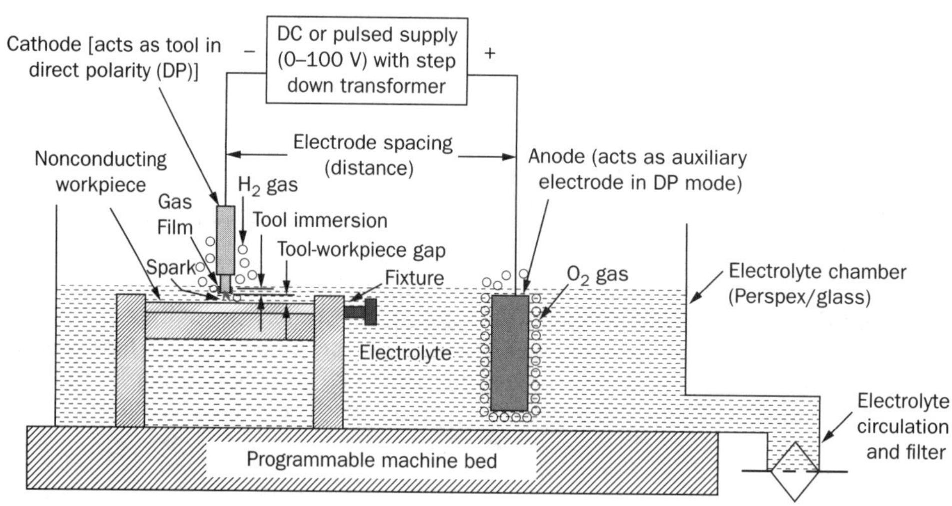

FIGURE 3.10.1 Schematic diagram of an ECDM setup.

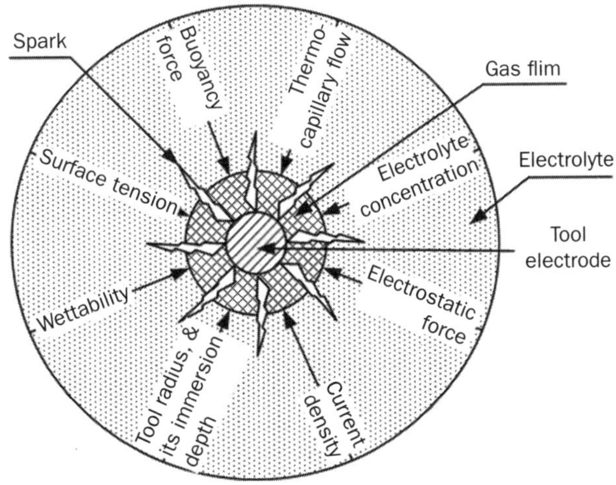

FIGURE 3.10.2 Parameters that influence gas film quality (Goud et al., 2016).

The formation of the gas layer and maintaining it are vital to the discharging process as it acts as the dielectric between the electrode and the electrolyte. There are a number of theories associated with the ECDM process that explain the significance of the gas layer. Apparently each gas bubble acts as a valve and the electrochemical discharge occurs due to its breakdown. The generation of electrolytic gas at the tool electrode is the main cause of occurrence of the discharge between the tool electrodes and the electrolyte. The switching effect in the electrical circuits may be the main factor that leads to the origin of sparks in ECDM. The mechanism of sparking is quite similar to the arc discharge in gases. The quality of machining in ECDM is mainly dominated by the gas film quality. There are a number of parameters which influence the quality of the gas film (shown in Fig. 3.10.2). These parameters include—buoyancy force, surface tension, thermocapillary flow due to temperature gradient between tool electrode and electrolyte, electrostatic force between bubble and electrode, wettability and nucleation sites on the tool electrode surface, current density, electrolyte resistivity, tool radius, and tool immersion depth.

Material Removal Modes in ECDM

There are several active mechanisms of material removal in ECDM. The major mechanisms include

(i) melting and vaporization due to sparking,

(ii) chemical etching at high temperatures, and

(iii) thermal stresses and micro cracking.

The predominant and most appreciable phenomenon is however the intensive local heating of the workpiece through sparking, leading to thermal erosion. Material gets removed due to intensive heating of the workpiece surface in the discharge zone that leads to melting and even vaporization. The local high temperature accelerates the contribution of chemical etching in this discharge zone. The etching occurs more predominantly through the grain boundaries which are characteristically weaker in structure.

Process Characteristics of ECDM

The ECDM process involves two major known phenomena—electro discharge as well as electrochemical actions. As mentioned earlier, however, the electro discharge action dominates the material removal in the process; the electrochemistry controls the kinetics of the process. The process involves a number of variables that influence the performance. The implementation of the process, although appears simple, needs a few specifications to be met. Precision control of the output of the process is possible, but differs significantly with the electrolyte-electrode materials. A few variants of the process, such as rotary ECDM and powder ECDM, are also being used effectively. The requirements for these variants are marginally different from the parent process. The main features and process characteristics of the ECDM process are illustrated in Table 3.10.1.

Types of ECDM

There are four common types of ECDM systems in practice. Implementation differs significantly, although the main principles of operation remain the same. These systems are illustrated briefly along with their schematic diagrams in Table 3.10.2.

Chemical Reactions in ECDM

The ECDM process is basically one of the applications of the controlled electrochemistry. Once the electrolytic cell is formed, which involves the tool (the cathode) and the working medium (the electrolyte), and suitable potential difference is created, a number of electrochemical reactions begin. The electrochemical reactions at the larger electrode and the electrolyte interface (anode) are different and result in generation of gases at the respective electrodes. The quality and stability of this gas film in the cathode are controlled at a desired level by suitably controlling the reaction kinetics in the system. Thus, these reactions are important continuation of the ECDM process. The common reactions occurring at both the metallic electrodes (M) in electrolyte can be expressed as follows [Eqs. (3.10.1 to 3.10.8)]:

$$M \rightarrow M^{z+} + Z^{e-} \tag{3.10.1}$$

$$M^{Z+} + Z(\mathrm{OH})^- \rightarrow M(\mathrm{OH})_Z \quad \text{(Insoluble metal hydroxide + precipitate)} \tag{3.10.2}$$

$$2\mathrm{H_2O} \rightarrow \mathrm{O_2} + 4\mathrm{H^+} + 4e^- \quad \text{(In acidic solution)} \tag{3.10.3}$$

$$4(\mathrm{OH})^- \rightarrow 2\mathrm{H_2O} + \mathrm{O_2}\uparrow + 4e^- \quad \text{(In alkaline solution)} \tag{3.10.4}$$

Reactions at the smaller electrode and the electrolyte interface while considering NaOH as the electrolyte can be expressed as shown in Eqs. (3.10.5) to (3.10.8):

$$2\mathrm{H^+} + 2e^- \rightarrow \mathrm{H_2} \tag{3.10.5}$$

Components/Condition	Type	Specifications	Optimal/Standard Settings
Polarity	Direct polarity (DP) Reverse polarity (RP)	The tool (cathode) is connected to the negative terminal. It is smaller in cross section than the anode (auxiliary electrode). The material removal (MR) takes place at the cathode. The cathode is larger in cross section and machining takes place at the anode.	DP is the most popular and widely used mode in ECDM process. MR using reverse polarity was reported to be faster in quartz.
Tool-electrolyte-workpiece gap	Distance (in mm)	The gap between tool tip and electrolyte and the gap between tool tip and workpiece is crucial.	Preferably the tool should be at a depth of ≈0.5–1 mm (0.020–0.039 in.) into the electrolyte. The preferred gap between tool and workpiece is in the range of 0–0.5 mm (0–0.020 in.).
Power mode	Direct current (DC) or pulsed DC.	The voltage values are ≥25 V up to 100 V.	50–70 V is the optimal voltage setting.
Transformer with a rectifier	Step down type.	Capable of converting 220 V AC mains supply to 100 V DC supply.	50–70 V is the optimal voltage setting.
Electrolyte	NaOH, KOH, NaCl, $NaNO_3$, NaF, HCL, H_2SO_4, etc.	20–30% concentration by weight is preferred.	NaOH is the most preferred and widely used electrolyte with ≈15–20% wt. concentration.
Electrolyte chamber	Made of Perspex[1]/glass or any other nonconducting material.	Fabricated as per requirement.	Nonconductive, nonreactive, and transparent material is preferred.
Settling tank	Made of Perspex/glass.	Fabricated as per requirement.	Nonconductive, nonreactive, and transparent material is preferred.
Filter	Standard industrial grade filter similar to the one used in honing machines.	Capable of filtering fine powders up to few nanometers.	The filter element should be standard and replaceable.

TABLE 3.10.1 ECDM Process Settings and Characteristics

Components/Condition	Type	Specifications	Optimal/Standard Settings
Work holding device	Made of Perspex/plastic or glass.	Fabricated as per the required work size.	It should have movable jaws, easy to hold and tighten the workpiece.
Tool (normally cathode)	Any conductive material (reduced to a fine diameter, resembling a needle tip).	End size (having the end diameter) of 50–1000 μm (0.002–0.039 in.) and tip length about 2 mm (0.079 in.).	Wear resistance materials such as stainless steel, tungsten, nickel, platinum, etc. are preferred.
Auxiliary tool (normally anode)	Any conducting material.	Size and volume are greater than the cathode.	Made of stainless steel or any noncorrosive metal; with cross-section diameter about 10 times higher than the tool.
Electrode spacing	It is the distance between the cathode and the anode.	Smaller spacing is preferred for higher MR and larger for lower tool wear (TW).	Optimum spacing is reported to be 40–60 mm (01.57–2.36 in.) for higher MR and 80–100 mm (3.15–3.94 in.) for lower TW criteria.
Machining table	The X-Y axis resolution of 1 μm (39 μin.), with range about ≥100×100 mm (3.94 × 3.94 in.).	A standard CNC table can be used or can be customized/fabricated using suitable stepper/servo motors.	Should be movable with precision, programmable preferred [computer numerical controlled (CNC) or any high-level language based].
Rotary ECDM fixture	Consists of a separate motor, fitted for imparting the rotary motion to the tool.	A small DC motor of capacity 10–60 RPM (revolutions per minute) is used. The speed can be varied using a regulator, as per the requirement.	Tool rotation of 25–35 RPM is considered stable, it improves the surface finish and MR; however, the sparking gets destabilized at higher RPMs.
Electrode cross section (c.s.) ratio	It is the ratio of cross section of the cathode to the cross section of the anode.	The tool (where MR takes place) is made smaller as compared to the auxiliary electrode, whose c.s. is 10–20 times larger.	For example in DP, a cross-section ratio of (cathode:anode) 1:10 is preferred.

Wire ECDM	The moving wire electrode is used for micro slotting applications on hard nonconductors and ceramics.	The wire material can be copper, stainless steel, brass, etc.	The preferred wire diameter size is 0.1–0.8 mm (0.004–0.031 in.), depending on different requirements.
Powder ECDM	Conductive powders and/or surfactants of different materials are added to the electrolyte in order to obtain enhanced finishing effect.	The powders of Al, graphite, silicon, Ti, etc. along with surfactants can be used to improve the surface integrity.	• Powder improves the finish and provides flat surfaces free of black carbon spots along with higher machining rates. • Improvement in surface finish (using Al + surfactant powder) up to Ra 0.172 μm (6.8 μin.) has been reported. • The powder ECDM improves over cuts and refines the micro cracks in the region.

[1]Perspex is a transparent polymethyl methacrylate thermoplastic. Lucite is a common brand name of it.

Table 3.10.1 ECDM Process Settings and Characteristics (*Continued*)

Type	Description	Schematic Sketch
Sinking ECDM	It uses shaped tools and finds application in micro machining of holes; 2D or 3D features. The main focus here is on the surface quality of side walls. The main components of these machines are power supply system, electrolyte supply system, electrolyte feeding system, machining chamber positioning, and vision system.	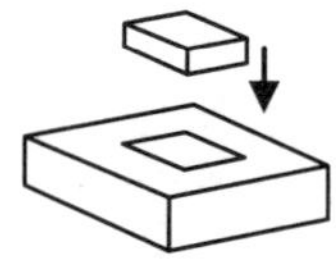
Drilling ECDM	This process uses a rotating tool electrode; mostly used for micro machining of holes. Fine drilling in nonconventional and/or difficult-to-machine materials has been achieved; however there are aspcct ratio limitations because of the inability of getting the electrolyte inside the cavity; hence maintaining the appropriate gap as well as gas film become issues.	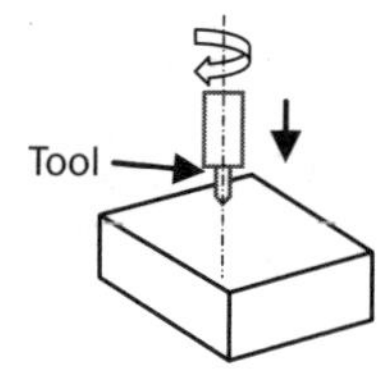
Traveling wire ECDM (TWECDM)	The TWECDM is capable of slicing large volumes without the need of forming electrode tools which are usually costly.	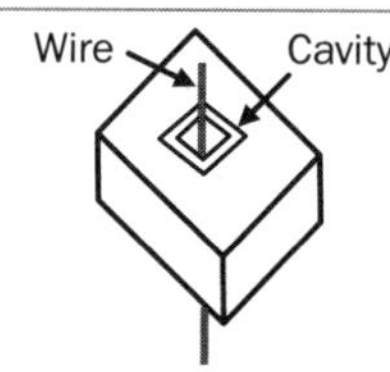
Milling ECDM	This is achieved by adapting a movement strategy similar to the conventional milling process using a simple micro sized electrode to produce a 3D cavity. It was first attempted using a tool speed of 0.5 µm/s (20 µin./s) and then extended for dimensions of 100 µm (0.0039 in.) wide and a few millimeters long. Different 3D profiles like micro channels, slots, key-ways, etc. can be produced with this system.	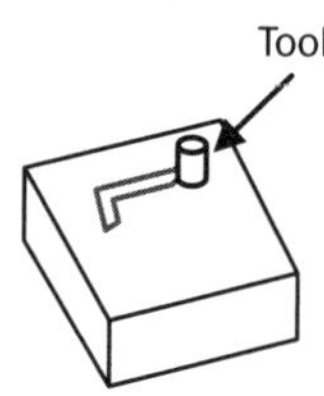

Table 3.10.2 Various Types of ECDM Systems

$$2H_2 + 2e^- \rightarrow 2(OH)^- + H_2\uparrow \quad \text{(Hydrogen bubble generation)} \tag{3.10.6}$$

$$Na^+ + e^- \rightarrow Na \tag{3.10.7}$$

$$2Na + 2H_2O \rightarrow NaOH + H_2\uparrow \tag{3.10.8}$$

These reactions are responsible for H_2 and O_2 generation at the cathode and anode, respectively.

Material (Glass)	Machining Conditions	Capability Indicators	Process Variant Used	Major Findings	Reported by
Glass (borosilicate)	Voltage, electrolytes, electrolyte concentration, and temperature were the variable parameters.	Machining time, hole expansion, and surface roughness	Micro drilling	Best surface finish achieved Ra ~0.08 μm (3 μin.) and MRR up to 1.5 mm/min (0.059 in./min), average hole dia. ~241 μm (0.009 in.).	Yang et al. (2001)
Glass (Pyrex)	Electrolyte: KOH Voltage: >30 V Current: 10 A	Mean current: 95–142 mA Depth: 100–300 μm (0.0039–0.012 in.) Gas films	Micro drilling	Hole dia: up to 463 μm (0.018 in.).	Cheng et al. (2010)
Glass (soda-lime, alumina)	Electrolyte concentration: 10–60% Duty factor: 60–80 Energy partition: 0.2–0.9 (R_w)	Material removal rate (MRR)	Micro drilling	MRR increases up to 30% due to increase in R_w in both glasses.	Bhondwe et al. (2006)
Glass (soda-lime, quartz)	Electrolyte concentration: 10–60% Applied voltage: 50 V Electrolyte: KOH, NaOH	Mechanism of material removal (MR) Crater size due to sparking	Micro drilling, Micro channel	MRR increases significantly as electrolyte concentration increases from 10% up to 45%, beyond which its role becomes insignificant. Experimentally showed average size of single crater to be ~ϕ300 μm (0.12 in.).	Goud and Sharma (2016)
Glass (soda-lime and borosilicate)	Voltage: 0–80 V Electrolyte concentration: 20–35% (wt.)	Overcut Penetration MRR	Micro drilling	Relationship in applied voltage to penetration and applied voltage to overcut.	Jain et al. (1999)
Glass (borosilicate alumina glass)	Voltage: 40–70 V Temperature: 40–70°C	MRR, machining depth and response surface equations.	Micro drilling	Abrasive-coated tools have shown improved performance.	Jain et al. (2002)

Table 3.10.3 Capabilities of ECDM with Respect to Machining of Different Materials

Material (Glass)	Machining Conditions	Capability Indicators	Process Variant Used	Major Findings	Reported by
Glass (borosilicate)	Voltage Current Electrolyte concentration Temperature	Machining time Hole expansion Surface finish	Micro drilling	MRR=1.5 mg/min (48 μoz) SR= 0.08 μm (3 μin.) Ra	Yang et al. (2006)
Borosilicate glass and quartz	Tool rotation: 20 RPM Machining depth: 0.5–3.0 mm (0.020–0.118 in.) Machining time: 1–6 min	Machining rate: up to 0.6 mm/min (0.024 in./min) Machining depth: 0.5–1.2 mm (0.020–0.047 in.)	Micro drilling	MRR improved using rotational tool. Thermal and spalling effects noticed.	Gautam and Jain (1998)
Alumina and quartz	Voltage: 50–70 V Temperature: 35–80°C Electrolyte (type)	MR: 122.7 mg in 70 min (0.000062 oz/min) Maximum machining depth: 1.35 mm (0.053 in.)	Micro drilling	Temperature found to have significant effect.	Jain and Chak (2000)
Alumina ceramic	Effect of glass content: 0–20% (wt.) Porosity: 0–30% (wt.)	MR: 40–110 mg (0.0014–0.0039 oz/min) Machining depth: 0.2–2 mm (0.0079–0.079 in.)	Micro drilling	The governing factor for MR and machining depth in ceramics is porosity.	Jain et al. (2010)
Alumina ceramic	Voltage: 70–90 V Electrolyte concentration: 20–30% (wt.) Electrolyte type: NaOH, $NaNo_3$, NaCl Tool tip geometry	Overcut= 0.33–0.56 mm (0.013–0.022 in.) for taper side wall, straight side wall, and curvature tools	Micro drilling	Most effective combination for high MRR and accuracy was 80 V and 25% NaOH.	Bhattacharya et al. (1999)
MMC (aluminum alloy-359, 20% SiC)	Current: 0.5–5 A Voltage: 20–120 V Pulse duration: 4–400 μs	Current/voltage waveforms and XRD	Micro drilling	Developed models and found capable of predicting the V_{bk} range as: 26–34 V.	Liu et al. (2010)

Kevlar-fiber-epoxy, glass-fiber-epoxy	Voltage: 65–80 V Electrolytic-conductance: 0.16–0.18 S ($mho.cm^{-1}$)	MRR	Micro drilling	MRR increases with rise in specific inductance and voltage and lowers with rise in tool diameter. MRR reported was ≈0.13 mg/min.	Tandon et al. (1990)
Quartz	Wettability of machining characteristics, gas film formation was studied.	SR and coalescence of gas films	Micro drilling	SR was found to profoundly affect the gas film and its coalescence.	Yang et al. (2010)
Glass (Pyrex-wafers)	Electrolyte: KOH Electrolyte concentration: 25% (wt.) Electrode: Cu, graphite Current: 10 A	Frequency; duty cycle of voltage pulse and MRR	Micro drilling	For less thermal damages; high frequency and low duty cycle ratio were preferred in fine-hole drilling up to 60 μm (0.0023 in.) diameter.	Kim et al. (2006)
Microscopic slides of soda lime glass	Using a robot, tool tip as ϕ 500 μm (~0.020 in.) S.S. needle and NaOH electrolyte with viscosities = 1.7, 4.5, 13, and 39 mPa.s (2.47×10^{-7} 6.53×10^{-7}, 1.89×10^{-6}, 1.44×10^{-11}, and 5.63×10^{-6} lbf.s/in.^{2})	Tool-workpiece gap, voltage, and tool speed were variables, and surface texture was measured	Micro channeling	A range of surface textures were developed and found to be viscosity dependent.	Jana et al. (2012)
Soda lime glass	Electrolyte concentration: 30% (wt.) Electrolyte: NaOH Electrolyte and tool speeds: <40 μm/s (0.0016 in./s).	Surface finish and wall boundaries were measured	Micro channeling	Smooth micro channels with undefined wall boundaries. At larger tool speeds, discretized and unsmooth surfaces with varying depths were found.	Didar et al. (2008)

Table 3.10.3 Capabilities of ECDM with Respect to Machining of Different Materials (*Continued*)

Material (Glass)	Machining Conditions	Capability Indicators	Process Variant Used	Major Findings	Reported by
Optical glass	Electrolyte concentration: 20% (wt.) Voltage: 65 V Tool rotation: 34 RPM Tool feed rate: 1.3 mm/min (0.051 in.) Stirrer speed = 100 RPM Stirrer location: an average distance of 100 mm (3.94 in) from tool tip.	Material removal and surface finish were measured and found to have enhanced effects through electrolyte stirring and tool rotation.	Micro channeling	Using tool rotation and electrolyte stirring MR up to 0.47 mg/min (16 μoz/min) and SR: 0.22–0.8 μm (9–31.5 μin.) Ra achieved.	Jawalkar (2013)
Glass	Electrolyte: NaOH Electrolyte concentration: 30% (wt.) Tool: SS 316L, 400 μm (0.0157 in.) dia Power: 40 V, 4 A	Drilling speed: 100 μm/s (0.004 in./s) Drilling depth: 425 μm (0.017 in.) (maximum)	Micro drilling	Higher drilling speed is attained at lower applied voltage (<30 V). Drilling speed at higher depth [>300 μm (0.012 in.)] is limited by the supply of sufficient electrolyte at the tool tip.	Wüthrich et al., (2006)
Glass (soda-lime and Pyrex)	Electrolyte: NaOH/KOH Electrolyte concentration: 20–40% (wt.) Work: 150 μm (0.0059 in.) thick	Contact force: mN Machining time: 500–2000 s	Micro drilling and micro structuring	Hole dia: 30 μm (0.0012 in.). Channel width: 40–46 μm (0.0016–0.0018 in.), Ra = 0.1–0.195 μm (4–8 μin.).	Cao Xuan et al. (2009)
Alumina ceramic, optical glass and quartz	Voltage: 45–90 V Frequency: 100–1000 Hz	MRR, voltage, and pulse current response: 3.8–27 A	Wire ECDM	MRR: 0.06 mm^3/min (3.7 $μin.^3$/min) for alumina ceramic with Ra = 3.5 μm (0.00014 in.).	Peng and Liao (2004)
Glass (Pyrex)	Abrasive conc.: 0–300 g/l Slit grits: 10–60 μm (0.00039–0.0024 in.) Power frequency: 30–100 Hz	Slit expansion: 0.024 mm (0.0009 in.) SR: 0.84–3.5 μm (33–138 μin.) Ra MRR: 0.03–0.09 mg/min (1–3 μoz/min)	Wire ECDM	Small grit size abrasive reduces SR, and refines micro cracks.	Yang et al. (2006)

Notations used in the table: MRR, material removal rate; MR, material removal; TWECDM, traveling wire electrochemical discharge machining; TWR, tool wear rate; TW, tool wear; V, voltage; V_c, critical voltage; g/l, grams/liter; V_{bk}, breakdown voltage; A, ampere; Hz, hertz; mm, millimeter; SR, surface roughness; mN, 10^{-3} newton; Ra, roughness average in micrometers; R_w, energy partition; MMC, metal matrix composite; SiC, silicon carbide; XRD, X-ray diffraction.

TABLE 3.10.3 Capabilities of ECDM with Respect to Machining of Different Materials (*Continued*)

Application Areas in ECDM

The most frequently found applications of the process are mentioned below.

- Micro drilling and development of micro channels.
- Reversible tubing connections for micro fluidics were made on 3D glass with leak-proof joints.
- Dressing of metal-bonded tools.
- Trepanning to produce 2.35-mm-diameter (0.093-in.) holes in quartz and 1.35 mm (0.053 in.) holes in alumina.
- Using Pyrex glass to make pressure sensors and acceleration sensors.
- Making silicon resonant angular sensors on Pyrex glass.
- Machining of Al_2O_3 ceramic through a variant of the process called—gas-filled electro discharge and electrochemical compound machining.
- Cutting fine slots and channels through wire ECDM process.

Capabilities of ECDM

Today ECDM is considered as one of the enabling technologies in micro machining. It forms the machining solutions for fabrication of 3D micro channels on hard materials like glass and quartz. Some key results depicting the capabilities of the ECDM process while fabricating micro channels and micro holes are listed in Table 3.10.3.

References

1. Bhattacharya B., Doloi B. N. and Sorkhel S. K., 1999, "Experimental Investigation into EDM of non conductive ceramic materials", *Journal of Material Processing Technology*, 95, 145–154.
2. Bhondwe K. L., Yadava V. and Kathiresan G., 2006, "Finite element prediction of material removal rate due to electrochemical spark machining", *International Journal of Machine Tools and Manufacture*, 46, 1699–1706.
3. Cao X. D., Kim B. H. and Chu C. N., 2009, "Micro structuring of glass with features less than 100 μm by ECDM", *Precision Engineering*, 33, 459–465.
4. Cheng C. P., Wu K. L., Mai C. C., Yang C. K., Hsu Y. S. and Yan B. H., 2010, "Study of gas film quality in ECDM", *International Journal of Machine Tools and Manufacture*, 50, 689–697.
5. Didar T. F., Dolatabadi A. and Wüthrich R., 2008, "Characteristics and modeling of 2D glass micromachining by SACE with constant velocity", *Journal of Micro Mechanics and Micro Engineering*, 18, 065016.
6. Gautam N. and Jain V. K., 1998, "Experimental investigations into ECSD process using various tool kinematics", *International Journal of Machine Tools and Manufacture*, 38, 15–27.
7. Goud M., Sharma A. K. and Jawalkar, C. S., 2016, "A review on material removal mechanism in electrochemical discharge machining (ECDM) and possibilities to enhance the material removal rate", *Precision Engineering*, http://dx.doi.org/10.1016/j.precisioneng.2016 .01.007.

8. Goud M. and Sharma, A. K., 2016, "A three dimensional finite element simulation approach to analyze material removal in electrochemical discharge machining", *Proc IMechE Part C: J Mechanical Engineering Science.* DOI: 10.1177/0954406216636167.
9. Jain V. K. and Chak S. K., 2000, "Electrochemical spark trepanning of alumina and quartz", *Machining Science and Technology*, 4(2), 277–290.
10. Jain V. K., Choudhury S. K. and Ramesh K. M., 2002, "On the machining of alumina and glass", *International Journal of Machine Tools and Manufacture*, 42, 1269–1276.
11. Jain V. K., Dixit P. M. and Pandey P. M., 1999, "On the analysis of electrochemical spark machining process", *International Journal of Machine Tools and Manufacture*, 39, 165–186.
12. Jain V. K., Singh M., Agarwal D. C. and Sidpara A., 2010, "Investigation into machining of alumina ceramics using ECSM process", Global Trends and Challenges in Design and Manufacturing, in *Proceedings of the AIMTDR Conference*, India, 235–240.
13. Jana D., Ziki A., Didar T. F. and Wüthrich R., 2012, "Micro texturing channel surfaces on glass with spark assisted chemical engraving", *International Journal of Machine Tools and Manufacture*, 57, 66–72.
14. Jawalkar C. S., 2013, "Investigations on performance enhancement of ECDM process while machining glass", Ph. D. Thesis, Institute of Technology Roorkee, Roorkee, India.
15. Kim D. J., Ahn Y., Lee S. H. and Kim Y. K., 2006, "Voltage pulse frequency and duty ratio effects in an electrochemical discharge micro drilling process of pyrex glass", *International Journal of Machine Tools and Manufacture*, 46, 1064–1067.
16. Liu J. W., Yue T. M. and Guo Z. N., 2010, "An analysis of the discharge mechanism in ECDM of particulate reinforced metal matrix composites", *International Journal of Machine Tools and Manufacture*, 50, 86–96.
17. Peng W. Y., Liao Y. S., 2004, "Study of ECDM for slicing non conductive brittle materials", *Journal of Material Processing Technology*, 149, 363–369.
18. Tandon S., Jain V. K., Kumar P. and Rajurkar K. P., 1990, "Investigations into machining of composites", *Precision Engineering*, 12, 227–238.
19. Wüthrich R., U Spaelter, Wu Y. and Bleuler H., 2006, "A systematic characterization method for gravity-feed micro-hole drilling in glass with spark assisted chemical engraving (SACE)", *Journal of Micromechanics and Microengineering*, 16, 1891–1896.
20. Yang C. T, Ho S. S. and Yan B. H., 2001, "Micro hole machining of borosilicate glass through ECDM", *Key Engineering Materials*, 196, 149–166.
21. Yang C. T., Song S. L. and Yan B. H., 2006, "Improving machine performance of wire ECDM by adding SiC abrasives to electrolyte", *International Journal of Machine Tools and Manufacture*, 46, 2044–2050.
22. Yang K., Chih P. C., Chao C. M., Cheng W. A., Jung C. H. and Biing H. Y., 2010, "Effect of surface roughness of tool electrode materials in ECDM performance", *International Journal of Machine Tools and Manufacture*, 50, 1088–1096.

CHAPTER 3.11

Micro Wire Electrical Discharge Grinding

LaRoux K. Gillespie

Manufacturing Consultant
Andover, Kansas

Electrical discharge machining (EDM) is used widely for cutting large parts as well as micro size features (see Chap. 3.10). The same process can be used for a form of grinding. There are two variations or applications that are described as micro wire electrical discharge grinding (μWEDG). Grinding normally implies taking a wheel made of abrasive material and using that to remove material from a part. With use the grains break down, revealing more grains that keep providing sharp cutting points. In μWEDG the electrical sparks of EDM break down a cobalt binder that holds grains in place, allowing more grains to be produced; therefore, when used to form grinding tools, EDM acts like normal wear in grinding.

But μWEDG also refers to what might be called EDM turning. The workpiece does not have to be a grinding wheel. It can be any electrically conductive material used for any purpose. The process can be used to produce micro ground shapes on finished parts as well as to make micro grinding tools that simply cannot be made by any other process because of the smallness and the brittleness of micro tools. For example, the process has been used since 1985 to produce electrodes for micro hole drilling, and it has been used to make electrical contacts as well as a variety of cylindrical shapes.

The process uses a fine-tuned wire EDM system dedicated to micro cutting. For example, in a typical application the wire is a very small-diameter brass wire 102 μm (0.004 in.) that advances past a wire guide which when the circuit is turned on causes minute electrical sparks to "eat" small particles from the workpiece (Fig. 3.11.1). When the intent is to make a micro electrode which will later make holes, the workpiece is typically a tungsten rod or wire which is turned axially. The cylindrical workpiece is fed axially downward while turning. By appropriate X-Y positioning, the tungsten workpiece is reduced to a very small diameter [25 to 50 μm (0.001 to 0.002 in.) for instance]. Because there is no physical contact with the workpiece, it can continue to be fed downward to make a very long aspect ratio part (or future tool).

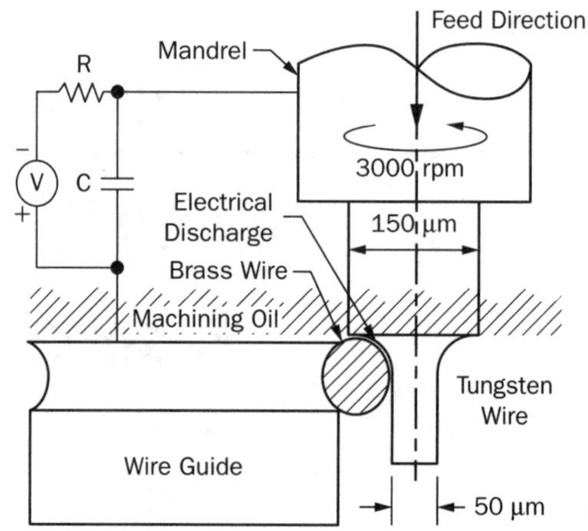

FIGURE 3.11.1 Schematic of micro WEDG process. (Courtesy American Society for Precision Engineering from Morgan, Shreve, and Vallance, 2003.)

A typical micro tool is produced in three steps on commercial machines. First the end is cut off perpendicular to the part axis, then the shape is roughed out, and then it is returned to finish the size. It is a slow process, but almost every micro manufacturing process is slow compared to large part machining. Straightness and roughness depend on machine accuracy and machining parameters, but roughness as low as 0.2 µm (8 µin.) and straightness of 4 µm (16 µin.) have been reported for 50-µm-diameter shafts 500 µm long (0.002 dia x 0.020 in.).

Polycrystalline diamond (PCD) and tungsten carbide tools as well as glass can be machined with µWEDG, making features as small or narrow as 8 µm (0.0003 in.). CVD diamond-coated tools are also machined with µWEDG. While glass can be machined this way, it is not a good choice for work materials.

The electrical spark produced in µWEDG erodes the conductive cobalt binder that holds PCD diamond grains together, allowing them to be cut as they were designed. The sparks also erode many glass compositions but with a different action. Figure 3.11.2 illustrates a multidisc µWEDG-made tool which is then used to make thin slots by EDM or to cut through leaving thin slices of the work material.

Users can make very long skinny electrodes via this process, which are then in turn used for unique electrical contact purposes, or to produce long aspect ratio holes, and tapered deep holes. With appropriate table or slide movements, it is possible to make square, triangular, or six-sided electrodes which are only 25 or 50 µm (0.001 or 0.002 in.) in diameter (Fig. 3.11.3). Reportedly, an electrode of only 4.3-µm (0.000169-in.) diameter has been produced by this process. By today's standards that would be the smallest diameter part that could be produced and it would be very difficult to make. A more common smallest size would be the 25- to 50-µm (0.001- to 0.002-in.) diameters.

By making dicing blades such as those shown in Fig. 3.11.2, one can make grooves of almost any depth and width, and they may be flat bottomed or v-shaped. It may be important to realize that this simple and easy to perform process of making slots can also make posts, hundreds of them standing above the surface. The user merely needs to grind grooves in one direction and turn the workpiece 90 degrees and cut

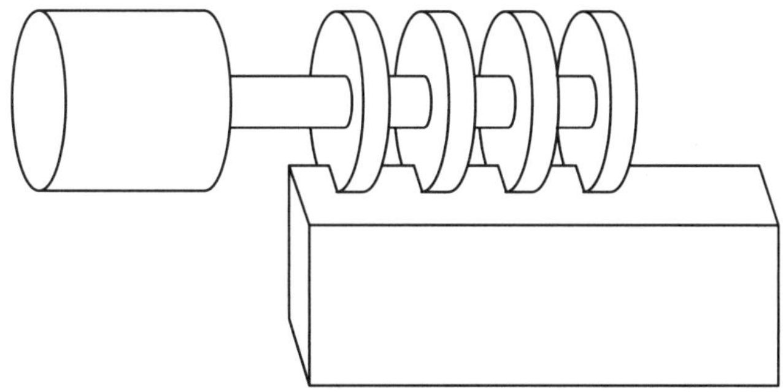

Figure 3.11.2 Example of μWEDG process used to make a rotating micro disk tool which makes multiple slots or slices thin sections from part material. (Courtesy C.-L. Kuo.)

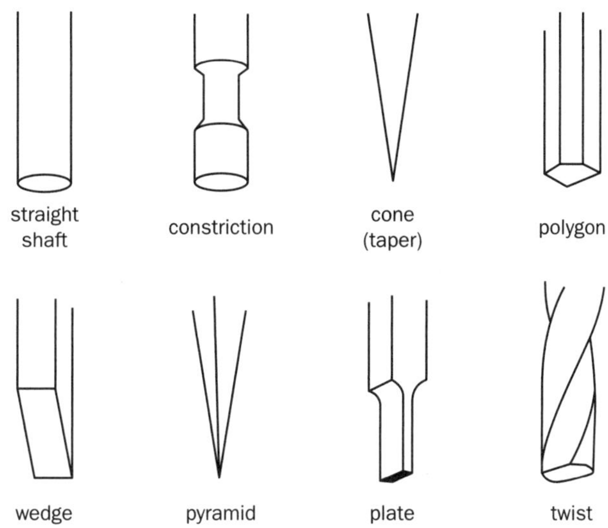

Figure 3.11.3 Example of tools made by μWEDG. (After Shao, 2015.)

more grooves in that direction and the result is a pattern of rectangular posts standing as deep as the workpiece was ground. The posts would be rectangular rather than round, but that may not be a limitation for many applications.

By machining an abrasive stick into a pencil-like shape, the tool can be used as a rotary grinding tool to grind micro slots. By using X-Y-Z tables the micro pencil then can grind 3D shapes including wavy grooves, and 3D outer configurations.

Micro grinding is a challenging process that requires a very high precision. Making the micro tools with μWEDG can be the easiest part of the operation.

In addition to making electrodes and grinding tools, μWEDG has also made minute cylindrical cutting tools by slicing flat faces off the cylinder in such a manner that desired shear angles are produced on the cutting tool. Stubby 60-μm (0.0024-in.) cutters have been made this way on PCD tools that were later used to "end mill" features.

While wire EDMing is used to produce most of the truly micro electrodes, at least one equipment manufacturer holds the future electrode up near a carbide block and then reverses the current so the tool is eroded to the diameter wanted rather than the carbide block. This produced a 25-μm (0.001-in.) electrode. This approach took 4 min to reduce a 381-μm-long (0.015-in.), 76-μm-diameter (0.003-in.) rod to 25 μm (0.001 in.) diameter.

While μWEDG realistically refers to producing cylindrical shapes, most users who have that capability will refer to any part on the equipment as μWEDG, even though the finished part has both cylindrical and other geometrical configurations. The part can be fed down without turning so that a flat face is generated, the part turned, and a similar step performed to make another surface. This truly is μWEDM, but users often find it simpler to refer to both processes as μWEDG or μWEDM.

Some commercial machines have the ability to μWEDG a tool, and on the same setup machine, the finished part without resetting either tool or part. There are companies with these capabilities and skills, but the equipment exists and details of actual operation are also published.

A number of useful references are presented below for readers wanting greater insight.

References

Han, F., Y. Yamada, T. Kawakami, and M. Kunieda, Investigations on Feasibility of Sub Micrometer Order Manufacturing Using Micro-EDM, *Proceedings American Society of Precision Engineering (ASPE) Annual Meeting*, 2003, vol. 30, pp. 551–554.

Hoffmeister, H.-W., and A. Wenda, Novel Grinding Tools for Machining Precision Micro Parts of Hard and Brittle Materials, Institute of Machine Tools and Production Technology Technical Paper, Braunschweig, Germany, 4pp.

Kuo, C.-L., and J.-D. Huang, Fabrication of Series-Pattern Micro-Disk Electrode and its Application in Machining Micro-Slit of Less than 10 μm, *Int. J. Machine Tools & Manufacture*, 2004, vol. 44, pp. 545–553.

Morgan, C., S. Shreve, and R. Ryan Vallance, Precision of Micro Shafts Machined with Wire Electro-Discharge Grinding, *Proceedings American Society for Precision Engineering*, 2003 Winter Annual Meeting, vol. 28, pp. 26–31.

Shao, B., Modeling and Simulation of Micro Electrical Discharge Machining Process, Mechanical (and Materials) Engineering, Dissertations, Thesis, and Student Research, Paper 92, 2015, University of Nebraska, Lincoln, Nebraska.

CHAPTER 3.12

Electron Beam Drilling

Kenneth Norsworthy

Owens Corning Ridgeview
Duncan, South Carolina

Electron beam drilling (EBD or EB drilling) is a thermal drilling process by which small through holes [in the range of 0.1 mm up to 1.0 mm (0.004 to 0.040 in.)] can be drilled at unusually high aspect (depth to diameter) ratios of up to 25:1, to a maximum depth of 9.0 mm (0.354 in.). This aspect ratio coupled with our high drilling speeds is what distinguishes this technology from all other thermal, chemical, and mechanical small-hole production technologies.

There are two different approaches to EB drilling: (1) percussion and (2) trepanning with the first involving single-pulse mode or perforation or multi-pulse shots. Trepanning, as the name implies, involves traversing the beam in a cylindrical fashion to cut the hole. Multi-pulse mode is generally not cost compatible with laser processes, but it may be used to improve wall surfaces.

EBD is performed inside a vacuum chamber. Under vacuum, a focused beam of electrons is accelerated up to two-thirds the speed of light. When the beam meets the surface of the material, it is vaporized as the beam burns through the remaining material. A special coating is applied to the back side of the material. As the beam breaks through the material the beam is switched off; the energy that enters the coating generates a micro explosion that blows the molten material outward to clear the hole. The resulting hole has a small radius at the entrance and a slight taper over the depth of the hole. The exit side of the hole has a defined edge without the radius as seen on the entrance side of the hole. This taper effect is a highly desired hole characteristic for filtering applications. The taper may be between 1 and 5 degrees.

EBD is a very cost-effective technology when part needs are from 10,000s to 1,000,000s of holes. However, since EBD is done in a vacuum chamber, parts with hole counts under 10,000 holes may be cost prohibitive due to the nonproductive machine time associated with vacuum chamber pump downtime.

Typical hole size diameter for EB drilling ranges from 0.05 to 1.5 mm (0.002 to 0.060 in.). Holes as small as 0.012 mm (0.0005 in.) in diameter have been EB drilled in 400 series stainless steel in less than 1 s. Holes less than 10 μm (0.0004 in.) may represent the lower limit for today's technology. Larger holes are drilled at a rate of up to 10 holes per second, while tiny holes can be produced at up to 150,000 per second!

Typical material thicknesses drilled range from 0.1 to 6 mm (0.004 to 0.236 in.). Holes can be perpendicular or inclined to the surface (up to 25 degrees). Typical tolerances for hole diameters are ±25 μm (±0.001 in.). The time required to make holes varies inversely with the size of the hole and the thickness of the material. Thus, the rate drops off to a "mere" 20 holes per second when EB drilling 5-mm-deep (0.197-in.), 0.7-mm-diameter (0.028-in.) holes.

Material hardness is not an issue, because EBD is a thermal process and can drill any grade of stainless steel, heat-treated tool steels, titanium, plastics, and ceramics as well as soft metals like copper foils. In fact, it can drill almost any material that can be processed in a vacuum environment. Because there are no mechanical forces on it, it can drill almost distortion-free even on thin foils. A very thin heat-affected zone may exist at the surface as well as some minute recast at walls near the top surface. It is typically much smaller than that produced by a laser or EDM.

Physical Part Size Limitations

Drilled cylinders from Ø200 up to Ø950 mm (7.9 to 37.5 in.) can be drilled to a maximum width of 1000 mm (39.4 in.). Many of the parts drilled are drilled on a cylinder. Since the machine is CNC programmed/controlled, it can drill any pattern or shape on a cylinder and then it can be cut into a flat sheet. The machine has a 5-axis manipulator which allows drilling round discs and angled holes as well.

Technology Applications

The size of holes that can be drilled coupled with the high aspect ratio is ideally suited for high-pressure filtration applications. Owens Corning, for example, uses EB drilling to make parts for its proprietary glass-making equipment as well as for other forms

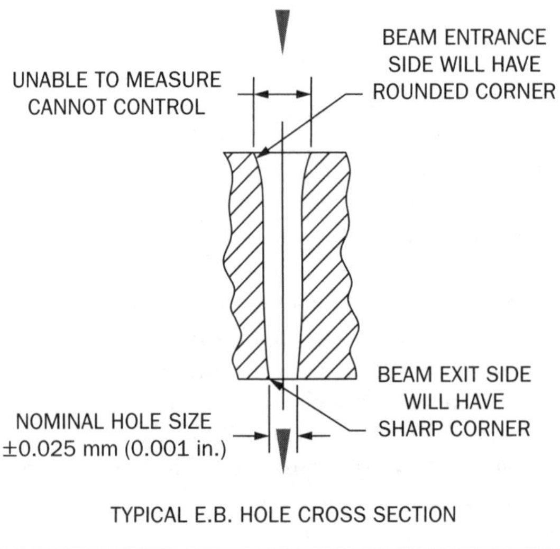

Figure 3.12.1 Typical appearance of an EB-drilled hole. (Courtesy Owens Corning Ridgeview.)

of contract work. Examples include filters for fiber glass insulation manufacturing (spinnerets, for example). Others use it for food and juice, process, pulp and paper, waste water, and recycled plastic filtration. Other applications are used for paint pigment particle milling, screening–particle separation, nonwoven products, and aerospace products. EBD has been used for gas metering orifices, fuel injectors, wire drawing dies, sapphire jewel bearings, light ray orifices, and space shot products as well as the more mundane applications.

In addition to holes, EB milling has been used to produce such features as 0.125-mm (0.005-in.) slots in ferrites, and slots or pockets from 0.025 to 10.0 mm (0.001 to 0.400 in.) although the majority of applications are for features on the smaller side of these dimensions. Tolerances for such features have been held to ±0.025 μm (±0.001 in.). Holes do not have to be round—slots, ovals, triangles, and many other shapes can be produced readily with this process. Straight walls, bowed-walls, large or small bell-mouthed holes, as well as intentionally tapered holes can be readily produced. Hole entrances are typically rounded, yet the hole exits are burr free and somewhat sharp. Figure 3.12.1 illustrates an EB-drilled hole.

There are only a few commercial houses that perform EB drilling, but some have been in business for decades. The process has been used in production since the 1950s and is well proven.

CHAPTER 3.13

Electron Beam Polishing

Masato Fukushima

Sodick, Inc.
Schaumburg, Illinois

Electron beam machine (EBM) polishing is a new surface processing technology that uses electron beam irradiation to remelt material surface to produce a smooth finish, mirror finishes in most cases [from 0.06 to 0.12 μm (2.4 to 4.7 μin.)], on a variety of machined surfaces. It is a nontraditional polishing process. It is being used in the biomedical, aerospace, and tooling industries as well as in a number of research efforts. Some specific material applications include 316 stainless steel, titanium, cobalt-chromium-molybdenum alloys, and several die steels and acrylics.

A high-power electron beam in repetitive microsecond blasts quickly heats and cools the surface. This rapid fusion and coagulation causes changes in the material surface. The electron beam affects roughly 5 μm (0.0002 in.) of the outer surface layer without damaging the original quality or part accuracy, and reforms the surface to a smooth finish (Fig. 3.13.1). In addition to making a smooth surface, micro cracks, chipping, or scratches irradiated with the electron beam during polishing become more water repellent and resistant to corrosion. This minimizes mold deposits from occurring in plastic injection molding.

Unlike traditional hand-polishing methods, there are absolutely no micro deviations in dimensional accuracy from spot to spot, and there is absolutely zero material removal. Consider for example any three-dimensional cavity. Hand polishing has a

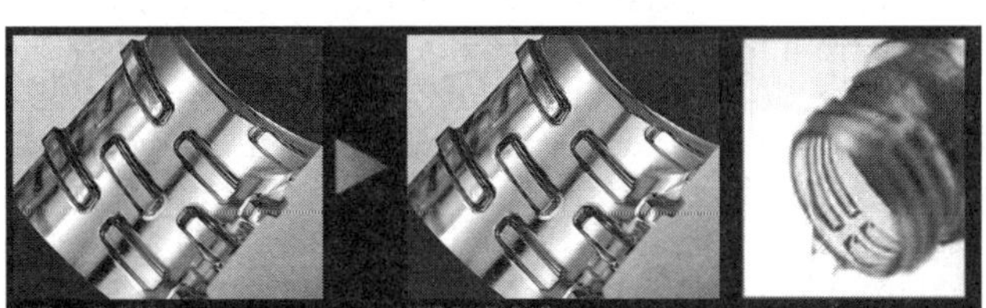

Figure 3.13.1 The threaded portion of screw cap molds requires a high finish to allow quick and damage-free release of the hot PET cap. Electron beam polishing takes the mold from a roughness of 0.56 μm (22 μin.) to 0.12 μm (4.7 μin.). With EBM polishing the time required fell from 3 h to 12 min.

FIGURE 3.13.2 This intricate three-dimensional mold is no challenge to polish so every part of the surface is uniformly smooth. Manual abrasive methods cannot duplicate this precision and repeatability.

difficult time reaching down into each portion of such cavities uniformly (Fig. 3.13.2). As a result, there is a variation in surface texture from one spot to another. All irradiated surface finishes evenly and uniformly to the same measurable accuracy throughout. This level of precision polishing has never been attainable with traditional polishing methods.

The electron beam itself is approximately 60 mm (2.36 in.) in diameter, but one machine is capable of irradiating up to a 300 × 300-mm (11.8-in.) workpiece surface area with programmable stage table movements. A typical maximum workpiece height is 100 mm (3.94 in.). The workpiece metallurgic reaction to the electron beam irradiation varies depending on the type of material and its composition. Generally, stainless steel (SUS) is the most compatible material for producing mirror-surface results. Carbide products also can be polished, but the mechanisms are different. By irradiating carbide with electron beams, the cobalt binder contained in tungsten carbide (WC) is diffused, which results in a hardened surface layer of solid tungsten, which makes it drastically easier to polish to a mirror-finish in just minutes.

EB polishing is highly effective for polishing molds, and also enhancing the surface quality of titanium and ceramic products in the medical industry, and to quickly improve the clarity of small acrylic resin products. When used on hot forging dies, tool steel electron beam polished surfaces allow nitride treatment such that the chromium binds with the nitrogen to produce a highly polished long-wearing chromium nitride surface that extends die life significantly. The chromium nitride atoms put the surface into a beneficial compressive stress that results in the extended product life. The exceptionally smooth surface from EBM polishing reduces friction, increases the ease of release from molds which can reduce mold cycle time, increases corrosion resistance, improves the sliding of press dies, and reduces gas clogging in minute vents or passages.

In production it reduces a 0.27-μm (11-μin.) surface on D2 alloy tool steel to a 0.18 μm (7 μin.) Ra over a large area (see Fig. 3.13.2) in 8 min. In another application the finish was reduced from 0.21 to 0.06 μm (8 to 2.4 μin.) Ra in only 40 min on STAVAX/SUS. The ability to reach into narrow slots or cavities is another unique aspect of EBM polishing (Fig. 3.13.3). The fact that the process can finish multiple miniature parts in a single load also makes the process economical for these high-precision repeatable finishes.

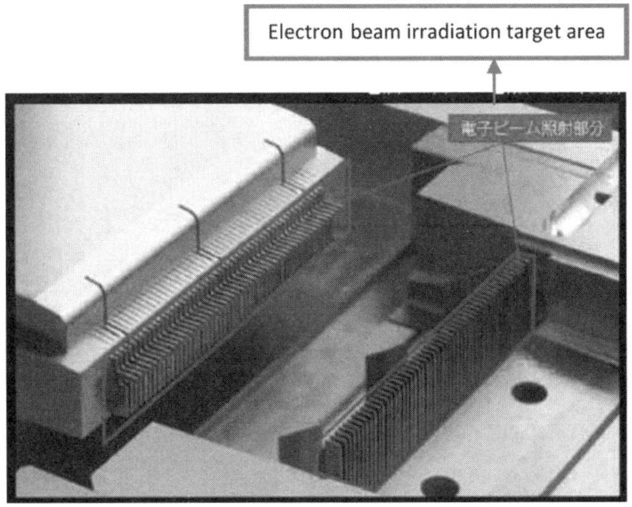

Figure 3.13.3 Electron beam polishing smoothens hard to reach surfaces and many details or parts in one setup.

In addition to polishing clean surfaces, EBM polishing can remove rust but will reform and seal the microscopic surface pores to reduce future corrosion. The process is used on cloudy acrylic resin products to remove cloudiness and make them transparent. By controlling the energy, intensity, and fusibility of the beam, sharp edges can be left sharp by the process. To prevent unintended exposure on adjacent surfaces, scotch tape is placed over those surfaces.

Hot forging dies also benefit from the treatment because after use at high temperatures the dies develop minute heat cracks on the die surfaces. EBM polishing reforms the material at the surface to extend die life.

In the micro manufacturing world, the exceptionally smooth surfaces offer higher success for removing thin-wall molded parts from the die since the friction forces are reduced.

The EBM polishing process can be used on most metals. Its most useful applications outside die polishing of conventional size products may be in the micro finishing world where small features must be polished and the small part or feature size allows rapid finishing.

CHAPTER 3.14

Designing for Chemical Mechanical Polishing

Jeffry J. Sniegowski

Sandia National Laboratories
Albuquerque, New Mexico

The Process

Chemical mechanical polishing (CMP), a planarization process first developed and used for the manufacturing of multilevel metal interconnects for high-density integrated circuits (ICs), has been adapted as an enabling technology in the fabrication of micro electromechanical systems (MEMS) or more generally, micro systems technology (MST). The CMP planarization technique alleviates many manufacturing issues introduced by topography generated during the fabrication of multilayer MST structures. Historically, this was specifically true with polysilicon surface micro machining, an additive manufacturing technology, which comprised as many as six added levels of polysilicon. Such a technology would not have been possible without the incorporation of CMP planarization between layers. For the designer, CMP not only eliminates design constraints introduced by nonplanar topography generated at intermediate fabrication steps, but also improves the overall quality of surfaces and potentially provides an avenue for integrating disparate process technologies. In general, CMP is applied to substrates onto or into which MST devices are constructed in a batch fabrication process similar to how ICs are manufactured. These substrates, commonly referred to as wafers, are most often round, thin, and comprised of single-crystal silicon and have diameters ranging from 100 to 300 mm (3.9 to 11.8 in.) with thicknesses of a few hundred micrometers to maybe a few millimeters (~0.012 to 0.250 in.). Commercial CMP tools are normally set up to handle such substrates. Nonstandard substrates can be CMP'd, but this may require specialized tooling at MEMS boutique foundries.

Most simply, CMP is a process to smooth and plane surfaces with the combination of chemical and mechanical forces. The CMP process produces both global and local planarization, as well as smoothing through relatively simple and quick processing. It can, in a way, be thought of as a hybrid of chemical etching and free abrasive polishing. Whereas mechanical grinding alone may theoretically achieve planarization,

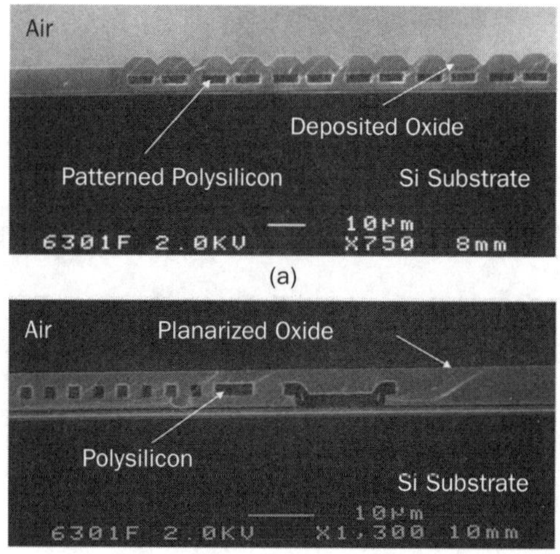

FIGURE 3.14.1 (*a*) Scanning electron microscope (SEM) cross-section image of a partially fabricated MST silicon substrate illustrating the uneven, severe topography before CMP planarization. Such topography is generated by the addition of another uniform, conformal layer of material onto a surface previously patterned with a design layer (such as the polysilicon layer shown here). (*b*) SEM cross section of a partially fabricated MST substrate post CMP planarization. Subsequent patterning of the oxide and addition of more layers can be accomplished without the interference from topography.

the surface damage is high as compared to CMP, and generally has no selectivity to materials. On the other hand, chemistry alone cannot attain planarization because most chemical etch reactions are isotropic.

CMP not only simply eases the design and manufacturability of MEMS devices by eliminating photolithographic and film deposition/etch issues generated by severe topography, but also enables far greater flexibility with associated designs and process complexity. First, CMP planarization includes topographic reduction on a local scale with the conversion of abrupt, vertical steps to much smaller steps that removes interferences and fabrication issues, while on a more global scale, it flattens transitions from large high areas to large low areas. Second, CMP can provide microscopic surface smoothing that produces a mirror-like surface at the nanometer scale.

Figure 3.14.1 exemplifies the basic function of the CMP process wherein an oxide surface having topographic bumps generated from an underlying polysilicon layer is planarized, removing the bumps in the oxide film and making it more appropriate for subsequent processing. This is accomplished by rotating a wafer under pressure against a polishing pad in the presence of a silica-based, alkaline slurry. The mechanical energy imparted to the abrasive slurry particle through pressure and rotation, in combination with materials-specific chemistry, causes high features to erode at a faster rate than low features, thereby planarizing the surface over time.

The motion and force of the MST substrate against the pad with a slurry containing microscopic particles causes mechanical abrasion, while the added chemistry to the slurry alters the material being polished on the surface of the wafer as illustrated in Fig. 3.14.2. The combined mechanical abrasion with chemical alteration is called chemical mechanical planarization or polishing (CMP). The removal rate of the material can

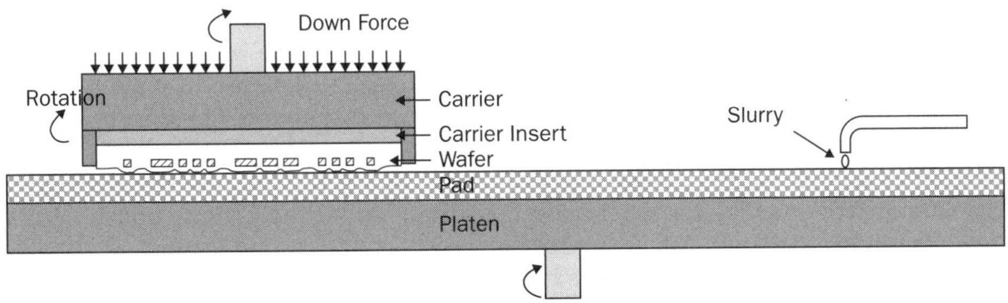

Figure 3.14.2 A schematic representation of a CMP process where an oxide surface with severe topography is planarized by rotating the inverted MST substrate under pressure against a polishing pad in the presence of a silica-based alkaline slurry.

be easily an order of magnitude higher with the simultaneous chemical and mechanical effects as compared to either one taken alone. Selectivity and polishing stoppage can be gained with the proper selection of chemistry and particles. Finally, the smoothness of the surface after polishing can also be optimized by using chemical and mechanical effects together.

Before proceeding further, it is emphasized and highly recommended that the designer ask questions of the chosen manufacturing facility regarding specific data on the fabrication output of their CMP processes that may impact the designer's MST product. These data may include step reduction (or print-through reduction), global flattening, and surface smoothness, in addition to which material systems they apply CMP. These items are more evident following the examples below. The goal is to provide a basic understanding of CMP and a quick guide to formulate appropriate questions by which to evaluate or compare multiple fabrication facilities. Details of the CMP process are beyond the scope of this chapter, as entire treatises are dedicated to the CMP process and its integration into a MST fabrication technology. One CMP reference is provided if the designer is inclined to further understand CMP [1].

Application of the Process

As an MST device is fabricated, layers of materials are added or removed, then patterned and etched with the device design. This cycle is potentially repeated numerous times, depending on how many levels are included in the design. Each cycle adds topography to the surface. The first layer on a pristine, flat substrate presents no problems, but as the subsequent layers are added, this topography introduces problems for manufacturing and eventually for the design function. The CMP process can be used to essentially re-flatten the surface topography, ameliorating both fabrication and design issues. This is the essential basic function of CMP. However, it can also provide more sophisticated functions that are described below.

In more general terms, there are three broad applications of the CMP process to MST fabrication. First, to planarize, or make flat, the layers, enhancing manufacturability and design. These can be the layers that will later form structural or functional materials in the final design, or can be the layers that are temporary support layers that are removed in the final steps of fabrication, exposing functional layers or producing free-standing mechanical structures. These temporary support layers are often referred to as sacrificial layers, and can be removed by various chemical means as a final step. Second, to touch polish a surface, producing a high-quality finish: this is important for applications such as optical devices. Third, to produce the design pattern in a layer (such as copper or tungsten) that cannot be patterned by the normal means of photolithography masking and etch. Such process is referred to as CMP damascene or dual-damascene. Damascene is basically the process of depositing the desired material into trenches having the desired design pattern and using CMP to polish away excess overlaying material.

As a starting point, from the designer's perspective three questions are proposed: (1) Do I need planarization at all? (2) If yes, does it only need to remove gross nonplanar structure? (3) Do I need a high degree of flatness and smoothness at either or both global and local scales? If the answer to (1) is no, then go design! If the answer to (2) is only the need for gross planarization, then most CMP processes offered will very adequately accomplish this. If the answer to (3) is yes, I need an optically flat and smooth surface, for example, it becomes imperative that the designer work very closely with the fabrication personnel to clearly understand what degree of flatness and smoothness is produced in the particular fabrication facility. If the facility has a well-developed CMP capability for MEMS, they may offer several CMP processes optimized to produce specific results on specific materials. It is hoped that the following further description of CMP planarization will readily guide the designer through these three questions. With CMP planarization, a change in design perspective occurs since the need to circumvent mechanical interference in a nonplanarized process is alleviated. Thus, more complex designs with greater functionality created from multiple levels are feasible.

The designer should inquire as to which layers in the fabrication technology are CMP'd and the relative thicknesses of those layers. Also, it is useful to inquire as to the purpose of each application of the CMP process. Whether it is for the purpose of planarization (impacts manufacturing and design), touch polish (impacts surface quality for design), and damascene (layer patterning, primarily only of interest to the fabricator). Although each specific application of CMP is optimized to a single purpose, there may or may not be cross-over benefit to the other impact areas (e.g., planarization CMP can provide high-quality surfaces, whereas a touch polish will not normally provide significant planarization). To first order, removal of mechanical interferences/obstructions and manufacturing issues is readily accomplished by a basic CMP process. However, where a high degree of flatness or ultrasmoothness is desired, the results are dependent on the specific material and CMP process. These can vary significantly foundry to foundry.

Enhanced Manufacturability of MST

The CMP process was first applied to an MEMS technology known as polysilicon surface micro machining, but can be applied to any MST wherein topography is introduced by the repetitive deposition and patterned etching of multiple films. Typically, at the completion of the MST fabrication, sacrificial layers (as their name suggests) are selectively removed, leaving the free-standing device structures in the form of channels, gears and linkages, or mirrors. Figures 3.14.3*a* to *c* both illustrate the essence of polysilicon surface micro machining and display the vertical topography which arises during a nonplanarized process. This topography can produce mechanical interference between moving parts and can complicates subsequent process steps. Figure 3.14.3*a* schematically

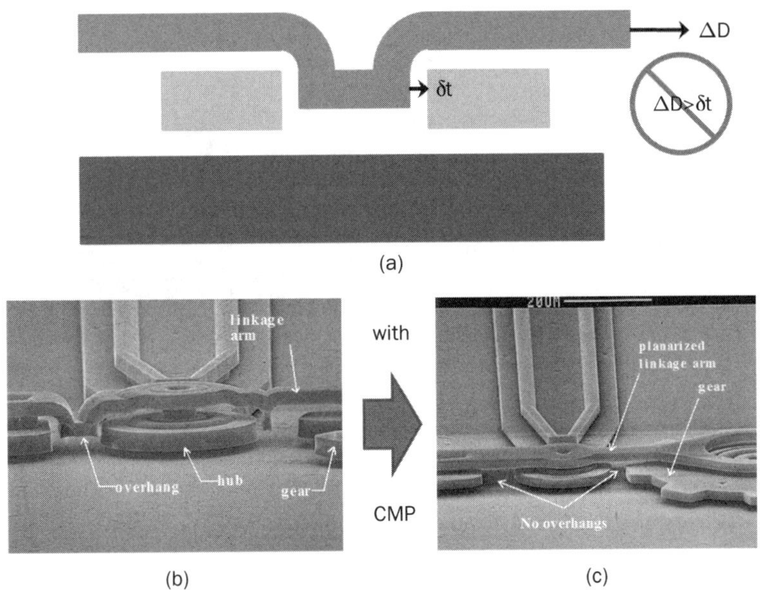

FIGURE 3.14.3 (*a*) A schematic illustration of an interfering protrusion generated in the upper, intermediate grey layer with a non-CMP process. In the final released device, under a displacement ΔD, the active MST linkage will stop in a collision with the underlying the middle, light grey layer. SEM image (*b*) is of a non-CMP-fabricated linkage while (*c*) is the same linkage fabricated with CMP. The protrusion interference is removed allowing a movement ΔD>δt and no longer needs to be considered during design.

illustrates this interference, where a completed linkage element is intended to move a lateral distance, ΔD, but is stopped at a travel of δt < ΔD due to the obstruction from a nonplanarized upper layer in the process.

The mechanical interference or physical obstruction arises when an upper layer of polysilicon must pass over the edge of a previously etched lower layer. An example of this interference is shown for a micro actuator in Fig. 3.14.3*b*, where the upper link runs over the edge of a gear producing an overhang feature that can interfere with the desired range of motion. This feature is due to the conformal deposition of the upper polysilicon film. Such link/gear interference can be alleviated by appropriate design of the micro actuator, as pointed out in Fig. 3.14.4 (left), or by planarization of the surfaces before subsequent deposition of additional films, as illustrated in the CMP-planarized joint in Fig. 3.14.3*c*, or the uncompensated design in Fig. 3.14.4 (right). Note such undesired obstruction also occurs if one is attempting to produce a fluid channel with a capping layer, wherein the intermediate layer is not planarized.

Higher-Order CMP Effects

CMP can optimize the top surface of a particular layer (e.g., a reflective metal surface). This includes smoothing the surface, eliminating or reducing print-through from the underlying layer topography, or improving the overall flatness on a larger scale. Print-through is defined as residual steps seen in the upper polysilicon surface from

Figure 3.14.4 (Left) SEM images of rack and pinion design with accommodation cuts (indicated by arrows) in the gear design for a non-CMP'd early process, while image on the right shows a similar design accomplished with CMP-planarized processing having no accommodation cuts. Note: the small regular holes seen are to expedite the final release etch for the device.

underlying layers following CMP. Regarding smoothing, CMP can produce surfaces with root-mean-square (RMS) roughness on the order of a few nanometers or less (e.g., 0.4 µin. or less), where a specific example is CMP reducing the RMS roughness of as-deposited polysilicon from 46 to 2.2 nm (1.8 to 0.09 µin.). To illustrate the most relevant higher-order CMP effects, schematic depictions with exaggerated dimension are shown in Fig. 3.14.5. These include surface polish and print-through (type A), dishing (type B), and bowing (type C) that may impact a design if not known or accommodated. Following these schematic illustrations, actual fabricated examples of these effects are provided in Figs. 3.14.6 to 3.14.8. Surface depressions, shown in Figs. 3.14.5*a* and *b*, can produce type A and type B print-throughs as a by-product of areas where the overlying film fills a patterned depression. Type A results when the width is less than the thickness ($w<t$) of the deposited film, whereas type B occurs as shallow dishing when $L>>t$. Normally, the touch polish is shallow and does not necessarily remove type A print-through if present in a design. This may adversely impact an optical mirror-type surface by introducing light-scattering artifacts. On the other hand, with planarizing polishing that removes significant thickness of material usually greater than or equal to H, type A features are simply polished out. However, this is not necessarily true of type B print-through having residual small steps at the edges of large depressions (or bumps) where the characteristic width of the depression (or bump) is much greater than the deposited film thickness ($L>>t$). Another even smaller effect can be a very slight bow (small convex curvature) of the surface over a planarized large bump as shown in Fig. 3.14.5*c*. There can also be a similar very slight dishing (small concave curvature) of the surface over a large planarized depression. (Note: this was not shown in Fig. 3.14.5*b* to avoid clutter.) In either case, if an ultra-flat surface is required, these effects need to be known to exist and inquired of the fabrication facility. However, if the planarized surface becomes part of a free-standing structure in the final device, other effects such as curvature due to residual mechanical stress can dominate the final surface curvature.

Print-through effects on the order of a couple hundred nanometers (~8 µin.) do not necessarily cause mechanical or manufacturing interference but may adversely affect optical elements, causing scattering or phase shifts. For a real example, Fig. 3.14.6 illustrates atypical surface profile of an overlying oxide pre- and post-CMP planarization.

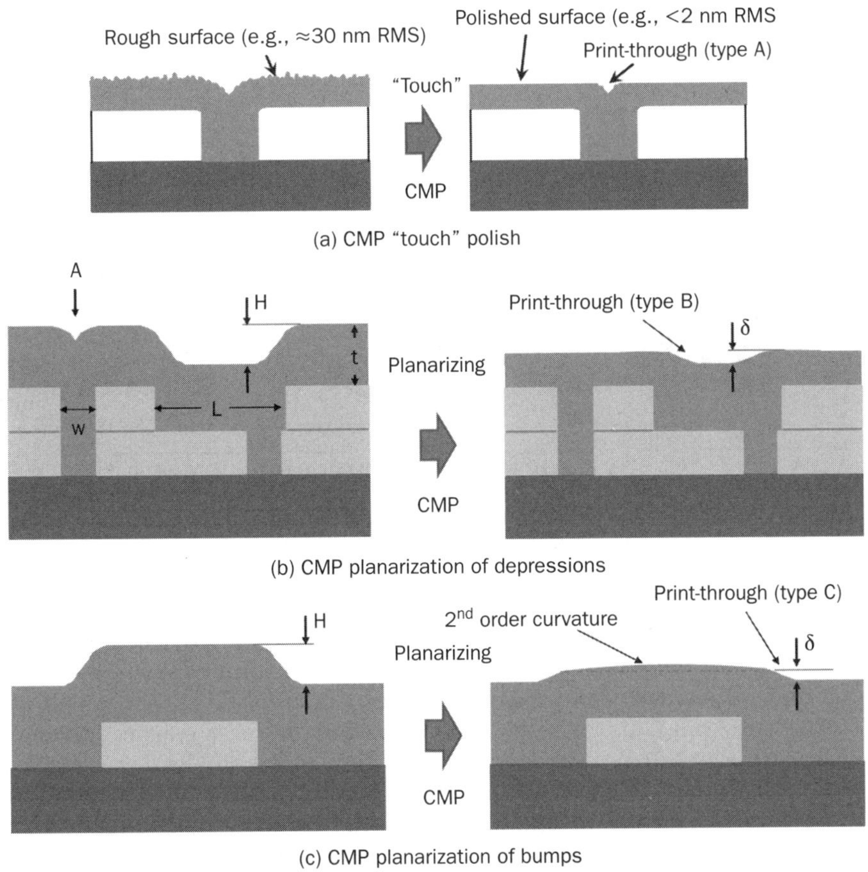

FIGURE 3.14.5 Schematic illustration of various second-order effects of CMP. Smoothing in the case of touch polishing in (*a*), while (*b*) and (*c*) illustrate effects that occur while CMP planarizing large areas of depressions or bumps, respectively. These help clarify the actual fabrication results shown in Figs. 3.14.6 to 3.14.8.

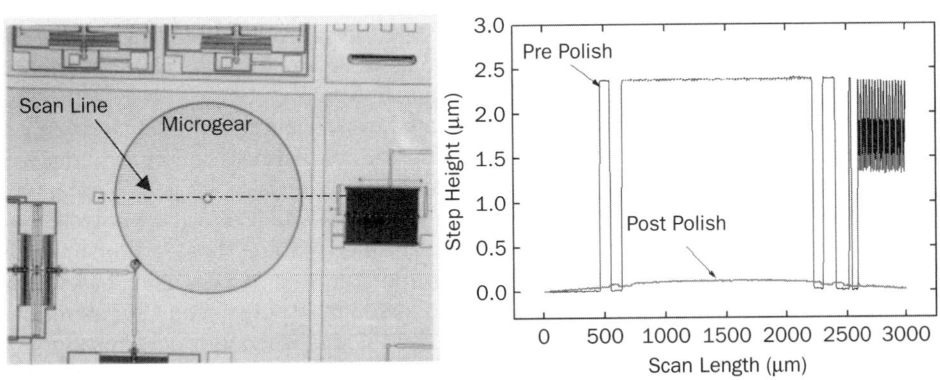

FIGURE 3.14.6 The profilometer scan on the right is across the gear in the optical micro graph on the left. Both the pre- and post-CMP topography heights of the oxide covering the gear are shown. The small <200 nm (<8 μin.) topography variation remaining over the micro gear is acceptable for micro mechanical applications.

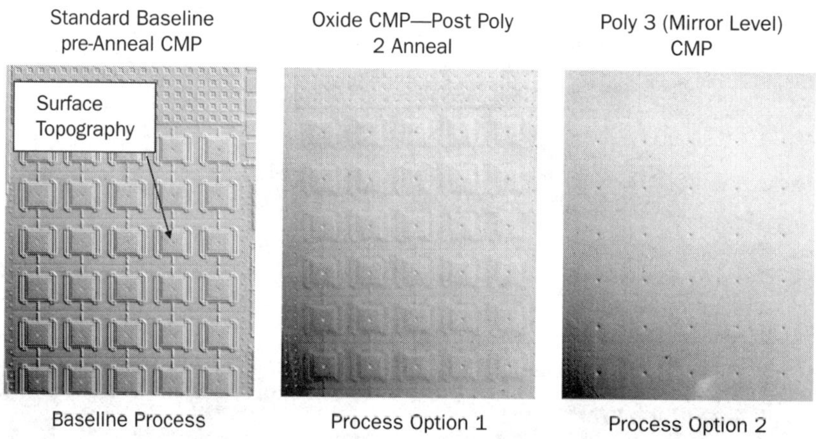

FIGURE 3.14.7 Nomarski mode microscope photographs of the polysilicon 3 layer for (left) baseline process, (middle) oxide CMP after the anneal step, and (right) baseline process with additional poly CMP step on poly 3 layer. The photographs give a good indication of the existing surface topography when viewed in the reflected-light Nomarski-interference contrast mode.

This oxide layer covers structures onto which additional structures may be built. One sees that the severe pre-CMP topography of 2.5 μm (98 μin.), which can lead to mechanical and manufacturing interferences, is eliminated. However, a relatively small [<200 nm (8 μin.)] global, residual bow remains, and even smaller [<100 nm (4 μin.)] localized print-through step (type C) still exists at structural edges. Specifically in the case of the global scale bow, it is good design practice to use CMP dummy structures to further reduce this residual. CMP dummies are simply extra, non-used structures consisting of all underlying layers that surround the active device areas. Dummies are used in all modern IC CMP processes to maintain a high degree of flatness, in that case to prevent manufacturing issues. IC foundries normally include these in the final design mask layouts through automated design tools. However, in the case of MST structures, it is critical that they are located well outside the range of motion of movable MST structures or the interior of channels or desired cavities. If the designer is not responsible for including CMP dummies in their MST design, it is imperative that the designer work with the fabrication facility/mask house to ensure there are no inadvertent design interferences introduced if they are automatically generated.

A final example is provided to illustrate how variations in CMP can highly impact demanding optical designs, where flatness and smoothness can be extremely critical. Figure 3.14.7 shows three differential interference contrast (DIC), or Nomarski mode, optical microscope photographs imaged after deposition of a final polysilicon micro mirror layer for the baseline process (left), modified oxide CMP step after an underlying polysilicon layer thermal anneal (middle), and modified process using baseline process up to the final micro mirror polysilicon followed by a polysilicon CMP planarization/smoothing step (right). As shown in Fig. 3.14.7, the baseline process produces a high degree of contrast and clearly shows the effect of the underlying pattern which has printed through to the front mirror surface resulting in increased topography. Such a result is unacceptable for micro mirror applications. The second and third photographs correspond to the revised process flows described above. The Nomarski mode

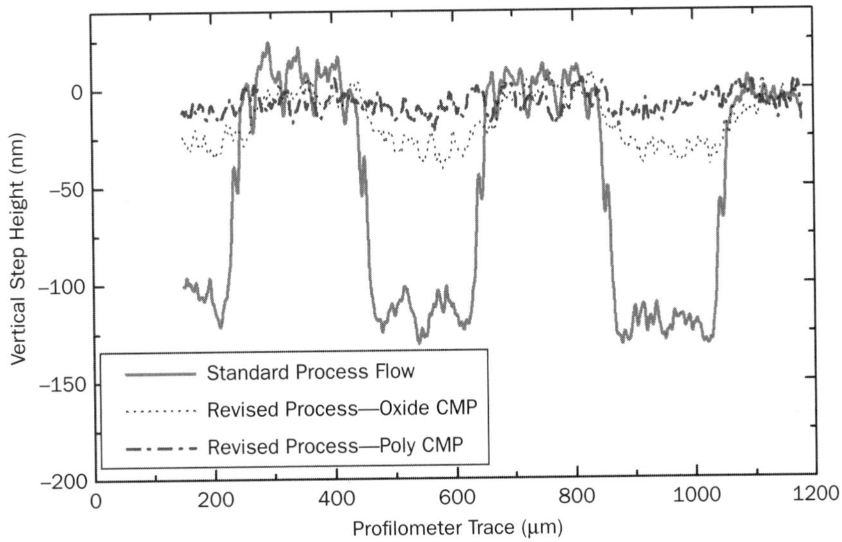

FIGURE 3.14.8 Profilometer measurement for the standard and revised process flows is shown. Both of the revised process flows show an improvement in planarity and produce acceptable flatness for optical micro mirror applications. The location of the profilometer scan is given in Fig. 3.14.7 (left).

photographs given in Figs. 3.14.7 (middle and right) show very significant topography reduction when compared to the baseline surface in Fig. 3.14.7 (left). Profilometer measurements for each of the process options and the baseline reference are given in Fig. 3.14.8. Both process alternatives provide acceptable $<\lambda/10$ wavelength flatness. The profilometer measurement that indicates the best result is with the polysilicon CMP process option (right) [Δ $step_{max}$ = 20 nm (0.8 μin.)].

From this example, it is imperative to request representative examples of the print-through reduction, global flatness, and surface smoothness from your specific foundry to determine whether or not they meet your design criteria.

CMP Limitations

The principle limitation of CMP is total material removal (i.e., height of topography removed). CMP provides a high-quality finish with little substrate damage, but it is relatively slow at material removal and, therefore, becomes an expensive process as film thickness increases. Typical film removal can range from submicrometer to several micrometers (<40 to 200+ μin.) in thickness. If significantly greater material must be removed, an initial removal using mechanical grinding or lapping followed by CMP to remove the damaged material and provide a high-quality finish is an alternative process. For example, mechanical grinding can have removal rates of approximately 100 μm (~0.004 in.) of Si in 6 min versus 0.65 μm (0.000026 in.) SiO_2 in 2.5 min with CMP. Mechanical polishing alone is unacceptable in many cases and, most specifically, for semiconductor IC processing due to excessive mechanical damage to the wafer surface. Although one may think that this is not essential to micro mechanical devices and structures, the damage can lead to mechanical stress gradients in layers that, upon release to form free-standing structures, generate undesired structural deformation (i.e., curvature of plates and beams). Again, strain gradients in layers and the resulting deformation are process parameters of which a designer must inquire at the foundry of choice.

Materials

CMP is typically used on silicon dioxide, polysilicon, and metal surfaces, which are all common materials in the IC industry. However, the basic technique is applicable to a wide variety of materials including polymers, semiconductors and their oxides, metals and metal alloys, composites, piezoelectrics, and many more. When used as a damascene process, it permits the use of difficult-to-etch metals such as copper, tungsten, or other materials of this character. The available materials are determined by the CMP capabilities developed at the specific foundry and, ultimately, the chemistries and pads provided by CMP vendors. If the particular foundry does not offer the specific material/CMP process of interest, it is recommended that another foundry be sought. Development of new material processes is expensive and time consuming!

Critical Process Parameters for the Designer

The designer can approach fabrication of their design from two directions. First, if the designer is already working with a specific foundry, identify what materials that foundry is capable of handling in their CMP process. If the materials are consistent with the requirements of the design, proceed to further understand the specific CMP process attributes, such as overall flatness dependency on design layout and final surface finish to see whether they also meet the design requirements. Second, if the existing material basis is not consistent with design requirements, determine whether the foundry is willing to develop the necessary material base. If not, proceed to locating alternate micro fabrication facilities that offer the necessary materials, and then vet their CMP process attributes. It cannot be overstressed that close interaction with the manufacturing personnel at the early stage of design is critical for favorable CMP outcome. In any case, knowing the detailed specifications regarding CMP performance at the specific facility chosen is critical. With the multitude of CMP tools, pads, chemistries available to a given foundry, and subsequent idiosyncrasies during process development and integration, it is important to discuss the detailed specifications of their CMP results. It is important to ask for process data that indicates both global and local planarization figures of merit, dependencies on layout density, and final surface quality, if it is an important design parameter.

Acknowledgments

Sandia National Laboratories is a multiprogram laboratory managed and operated by Sandia Corporation, a wholly owned subsidiary of Lockheed Martin Corporation, for the U.S. Department of Energy's National Nuclear Security Administration under contract DE-AC04-94AL85000.

Reference

1. *Chemical-Mechanical Planarization of Semiconductor Materials,* M. R. Oliver (Ed.), Springer Series in Materials Science, 2004, ISBN 3-540-43181-0.

CHAPTER 3.15

Micro Ultrasonic Machining

LaRoux K. Gillespie

Manufacturing Consultant
Andover, Kansas

Ultrasonic machining (USM) uses a vertical tool and an abrasive slurry to machine very hard materials. Typically, mechanical vibrations at an ultrasound frequency of 20 to 40 kHz move the tool and slurry up and down against the workpiece. The tool never touches the workpiece, but it forces the abrasive slurry to pound away below the tool. The constant hammering at these high speeds erodes the material in the shape of the tool's cross section. The tool vibrates at an amplitude of a few micrometers (Fig. 3.15.1). The slurry is a mixture of fine abrasive particles [usually in the range of 0.5 to 5 μm (20 to 200 μin.)] and water or oil. It is essential, however, to keep fresh slurry beneath the tool, which requires more effort for micro features than for larger applications. Microscopic particles might not seem like they would provide much cutting, but the ultrasonic action propels the particles in some conditions so they act like they are 150,000 times heavier than they actually are.

In addition to being ideal for very hard workpiece materials, the process does not alter mechanical, thermal, chemical, or electrical properties of the workpiece. As a general rule there are no added internal stresses from the process. Thus the part's fatigue strength of the material does not change. Unlike conventional machining, there are no burrs or deposits left on feature edges. A small radius is normally produced there. For glass some chipping can occur.

While the process can also be used as a disc-grinding process to improve surface finishes, it is not clear whether this is a major benefit. While its largest use is to machine internal features, it also can as easily contour the outer dimensions. Die-sinking is one long-used application as well as contouring complex features.

USM Shapes and Tools

The tools can have any cross section, but they are often round because most often they are used to drill micro holes. Tools are made of austenitic stainless steel and they can be

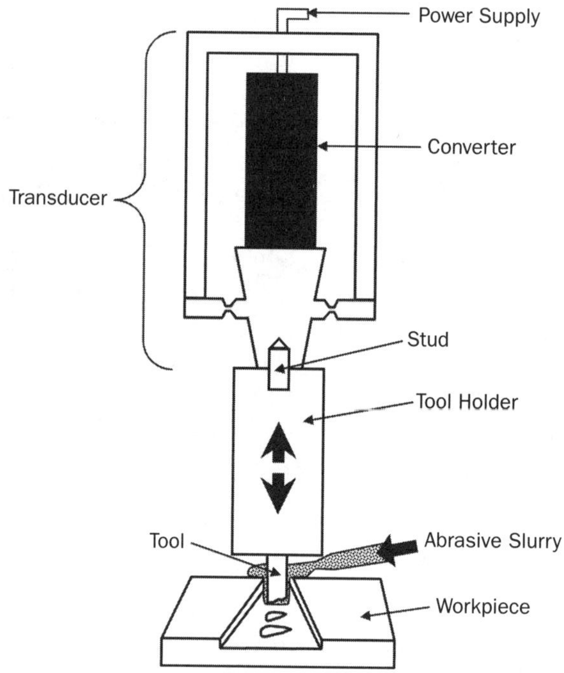

FIGURE 3.15.1 Schematic of micro USM process. (Courtesy Bullen Ultrasonics.)

hollow tubes. Micro USM produces holes ranging in size from 20 to 400 μm (0.0008 to 0.016 in.), but the process can produce very large holes or features if desired. The micro holes normally are round, but square- and triangular-shaped ones are also made. Depth accuracies of 10 μm (0.0004 in.) are held.

Arrays of pillars 280 μm (0.011 in.) in diameter and 6000 μm (0.236 in.) deep have been made in lead zirconate titanate (PZT). These were made with a tool having an array of 300-μm-diameter (0.012-in.) holes. In essence, the tool looked like an empty honeycomb structure.

One of the smallest applications has been a hole of 15-μm (0.0006-in.) diameter, 32 μm (0.0013 in.) deep. There is no known lower limit to what can be produced, but making tools smaller than this is a challenge. One researcher has drilled holes only 5 μm (0.0002 in.) in diameter in silicon and soda lime glass, but that is an exception to what might be called normal micro practice.

The process can machine cavities of any shape as long as the shape can be viewed totally from the top of the part. The tools can have tapered walls. Slots either blind or through are made: rectangular slots 47 μm (0.0019 in.) wide, by 500 μm (0.020 in.) long and 60 μm (0.0024 in.) deep for example. These are not limitations, but are representative of what can be done with appropriate equipment. Slots do not have to be rectangular. Serpentine slots can also be cut.

Glass-mica (Macor™) ceramic plates have been cut 34 μm (0.0013 in.) deep with a multi-pillar tool at 18 μm/min (0.0007 in./min.). Spiral shape micro actuators have also been made from bulk PZT.

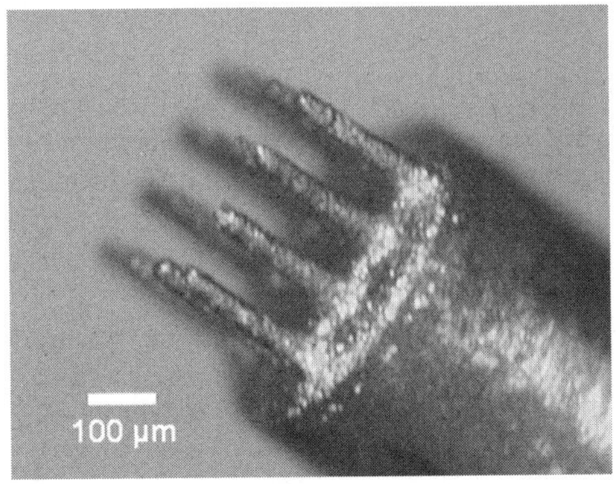

Figure 3.15.2 Example of micro USM tool having several pillars for producing holes. The micro diameter tool has 16 pillars of 25-μm (0.001-in.) diameter. This tool was able to make four sets of holes. The ends of the tools wear faster than the sides as shown here. (Courtesy Egashira, 2004.)

The maximum recorded aspect ratio for micro USM was for a 60-μm-diameter (0.0024 in.) tool drilling 500 μm (0.020 in.) deep (aspect ratio of 8.3). Deeper details may be possible. One application used 16 pillars 18 μm (0.0007 in.) in diameter as a single tool to make 16 holes. The challenge is to make the tool (see Chaps. 3.7 and 3.11). Another example is an array of pillars 280-μm (0.011-in.) diameter cutting to 6000-μm (0.236-in.) depth. Larger diameter tools clearly can go much deeper than micrometer size tools. This latter example has an aspect ratio of 21 (Fig. 3.15.2).

A novel way to make multiple micro slots has been patented by Dr. Shinozuka of Yokohama National University. Stainless steel shims of the desired slot thickness are bonded between layers of tissue-covered epoxy. The tissue-epoxy thickness is the separation distance of each slot. With the shim thickness facing down as the tool, the abrasives wear out the tissue-epoxy layer, at the end leaving only the shims to do the cutting. The tissue-epoxy acts as a spacer and as a means to add strength to the overall tool that would not be possible without some filler.

Workpiece Materials

USM only works best on hard materials. Ceramics and hard metals are the typical applications. Many silicon wafers can be machined rapidly and an array of holes can be drilled or they can be drilled one at a time. Most holes are through holes, but the process also makes blind holes with no difficulty. Glass of all types, quartz, and silicon are easy targets for the process. Titanium is another metal that responds to USM. Alumina, beryllium oxide, canasite, composites, ferrite, glass, single crystal materials, polycrystalline diamond compact, sapphire, piezoceramics, silicon carbide, silicon nitride, zirconia, graphite, glass carbon, and stainless steels have all been machined by USM. Most stainless steels however are too soft to work well. As a specific example of the process, polycrystalline diamond particles ranging from 0.37 to 3 μm (15 to 118 μin.) are used for micro work on single crystalline ⟨100⟩ silicon and fused quartz.

Equipment

Micro USM machines can be purchased and they are also often made for specific in-house use. For the micro level work, some equipment designers have combined wire electro discharge grinding (WEDG) with USM to produce tools that leave micro holes as small as 20 μm (0.0008 in.) in diameter. For the very small holes like these, hole depths may be limited to 50-μm (0.002-in.) depth.

The equipment is relatively low cost and simple to use.

In most large-part instances, the tool is fed down into the workpiece, but in several micro feature applications the workpiece setting in a tank is fed upward to the tool.

There is a vertical force between tool and workpiece for the micro cutting. It can be as low as 0.05 to 0.1 N (0.011 to 0.022 lb).

Process Variations

There are two variations of conventional USM: (1) the tool is not rotated and (2) a rotating tool is used. The latter is primarily used for round holes. Any other tool shape would by default produce a round hole since the tool's rotation generates a circle.

In addition to conventional USM, some applications have used a rotary grinding tool that eliminates the need to use a slurry. A liquid coolant is still used and its main application is to flush workpiece and tool debris out of the cut area so more cutting can take place. This latter variation is sometimes referred to as rotary ultrasonic machining (RUM) but the conventional approach also can use rotary motion. The RUM variation cuts by the hybrid action of USM and conventional grinding. Electroplated diamond tools are used with grit sizes of 30 to120 μm (0.0012 to 0.0047 in.). Tool tip shapes are normally cylindrical, but conical- or ball-ended tools are also simple to make. This later variation (RUM) is also referred to as ultrasonic impact drilling and ultrasonic vibration-assisted grinding.

Ultrasonic-assisted operations at the micro level include micro ultrasonic vibration lapping to improve the precision of micro EDMed holes, and USM-assisted turning has reduced machining time, workpiece residual stresses and strain hardening, and improved workpiece surface quality and tool life on some materials. Combining EDM and ultrasonics so they both operate at the same time at the micro level has also been shown to improve capability, although equipment makers have not incorporated this into their commercial machines yet.

There is some data to say that by using electrorheological fluids as the slurry carrier chipping is reduced in some instances.

In a conventional approach to high-definition USM (HD-USM), it has been shown on fused silica that the minimum machining rate was 10 nm/s (0.4 μin./s) while initially holding the tool off the workpiece by 40 μm (0.0016 in.). An average surface roughness (Sa) of 30 nm (1 μin.) was achieved by machining with 10-nm (0.4-μin.) diamond abrasive particles in the slurry. This is reportedly ≈7 times smaller than typical conventional micro USM (Viswanath, 2014).

USM makes features in hard materials, and as reported here to improve surface finishes. It can also be used as a trimming tool for those same hard materials and as a means to radius top edges of features.

References

Egashira, K., T. Taniguchi, H. Tsuchiya, and M. Miyazaki, Microultrasonic machining using multitools, *Proceedings of the 7th International Conference on Progress Machining Technology (ICPMT '04)*, December 2004, pp. 297–301.

Jain, V., A. K. Sharma, and P. Kumar, Recent Developments and Research Issues in Microultrasonic Machining, *ISRN Mechanical Engineering*, 2011, vol. 2011, Article ID 413231, 15 pages (doi:10.5402/2011/413231).

Sarwade, A., "Study of Micro Rotary Ultrasonic Machining" Industrial and Management Systems Engineering, Dissertations and Student Research. Paper 6, University of Nebraska, 2010.

Sun, X. Q., T. Masuzawa, and M. Fujino, Micro ultrasonic machining and self-aligned multilayer machining/assembly technologies for 3D micromachines, *Micro Electro Mechanical Systems, 1996, MEMS '96, Proceedings. An Investigation of Micro Structures, Sensors, Actuators, Machines and Systems. IEEE, Proceedings the Ninth Annual International Workshop on*, San Diego, CA, 1996, pp. 312–317 (doi: 10.1109/MEMSYS.1996.494000).

Viswanath, A., T. Li, and Y. Gianchandani, High resolution micro ultrasonic machining for trimming 3D microstructures, *J. Micromech. Microeng.*, 2014, vol. 24, 065017 (http://iopscience.iop.org/0960-1317/24/6/065017).

CHAPTER 3.16

Cylindrical Micro Grinding

Jim Boldig

Custom Wire Technologies, Inc.
Port Washington, Wisconsin

Process

Cylindrical grinding's biggest advantage is its ability to provide consistent, precise, fine-finish micro parts. It can do that on short, stubby configurations and depending on part design, even parts as long as 460 mm (18 in.). Micro grinding can produce square and triangular and even six-sided shapes as well as perfectly round ones. Centerless grinding, one form of cylindrical grinding, is unique in the ability to provide superior tolerance-holding capabilities, sometimes holding to tighter tolerances than a wire manufacturing mill can hold. Its ability to improve the roundness and tolerance of cylindrical stock is why it is often used as the first operation for any part made on a CNC screw machine or chucker lathe. By making the beginning stock round, every part made on subsequent will be more accurate and more repeatable. Working with diameters as small as a human hair requires great insight into the mechanics and machine capabilities, far beyond that of more conventional-sized part.

Micro grinding is a good choice for accurate micro parts, while turning on lathes or other processes may not be able to produce diameters less than 0.076 mm (0.003 in.) without damaging the part. This is particularly true if the micro parts are long. Micro grinding is repeatable over large volumes with consistency. C_{pk} values (a measure of repeatability) of 1.33 or higher are not unreasonable.

Grinding with an abrasive wheel can produce very fine finishes [down to 2 μm (8 μin.)] and very precise dimensions. It is typically a much better method for machining very hard materials than milling or turning. Moreover, the process works better when taking small depth of cuts, like reducing a wire diameter by only 5 μm (0.0002 in.). Grinding can be a slow process to achieve the appropriate surface finish or diameter tolerance, however, it can produce large volumes with relative ease. In machining processes, such as milling, the tool has only a few cutting surfaces, whereas a grinding wheel has hundreds (if not thousands) of abrasive points which slowly remove material with each revolution of the wheel.

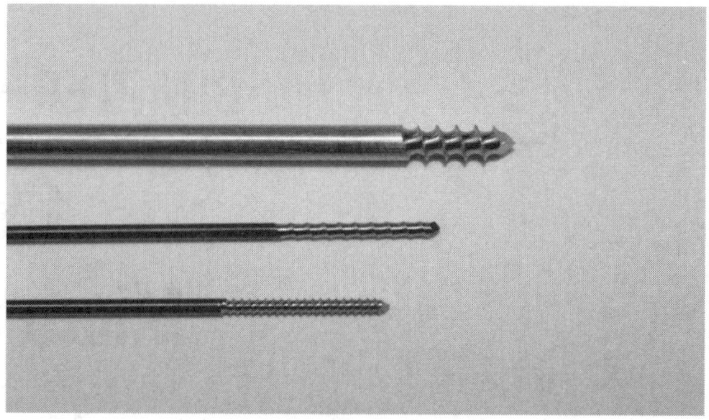

Figure 3.16.1 Some typical micro ground medical parts.

Cylindrical grinding is a method which uses more than just cylindrical-shaped material. Square and even hexagon-shaped bar stock can be processed using the cylindrical grinding method with the correct tooling.

Characteristics and Applications

Ground micro parts are used in a variety of industries, including mold making, medical instruments, watch-like applications, electronics, defense, aerospace, and metrology. Orthopedic implants, inspection rods, and mold release pins are some very common applications. Figure 3.16.1 shows a few parts produced for the medical industry, one of the biggest beneficiaries from the small sizes of parts micro grinding can produce. Minimally invasive surgical procedures and their devices continue to demand smaller and smaller ground parts.

Micro Size

Starting material stock used in micro grinding ranges from 0.152 to 3.80 mm (0.006 to 0.150 in.). These sizes are usually limited by the capabilities of the equipment, not the grinding process. With finished part diameters less than the thickness of a human hair [0.076 μm (0.003 in.)], a microscope or vision system must be used to view and/or inspect these features. Finished part diameters down to 0.050 mm (0.002 in.) can be successfully produced with great repeatability. However, small starting-material stock [≤0.25 mm (≤0.010 in.)] can be difficult to load into the machine, because of the extreme flexibility of the material at these sizes. Machine operators may be required to load the material into the machine by hand in these small sizes. In these instances operators may experience difficulties seeing the material or loading it into the small-sized tooling. (Think of a hair size wet noodle 3 feet long going into a hole one-tenth the size of a needle eye.)

Some parts begin with a starting material size that is 5 to 7 times larger than the smallest diameter of the wire. However, a tradeoff exists with these small diameter parts. These small parts require a precise, careful machine setup. With such setups, diameters less than 0.05 mm (0.002 in.) can be achieved, but they will require slower speeds than larger diameters. Additionally, the diameter and length tolerances play

FIGURE 3.16.2 Challenging micro medical part, small, multiple diameters, long and precision.

a major role in the production of a part. Greater precision is necessary when grinding a part diameter to ±0.005 mm (±0.0002 in.) versus a part ground to a tolerance of ±0.013 mm (±0.0005 in.). A careful watch over in-process inspections is required to ensure parts are not falling outside of such tolerance windows. Figure 3.16.2 is an example of a challenging micro ground part. It has two diameters and unlike what many think of as micro it is extremely long. It is one aspect of the micro medical world.

A unique advantage of some grinding equipment is the ability to grind parts longer than 3.0 m (118 in.). Theoretically, if the ability to support the wire on the in-feed and exit side of the machine exists, a part could be made as long as one deems necessary. Finished parts can be very short or very long as shown in Fig. 3.16.2. In addition, they can be very long with some very short precision dimensions also. For example, a taper ground on a wire can occur with a sharp angle over 0.5 mm (0.020 in.). Tapers which gradually increase over 0.5 m (20 in.) are possible too—high precision over long distances on micro diameters. There are limitations when it comes to length. The part being ground must be manipulated in some form, either rotating or traversing. This usually requires additional material outside of the length of the finished component. Depending on the length of the part and the ground features, the part may have to be cut from a longer stock to allow the machine to move and grind the features. This distance is dependent on the type of machine, but it could range from 150 to 460 mm (6 to 18 in.) of additional material. Other features which can be ground into these parts include flats, radii, and threads. Special equipment may have the ability to rotate the part to turn a round wire into a triangular, square, or even hexagonal shape. Equally spaced flats, similar to ones for chucking a twist drill into a spindle, are possible with some additional engineering and setup time.

Materials

The most common ground metals for micro parts are stainless steel, titanium, or nickel titanium (nitinol). Other materials such as plastics, steels, ceramics, or carbides can be ground as well but are less common. Each material is unique and requires a different approach to grind successfully. For example, titanium or nitinol grinding requires different programming parameters than stainless steel would require. These parameters could be rotational speed of the wheel, rotational speed of the part, linear feed speeds, type of coolant, type of wheel, etc. All of these variables are unique to each part and can change among the same material and even change with material of the same size.

300 series stainless steel is the most common material in the medical industry. The most common grade being ground is 316LVM. It has great corrosion resistance, strength, and biocompatibility. When compared to titanium alloys, 316LVM is a more economical choice as well. Depending on the temper condition of the material, ultimate tensile strengths range from 600 to 2000 MPa (90 to 300 ksi). 304 stainless steel is another common material, with properties similar to 316LVM.

Titanium has advantages over stainless steel because it is lightweight, and has outstanding corrosion resistance. In the area of medical devices, titanium is fully biocompatible. Specific grades of titanium, especially commercially pure titanium, promote growth of bone into the material, which helps stabilize implants in the body. Titanium ultimate tensile strength ranges from 1000 to 1450 MPa (145 to 210 ksi).

Nitinol is a unique material in its shape properties, of which there are two different types. *Shape memory nitinol* allows it to be deformed at one temperature (room temperature for instance), and then return to its original shape when heated to its transformation temperature. *Super elastic nitinol* is a very flexible material which can be manipulated to many different shapes and will return to its original shape without the need for heat (assuming it has not been deformed beyond its yield point). A variety of grades are available in both shape memory and super elastic. They both have similar ranges of ultimate tensile strengths between 1400 and 1700 MPa (200 and 250 ksi). However, the transformation temperatures are usually a more important component of nitinol, ranging from <0°C to 22°C (32°F to 72°F) for super elastic alloys and 22°C to >85°C (72°F to 185°F) for shape memory alloys.

Tolerances

In the micro grinding process, tolerances are dependent on many variables, including the skill of the setup operator and the capabilities of the machine itself. Ground parts may feature diameters with tolerances in the range of ±0.013 to ±0.025 mm (±0.0005 to ±0.0010 in.). These are common and most precision grinding equipment will have the ability to reach such specifications. When tighter tolerances are specified, such as ±0.0025 to ±0.0076 mm (±0.0001 to ±0.0003 in.) greater precision equipment must be used and experimentation with grinding parameters is required. Achieving tight tolerances is more costly, more challenging, and less likely to *consistently* be produced, but such tolerances can be achieved for many designs. While tighter tolerances can be achieved than listed here, it takes exceptional operators, temperature control, and ultraprecision machines. The smaller the starting part diameter the more difficult it becomes.

Similarly the tighter the length tolerances, the more precise the setup must be. Normal operating tolerances for ground features can range from ±0.50 to 1.00 mm (±0.020 to 0.040 in.). If tighter tolerances must be achieved, it is possible to grind to ±0.13 to ±0.25 mm (±0.005 to ±0.010 in.). Grind length tolerances are dependent on the traverse speed of the equipment being used. The faster the part traverses past the work wheel, the more open the tolerance should be. If precision length tolerances are needed, the workpiece should be slowed to a manageable rate where length tolerances of 0.05 mm (±0.002 in.) can be achieved.

Normal operating conditions can achieve a surface finish of 8 µm (32 µin.) Ra. However, surface finish is dependent on more variables than other areas of the part. Generally, it depends on wheel compositions, speeds, and coolant types. Surface finishes as fine as 2 to 4 µm (8 to 16 µin.) Ra are also feasible options. A finish as fine as 1 µm (4 µin.) Ra may be possible with extra time and correct program parameters.

Angular tolerances are more dependent on the resolution of the equipment than other tolerance groups. Common tolerances are ±2.0 to 5.0 degrees for nearly all angular features. If a certain dimension needs to be held to a tighter specification, tolerances of ±0.5 to 1.0 degree can be maintained.

Practicality is a basic rule of manufacturing. Do not specify high precision where it is not needed. Is it necessary to hold an angular tolerance of ±0.5 degree or an overall length tolerance of ±0.010 in. for the part, or can it be more generous? Tight tolerances are not necessary for certain features and drive up the cost of parts. Use tight tolerances only when necessary.

Design Recommendations

Figure 3.16.3 demonstrates geometries that are possible and not possible for micro cylindrical grinding. Geometries similar to the one on the left cannot be ground with this process, and when the part diameter is truly micro no grinding process can do it. The one on the right can be ground. There are alternative methods of manufacturing parts such as that on the left, but cylindrical grinding is not one.

A grinding operation will usually leave a burr on the part being ground. This isn't the case for all parts but is common for parts which exhibit sharp corners where flats or steps occur. While measures are taken to avoid burrs, they are unavoidable in some cases. This is especially the case when grinding hypodermic tubing with features which expose the inner diameter of the tubing. If burrs are a concern for the final product, a secondary finish process should be specified or noted to be burr free. Electropolishing is a common finishing process which can help eliminate small burrs while enhancing the appearance of the part. Other processes such as deburring with a bristle brush may be required to remove any burrs.

Sharp corners and transitions are possible with correct tooling but a maximum radius should be identified. The sharpest corner will come from the very first part and subsequent parts will have a small radius in the corner as the wheel begins to wear. A maximum radius callout will allow the manufacturer to ensure they are within specification (Fig. 3.16.4).

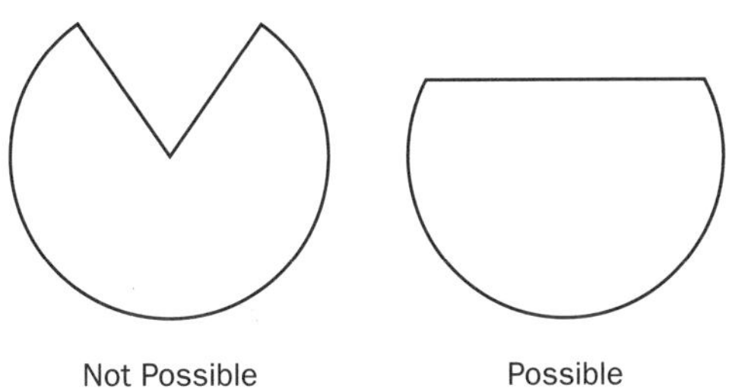

FIGURE 3.16.3 Geometries that are possible and not possible in cylindrical grinding.

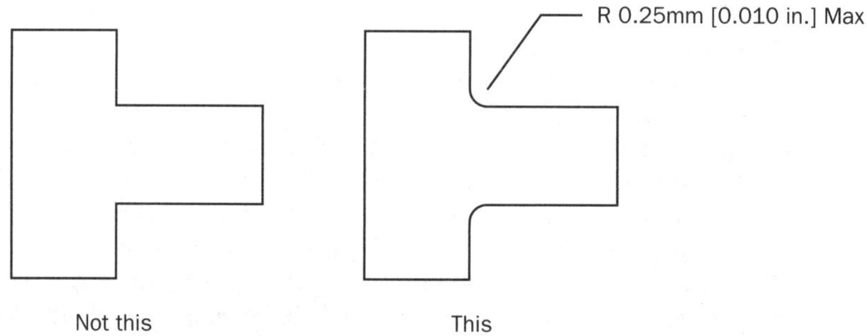

Figure 3.16.4 Specify radius for internal corners.

Specific thread callouts meeting ANSI or ISO standards should be avoided where possible. Some micro grinding equipment cannot grind threads into components. This is usually left for thread-forming equipment. Although a specific thread is possible to grind, some applications do not need a standard thread. These cases may be using the thread for retention in a bone or help advance a drill through the bone, but may not need to mate with another threaded component. Specifying a minor diameter and pitch can result in simple, acceptable results.

Complete drawings specifying form and function help the manufacturer understand the part purpose. Size, material, and finish are three key characteristics which should be included on any drawing. If a range of sizes or multiple materials are acceptable for the end product, these should be communicated to the manufacturer. Critical dimensions to be inspected should be clearly identified. Tolerances and reference dimensions should be identified as well.

Economics of Micro Grinding

The micro grinding process can be used in large-volume production runs or in prototype and development projects. Lead times for any component usually depend on raw material availability. Depending on volume and complexity, normal lead times range from 2 to 4 weeks with in-stock material. There are options available to expedite the process to 10 business days or less for an additional fee. If material is not in-stock and must be purchased from the mill, lead times can be extended by 3 to 6 weeks depending on the specific material being ordered. Stainless steel usually requires 3 to 4 weeks, while nitinol or titanium may need 5 to 6 weeks lead time.

Factors to be considered in the cost of manufacturing a part using the grinding process include the cost of raw material, setup time, production time, and inspection time. Other factors include complexity, expedite fees, and tooling. Raw material costs change frequently, but stainless steel grades can range from \$20 to \$50 per pound. Titanium and nitinol are generally more expensive. Simple machine setups may take only 1 h to complete while longer precision setups could be in excess of 4 h. Production times depend on the geometry of the parts, normally ranging from 2 to 4 min. It is possible for parts to take 10 to 20 min, although uncommon. Tooling costs generally don't vary, normally ranging from \$250 to \$500 per piece for each wire size. However, each wire size may need three identical pieces to complete the job. Tooling may not last and could be consumed during a large-volume order. Complexity and expediting are determined on a case-by-case basis. They normally include engineering time, break-in fees, development time, etc. Part costs are widely varied. Large-volume orders will have a discounted price versus lower-volume projects. Production volumes normally range from 100 to 500 pieces. Large production orders of 100,000 to 1,000,000 pieces are routinely produced. Prototype volumes range from 1 to 25 pieces. Parts which require little to no operator involvement benefit from more pricing discounts. Some parts can't eliminate an operator, but large volumes can help offset the higher starting costs.

CHAPTER 3.17

Grinding with Mechanical Micro Tools

LaRoux K. Gillespie

Manufacturing Consultant
Andover, Kansas

Introduction

The manufacture of micro sized parts and micro features in some instances can be made by conventional size tools and machines, but in many situations it requires micro sized tools. Being micro sized they bend and break far too easily and require extreme care to produce. Cylindrical grinding of minute cylinders is performed on several commercial cylindrical grinders (see Chap. 3.16). As an example of a conventional size tool grinding micro features, dicing blades having a thickness of only 15 μm (0.0006 in.) are made with a wheel diameter of 56 mm (2.2 in.) (Fig. 3.17.1). Tools such as this can produce narrow channels and ledges, as well as dicing material. This is not a micro wheel, but it makes micro cuts with exceptional smooth sides. Grinding of configurations more complex than cylinders is described in this chapter using micro end mill-like tools to grind micro features (Fig. 3.17.2).

Cutting micro features requires very small grains that wear predictably while removing very small amounts of material. That has to happen without smearing workpiece metal all over the grinding wheel. Diamond and cubic boron nitride (CBN) are the two most common wheel materials because of their high grain strength. Grinding of any size part is slow compared to other cutting methods and that is particularly true at microscopic conditions. The fact that it can produce mirror finishes on ceramic surfaces makes it particularly attractive for some products.

Making the Tools

Research has indicated that the practical lower limit for grain size in the micro pencil sizes is about 30 μm (0.0012 in.). The PCD or CVD tools are made by coating diamond particles in layers onto cemented carbide shafts. The commercial tools are straight

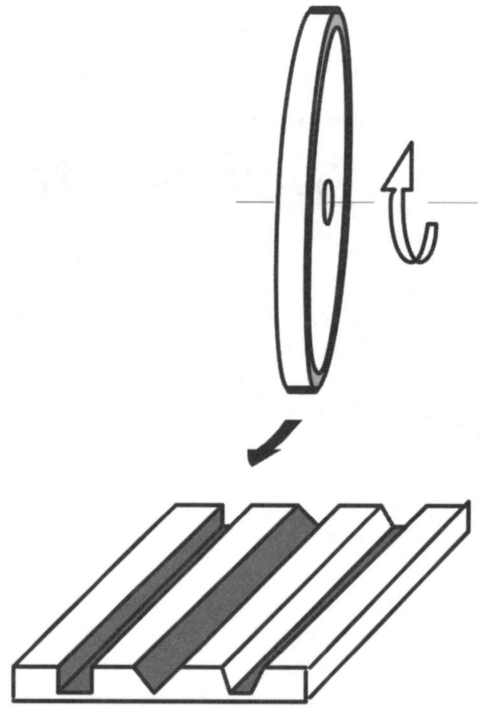

FIGURE 3.17.1 Dicing wheel-like grinding wheels are used to produce micro channels and intersecting grooves. (Reproduced with permission from Hoffmeister and Wenda, 2000.)

sided, but other shapes can be made, since they can be coated onto shaped shafts readily or shaped by μEDM or μWEDG breakdown of bonds in specific shapes.

Machines for Micro Grinding

When the cutting tool (wheel) is less than the diameter of a human hair and precision surfaces are required, very high-precision machine tools are needed. Machines with this ability have resolutions of 1 nm (0.003 μin.) and ultrastiff design. Electrolytic in-process dressing is used to keep the sharp grains at appropriate protrusions. Spindles need to provide rotational velocities of 40,000 to 200,000 rpm to give needed cutting speeds for a variety of materials and conditions. Tools will be fed on both sides at rates of 10 mm/min (0.040 in./min).

Capabilities of the Process

Micro grinding is typically limited to the hard metals and ceramics, particularly the brittle ceramics. With fine abrasives micro grinding has achieved surface finishes of 10 nm (0.4 μin.) Ra. It has ground intersecting grooves as narrow as 20 μm (0.0008 in.), leaving rectangular posts.

The typical smallest "pencil" style commercial tool is a micro CVD diamond-coated grinding wheel, 50 μm (0.002 in.) in diameter. "Pencil" style wheels refer to a cylindrical

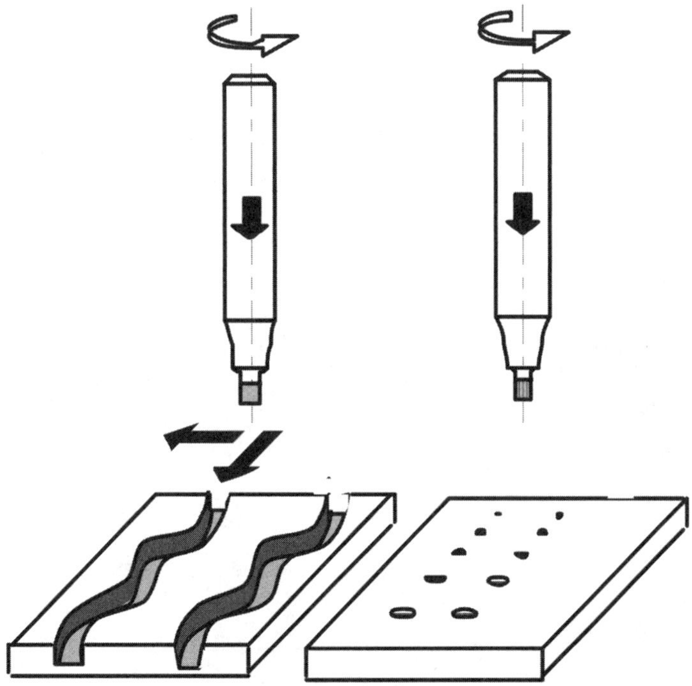

Figure 3.17.2 Micro pencil-like grinding is used to produce slots, cavities, and external sculptures for micro products. (Reproduced with permission from Hoffmeister and Wenda, 2000.)

tool similar to a small-diameter milling cutter. Because the hair-like tools bend easily, they are limited to grinding at very shallow depths. They face the same limitations as micro milling tools do.

To produce a 10-nm (0.4-μin.) surface roughness the depth of the cut must be less than 100 nm (4 μin.), an amazingly small cut for those individuals used to cutting conventional parts. Using micro grinding silicon, ferrites, glass, steel, and ceramics have been ground to mirror finishes [0.2 μm (8 μin.)].

The smallest geometries that can be ground are a function of area to be ground, and the grain size that can be incorporated into the wheel. A 76-μm (0.003-in.) grit, for example, will not allow a smaller geometry than a 76-μm (0.003-in.) groove or corner radius to be cut. Smaller abrasive size will considerably slow grinding and raise the grinding forces since a larger surface area is in contact. Smaller grains can be used however.

For glass, silicon, and sintered carbide, a 0.35-mm-diameter (0.014-in.) grinding pencil makes problem-free cuts at 130,000 rpm. Silicon channels are ground with 100 μm (0.004 in.) PCD tools at 40,000 rpm and 0.5 mm/min feed at depths of 2.5 to 4.0 μm (0.0001 to 0.00016 in.). In addition to conventional cutting conditions, users may need to employ force dynamometers to prevent tool breakage and employ an intense flushing of chips from the cutting area. Forces at which the tool will break can be predicted by the Hänchen-Decker method and range from 1 to 2 N (0.22 to 0.45 lb). Tool life may be as short as 5 min of cutting action if forces fall above the calculated values.

Research has shown that in Pyrex glass and stainless steel channel, grinding needs to leave at least 0.5 μm (0.000020 in.) wall between channels to prevent damage with

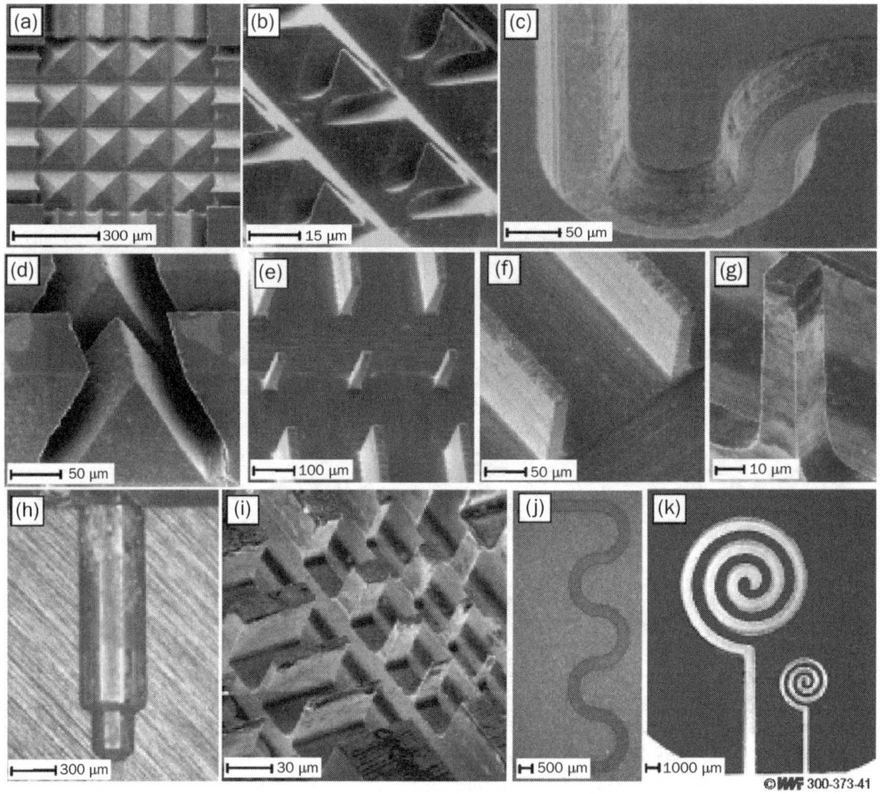

FIGURE 3.17.3 Micro features ground silicon, glass, zirconium oxide, aluminum oxide, and sintered carbide. Serpentine channels were cut with pencil style tools while most of the straight channels were cut with dicing wheel style blades. (Reproduced with permission from Hoffmeister and Wenda, 2000.)

this grinding. Grinding stainless steel with these tools leaves burrs along the wall edges. While they can be removed by flat "sanding" with fine abrasives, some burrs may be turned into grooves (see Chap. 3.34 for other comments on burrs).

A wide variety of ground geometries have been produced. The tools used may have V-shape bottoms or be ball shaped as well as flat ended. These minute tools have made a variety of channels, posts, ribs, serpentine cutouts, angled slots, and other three-dimensional shapes in silicon, glass, zirconium oxide, aluminum oxide, germanium, and sintered carbide (Fig. 3.17.3). All such features can be produced in sizes as small as 25 to 100 μm (0.001 to 0.004 in.), but tool life is short, and the challenges are high. Success is much easier and more consistent with larger diameter tools. For most general uses with good success, tools should be at least 0.25 mm (0.010 in.) in diameter and 0.5-mm-diameter (0.020-in.) tools will be preferable.

Because tool breakage and edge quality are such challenges, users are turning to laser heating of the work area just before the grinding tool advances to make the cut. This reduces cutting forces on the tool, improves machinability, and results in better surfaces without metallurgical damage. This requires equipment that combines precision motion and both cutting and laser capabilities, but it greatly enhances the potential for success, as well as faster production.

Very few manufacturing facilities have the capability to grind with 50-µm (0.002-in.) pencils, but the technology does exist in U.S., Japanese, and German industry. Germany's Institute of Machine Tools and Production Technology (Braunschweig, Germany) is a leader in developing tools for these applications.

Other Processes

When making micrometer size channels, grooves, cut outs, and microscopic shapes, users may be better advised to pursue some of the advanced nontraditional processes. Wire EDM (WEDM), wire electrical discharge grinding (WEDG), wire electrochemical discharge grinding (ECDG) are other processes that can make similar geometries to that produced by conventional grinding. Micro EDM is described in Chap. 3.7 and WEDG in Chap. 3.11. Micro electrochemical EDM is described in Chap. 3.10 and micro turning in Chap. 3.5. Micro laser machining may be able to produce some geometries normally produced by micro grinding and that process is covered in Chap. 3.6. Micro milling has produced slots only 50 µm wide (0.002 in.) but it too faces the same breakage and quality issues that micro grinding does.

Micro wire electrochemical grinding (WECG), which is not widely known or used at micro levels, can be used after WEDG grinding to smooth the surface, but there is little published information to assist users.

References

Hoffmeister, H.-W., and A. Wenda, Novel Grinding Tools for Machining Precision Micro Parts of Hard and Brittle Materials, *Proceedings ASPE Annual Meeting*, 2000, vol. 22, pp. 152–155.

Masuzawa, T., and H. K. Tönshoff, Three-Dimensional Micromachining by Machine Tools, *Annals of the CIRP*, vol. 46, no. 2, 1997, pp. 621–628.

Morgan, C., R. Ryan Vallance, and Eric R. Marsh, Micro-machining and Micro-grinding with Tools Fabricated by Micro Electro-discharge Machining, *International Journal of Nanomanufacturing*, 2006, vol. 1, no. 2, pp. 242–258.

CHAPTER 3.18

Micro Coining

LaRoux K. Gillespie

Manufacturing Consultant
Andover, Kansas

Coining has been used for over 2000 years to make products and features on products. The material is placed in a confined die and subjected to a sufficiently high pressure to induce plastic flow on the surface and just below the material surface. Monetary coins and medals are the two most well-known examples, but there are many other industrial products that also utilize coining. It is a rapid strike process with many parts per minute made. It can provide surface features and edge rounding in one stroke, and in certain designs can also provide desired metallurgy aspects such as surface hardening. Unlike other processes, it can produce desired features on both sides of the part at the same time and the process can produce flat surfaces or patterned surfaces. Nickel alloys, molybdenum, and other exotic materials may be difficult to machine in micro sizes, but they can be coined easily. While most coined parts are metal, that is not a requirement. Paper and plastic can be readily coined as can ceramics in the green state (i.e., not hardened).

At the microscopic level, however, several issues pop up that are not an issue for larger parts. When the part is only 2 to 10 grains thick, the metallurgical characteristics seem to defy logic. With thicker and larger parts coining causes plastic flow via dislocation movement, but in micro forming metal movement involves much more of changing individual grains rather than the dislocation spaces between grains and that affects the required forces and the "flowability" of metal into the dies. The result is often incomplete filling of dies compared to conventional parts. Stated simpler—it is a challenge to coin features having only a few grains to move.

In some cases the Hall-Petch relationship holds at micro scales and minute numbers of grains, but at others it appears to not. This relation shows that the flow stress[1] increases as the reciprocal of the square root of grain size. Specifically, the relation seems to hold when the part size is much larger than the grain size. Larger grain sizes will flow better.

---

[1]Flow stress is also referred to as the true stress-strain curve. Flow stress is the instantaneous value of stress required to continue plastically deforming the material—to keep the metal flowing.

Coining is of interest to micro part manufacturers not only because it can make complete parts in one hit of a forging die, but it can also produce micro features such as channels and similar geometries for micro fluidics and other functions much faster than other processes. Embossing is a major process for many of these applications and coining and embossing share the same basic issues (for roll-to-roll embossing see Chap. 3.26[2]). As an example, coining can easily make channels for heat exchangers, fuel cells, and chemical micro reactors with 0.4-mm-wide (0.016-in.) channels with 0.4-mm-wide (0.016-in.) ribs between the channels in 0.5- to 0.8-mm-thick (0.020- to 0.031-in.) sheets of aluminum, copper zinc, and chrome nickel. Then these channeled sheets can be stacked to make multilevel assemblies—micro features in larger assemblies.

Coining requires much higher forces when the parts/features are fully constrained, and a slight chamfer near edges may allow better material movement in the center of the part/feature.

While many micro products are based on straight-line grooves or channels, coining can impress spirals or many other configurations as well. Bonding pads, covers, heat sinks, lead frames, tabs, terminals, and other components for micro electronic assemblies and packages are coined in materials such as Kovar, molybdenum, tungsten, and copper. Solder and brazing performs, including gold-tin ones are regularly coined for micro electronic joining applications. Coining also can be used to imprint almost nano size posts, stud bumps, and related surface effects that are later used for electrical contacts or other applications. Such application may be called imprinting or embossing rather than coining, but the same basic technology is employed. In this case the challenge becomes making the die that is used to produce the mirror image on the part.

High-strength steel, 304 stainless steel, and titanium micro sheets have been coined, some at room temperature and others for easier flow at hot temperatures. Copper is one of the easiest to coin and it provides ease of fabrication, conductivity, and wide application for micro parts. Ultrasonic vibration helps provide a smoother surface in coining. Features as small as 1 μm (39 μin.) have been made by silicon dies in aluminum without any use of heat or other energy, but this is the smallest size known and not a commercial application. The commercial lower limit on size is at least five times this value. Laser-assisted micro coining has been done using sapphire tools to allow the laser to pass through to the stainless steel part, but again that is not a commercially viable application yet.

Because the grain structure, sheet thickness, initial geometry, and feature geometry all influence the end product, both experimentation and simulation are required for success for materials less than 1 mm (0.040 in.) in thickness. The references cited in this chapter provide a good start for understanding the conditions necessary. From a design standpoint draft greatly improves the formability of micro size features. When channels are coined, the ratio of channel width to grain size is a critical factor in forming performance. As this size ratio falls below 10, the reduced number of grains within the field causes reduced degrees of freedom of slip.

Commercial machines are available to make coined parts and features down to about 0.5 mm (0.020 in.) and they can hold tolerances of ±5 μm (±0.0002 in.). The few

[2]Coining involves changing the surface of parts, and so does embossing. Cold heading can make the same parts, but cold heading typically refers to changing the geometry of the material from a wire into a shape. All three can produce similar surface effects, but coining typically does not change the basic part configuration, just top and/or bottom surfaces.

manufacturers who coin smaller parts and thinner parts have made their own machines for doing so and these machines are kept hidden from public view. From a design standpoint the key message is that once the ratio of feature size to grain size falls below 10, higher forces and less material movement will occur.

References

Hirt, G., and B. Rattay, Coining of Thin Plates to Produce Micro Channel Structures, Initiatives of Precision Engineering at the Beginning of a Millennium, *Proceedings of 10th International Conference on Precision Engineering*, 2001, Springer Verlag, 2002, pp. 32–36.

Neugebauer, R., A. Schubert, J. Kadner, and T. Burkhardt, High Precision Embossing of Microparts, Advanced Technology of Plasticity, vol. II, *Proceedings of the 6th ICTP*, Springer Verlag, Berlin, 1999, pp. 921–926.

Kim, G.-Y., M. Koç, and J. Ni, Experimental and Numerical Investigations on Microcoining of Stainless Steel 304, *Journal of Manufacturing Science and Engineering*, vol., 130, August 2008, pp. 041017-1–041017-6.

Raulea, L. V., L. E. Govaert, and F. P. T. Baaijens, Grain and Size Effects in Processing Metal Sheets, *Proceedings of the 6th ICTP Conference on Technology of Plasticity*, Nuremberg, Germany, 1999, pp. 939–944.

Nam, J. S., S. W. Lee, and H. S. Kim, Investigations into the Size Effect on Plastic Deformation Behavior of Metallic Materials in Microcoining Process, *International Journal of Precision Engineering and Manufacturing*, vol. 15, No. 1, pp. 5–11.

CHAPTER 3.19

Magnetic Abrasive Finishing

Hitomi Yamaguchi

University of Florida
Gainesville, Florida

The Magnetic Abrasive Finishing Process

Magnetic abrasive finishing (MAF) deburrs and polishes by moving abrasives in a mass of ferromagnetic particles via a magnetic field or by employing a magnetic abrasive, a composite of iron and abrasive, as a "cutting tool." Aluminum oxide magnetic abrasive 80 μm (0.0031 in.) mean diameter overall with alumina particles smaller than 10 μm (0.0004 in. diameter) is a popular composite tool. Figure 3.19.1 shows a schematic of the MAF process for a ferromagnetic component using magnetic abrasives. When magnetic abrasives are introduced into a magnetic field, the magnetic abrasive is magnetized and follows the lines of magnetic force to form particle chains which act like a brush. When a workpiece is inserted into the media, the chains conform to the workpiece surface and push against it with magnetic force. By moving the magnet (rotating or translating), the particle "brushes" follow the magnet and drag the abrasive across the surface. This relative motion between the abrasive and workpiece surface smooths the surface by removing material.

This process has been used to finish the internal surfaces of capillary tubes, stents, catheter shafts, needles, prosthetics, cutting tools, turbine blades, airfoils, optics, sewing machine, watch and camera parts, and other small components and features. Surfaces of ground stainless steel parts can be finished to around 10 nm (0.4 μin.) Ra [1], and surface finishes of precision components (e.g., optics, wafers, etc.) can be reduced to less than 1 nm (0.04 μin.) rms or Ra [2, 3].

The magnetic force acting on the magnetic abrasive is a major factor of the finishing force in MAF. The magnetic force acting on a magnetic particle in a nonuniform magnetic field is generally expressed as follows:

$$F = V \chi H \cdot \nabla H \tag{3.19.1}$$

where V is the volume of the magnetic particle, χ is the magnetic susceptibility of the particle, H is the magnetic field intensity, and ∇H is the gradient of the magnetic field

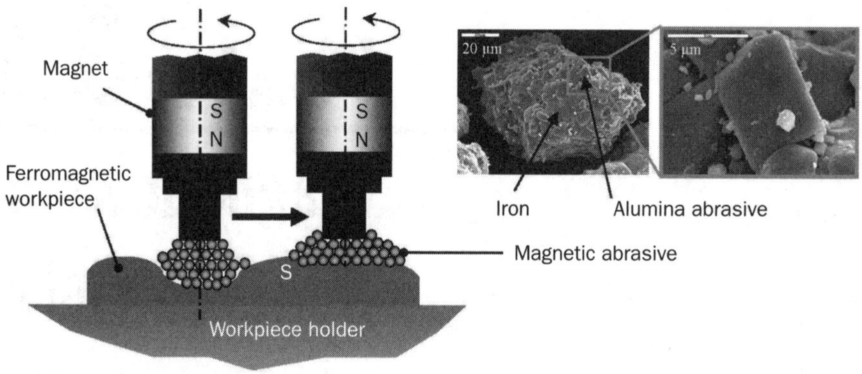

FIGURE 3.19.1 MAF processing principle for ferromagnetic workpieces.

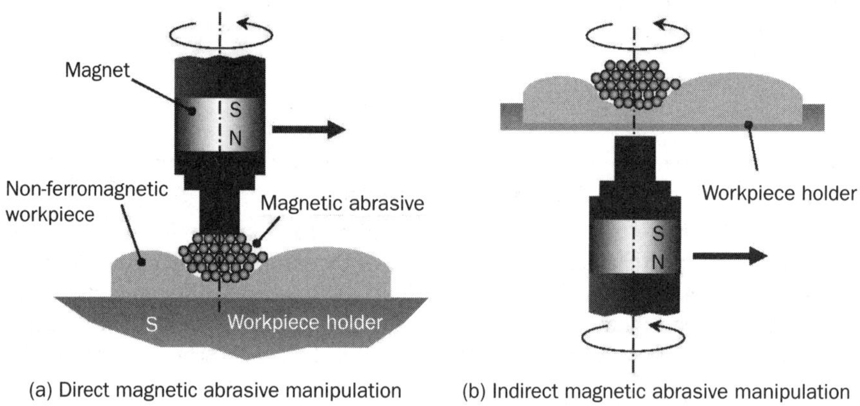

FIGURE 3.19.2 MAF processing principle for non-ferromagnetic workpieces.

intensity. While the first two terms in Eq. (3.19.1), V and χ, are related to the particle, H and ∇H are dependent on the magnetic field at the finishing area. Therefore, generating a sufficient magnetic field (described by $H \cdot \nabla H$) at the finishing area is key to designing an MAF process. The magnetic field generator can be either an electromagnet or a permanent magnet. For permanent magnets, neodymium rare earth (Nd-Fe-B) permanent magnets are typically used because of their high coercive force and saturation magnetism. Because mild carbon steels have high magnetic susceptibility χ and are easily machined, mild carbon steel pole tips are attached to the magnets to adjust the magnetic field at the machining area.

When the particle brushes suspended in a magnetic field meet an obstacle, the particles reorient themselves within the brushes, which allows MAF to conform to the target surface (Fig. 3.19.1). In addition, the "brush" stiffness can be adjusted by changing the magnetic field or particle size. These are distinct characteristics of the particle brushes, which allow finishing of complex geometries and various materials.

The magnetic properties of the workpiece also influence $H \cdot \nabla H$ at the finishing area. Figure 3.19.2 shows schematics of MAF processes for non-ferromagnetic workpieces. While the magnetic flux flows *into* a ferromagnetic workpiece (Fig. 3.19.1), the

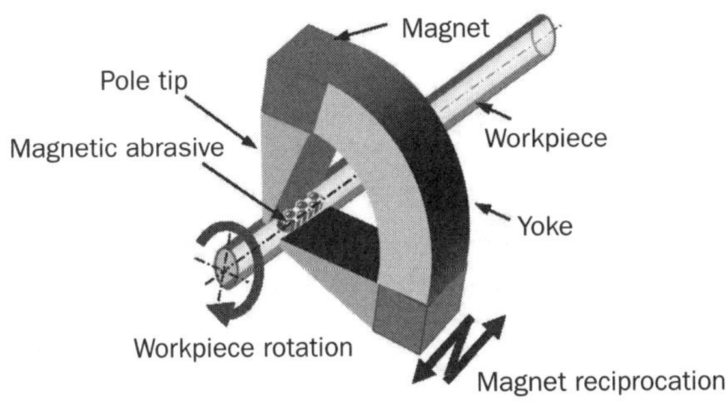

FIGURE 3.19.3 MAF processing principle for internal finishing of rotatable workpieces. (Courtesy ASME [4].)

magnetic flux flows *through* the non-ferromagnetic workpiece. A ferromagnetic workpiece holder under the non-ferromagnetic component is necessary to obtain a magnetic field to generate sufficient magnetic force on the particle for finishing the target surface. Practically speaking, the magnetic flux density at the finishing area should be greater than 0.3 T.

The unimpeded magnetic flux flow through the non-ferromagnetic workpiece provides distinct advantages of MAF. Figure 3.19.2*b* shows surface finishing of a non-ferromagnetic workpiece in which the magnetic abrasive is manipulated from the opposite side of the target surface. With a stationary workpiece, the magnetic field can be moved to drag the magnetic abrasive by magnetic force and generate motion relative to the workpiece. Alternatively, relative motion between the workpiece and the abrasive can be achieved by moving the workpiece while the field and the abrasive remain stationary. The MAF process can be applied to tubular workpieces by manipulating the particles from outside the workpiece.

Figure 3.19.3 shows a schematic of MAF internal finishing of non-ferromagnetic tubes. Magnetic abrasive is introduced into the tube and suspended by a magnetic field generated by magnets outside the tube. As the tube is rotated at high speeds, the magnetic abrasive moves relative to the tube surface. This generates a smooth surface consisting of long cutting marks. By translating the magnet parallel to the tube axis, the magnetic abrasive follows the magnet motion and extends the finished area, enabling the internal surface finishing of long, slender tubes.

Figure 3.19.4 shows a photograph of a representative MAF machine using this principle. By this approach, MAF has successfully internally finished a variety of non-ferromagnetic tubes (e.g., needles, catheter shafts, elbows, and other piping systems) made of stainless steels, aluminum alloys, copper alloys, and ceramics.

Characteristics and Applications of the Resulting Product

Table 3.19.1 shows the classification of MAF by its various process parameters and lists some applications for each type. The MAF process uses either a static magnetic field or an alternating magnetic field (such as rotating magnetic field). When using static magnetic fields, the magnetic abrasive is attracted by the field and pushed against the target surface. Relative motion between the magnetic abrasive and workpiece is obtained by physically moving either the workpiece or magnetic field generator (permanent magnet or electromagnetic coils).

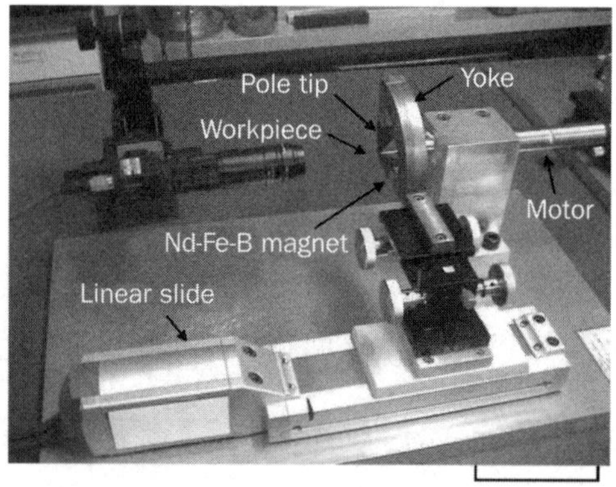

Figure 3.19.4 Photograph of finishing machine. (Courtesy ASME [4].)

Magnetic Field	Magnetic Tool	Abrasive	Applications
Static magnetic field	• Particles • Flakes • Pins • Blocks • Rods • Magnetorheological fluids (MRF) • Magnetic compound fluids (MCF)	• Magnetic abrasive • Alumina • Silicon carbide • Cerium oxide • Chromium oxide • Diamond • etc.	Surface finishing • Plates, wafers • Cylinders • Straight and bent tubes • Bellows • Curved surface, lenses • Balls • Complex geometries • Free-form dies • Medical implants (stents, catheter shafts, knee prosthesis, etc.)
	• Magnetic fluids (MF) • Permanent magnets		Edge finishing
Rotating/ linearly traveling magnetic field	• Permanent magnets • Metastable stainless steel pins		• Cutting tools, drills, reamers • Burrs produced by cutting, grinding, or laser machining

Table 3.19.1 Classification of Magnetic Abrasive Finishing

A representative MAF-processed surface of a 304 stainless steel tube [⌀1.0 × ⌀0.8 × 60 mm (0.040 × 0.031 × 2.362 in.)] using a mixture of 20 wt% of magnetic abrasive and 80 wt% of iron particles [150 μm (0.006 in.) mean diameter] reduced surface roughness Rz from 1.5 to 2.0 μm and from 0.3 to 0.5 μm (12 to 20 μin.) [4]. The MAF-processed surface generally consists of an accumulation of long scratch marks perpendicular to the axis. Typical operation time per part ranges from 1 to 5 min.

As shown in Table 3.19.1, MAF uses various magnetic tools, including powders, flakes, pins, rods, blocks, etc. Magnetorheological fluids, magnetic compound fluids,

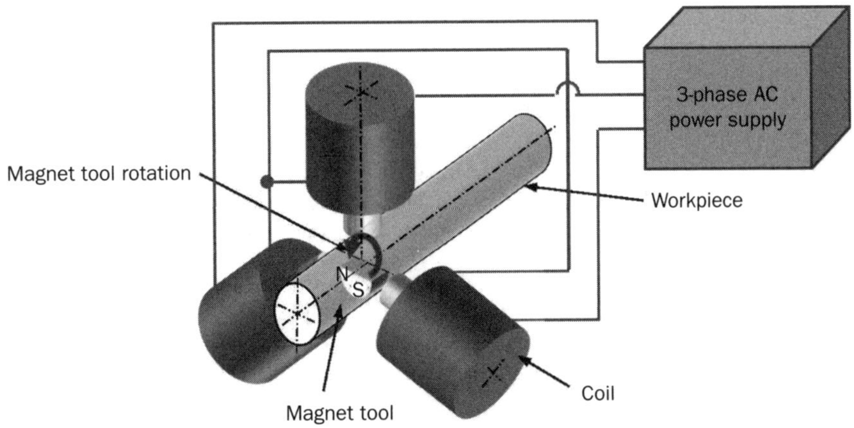

FIGURE 3.19.5 MAF processing principle for internal finishing using rotating magnetic field.

and magnetic fluids can also be used as tools. However, the most common tools are iron particles because iron has high magnetic susceptibility χ and generates strong magnetic force. Moreover, a wide variety of iron particles is available on the market. The particle size is chosen to achieve the desired material removal rate and surface finish and varies from several micrometers to a few millimeters. As shown in Eq. (3.19.1), increasing the magnetic particle volume V increases the magnetic force acting on the particle. The use of larger particles increases the material removal rate. The selection of magnetic and mechanical properties, geometry, and sizes of the magnetic tools alters the magnetic force on the tools and their behavior during finishing.

The abrasive type and size also influence the finishing characteristics. The ferromagnetic component (iron) generates magnetic force, and the abrasive component (aluminum oxide) machines the target surface. If the workpiece requires a different abrasive, such as silicon carbide, the desired abrasive can be simply mixed with various sizes of magnetic tools (e.g., iron particles) instead of fabricating composite abrasive particles. The magnetic tools press the abrasive grains against the target surface by magnetic force. As the magnetic tool motion slides the abrasive over the target surface, the abrasive grains, machines, and finishes the surface.

The MAF processes using alternating magnetic fields are generally called *magnetic field-assisted finishing* processes. Figure 3.19.5 shows a schematic of internal finishing of tubes using a rotating magnetic field. Three coils are installed 120 degrees apart and are connected to a three-phase AC power supply. When a three-phase exciting current is supplied to the coils, a rotating magnetic field is generated. A permanent magnet wrapped by an abrasive tape or belt is inserted into the tube and driven by magnetic force to rotate along the inner surface of the tube.

Metastable stainless steel tools (pins or cut wires) introduced into an alternating magnetic field move in response to the alternating magnetic field. When a *non-ferromagnetic* workpiece is inserted into the mass of moving pins, the pins collide with the workpiece and machine the workpiece surface. The processed surface consists of isotropic short cutting marks and indentations of the pins. This method is useful not only for surface finishing but also for edge finishing. This method is known as *magnetic polishing* or *magnetic barreling* and is used for surface and edge finishing of complex components such as those used for watches or jewelry.

Tools introduced into a hollow workpiece in an alternating magnetic field collide with the inner surface of the workpiece in response to the alternating magnetic field. The machined surface in this case is also an accumulation of isotropic short cutting marks rather than long directional cutting marks.

The geometrical, magnetic, mechanical, and chemical properties of the workpiece are keys to determining which MAF method in Table 3.19.1 is most suitable for each application. The machine configuration and processing conditions should be designed according to the desired material removal rate and surface conditions of the target surface.

Materials Suitable to MAF

MAF has been applied to metals (steels, aluminum alloys, copper alloys, stainless steels, cobalt-chromium alloys, nickel alloys, magnesium alloys, etc.), ceramics (glass, quartz, aluminum oxides, silicon nitrides, tungsten carbides, cermet, etc.), silicon, plastics, etc. The magnetic properties of the workpiece are critical to MAF process design because they influence the magnetic field at the finishing area. For example, the following three cases are *not* potential MAF applications because of the difficulties in generating a sufficient magnetic field at the finishing area.

- External finishing of thick non-ferromagnetic components

  The magnetic force acting on two pieces of matter is inversely proportional to the square of the distance between the two objects. Although a ferromagnetic steel component can increase the magnetic field intensity, it is difficult to obtain magnetic fields sufficient for finishing with thick non-ferromagnetic workpieces.

- Internal finishing of ferromagnetic components

  Because magnetic flux travels into ferromagnetic material, the inside of a ferromagnetic tube in a magnetic field is free of any magnetic field, and the magnetic abrasive magnetically adheres to the inner surface of the tube. When the tube is rotated at high speed, the magnetic abrasive rotates with the tube, and there is no relative motion between them. If the magnetic field is strong enough to magnetically saturate the workpiece, however, MAF is applicable. Note that this approach works in practice only for limited cases, such as tubes with very thin walls (less than 1 mm) [5].

- Magnetic barreling of ferromagnetic components

  Ferromagnetic components are magnetized in the field and attract the magnetic tools, which prevents collisions of the tools with the workpiece surface so the MAF process fails.

Specific Design Recommendations

The design of the magnetic field is the most critical concern for successful application of MAF. This field determines the magnetic force acting on the particle brushes (finishing force) and the motion of the particle brushes. The magnetic field generators can be either permanent magnets or electromagnets. Permanent magnets are more compact than electromagnets and do not require additional power to generate the magnetic fields. In addition, various kinds and sizes of permanent magnets are commercially

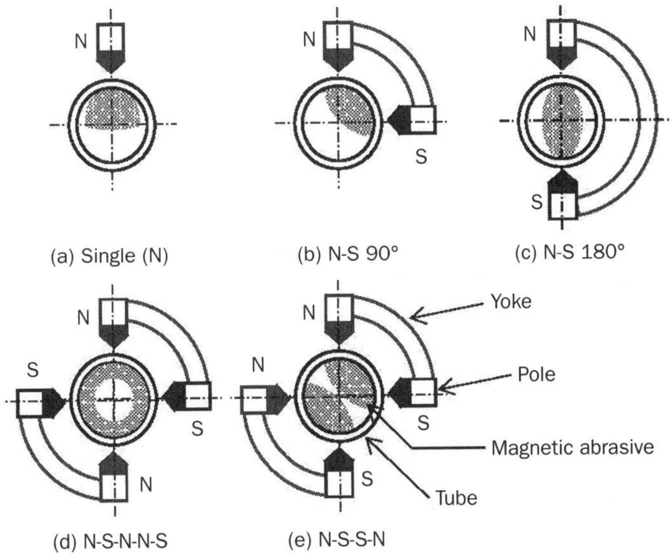

FIGURE 3.19.6 Arrangement of magnetic poles. (Courtesy The Japan Society of Mechanical Engineers [6].)

available. Permanent magnets are more suitable for cases in which the magnetic field generator needs to be rotated or translated at high speeds. As a result, the use of permanent magnets is often more economical than the use of electromagnets. On the other hand, electromagnets can easily generate both static and alternating magnetic fields, and the fields can be tightly controlled by adjusting the electromagnet exciting current.

The magnet arrangement is more critical in internal finishing than in external finishing because the magnetic abrasive is not directly in contact with the magnet in the internal finishing process. Using the internal finishing process as an example, Fig. 3.19.6 shows five magnet arrangements and the corresponding magnetic abrasive distribution for internal finishing of 304 stainless steel tubes [20 mm OD and 18 mm ID (0.787 in. OD and 0.709 in. ID)]. The magnetic abrasive follows the lines of magnetic force. While the single-pole case (Fig. 3.19.6*d*) generates the lowest magnetic force due to the low value of $H \cdot \nabla H$ at the finishing area, the N-S 90-degree case shows the highest magnetic force. The pressure (magnetic force per area) ranges from 20 to 60 kPa (2.9 to 8.7 psi) in the internal finishing using iron particles [510 μm mean diameter, 4 g (0.020 in. diameter, 0.009 pounds weight)] and magnetic abrasive [80 μm mean diameter, 1 g (0.0031 in., 0.0022 pounds)] [6]. In the case of Fig. 3.19.6*d*, the ferromagnetic particles adhere along the inner surface of the tube, and the magnetic abrasive does not move relative to the tube surface, regardless of the tube rotation or pole tip rotation, and no surface finishing action is performed. Except for this, all arrangements showed relative motion and performed surface finishing.

For larger diameter tubes, more magnets are needed to generate a strong magnetic field and sufficient magnetic abrasive motion. When MAF is applied to smaller tubes, the magnets must be placed close to one another, which raises difficulties in obtaining a sufficient magnetic field at the finishing area inside the tube. In practice, pole tips are attached to the magnets to concentrate the magnetic field at the finishing area.

The equipment for MAF is relatively simple and low cost, but is usually designed for the specific application using finite element magnetic field analysis to obtain the desired magnetic field.

References

1. Fox, M., Agrawal, K., Shinmura, T., Komanduri, R., 1994, Magnetic Abrasive Finishing of Rollers, Annals of the CIRP, 43/1:181–184.
2. Kordonski, W., Golini, D., 1999, Fundamentals of Magnetorheological Fluid Utilization in High Precision Finishing, J. Intell. Mater. Syst. Struct., 10:683–689.
3. Yamaguchi H., Yumoto K., Shinmura T., Okazaki T., 2009, Study of Finishing of Wafers by Magnetic Field-assisted Finishing, J. Adv. Mech. Des., Syst. Manuf., 3/1:35–46.
4. Yamaguchi, H., Shinmura, T., Ikeda, R., 2006, Study of Internal Finishing of Austenitic Stainless Steel Capillary Tubes by Magnetic Abrasive Finishing, J. Manuf. Sci. Eng., 129/5:885–893.
5. Yamaguchi, H., Shinmura, T., 1995, Magnetic Abrasive Finishing of Inner Surface of Tubes, Proceedings International Symposium for Electromachining-XI, 963–976.
6. Yamaguchi, H., Shinmura, T., Kaneko, T., 1996, Development of a New Internal Finishing Process Applying Magnetic Abrasive Finishing by Use of Pole Rotation System, Int. J. Jpn. Soc. Prec. Eng., 30/4:317–322.

CHAPTER 3.20

Designing for Micro Abrasive Waterjet Machining

Don Miller

Finepart Sweden AB
Bollebygd, Sweden

The Process

Micro abrasive waterjetting is a relatively new method for machining miniature parts. It is a general purpose machining method, in that it cuts virtually all engineering materials. Micro abrasive waterjet machining is a technology that ablates microscopic amounts of material in nanoseconds with a frequency of megahertz without causing subsurface changes in workpiece material. A micro abrasive waterjet machining platform can have a positional accuracy better than 2 μm (0.0001 in.) and some materials up to 10 mm (0.4 in.) thick can be cut with an accuracy of 10 μm (0.0004 in.) and an edge finish of ≈1 μm (40 μin.) Ra. Fully enclosed and located in a clean workshop environment, an operator can get on with other business activities while monitoring operations on a smart phone.

Relationship to Other Micro Cutting Methods

Miniature parts typically have features with dimension less than 300 μm (0.012 in.). Machining features less than 300 μm by mechanical methods becomes increasingly difficult as feature sizes approach 100 μm (0.004 in.). This is due to a number of factors with low tool flexural stiffness being particularly important in limiting the amount of force that can be applied through cutting edges to a workpiece. Avoiding the need to apply forces to a workpiece makes non contact machining the main means of machining miniature parts. Non contact machining methods ablate microscopic amounts of material at high frequency and impose negligible forces on workpieces. Ablation can be by the energy in photons being transferred to a material electron to break molecular bonds or to cause vaporization/explosion of material, by sparks that melt and vaporize electrically conducing material, or by particle impacts that generate chips of material.

The three non contact micro cutting methods that ablate material at a sufficient rate to be used extensively are:

- Ultrashort pulse lasers—the dominant micro machining method
- Wire and micro wire EDM—particularly for electrical conducting hard and superhard metals and ceramics
- Micro abrasive waterjets—a machining method that has been known since the early 1990s but not seriously commercialized until 2010

When machining metals the three methods thermally ablate material. Subsurface changes in material properties are avoided with ultrashort pulse lasers operating with pulses in the low femtosecond region. In the case of abrasive waterjets, water droplets traveling at supersonic velocities in a cutting jet effectively provide infinite cooling and prevent subsurface heat affects. Abrasive waterjets are generally referred to as a cold cutting method.

Abrasive Waterjet Generation and Cutting

The key technology of the abrasive waterjet industry is the transfer of momentum from a high-speed waterjet to abrasive particles within a cutting head's focus tube. Figure 3.20.1 shows the general layout of an abrasive waterjet cutting head that entrains

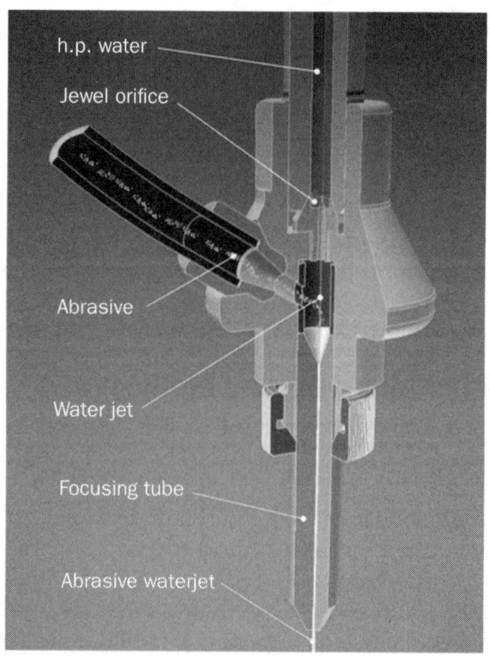

FIGURE 3.20.1 Micro waterjet entrainment cutting head.

abrasive carried to a cutting head in airflow. Water at 4000 bar (58,000 psi) or so is passed through a restriction (jewel orifice) made from diamond, ruby, or sapphire to generate a waterjet traveling at over twice the speed of sound. A waterjet traverses a chamber in a cutting head and enters the bore of a focus tube, thereby causing a vacuum that induces abrasive particles in a carrier fluid to flow into the chamber and be entrained into the focus tube bore. In a focus tube, bore particles are accelerated to about 600 m/s (2000 ft/s) they leave the focus tube as a focused cutting jet.

At exit from a focus tube, abrasive particles occupy about 1% of the volume of a cutting jet and water droplets about 10%. The remainder of a cutting jet is air when abrasive particles are carried to a cutting head in airflow and water vapor when particles in a water-based slurry are entrained into a waterjet. Typically abrasive particles have mean diameters a quarter of the focus tube diameter. A micro abrasive waterjet 200 μm (0.0078 in.) diameter would use under a kilogram (2.2 lb) of abrasive per hour.

Particle mass depends on the cube of the particle diameter (X^3) whereas water and particle mass flows through a cutting head depend on cutting jet diameter squared (X^2). This results in the number of particles impacting on a workpiece per second increasing as cutting jet diameters decrease. Micro machining with an abrasive waterjet involves several million ablation events per second (MHz) that last nanoseconds. Although several million impacts occur per second, each impact is essentially an isolated event as particles are spaced out many diameters apart longitudinally and each particle occupies less than 10% of a cutting jets cross section. In reality 80% of the most commonly used abrasive (garnet) particles are broken down to some extent in a focus tube so larger particles are accompanied by numerous smaller particles.

Chip formation when cutting ductile materials occurs by the plowing action of particles and by secondary removal or ridges caused by prior particle impacts. In the case of brittle materials, if the impact force is insufficient to cause cracks to propagate along grain boundaries, or existing faults within grains, then ductile material removal processes prevail. If impact forces are sufficient to cause propagation of a crack network, then individual grains are removed. Because of the short timescales involved, cracks do not propagate into the bulk of the material.

The interaction between a cutting jet and a workpiece is complex, with the cutting jet leaving a cut lagging behind the jet entry into a cut. The process inherently causes striations on cut surfaces, which are minimized or essentially eliminated by cutting at a slower speed (when the material is not too thick for the cutting jet diameter). When changing cutting direction and moving around curves, cutting jets have to be slowed to bring the cut face essentially vertical. Increasingly, 3-axis machining centers have some limited 5-axis motion that is used to compensate for cutting jet deviation through a workpiece and to reduce cutting times by up to 40%. Limited 5-axis motion can virtually eliminate cutting edge taper and provide for release angles when cutting out punch tooling and other applications requiring defined edge taper. Figure 3.20.2 illustrates parts cut using limited 5-axis motion.

An abrasive cutting head both generates and focuses a cutting jet. This is in contrast to a laser where a laser beam is generated remotely from a cutting head and cutting head optics focus a beam onto or in the workpiece. After a relatively short distance from the focal point, a laser beam energy density drops dramatically which is not the case with an abrasive waterjet. Having a focused cutting jet has advantages and disadvantages. A major advantage is a micro abrasive waterjet will cut much thicker materials than an ultrashort pulse laser. The major disadvantage of a persistent cutting jet is

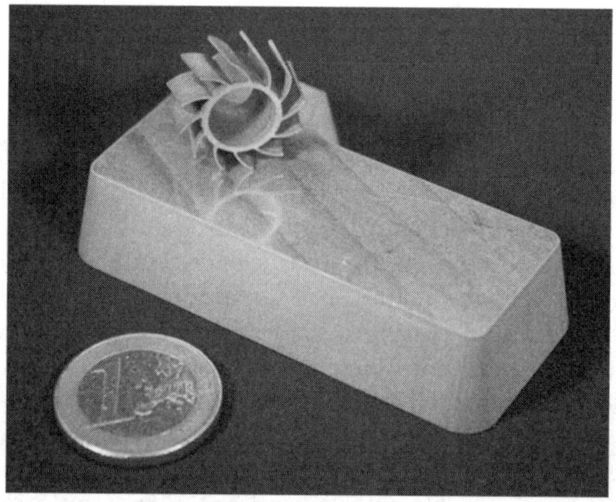

FIGURE 3.20.2 Examples of 5-axis cutting of 15-mm-thick (0.590-in.) aluminum with 14-degree release angle. Turbine cut from 8-mm-thick (0.315-in.) tool steel—blades and interior hole have 12-degree taper. (Courtesy: KTH University Stockholm/Finepart.)

that it retains damaging and dangerous cutting capabilities over substantial distances, which can be as high as a 1000 jet diameters. Even after cutting through a workpiece, a jet retains considerable cutting energy and the jet can be deflected substantially from vertically downwards. Containing and dissipating the residual energy in cutting jets is achieved by having a tank of water above which a workpiece is located. When a cutting head is mounted on a 5-axis motion system the greater the angling of the cutting head from the vertical the smaller the area of a machine over which cutting can take place. Catching the abrasive and workpiece material chips in a water tank has benefits when cutting high-value materials, such as gold, as the spent abrasive with its valuable material can be sent for refining.

A cutting jet is a "floppy cutting tool" and its interaction with workpieces in milling is complex as the cutting intensity drops from the center of a cutting jet to the jet edge, at right angles to the cutting head movement. Not only is the tool "floppy" but it is not possible to directly monitor the cutting process because of ricocheting abrasive particles, water droplets, and water vapor enveloping the milling region. Impressive milling operations are carried out with abrasive waterjets but a very high level of predictive computer modeling is required. Extensive milling studies are still needed to improve abrasive waterjet milling processes.

Micro Abrasive Waterjet Machining Centers

The design of a state-of-the-art micro machining center follows practices adopted for micro milling and micro wire EDM machine tools of using linear motors on the X-Y axes to provide rapid precise movements with a life expectancy, while maintaining machining accuracy, in excess of 10 years. A high-precision ball screw is used on the Z axis to prevent cutting head movement following power failure. Figure 3.20.3 shows a micro abrasive waterjet machining center.

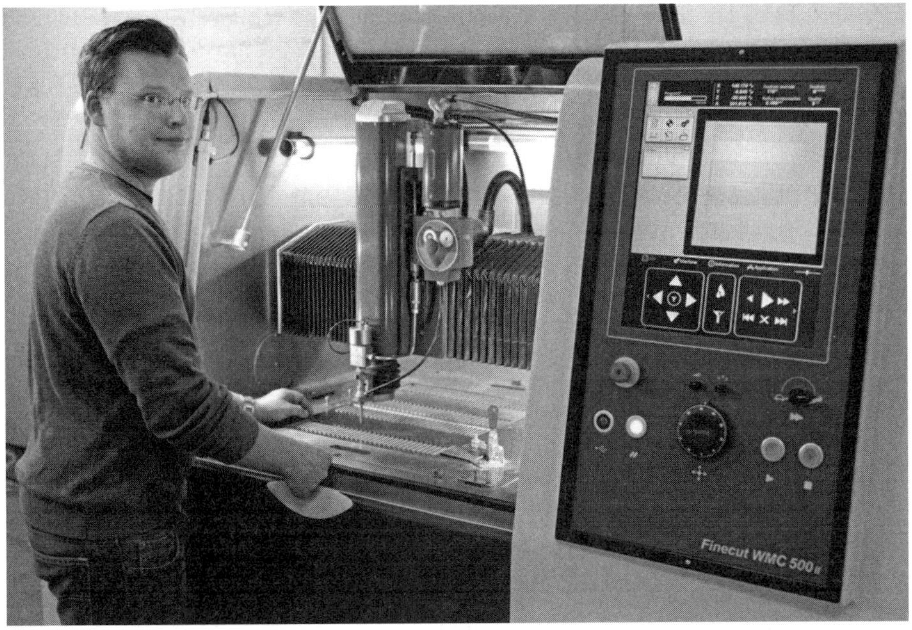

FIGURE 3.20.3 Abrasive waterjet machining center.

Typical requirements and operation of a micro abrasive waterjet machining center include:

- A cutting envelope XYZ of 500 × 500 × 80 mm (19.7 × 19.7 × 3.15 in.)
- Linear motors on the X-Y axes with free-floating absolute invar scales, with nanometer interpolation providing a repeatability over the working area of ±3 μm (0.0002 in.)
- A state-of-the-the-art multi-axis controller (e.g., FANUC's 31i-B)
- Cutting head to generate 300- and 200-μm-diameter (0.012- and 0.008-in.) cutting jets, with the option for future fitting of a cutting head to generate 100-μm-diameter (0.004-in.) cutting jets
- A rigid machine structure with workpiece fixturing to the machine structure
- Suspended jet catcher tank to avoid vibration transmission to the machine structure
- Enclosed cutting environment with safety interlocks

Workpiece Holding

Low cutting forces allow relatively simple fixturing to be used for many jobs. As cutting jet diameter decreases, capillary forces become more important and avoiding cut miniature parts moving into the path of a cutting jet becomes challenging. To prevent loss of machining reference during work transfer, more sophisticated jigs are used when workpieces are transferred between machining centers. For instance, a wire EDM job shop may use a micro abrasive waterjet machining center to do 2D cuts on a workpiece and drill starter holes for the wire. The workpiece is then transferred in its jig to a wire EDM machine for high-precision 3D cuts.

Human Machine Interface and Control System

The success of MAW machining center is intrinsically linked to the human machine interface (HMI). Not only is it the operator's means of interacting with and programming the machining center, but it is also the means by which the manufacturer and the control system supplier can access and support the machine over the internet. In effect an operator has a very high level of support from experts in abrasive waterjet and control technology and access to updates to the HMI and control system.

Over 25 years of operating experience with conventional abrasive waterjets has gone into the design of HMIs for micro abrasive waterjet cutting. An HMI provides intuitive program learning, touch screen programming, a library of abrasive waterjet functions, and facilities to easily correct mistakes.

A technical competent person can expect to operate a micro abrasive waterjet machining center after a short period of training. As with any machine tool operating experience is fundamental to building the knowledge to use a machining center in innovative ways and to its full potential.

Comprehensive CAM Software

Developing comprehensive CAM software for abrasive waterjet cutting has been relatively easy compared with other machining methods because:

- Abrasive waterjets produce a focused cutting jet, thus there are no dependent tool or cutting beam parameters to vary as there are with other machining methods
- Cutting by particle erosion is a continuum process down to micrometer diameter cutting jets, so cutting jet erosion rates are simply related to cutting jet diameter
- Only a limited amount of straight-line cutting performance data is required to characterize individual materials, with the result that CAM databases contain information on most engineering materials
- Adequate data exists to predict how cutting speed needs to vary along curved cutting paths and at changes in cut direction
- Simple algorithms have been developed to predict cutting speeds, given the machine operating parameters, the material, the material thickness, and the required tolerances and surface finish

The continuum nature of the erosion process means CAM software originally developed for macro abrasive waterjet machining applies to micro machining. This is in contrast with other machining methods, both mechanical contact and non-contact, where the material removal mechanisms change between macro and micro machining.

Typically CAD/CAM software accepts input from most types of drawing data files and produces machining instructions in G code, given the:

- Material to be machined
- Material thickness
- Cut edge quality, which can be specified at any point along a cut path
- Any changes from the prior set jet piercing parameters and jet lead in and lead out locations and form

Machining center operating parameters will already be stored in the software, but these can readily be varied.

Ideal Job Shop Micro Machining Tool

Micro abrasive waterjets meet the requirements of job shops which must provide precision miniature parts in different materials in small to large lot sizes on a just in-time basis. Their particular benefits are:

- High-productivity machining of miniature parts
- Cold cutting of metals, ceramics, glasses, composite, and layered materials with quality edge finishes that are free of heat effects and cracks and with minimal burrs
- Processing of high-tech materials that are difficult or uneconomic to process by other methods
- Cutting materials with thicknesses from tens of micrometers (<0.001 in.) to 15 mm (0.059 in.) depending on cutting jet diameter
- Minimal force exerted on workpieces allowing high aspect ratio ribs/slots
- From part design to finished part in minutes to a few hours
- Cost-effective from a single part for R&D to thousands of parts in production
- User-friendly, PC-based CAD/CAM software for 2D and 3D machining

Figures 3.20.4 and 3.20.5 show typical parts machined with micro abrasive waterjets. Abrasive waterjets can be operated to cause minimal burrs which is a major advantage for medical implants, such as shown in Fig. 3.20.4, as secondary machining operations can be avoided to remove burrs. The application of minimal force to workpieces allows thin webs to be reproduced as shown in Fig. 3.20.5.

Table 3.20.1 presents indicative examples of straight-line cutting speeds for a 200-μm-diameter (0.008-in.) cutting jet. The abrasive waterjet industry has adopted a scale of 1 to 5 to characterize surface finishes, with 1 being a separation cut which would have deep striations to 5 which is a good quality surface free of striations. This scale has been extended to 9 to accommodate higher quality surfaces produced by micro abrasive waterjets. In Table 3.20.1 two cut surface qualities are given equivalent to cut surface qualities of 3 and 9. A separation cut quality 1 speed would be roughly twice that of quality 3. The CAM software modifies the straight-line cutting speed dependent on how close a cutting jet is to a change of direction and on a curve's radius; the smaller

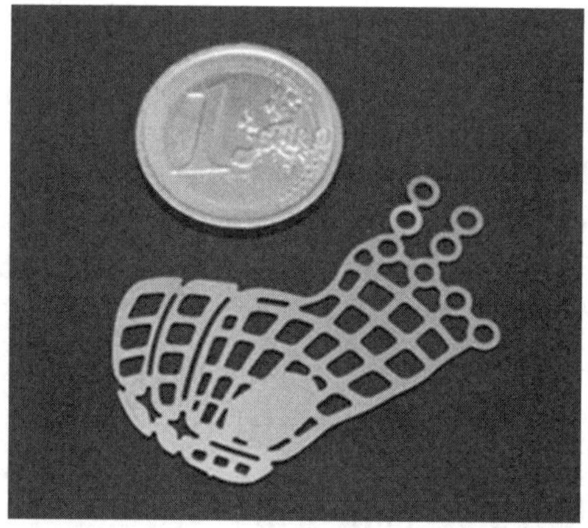

Figure 3.20.4 Medical implant.

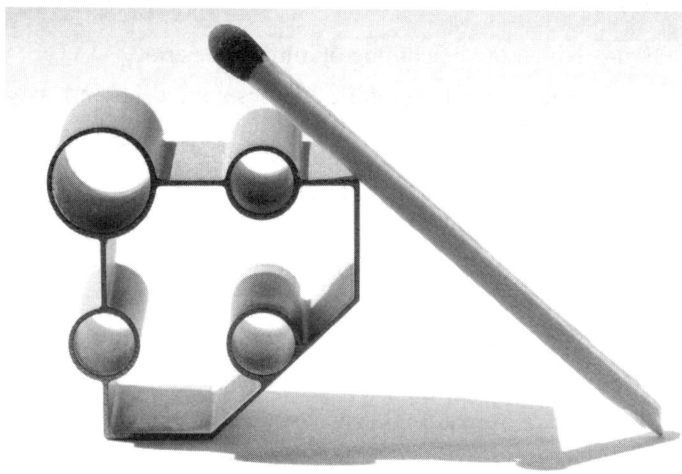

Figure 3.20.5 Example of cutting thin webs.

the radius the slower the cutting speed. At the quotation stage the CAM software can be used to compute the cutting time and the job cost.

Cutting parameters scale roughly linearly, so a 300-μm (0.012-in.) cutting jet would cut the materials listed in Table 3.20.1 1.5 times faster for the same thickness and cut materials 1.5 times thicker at the same speed.

Future Developments

The development of micro abrasive waterjets has been delayed by problems caused by Van der Waals, electrostatic, and capillary forces acting on particles less than 50 μm (0.002 in.) in mean diameter. Such particles have a large surface area relative to their

Material	Material Thickness (in.)	Medium Quality [Cutting Speed (in./min)]	Highest Quality [Cutting Speed (in./min)]
Stainless steel	0.03	3.0	1.2
Stainless steel	0.10	1.2	0.5
Titanium	0.03	5.0	2
Titanium	0.10	1.6	0.6
Inconel	0.03	3.1	1.2
Hardened tool steel (60RC)	0.03	1.0	0.4
Gold	0.03	6.0	4
Carbon fiber	0.03	24	4
PEEK	0.10	3.6	1.4
SiC ceramic	0.05	0.4	0.12
Glass	0.10	12	4

Table 3.20.1 Examples of Micro Waterjet Machining Capabilities

Figure 3.20.6 Experimental test piece cut with entrained slurry 125-μm (0.005-in.) cutting jet diameter.

mass. They flow in stick/slip mode which makes them difficult or impossible to meter, and when carried in airflow to a cutting head, static electricity causes the particles to clump and block passages. Developments are now taking place to entrain abrasive as a slurry for generating cutting jets down to 40 μm (0.0016 in.). Figure 3.20.6 shows a test part cut with a cutting jet generated by entraining abrasive slurry. Cutting jets below 40 μm (0.0016 in.) in diameter will probably not be practical for some time because of problems of guaranteeing that oversized particles do not reach cutting heads and block focus tubes.

As cutting jet diameters decrease it becomes practical to make focus tubes from diamond, and this will allow the economic cutting of hard ceramics which cannot be economically cut using conventional focus tubes made of tungsten carbide.

Compared to ultrashort pulse lasers, with their vast array of applications and numerous process variables, abrasive waterjets with just a few variables may seem low tech. However, what abrasive waterjets are good at, which is controlled ablating of virtually all materials, mean they have capabilities that no other machining method can match. This is particularly true for many micro machining applications, and since we live in the age of miniaturization of products and devices, micro abrasive waterjets will be a manufacturing process of choice for decades.

CHAPTER 3.21

Photochemical Machining for Micro Parts

David M. Allen

Cranfield University
Bedford, United Kingdom

Process and Technology

Photochemical machining (PCM), also known as photoetching, photomilling, photofabrication, and chemical blanking, plays a valuable role worldwide in the production of both precision parts and decorative items. PCM has flourished, particularly in the last few decades, by producing micro parts for the development of micro systems technology (MST), micro electromechanical systems (MEMS), micro air vehicles (MAVs), micro electronics, portable communication and miniaturized devices, instrumentation, micro surgery, and implantable bio-devices.

The process involves photoresist technology to enable imaging of the required part (usually via a phototool set) and etching technology to chemically machine the part. The process is similar to those processes used in the fabrication of printed circuits and silicon chips. PCM is usually carried out by subcontractors as the process involves the use of automated photoresist coating and etching machines, chemical processing equipment, and a great amount of in-house technical and chemical know-how. The PCM business turnover in 2000 was calculated at approximately US$ 6 billion. A schematic of the PCM process is shown in Fig. 3.21.1.

PCM is usually applied to fabricate flat parts which can then be folded or formed, if 3D parts are required, by secondary processing. PCM has also been applied to surface etching of three-dimensional parts such as cylinders and spheres.

Characteristics, Applications, and Limitations of the Resulting Product

Characteristics and main advantages of PCM include

(1) Low-temperature processing that does not affect the physical or chemical properties of the starting material.

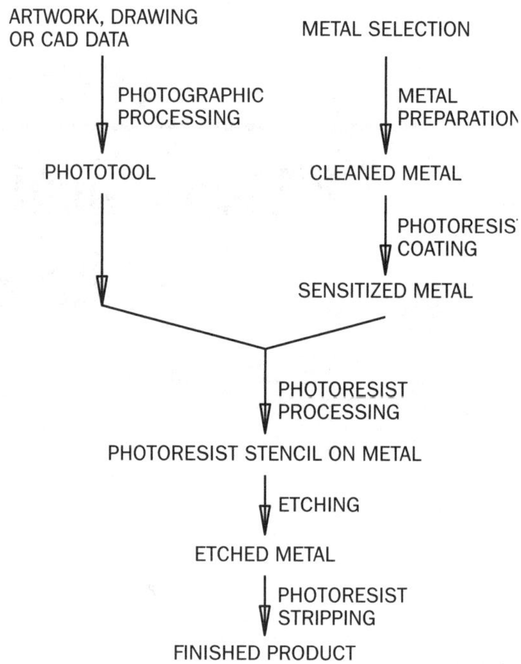

FIGURE 3.21.1 Schematic of the PCM process.

(2) The ability that if you can draw it, then PCM can make it, subject to the limitations discussed below.

(3) Extremely short lead times. Parts can be manufactured within 24 h of receiving the part CAD file. Field testing can therefore be carried out rapidly and, if part modifications are required, these will be inexpensive and also quick to make.

(4) Micro parts made by PCM *will not require deburring* as burrs are not formed. This is a very positive advantage as micro burr removal methods have been the subject of much heated debate and the best solution is to avoid their formation in the first place!

(5) Micro parts may be produced separately or tabbed into a sheet for ease of handling, automated inspection, or assembly.

The main limitations of PCM, in terms of material thickness, T, are

(1) Minimum hole diameter, $\Phi_{min} = 1.1T$ but, for very large volumes of product using automated roll-to-roll processing of materials, Φ_{min} may be reduced to $0.8T$.

(2) An upper thickness limitation of 2 mm (0.080 in.) although 6 mm (0.236 in.) is possible for low-resolution parts.

(3) Some deviation from a classical straight profile at the edges of parts, usually controlled to $<0.25T$. The ability to vary edge profile can be a positive advantage for PCM and typical profiles are shown in Fig. 3.21.2. As the etchant penetrates

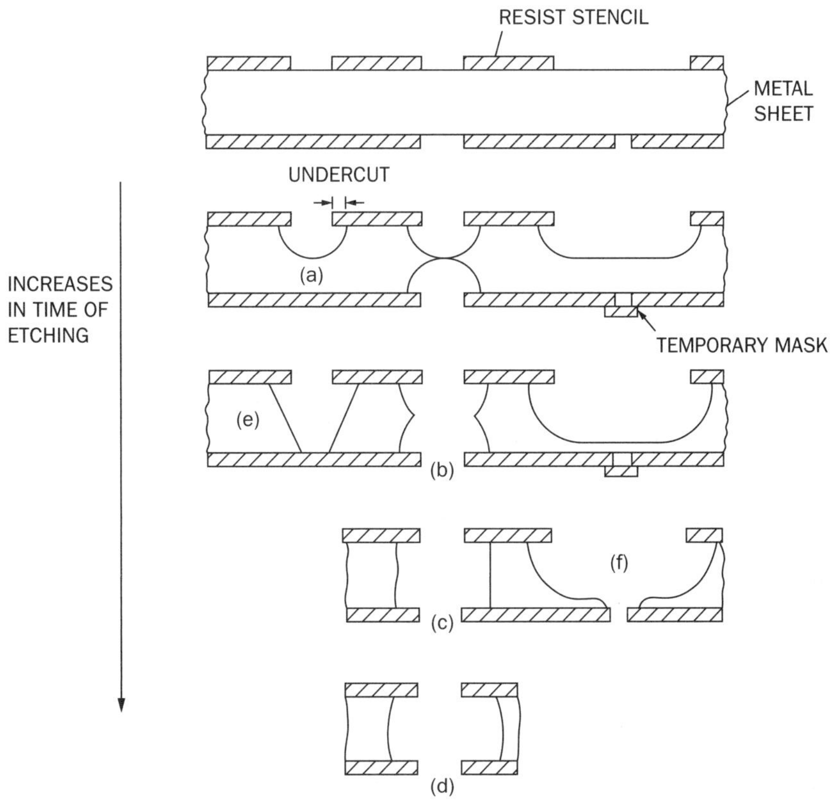

Figure 3.21.2 Edge profile development as a function of etching time showing (*a*) half etch, (*b*) biconvex, (*c*) straight, (*d*) biconcave, (*e*) sharp, and (*f*) tapered profiles.

through the thickness of the material, it also etches sideways beneath the photoresist coated on the metal surface. The etch factor is defined as the depth of etch divided by the undercut (see Fig. 3.21.2) and is used by the subcontractor to modify the part CAD file with suitable etch compensations to achieve a correctly dimensioned part after etching.

Note the additional capabilities:

(1) Produces surface-etched features at the same time as perforations, providing process cost advantage of carrying out two different processes (milling and piercing) simultaneously by a single chemical process rather than two physical processes.

(2) Produces sharp edges (see Fig. 3.21.2 profile *e*) by single-sided etching. These profiles have been utilized, for example, in surgical scalpel production.

In general, a dimensional tolerance of $\pm 0.15T$ is an accepted estimate of tolerance on etched dimensions, but this will vary according to the metal, sheet size, etchant chemistry, and equipment involved.

Hole-center to hole-center dimensions are independent of etching time and reflect the accuracy of manufacturing the phototool from the CAD file. Tolerances are of the order of ±0.05% of the dimension concerned, for example, 50 ± 0.025 mm (2 ± 0.001 in.) is typical.

The applications of PCM are many and various. A selection of parts from different technologies applicable to micro parts manufacture include

- Optical applications: iris leaves (spring steel); encoder discs and light attenuators (stainless steel and aluminum); graticules (brass); bespoke apertures (tantalum)
- Mechanical applications: washers, gaskets, grids, filters, shims, diaphragms, levers, links, brackets (steels and brasses)
- Electronic applications: lead frames and connectors (copper alloys); emitter contacts (molybdenum)
- Medical applications: stents (shape memory alloys and magnesium), filters and probes (stainless steels and titanium)

Finishing processes such as anodizing, chemical blackening, chemical polishing, electroplating, painting, and powder coating can be applied to PCM parts to increase their functional versatility.

Economics

Every manufacturing method has its advantages and disadvantages and should be selected accordingly. If the part to be produced is at the limit of the process capability of the method selected, the yield of acceptable products will be poor and will be reflected in high-production costs. Other manufacturing processes, which may be in competition with PCM, are:

(1) Photoelectroforming (PEF) involving electroforming through a photoresist stencil

(2) Wire electro-discharge machining (WEDM) or wire spark erosion

(3) Laser beam machining (LBM)

(4) Stamping or fine blanking and piercing

In deciding which method to use for part production, the technical aspects of the method should be checked to ensure the selection of appropriate techniques for the material used and its thickness as well as part specifications in terms of edge profile, flatness, and dimensions. Manufacture of a part by PCM requires consumables such as cleaners, photoresists, etchant, and strippers to be included in the part cost (P_E) which is therefore relatively high. Phototooling also needs to be made but this involves relatively low costs (A). These economic factors contrast strongly with those of stamping where a high tooling cost for the manufacture of a punch and die (D) is associated with a very low part cost (P_S). It is apparent, therefore, that PCM is cheaper for low-volume production while stamping is cheaper for high-volume production. The break-even quantity (Q) is dependent on the geometric complexity of the part and is calculated from the formula:

$$Q = (D - A)/(P_E - P_S) \tag{3.21.1}$$

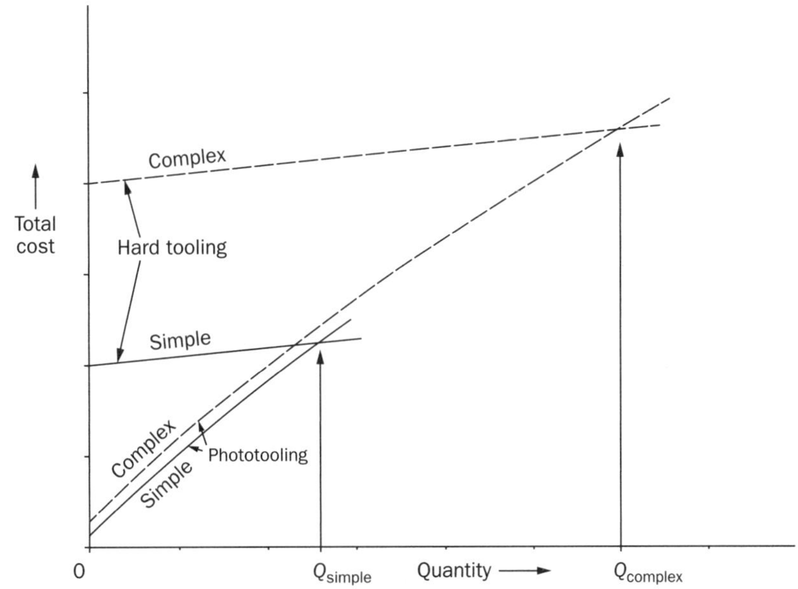

Figure 3.21.3 Typical economics of the production of simple and complex parts by PCM and stamping showing break-even quantity (Q) increasing with part complexity.

Assuming PCM and stamping produce parts to identical technical specifications, PCM becomes increasingly economic as part complexity increases. This is illustrated in Fig. 3.21.3.

In general, the factors influencing the most economical method of production are part complexity, batch size, and part thickness. Large-volume production tends to favor stamping; very low volumes favor WEDM; high complexity favors PCM; thicker components favor LBM; thinner components favor PCM; and extremely thin [<0.025 mm (0.001 in.) thick], high-resolution components in *metals only capable of being electroplated* favor PEF.

Materials Suitable for This Process

It should be noted that PCM is normally applied to metal foils and sheets, most commonly stainless steels and brasses. Some polymers such as Kapton® (a polyimide) can also be etched, but this is a relatively rare process in comparison to metal etching. However, as the etching process involves accelerated but controlled corrosion, more corrosion-resistant metallic materials require more aggressive etchants to be used. Thus, Table 3.21.1 shows a selection of metallic materials that are easily etched in standard etchants such as ferric chloride, cupric chloride, and dilute mineral acids and those that require more aggressive etchants such as mixed or concentrated acid formulations and alkaline etchants based on sodium or potassium hydroxide. Depending on the designer's choice of metal for the micro part, the PCM subcontractor will choose the most appropriate etchant for optimum product yield.

Good	Good to Fair	Fair to Poor	Poor
Copper	Precipitation-hardening stainless steels	Molybdenum	Tungsten
Copper alloys	Martensitic stainless steels	Nichrome (Ni, Cr)	Hastelloy C
Zinc	Ferritic stainless steels	Vanadium	Titanium
Carbon steel	Austenitic stainless steels	Chromium	Rene 41
Kovar (Fe, Ni, Co)	Inconel alloys (Ni, Cr, Fe)	Gold	Niobium
Nickel	Hastelloy B	Lead	Tantalum
Nickel alloys		Manganese	
Magnesium		Rhenium	
Aluminum		Zirconium	
Anodized aluminum			

Table 3.21.1 Etchability of Metals

Feature	Material; Thickness	Design Considerations
Two 1.0125 ± 0.0125-mm-diameter (0.0405 ± 0.0005-in.) holes	Spring steel; 0.050 mm (0.002 in.)	A very flat, distortion-free and burr-free camera shutter blade was required prior to polymer coating. The etched edge profiles of the holes were required to be biconcave.
Arrays of conical holes 0.05 mm (0.002 in.) in diameter with hole-center to hole-center distances of 0.3 mm (0.012 in.)	AISI 316L stainless steel; 0.050 mm (0.002 in.)	A corrosion-resistant filter was required comprising an array of tapered profile (conical) holes.
Arrays of 0.3-mm-diameter holes	Brass; 0.020 mm (0.0008 in.)	A contact strip was required for a piezo-mechanical actuation device.
Arrays of 0.1-mm-wide (0.004-in.) hexagonal holes	AISI 304 stainless steel	An insect-proof cylindrical mesh was required for instrument sensor protection.
Two 0.2 mm (0.008 in.) circular holes, an oval recess 0.073 mm (0.003 in.) deep with a central 0.1-mm-diameter (0.004-in.) hole and a drainage channel 0.031 mm (0.0012 in.) deep connected to one of the larger holes.	Ni-Span-C®; 0.1 mm (0.004 in.)	A small prototype blood pressure transducer body with outside dimensions approx. 2.3 × 0.7 mm (0.09 × 0.028 in.) was required to sit within a vein or artery and be placed in situ by means of a hypodermic needle.
A two-sided phototool was registered such that when concentric grooves displaced relative to each other were etched from two sides, it produced a corrugated foil approx. 0.015 mm (0.0006 in.) thick	AISI 316L stainless steel; 0.025 mm (0.001 in.)	A prototype circular corrugated pressure sensitive diaphragm 6.9 mm (0.27 in.) in diameter was required for inclusion in a radio capsule that is swallowed to monitor alimentary tract pressure.
An 8-mm-ID (0.315-in.) cylinder of intersecting wires 0.15 mm (0.006 in.) wide	Magnesium; 0.25 mm (0.01 in.)	A prototype nontoxic biodegradable stent was required that dissolved in blood.

Table 3.21.2 Examples of Micro Features Produced by PCM

Specific Design Recommendations and Issues

The designer can use standard CAD software packages to define the required micro part. PCM subcontractors will give recommendations for acceptable file formats but generally accepted standard formats include .dwg and .dxf files. The subcontractor will then adapt the file to make the phototooling set and optimize product yield by building in etch compensation allowances dependent on the metal thickness of the part required.

It is highly recommended that the designer has an in-depth discussion with the PCM subcontractor when submitting the part CAD file. Some subcontractors specialize in etching particular metals (e.g., aluminum and its alloys) or polymers and have the expertise that other subcontractors may not possess. Any potential issues arising can then be discussed. It should be remembered that the subcontractor has the experience of producing a very wide range of parts in a variety of metals and thicknesses (see Table 3.21.2) and is therefore able to give the designer the part required on time, and at the agreed price, once the designed part has been accepted for production by PCM. Once the first prototypes have been produced to the designer's satisfaction, no further changes in process steps are required and PCM is then ready for mass production of the parts if required in high volumes.

CHAPTER 3.22

Micro Molding Overview

Aaron Johnson

Accumold
Ankeny, Iowa

In its simplest definition, micro molding is a manufacturing process that is just like general injection molding, but the much smaller components create new challenges and approaches. Micro molding can be defined by the size of a part, but also can refer to parts with micro features or micro tolerances. Conventional parts with micro features may require the same considerations as micro-size parts (Fig. 3.22.1).

Like general molding, micro molding is the process of melting plastic, introducing the molten material into an injection system, and forcing it to fill the desired shape. It is sometimes believed that because these molded parts are micro in size the general laws of molding do not apply. They do.

That being said, the differentiation between where general modeling ends and micro molding begins isn't always clear. Geometry, material, quality, features, tolerances, and/or annual unit demands can each have a profound effect on the success of any given project. As technical as the micro molding process is, it is incredibly situational. Parts that are needed for micro optics may be quite dissimilar to those for electronic control panels. The key element of micro molding is that it is a process designed to quickly and economically produce thousands, millions, and even billions of parts

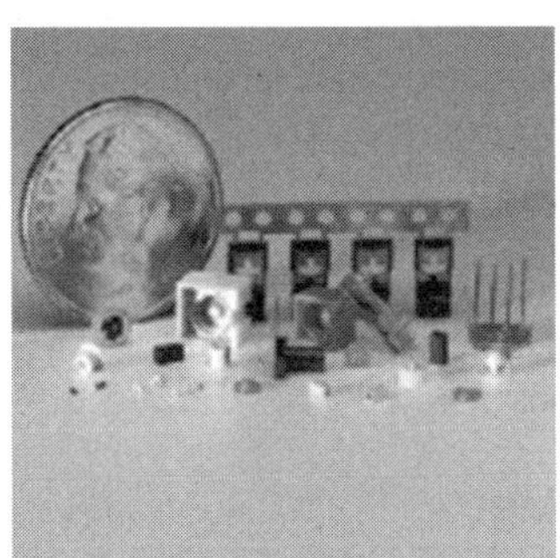

FIGURE 3.22.1 A variety of micro molded examples. It is often better to show the capabilities than define them.

with the same precision. As such, it is ideally a high-production process, but because of its speed and reproducibility, it also can be appropriate for quantities in the low thousands.

Applications for Micro Molded Parts

Micro molded parts can be found virtually everywhere: in electronics, cell phones, telecommunication equipment, surgical tools, hearing aids, medical devices, antennas, connectors, and more. The tiny lenses in these micro molded parts are frequently used in micro sensors, medical diagnostics, endoscopic tools, and minimally invasive surgical instruments. The auto industry relies on sensitive high-tech parts such as steering modules and sensors that aid in a vehicle's safety, handling, and fuel efficiency. And, given the exacting environments in which military and aerospace parts must function, these industries demand that components be developed with micro tolerances, metrology, and security in mind.

Individual industries often require strict production protocols, which add to the complexity of presenting all the needed design consideration. Today's medical devices, for example, require medical-grade thermoplastics, clean room production, and lot traceability. Products with micro electronics must be built with wafer-thin and feather light components.

Types of Micro Molding

Common forms of micro molding include:

- Conventional molding of a single material
- 2-shot molding (two different materials)
- Insert micro molding
- Lead frame micro molding
- Micro overmolding
- Continuous molding

Small, Miniature, and Micro

Micro-size molded parts can range from approximately a half inch (12.7 mm) in size to those with features measured in micrometers (Fig. 3.22.2). "Small" molded parts are slightly larger than micro molded parts. They can be approximately a half inch to about 3 in. (12.7 to 76.2 mm) in diameter. Like micro molded parts, these require an exceptional degree of precision and quality during production. Small molded parts may have micrometer tolerances or they can have micro-size features measured in micrometers. There is a distinction in the complexity of these small parts that separates them by size from what would be considered more "commodity" level parts.

Two-Shot Micro Molding

In this specialized process, two different plastic materials are shot in sync and in one single operation (Fig. 3.22.3). When appropriate to the component configurations, this type of process can reduce labor, create a more robust part configuration, and help

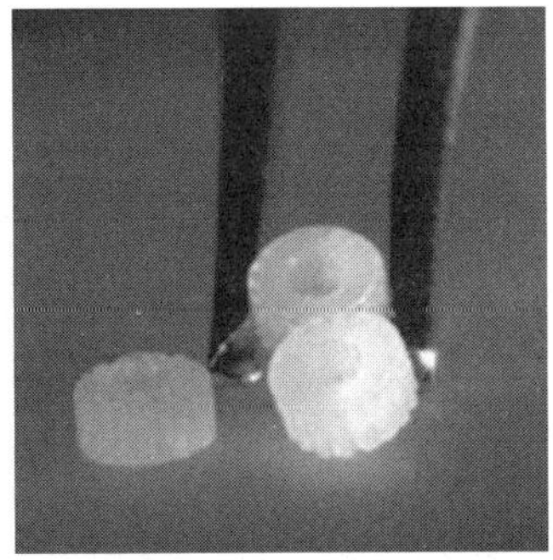

FIGURE 3.22.2 These micro molded rotors held by tweezers have 18 tiny teeth around the rotor's 2.34 mm (0.092 in.) diameter.

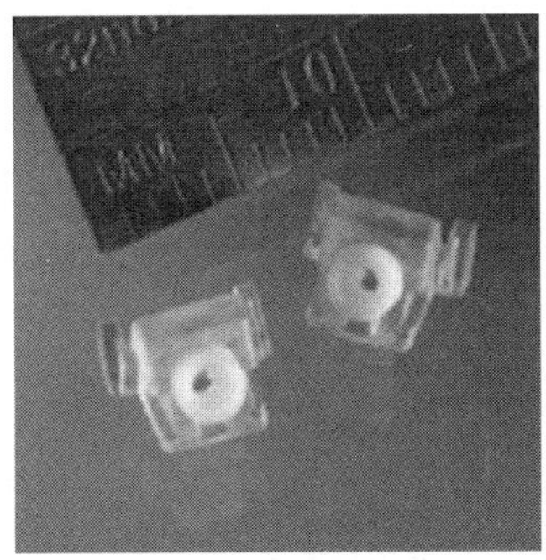

FIGURE 3.22.3 The hard, clear ABS part here was combined with a soft, white TPU material to create a perfect bond. The 2-shot molding process saved manufacturing resources and replaced the assembly and gluing process.

contain costs. Again, the common distinction from standard molding is the complexity of performing this process on a micro-size scale.

Insert Micro Molding

Insert micro molding provides the capability to mold a wide variety of materials—such as metal or glass—within or around plastic parts. Similar to 2-shot molding, this process not only reduces handling and assembly, it can solve critical issues in manufacturing. Capturing small- or micro-size inserts often requires a custom molding and/or automation system to handle the delicate media (Fig. 3.22.4).

Lead Frame Micro Molding

Lead frame packaging technology requires high-quality metal lead frames that produce exceptionally intricate designs. Plastic can be molded on and around all different types of metal, tin, copper, silver, and gold to form an airtight seal. Some of these processes produce completed parts, while other components are produced and sent to a general manufacturer for further processing.

In this process, a blank metal strip, or "lead" strip, is articulated through the mold during processing and plastic is molded over the metal to form a new component (Fig. 3.22.5). This process usually involves fabricating a system to feed the metal

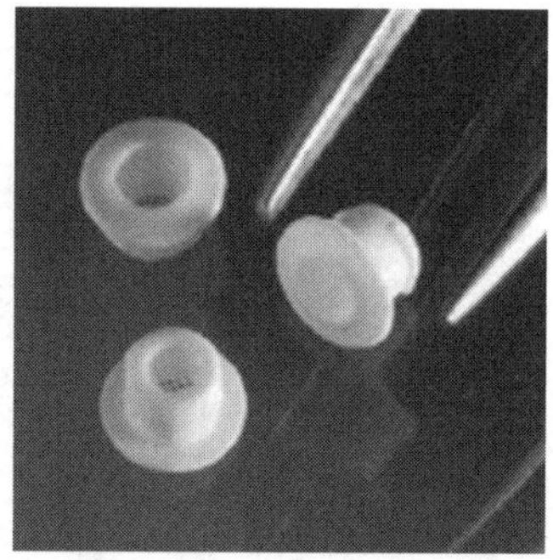

FIGURE 3.22.4 Fragile fabric filter insert molded into a plastic frame. The mold had to be built so precisely that the filter was not damaged during the molding process and the tight tolerances of both the filter and the frame were retained.

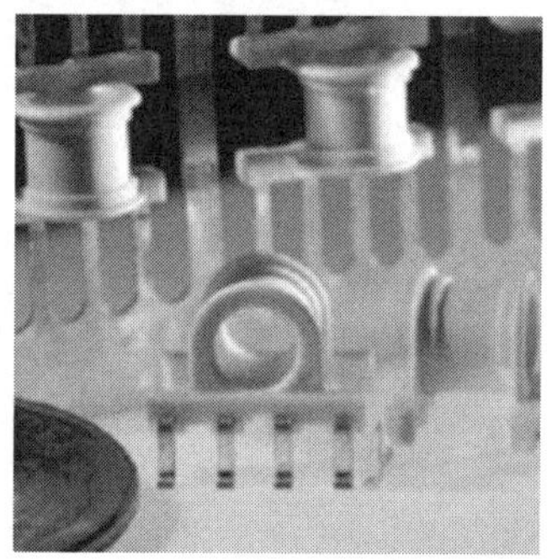

FIGURE 3.22.5 This bobbin was molded in a multi-cavity mold on a continuous strip. The cavity-to-cavity variation averaged between 5 and 7.6 μm (0.0002 and 0.0003 in.).

lead frame into the machine on a continuous reel. This allows the molded parts to be re-reeled after molding and shipped to the customer for further manufacturing.

Lead frame molding has huge advantages for high-volume manufacturing because it is designed to be received into the customer's automation system with little to no handling required. This process also is duplicated for single components when they are placed in a tape-and-reel packing system after molding. Here again, these reels are designed to provide a carrier for automation during the next phase of manufacturing.

The main distinctions between lead frame and micro molding come not only in the molding side of the equation but also in the lead strip stamping design and quality. The consistency of the lead strip in terms of flatness or feature locations is key, especially if there is little plastic surface area to work with and/or the expectation of the tolerances is very tight, as in a few micrometers (a few 0.0001 of an inch).

Micro Overmolding

Overmolding often refers to a process that molds plastic "over" a part made of plastic, metal, or other material and, in some fashion, binds the two parts into one (Fig. 3.22.6). This process is synonymous with insert molding and is often used interchangeably.

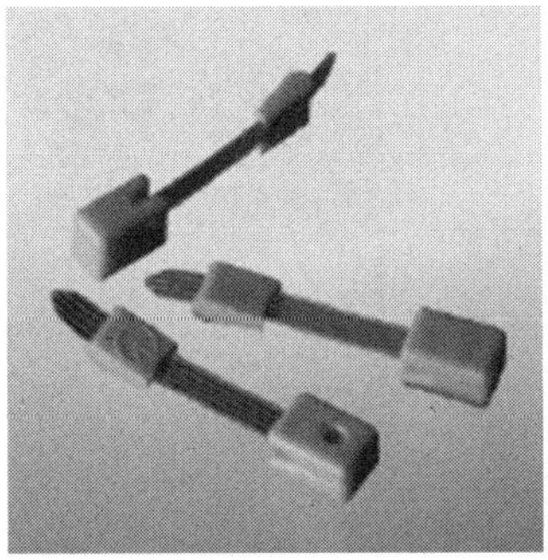

FIGURE 3.22.6 A very sophisticated overmolding product in which delicate flex-circuit leads were inserted, held, and shut off around the mold to form a very tight seal without damaging the flex-circuit or the connector.

Insert molding, lead frame molding, and overmolding are all variations of the same basic process.

Enhancing Success in Micro Molding

In virtually every case, a micro molded project starts with a design for manufacturability (DFM) conversation. It truly is essential.

In a general molding project, it is common for a designer to select a molder, upload a 3D file to the molder's website, and receive a quote in moments. With micro molding, however, the relationship between the geometry, the material, and the quality requirements is too complex for simply a computer-generated response. Until recently, most flow analysis tools would suggest that micro molded parts were not even moldable. In the micro molding process, the DFM conversation between the designer and the micro mold manufacturer works through these topics.

Geometry and Material Selection

The DFM will begin with the ideal design for the component and a discussion about what is possible through micro molding.

General molding rules apply for micro molded parts, but there are many more constraints. The mold needs a gate, a way to eject the part, and even draft on micro-size molded parts. For micro work, wall thinness, length, and depth are critical issues. Material needs are important for a part of any size, but they become even more important for micro parts because they shed temperature differently, have less surface to bond, and could fail because of a single bubble. Normal molding stray particles cannot be tolerated in micro features.

Typically, a material is chosen for its inherent properties such as strength, heat resistance, or biocompatibility. However, material data sheets can be misleading since many of their molding suggestions for gate size, flow capability, or other features are based on much larger size moldings. Through the DFM, the designer can tap into the experience of the micro molder and be better equipped to identify materials that will satisfy the project goals.

Figure 3.22.7 LCP part.

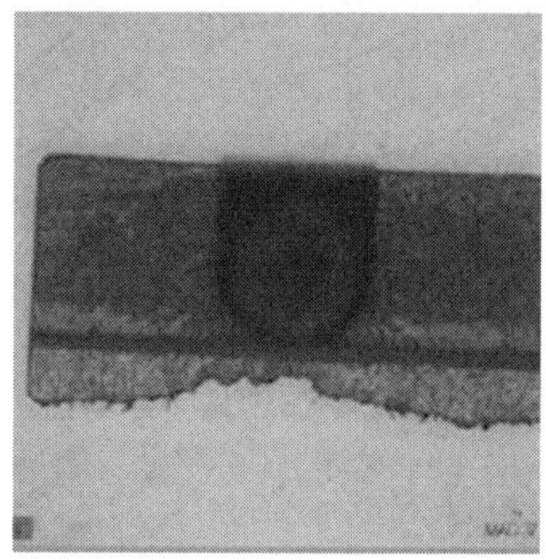

Figure 3.22.8 Same part made from PEEK.

In conjunction with part geometry and material, the question of how small one can mold is always the next question. Because the described micro molding process is very situational, it is virtually impossible to express what can be accomplished by means of a list or chart, but the following summarizes some key issues.

Materials

Virtually any thermoplastic can be processed for micro molding. However, there is a direct correlation between the feature size and the material selected. For example, liquid crystal polymer (LCP), polyethylene (PE), polypropylene (PP), and polyoxymethylene (POM) are examples of materials capable of thin-wall, high aspect ratio features. Others, like polyetherimide (PEI) or polyether ether ketone (PEEK), will struggle to fill. Figure 3.22.7 shows an LCP-molded part with an extended wall thickness of 76.2 μm × 32.4 m (0.003 × 1276 in.) in length. Figure 3.22.8 is the same part design-molded with PEEK. This type of outcome is a common variance with micro-size features between different material selections.

Part Size

Figure 3.22.9 is an example of a micro molded part (tiny spot in lower right) next to a standard sample bar with a common gate recommendation. It is molded of LCP and measures 380 × 360 × 300 μm (0.015 × 0.014 × 0.012 in.). An individual part weighs

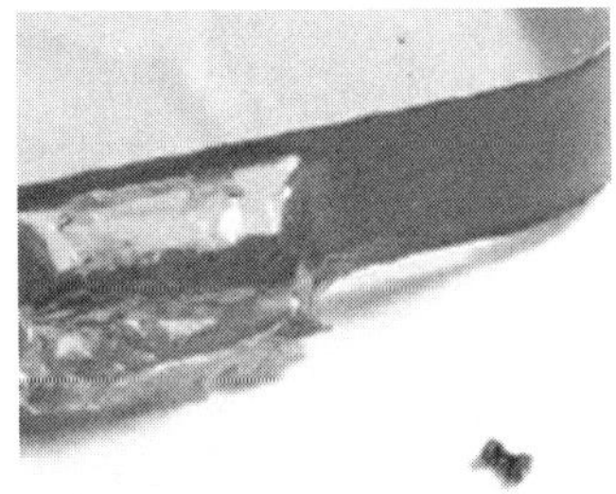

FIGURE 3.22.9 LCP micro part (the tiny spot in lower right).

FIGURE 3.22.10 PEI lens next to penny.

0.000085 g. Part size also is limited by the relationship between the material selection and geometries.

Feature Size

Feature size can be described either as a geometric shape or as a surface detail. Figure 3.22.10 shows a part PEI-molded that has a 100-μm-diameter (0.0039-in.) aspheric lens 125-μm (0.0049-in.) base radius with a 3-μm (0.0001-in.) tolerance critical to part functionality. The basic part dimensions are 0.700 × 1.00 × 1.700 mm (0.028 × 0.040 × 0.067 in.). Figure 3.22.11 shows a PEI-molded part with surface features down to 0.5 μm (0.000019 in.) in size. Size also is limited by the relationship between the material selection and geometries.

Challenges

Some reasons why micro molding is so challenging and demands in-depth reviews include:

- Boundary layer friction is higher in micro-size channels and walls
- Boundary layer undergoes high shear stress

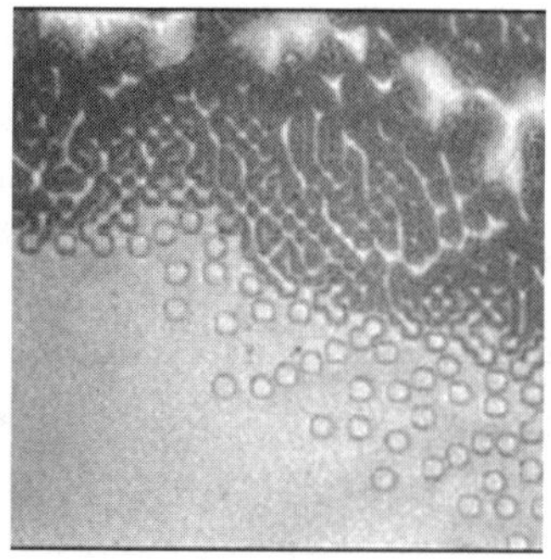

FIGURE 3.22.11 PEI part with 0.5-μm (20-μin.) surface details.

- Melt slip along walls depends on dimensions and material
- Plastic materials freeze (cool) much more quickly in thin walls
- Much higher pressures required to force melt through micro channels to part
- Fill times measured in milliseconds
- Ejection without damage of micro parts is difficult
- Molds must be cleaned of microscopic material
- Shrinkage is difficult to predict
- Material being pushed into mold must be metered in milligram or microgram shots
- Mold temperatures must be held very closely
- Final part handling and shipping constraints

Quality and Critical Features

Often, the tolerances on the critical dimensions of micro molded projects are set at 25 μm (0.001 in.) or less. But—especially with micro optic components—dimensional tolerances of ±2 to 3 μm (0.000078 to 0.000118 in.) can be held.

CHAPTER 3.23

Micro Metal Powder Injection Molding

LaRoux K. Gillespie

Manufacturing Consultant
Andover, Kansas

Metal powder injection molding combines traditional powder metallurgy (P/M) with plastic injection molding. Injection molding is one of the lowest cost processes for mass producing parts, often producing parts for pennies. Metal powder, ceramic, intermetallic compounds, and composite powders can all be injection molded. The metal powder process mixes metal powders with a large amount of polymeric or wax binder and the combination is forced by a rotating screw into the mold. The resulting green compact has the desired shape and is strong enough to handle, but the metal is not fully fused together. The organic components are removed from the part after the part is taken from the mold (by pyrolysis or solvents). The resulting all-metal component then is sintered to fully fuse the metal and provide the necessary strength. While conventional metal injection molding (MIM) can claim 95% of full density, part geometry and features size can reduce that for micro parts.

Micro dimensions challenge injection molding because it is difficult to push the injected materials completely into narrow channels and cavities. Metal mold walls quickly chill the incoming fluid which restricts further flow and leaves some areas incompletely filled. This becomes a larger problem when high aspect ratios are to be filled. As the materials fill the molds, some air or gases have to escape and narrow passages restrict that removal. Removing the delicate part from the mold can be a challenge without damaging it, and handling it at each stage requires great care. Using finer powder particles helps the flow, but the high specific surface area on nano size powders increases viscosity rapidly.

It is easier and less expensive to make micro products via mold inserts rather than to cut micro delicate features in a large mold. These inserts, made of nickel, nickel iron alloys, silicon, tungsten-cobalt, and other materials can readily be removed and replaced when worn or otherwise inadequate. Silicon and plastic inserts allow better thermal flow and prevent sudden freezing of the molded materials.

Micro MIM Materials

The most common metallic powders used for the MIM process include low-alloy steels, high-speed steels, stainless steels, tool steels, superalloys, intermetallics, magnetic alloys, Kovar, hard metals, aluminum, copper-tungsten, molybdenum-copper, and titanium. Copper is also an available powder in both micro and nano sizes. The nano sizes result in different production effects than the micro sizes. Typical particle diameters of 316L stainless steel powder are 10 and 2 μm, and other materials have similar size ranges.

The material used with the submicrometer powders to facilitate flow for fine details can consist of water-soluble binder PEG and PMMA or just wax. The PMMA binder can be removed from the parts faster and easier than some of the other binders.

Parts and Features

One of the best examples to define the capability of the process is the production of a planet gear assembly. The gears included a 24 tooth, module: m = 0.07 mm gear [1 mm (0.040 in.) in diameter] made from 17-4PH stainless steel. A water-atomized powder [D50 = 2 μm (79 μin.)] and oxymethylene-based binder was used. The green compacts were debound after molding at 600°C for 2 h in N_2 gas, and then sintered at 1150°C for 2 h in argon gas. The parts were then age-hardened at 480°C for 1 h. The resulting parts were a class 7 gear. Grinding the minute gear rather than molding it produced a class 5 gear; therefore, the resulting accuracy was less than the ground gear, but the production rate was much faster.

For practical limits on part size, this working gear probably represents the smallest functioning (rotating, mating, turning) size appropriate for the process today.

Figures 3.23.1 and 3.23.2 illustrate a wide variety of some of the smaller micro parts made by powder metal injection molding.

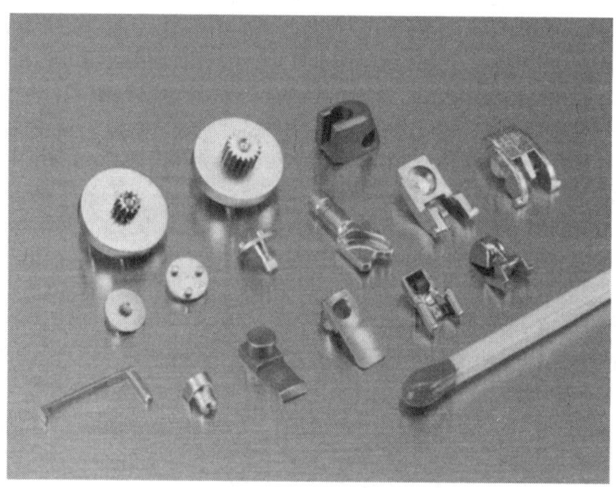

FIGURE 3.23.1 A sample of micro powder metal injection molded parts. (Courtesy Parmaco-Metal Injection Molding AG.)

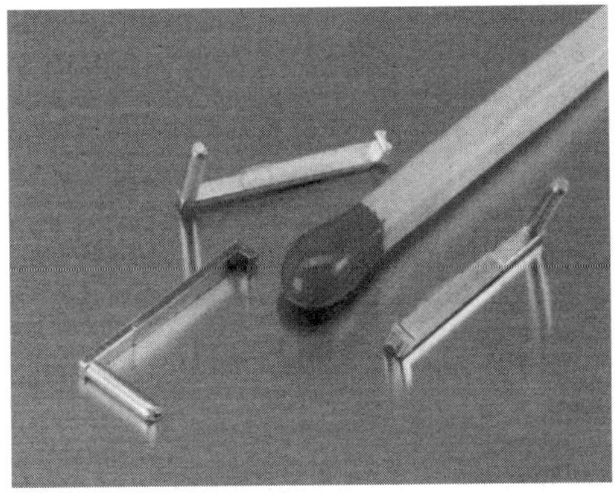

FIGURE 3.23.2 Slender micro metal powder injection molded parts. (Courtesy Parmaco-Metal Injection Molding AG.)

Using plastic molds, it is possible to make pillar arrays more successfully than with metal molds. The plastic allows the metal to cool faster at the end of the shot which helps final properties, and the plastic mold allows demolding without damaging the pillars. Pillars as small as 50 by 150 μm (0.002 × 0.006 in.) tall have been produced with 3-μm (0.00012-in.) 316 stainless steel particles. These pillars are made in arrays and they in turn can become tools for other micro processes, or serve as heat dissipaters, or flow channels.

Some of the smallest parts produced by the process have been 350 μm (0.014 in.) in diameter after sintering, but these were research parts, not production-worthy applications. Sintering shrinks most dimensions and that can be as much as 30% to 40% of the molded part dimension. That is well above shrinkage on most inch-size components. Shrinkage of micro features depends on the feature dimensions and on the process parameters. Thin sections exhibit higher relative shrinkage than thicker sections.

Surface roughness of 0.4 μm (16 μin.) Ra and an Rz of 3 μm (118 μin.) have been produced with the metal process, but that depends on powder size, binder, pressure, sintering, and other parameters.

Equipment

Several manufacturers make micro metal injection molding machines. One even allows 2-shot molding, so metals having a different property can be added to the primary component. That is not a common practice, but it can be done. At least a handful of companies provide micro level manufacturing for other companies.

Design Considerations

Because the resulting parts are not as solid or dense as parts made from solid stock, it is essential that designers specify and manufacturers monitor the resulting part strength or performance of metal injection molded parts. It is even more critical for micro parts since the opportunity for minor changes can make a large impact on microscopic features. Designers need to state the allowable density and strength of parts. For many applications, a test sample like a dog bone tensile test sample is molded at the same conditions for quality control sampling. While it is also appropriate for larger parts, it will be critical for success to do it for micro features and micro parts.

Because of part and feature shrinkage after sintering, experimentation is required to obtain the desired final dimensions.

Micro structure arrays are more difficult to release from the molds than non-array parts, in part because there is more surface area to adhere to the mold and in part because the pillars are generally very minute and more subject to damage.

Most of the design restraints for large part powder metal injection molding and for powder metal forming apply to the micro world as well (see Bralla's *Design for Manufacturability Handbook* for numerous examples). Unfortunately, dimensional data for the larger parts just do not apply at the micro level because of the differences in particle size used, binder differences, shrinkage differences, and geometry often required for micro parts.

Some *general guidelines* for micro features include

- Provide one flat surface on the part for support during sintering.
- Do not design very sharp corners or edges [use radii >0.05 mm (0.002 in.)]
- Holes smaller than 0.1 mm (0.004 in.) diameter will be difficult to produce.
- Walls thinner than 0.1 mm (0.004 in.) will distort and fail to fill.

References

Bralla, J. G., *Design for Manufacturability Handbook*, 2nd ed., 1999, McGraw-Hill, New York, pp. 167–176.

Imgrund, P., A. Rota, and F. Petzoldt, Manufacturing of multi-functional micro parts by two-component metal injection moulding, *The International Journal of Advanced Manufacturing Technology*, 2007, vol. 33, Number 1–2, pp. 176–177.

Islam, A., N. Giannekas, D. M. Marhöfer, G. Tosello, and H. N. Hansen, The shrinkage behavior and surface topographical investigation for micro metal injection molding, *AIP Conference Proceedings*, 2015, 1664, 110007 (DOI:http://dx.doi.org/10.1063/1.4918482).

Nishiyabu, K., Micro Metal Powder Injection Molding, Some Critical Issues for Injection Molding, *InTech*, http://cdn.intechopen.com/pdfs/33647/InTech-Micro_metal_powder_injection_molding.pdf (DOI:10.5772/34498).

Son, S.-J., Y.-S. Cho, and C.-J. Choi, Advanced micromanufacturing for high-precision micro bearing by nanopowder metallurgy and LIGA processing, *Reviews on Advanced Materials Science*, 2011, vol. 28. pp. 190–195.

CHAPTER 3.24

Micro Stamping

LaRoux K. Gillespie

Manufacturing Consultant
Andover, Kansas

Micro parts and micro features can be stamped in pressing operations with punch and dies until the required features reach too minute a dimension. For over 50 years simple micro shims only 25 μm (0.001 in.) thick have been stamped out; 200,000 of these can be mashed into a normal finger-size sewing thimble. Similarly, another stamped micro part is so small that 5000 of them can safely be placed in only half of a normal prescription medicine bottle.

The definition of what constitutes "micro" stampings varies widely. For many of the most minute stamping operations, 2 mm (0.080 in.) or smaller parts are micro parts and 0.1 mm (0.004 in.) features are micro (Fig. 3.24.1). Larger parts might also be defined as micro, but at least these examples provide a range of what can be done as guidance for designers.

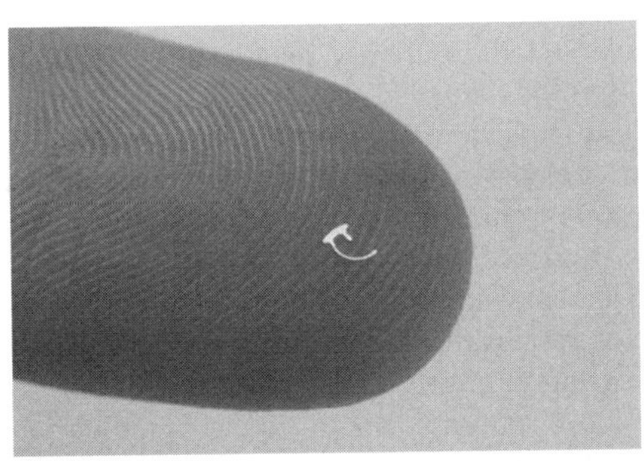

Figure 3.24.1 Example of micro stamped part. (Courtesy Top Tool Co.)

Stamping for many individuals implies just cutting a flat piece with a punch and die.[1] Stamping, however, also can include bending, coining, skiving, cold heading, and drawing to make three-dimensional configurations. Blanking refers to cutting out a specific shape, and piercing involves driving a punch through the shape to make holes or internal cutouts. Bending, curling, and flattening are forming operations, and they can be part of a stamping operation, of a continuous line that starts by piercing and blanking.

This chapter is devoted to blanking, piercing, and bending. Micro coining is discussed in Chap. 3.18.

The Impact of Part Design

Blanking out a micro part can be an easy and rapid process (200 to several thousand per hour) when the configuration is simple (micro flat shims for example). But when the part configuration requires several near-intersecting features, the task becomes much more complicated. When the material thickness is only 25 μm (0.001 in.), the challenge is also compounded. In both instances there are less than a handful of companies with the expertise to successful meet tight tolerances of complex configurations.

Simply stated, when parts are smaller than 1 or 2 mm (0.040 to 0.080 in.) and features approach 100 μm (0.004 in.) wide or diameter, there are much better approaches than stamping them out. Simple doughnut shims and similar simple configurations can be stamped out, but more challenging shapes should not be attempted, and if they are it should only be with one of the most knowledgeable leaders in the business.

Consider a series of micro holes punched, leaving only a few thousandths of an inch between them as remaining stock. The grain size for most metals falls in the range of 1 to 150 μm (39 μin. to 0.0059 in.) or larger; thus, there are only a few grains between holes. Because the grains are not uniform, some distortion of the hole geometry will occur.

The same is true for ultrathin metals which can be as thin as 12.5 μm (0.0005 in.). The grains may be smaller than the punch radius which causes a change in the cutting action from shearing to tensile failure. That can distort the dimensions also as well as increasing the forces required to cut.

Whether for blanking, piercing, or micro forming of thin metals, operations that only consist of four or five (or fewer) grains will result in higher variation than if more grains are involved. Stated differently, plastic deformation behavior is defined by only a few grains located in the deformation regions and important size effects occur. Smaller grains provide more consistent sheet metal components.

It is more than grain size alone that is significant. Grains are not uniform. Their distribution and crystal orientation within the sheet affect shearing. While the number of grains in the cross section decreases with respect to thicker components, the anisotropy of the grains also leads to significantly more variation in results. As a result of this, slip and coordination of grains are more difficult during the shearing which ultimately leads to increase in shearing strength. The resultant geometry of micro hole cross sections from large grains and grain anisotropy causes large variance in blanked edge profile, distribution of rollover, burnish, fracture, and burrs when a large grain size is present.

[1]In the micro manufacturing world, "stamping" can also refer to using rubber-like stamp pads to emboss soft materials.

The ratio of punch-to-die blanking clearance to grain size is also a significant factor, much larger than in normal size components. It is difficult to reduce the clearance in proportion to the grain size.

Despite the challenge, stamped micro parts are used in micro electronics, in everyday items like pacemakers and hearing aids, and in some neuroelectrodes used for pain management. They are found in computer read-and-write heads, inkjet printers, pressure sensors, connector pins, resistor caps, contact springs, and chip lead frames.

As an example of the typical smallest features stamped out, electronic lead frames are blanked and pierced in copper and FeNi sheets of only 150 µm (0.0059 in.) thickness. Long leads (up to 220 per part) only 100 µm (0.004 in.) wide are stamped out on 168 µm (0.0027 in.) pitch at rates of 160 to 2000 parts per hour. The very complex and thinnest ones require the longer times. The 168 pitch with 100-µm-wide (0.004-in.) leads, leaves 68 µm (0.0027 in.) between the lead fingers (i.e., open space). Millions of these lead frames are stamped out. The investment in tooling and process is high but is recovered by the large quantity made and the relatively high-speed operation.

Micro punching has successfully produced micro hole arrays of a 50 × 4 matrix of 0.6-mm-diameter (0.024-in.) holes in 200-µm-thick (0.0079-in.) brass (all done in 120 s). Similar arrays have been produced with 150 µm (0.0059 in.) punches in 80 µm (0.0031 in.) brass. Diameters can be held to ±1 µm (±39 µin.) and positions to ±3 µm (±118 µin.) in this material with these sizes. Holes were spaced 1000 µm (0.039 in.) apart in x and y dimensions, leaving a 400-µm (0.016-in.) wall between them.

The lower limit to spacing between stamped holes is constrained by the need to support the punches and dies from moving. While micro EDM can produce holes leaving a wall thickness of only 100 µm (0.004 in.), this leaves little room to support the forces from the punches.

The smallest practical limit to punched holes is about 50 µm (0.002 in.) and that is in sheets only 25 to 50 µm (0.001 to 0.002 in.) thick (aspect ratio of 0.5 to 1.0). This aspect ratio is somewhat universal for micro stamped features and should be considered a logical design limit for thin fingers as well as small holes.

Stamped part tolerances can be as low as 0.01 mm (0.004 in.), and a few stamped parts are made with materials as thin as 0.013 mm (0.0005 in.). Machining cannot produce parts that thin unless they are stacked many layers high, so stamping is a lower cost process when many parts are to be produced.

While punches have been made as small as 50 µm (0.002 in.) in diameter for experiments, a more reasonable lower limit size is 80 to 100 µm (0.0031 to 0.0039 in.) diameter. As noted above for the lead frames, fingers should also be roughly 100 µm wide or bigger. Distance between features as noted for the lead frames may be as small as 68 µm (0.0027 in.), but wider punches provide longer life and less feature distortion.

The above discussion was limited to stamping out or blanking parts. Micro parts can be formed by bending in presses just as their larger components are, but the challenge again is the impact of grain size and the ability to get micro tooling appropriate to the need. The standard practice for larger parts is not effective for the micro parts. It requires considerable experimentation on highly uniform material to produce consistent results in bending. Lead bending studies illustrate that spring back when bending lead frame fingers is larger when bending along the material's rolling direction that is perpendicular to it, so grain orientation as well as size and number of grains are all important factors.

While stamped parts typically begin with thin sheets or even foils, it is also possible to flatten wire to begin the process and it can be done in a continuous line.

Deep drawing is a stamping process, and one for which there is a growing interest for micro parts. Deep-drawn cups 1 mm (0.039 in.) in diameter and 0.4 mm (0.016 in.) deep have been formed in 20-μm (0.0008-in.) aluminum sheet. Shimizu et al. (2012) have drawn cups as small as 0.45 mm (0.018 in.) diameter in 20-μm-thick (0.0008-in.) 304 stainless steel, phosphor bronze, and pure titanium. That appears to be the lower limit possible today in these materials.

Cold heading takes a wire or similar cylinder and mashes the end to make a head. The materials common to this process include copper, gold, silver, and gold and silver alloys, brass, platinum, tantalum and the alloys of these latter two, titanium and its ductile alloys, nickel and its alloys having more than 20% ductility, iron, steels, stainless steels, and Kovar. Brittle materials are not amenable to cold heading.

Cold heading has been used commercially on wire just 0.3 mm (0.012 in.) in diameter. Parts can be cold-formed at rates of 50 to 400 parts/min. The process can work on smaller wire, but commercial production equipment does not exist. At least two major companies have made their own equipment for smaller work.

Several researchers are exploring the field of micro forming to help advance each of the areas discussed.

Materials

Today's micro stamped parts are made from platinum, platinum iridium, gold, niobium, stainless steel, beryllium copper, titanium, nitinol, and Paliney. While most users will stamp metal parts, Mylar and other polymer sheets can also be blanked and pierced to micro dimensions.

Blanking and piercing, like machining, produces burrs on the cut edges, and unless kept very small these burrs are entirely unallowable. While the flat stock allows some simple flat-work deburring processes, it may be very difficult to remove them without distorting some features or removing stock from the surfaces. Deburring formed micro parts can be extremely difficult.

Design

Most of the design knowledge for the very micro features is held by commercial companies who for proprietary reasons prefer to keep it within company walls. They work with designers to jointly reach acceptable designs, but do not want to lose their competitive advantage formally providing that in public ways. As a result, there is little formal guidance for designers other than understanding the state of the art of lower limit possibilities.

It is essential in micro stamping to provide as much distance from edges as possible for holes, cutouts, bends, etc. Doing so helps reduce distortion of the surfaces around edges. Heat treatment also will cause distortion and special attention must be paid to keeping stock stable during heat treatment.

Designing features at the micro level requires an extensive understanding of material properties and manufacturing forces and stresses. In some instances the expertise is so critical that only a single material can deliver the desired size and shape to specified tolerances. This requires designers to consider materials other than the first choice. Stated differently, success in micro stamping requires an extensive understanding of metallurgy, stresses, and geometry influences on distortion. Unfortunately, there are few design guides for micro features and few experts in this area.

- Experimental studies and analyses of the starting material and finished part are required at the micro levels.
- From a design aspect, designers need to balance the number of grains and part thickness carefully before committing to a design. That means exact specification of material properties.
- As for many micro processes, designers need to consult manufacturers before finalizing designs in order to prevent unmakeable products. Micro stamping is extremely challenging.
- Micro parts of precious metals although small and stamping being the lowest cost process for many configurations, at over $1000 per ounce, parts can still be expensive.

Reel-to-reel chemical etching, or the lower volume conventional sheet at a time photochemical etching, generally is a better choice for very narrow fingers and lace-like components. The lithographic photoetching process produces no forces and no burrs and can produce thousands of parts per hour while holding tight tolerances. It can produce thin fingers less than 25 μm (0.001 in.) wide, while blanking and piercing operations typically cannot leave fingers thinner than 100 μm (0.0039 in.) and that is not possible on many configurations. Chapter 3.21 discusses this process.

For a low quantity of parts, micro EDM or micro WEDG are more appropriate choices for many geometries. Chapters 3.7 and 3.11 discuss these processes.

Several references are cited below for more detailed understanding of the metallurgy issues.

References

Geiger, M., F. Vollertsen, and R. Kals, Fundamentals on the manufacturing of sheet metal micro parts, *CIRP Annals*, vol. 45, no. 1, 1996, pp. 277–282.

Joo, B. Y., S. H. Rhim, and S. I. Oh, Micro-hole fabrication by mechanical punching process, *Journal of Materials Processing Technology*, vol. 170, no. 3, 2005, pp. 593–601.

Razali, A. R., and Y. Qinb, A review on micro-manufacturing, micro-forming and their key issues, *Procedia Engineering*, vol. 53, 2013, pp. 665–672.

Shimizu, T., M. Yang, and Ken-ichi Manabe, Chapter 5: Impact of surface topography of tools and materials in micro-sheet metal forming, in *Metal Forming—Process, Tools, Design*, Mohsen Kazeminezhad (ed.), Published by InTech, 2012 (http://dx.doi.org/10.5772/48296).

Xu, J., B. Guo, C. Wang, and D. Shan, Blanking clearance and grain size effects on micro deformation behavior and fracture in micro-blanking of brass foil, *International Journal of Machine Tools and Manufacture*, vol. 60, September 2012, pp. 27–34 (doi:10.1016/j.ijmachtools.2012.04.001).

Xu, J., B. Guo, and D. Shan, Size effects in micro blanking of metal foil with miniaturization, *The International Journal of Advanced Manufacturing Technology*, vol. 56, no. 5–8, 2011, pp. 515–522.

Xua, J., B. Guo, D. Shan, C. Wang, J. Li, Y. Liu, and D. Qu, Development of a micro-forming system for micro-punching process of micro-hole arrays in brass foil, *Journal of Materials Processing Technology*, vol. 212, 2012, 2238–2246.

CHAPTER 3.25

Designing for Micro Hot Embossing

Jacobo Paredes

CEIT and Tecnun (University of Navarra)
San Sebastián, Spain

Marc Chooljian and Dorian Liepmann

University of California
Berkeley, California

The Process

Micro hot embossing technique takes advantage of the relatively low glass transition temperature (T_g) of many of the thermoplastic materials to conform a shape to a master stamp. This process is a well-known process widely used in industry for many devices, together with other techniques such as roll-to-roll or injection molding. However, hot embossing is better suited for prototyping or low-volume production. It is a very affordable and versatile technique that covers the gap between research and large-scale production.

On top of that this technique presents some advantages over the other mass production techniques, such as less polymer stress due to a reduction in material shrinkage and residual stresses, and smaller temperature cycle. This of course comes with a price: fewer cycles per time unit using a less-established technology.

The term "micro" refers to structures that are in the range of a few micrometers to hundreds of micrometers, although this method is certainly used in bigger sizes as well. The resolution of this technique goes nearly to the nanometer scale faithfully replicating the mold structures.

This process involves two main steps: (1) design and fabrication of the mold or stamp and (2) the hot embossing per se of the features onto the plastic material. Device design should take into account some considerations about the limitations of fabrication methods that will be described in subsequent paragraphs.

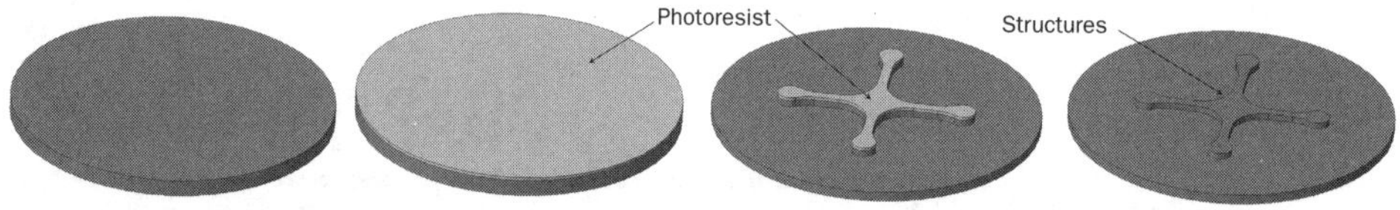

FIGURE 3.25.1 Schematic of the fabrication process of silicon masters for hot embossing.

Fabrication of Molds for Micro Hot Embossing

The mold is typically fabricated by means of micro technologies due to the high-resolution requirement [in a wide range of nanometers to tens of micrometers (fractions to tens of millionths of an inch)]. The device layout is generally printed onto a photolithographic mask and then transferred to a photoresist spin coated on the substrate. Fabrication methods depend on the substrate chosen for the mold. Silicon, which is a very common material used for the inserts, can be micro machined with reaction ion etching (RIE) and other methods. Figure 3.25.1 presents a brief schematic of the fabrication of silicon inserts for hot embossing.

There are other methods and materials suitable for this purpose such as nickel electrode position on stainless steel wafers, casting of PDMS [poly (dimethylsiloxane)] on an SU-8 mold (SU-8 is an epoxy-based photoresist used for generating structures usually over a silicon substrate), etc. High-precision machining systems may also be used for this purpose, achieving high-strength molds with long lifespan. The properties of each mold will be different and therefore their desirable uses.

Micro Hot Embossing of Thermoplastics

The mold is then used for transferring the features to the polymer target. Therefore, it is necessary to understand how processing parameters such as time, temperature, and pressure affect both embossing and demolding. The optimum conditions to perform these two steps of the process follow.

The glass transition temperature of a thermoplastic depends not only on the temperature, but also on the pressure at which the plastic is exposed. Increasing the pressure will raise the T_g of the polymer and its viscosity. Therefore, it is necessary to apply greater temperatures [in the order of 10°C (18°F)] to ensure that the polymer continues to flow under pressure. Another consequence of this behavior is that the increase in viscosity will require more time for the polymer to accommodate to the features.

Due to the mismatch of thermal expansion between the mold (almost negligible) and the plastic (orders of magnitude of variation), users need to account for the fact that the final product could differ slightly from the mold. There is no equation or general law for calculating this effect; it must be found for each different scenario.

Figure 3.25.2 represents the summarized sequence of this process in a hot press or embosser. It consists of five main steps: (1) make contact between the mold and polymer sheet, (2) heat up the platens of the embosser above the T_g of the plastic [at least 10°C (18°F) higher], (3) apply pressure, (4) cool down below the T_g and release the pressure, and finally (5) demold to separate the mold and the plastic, then start the process again.

There are some specific characteristics to each step that should be taken into consideration when designing or performing this method. Figure 3.25.3 depicts a

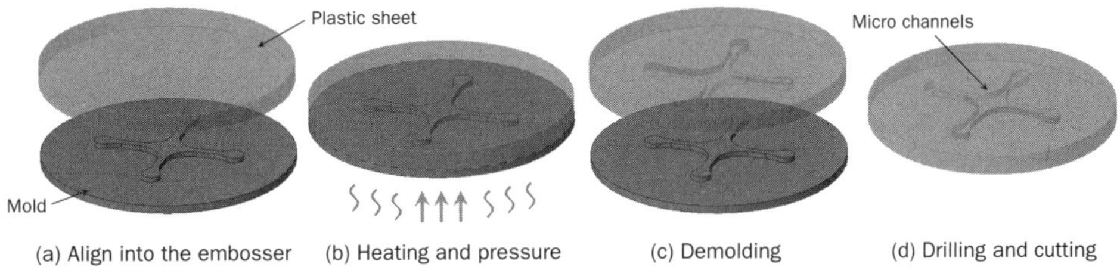

FIGURE 3.25.2 Schematic of the process of micro hot embossing.

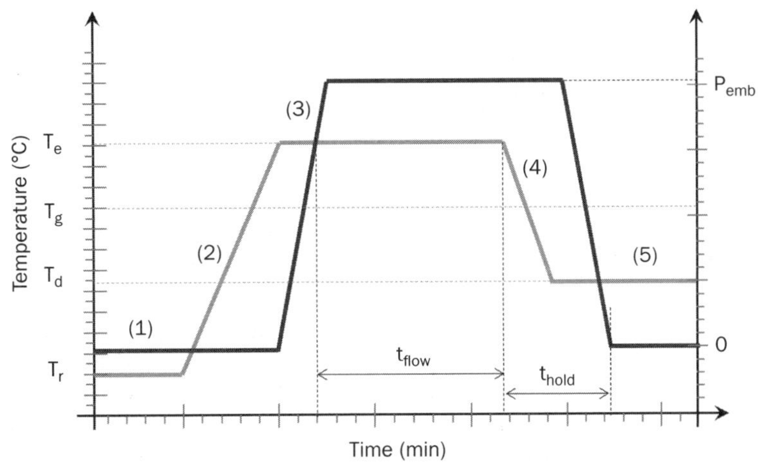

FIGURE 3.25.3 Temperature and pressure applied on the plastic during the embossing process. T_e is embossing temperature, T_d demolding temperature, T_r room temperature, and P_{emb} the pressure at which the pieces are pressed.

temperature-pressure graph of an ideal embossing process. The total duration of the process known as the cycle time will set the number of processes over time.

1. Bringing the mold and plastic together after heating the platens is an important step to avoid air trapping and improving homogeneity. This situation is especially important when the embosser is not equipped with a vacuum chamber. In this case it is recommended to wait until the heated air escapes the interface before applying pressure and then pursue the next step of the process.
2. Heating and pressure *ramps* depend on the specifications of the embosser and don't represent a critical configuration of embossing.
3. The period of time in which we lead the *plastic flow* (in the viscous state over the T_g) is important to ensure that all features are correctly transferred. Pressure values are typically in the range of 5 to 40 MPa (525 to 5800 psi).
4. *Pressure* is held while cooling the system below the T_g so that the plastic keeps the shape of the mold. This period of time is marked in the figure as t_{hold}. After that pressure is released, pieces can then be separated. Cooling is very important together with the next step due to thermal stresses caused by the shrinkage

Material	Acronym	α_L (10^{-6} m/mK)*	Available Fabrication Technique
Silicon	Si	2.56	Wet and dry etching
Glass	Glass	3.2	Wet and dry etching
Stainless steel	SS	10–17.3	Micro machining, electrodeposition
Nickel	Ni	13	Electrodeposition
Cooper	Cu	17	Micro machining, electrodeposition
Aluminum	Al	23.1	Micro machining
Polydimethylsiloxane	PMDS	3	Casting on resists (i.e., SU-8)

*1×10^{-6} m/mK = 0.555×10^{-6} in./in.°F.

Table 3.25.1 Characteristics of Commonly Used Molding Materials

mismatch between plastic and mold. Also, it is important to design feature sizes according to the coefficient of thermal expansion (α_L) (see Tables 3.25.1 and 3.25.2 for the specific values of common materials).

5. The *demolding* temperature and the operation is one of the most critical steps to ensure a perfect outcome. Many types of defects can be caused due to the high adhesion between the plastic and the mold: distorted features, localized sidewall bulges, mold features pulled out and embedded in the plastic, etc. Approaches that help reduce these effects are: increase the draft angle of the structures whenever it's possible; demold at high temperature using separation flags, air knife, or other specific methods that prevent mechanical deformations. However, there is no general law or equation to provide the optimum temperature for demolding; it will depend on the materials as well as the feature sizes and designs.

Typical Applications

Micro hot embossing has become a very popular technique over the past decades for fabricating thermoplastic micro devices. This technique is especially attractive for micro fluidic devices and optical micro components due to the suitable characteristics of polymers for these applications. There is a wide range of different thermoplastics that show very different characteristics.

Micro fluidic devices include lab-on-a-chip devices, micro total analysis systems, point-of-care devices, cellular culture platforms, chemical and molecular assays (like PCR or electrophoresis), etc.

Most common requirements are transparency and chemical compatibility (biocompatibility, ease of sterilization, etc.) with a long lifespan. Also, the sizes for these kinds of applications are in the range of tens to hundreds of micrometers (thousandths to tens of thousandths of an inch) for the layout and tens of micrometers (thousandths of an inch) in height. Most common designs are two-dimensional; however, it is possible to build three-dimensional structures such as mixers, valves, chambers, etc. (using multilayer approach). Channel cross sections are usually square or rectangular but it is possible to have them semicircular.

Thermoplastic	Acronym	T_g (°C)	α_L (10^{-6} m/mK)*	Properties
Polyethylene	PE	–125	108–200	Flexible, economical, chemical resistant
Polypropylene	PP	0	100–200	Flexible and economical
Polyamide	PA	60	110	Good mechanical resistance
Polyethylene terephthalate	PET	70	59.4	Low friction
Polyvinylchloride	PVC	80	50.4	Economical
Polystyrene	PS	95	70	Transparent and economical
Acrylonitrile butadiene styrene	ABS	80–125	73.8	Mechanical resistance
Poly(methyl methacrylate)	PMMA	105	75	Good transparency
Polyether ether ketone	PEEK	143	47	Mechanical and chemical resistance
Polycarbonate	PC	147	70.2	Transparent and mechanical resistance
Cyclo-olefin polymer	COP	123–156	60–70	High light transmission
Cyclic-olefin copolymer	COC	78–178	60–70	High light transmission
Polytetrafluoroethylene	TEFLON	250	112–125	Excellent chemical resistance

*1×10^{-6} m/mK = 0.555×10^{-6} in./in.°F.

TABLE 3.25.2 List of the Main Thermoplastics Used in Hot Embossing

Optical micro components include micro lenses, waveguides, filters, etc. In this case the optical properties of the selected plastic are critical. Many cases require chemical compatibility and durability as well. Depending on the wavelengths, light can be a degradation source for some materials and therefore something to consider for these applications.

For many applications it is possible to integrate components of both groups (micro fluidics and optics) to analyze contents of a micro fluidic channel directly through the optical components.

Surface finish and tolerances are important since small deviations in the feature sizes can affect the results of experiments. Both finishing aspects (finish and tolerance) depend on the quality of the mold and the plastic material used for that specific application. Regarding this point, it is also particularly interesting that one can also change the hydrophobic characteristics of the surfaces (this could be achieved by embossing combinations of small features in the micro nano scale).

Materials Suitable to This Process

Tables 3.25.1 and 3.25.2 describe some common materials used for master fabrication and for hot embossing for different applications.

Materials for Mold/Stamp Fabrication

The main requirements for any material to be used as a substrate for fabricating a mold for hot embossing are: (1) structural stability of the features in the temperature range of use; in other words, the T_g of such material should be considerably higher than that of the thermoplastic intended to conform. This will ensure the durability for many embossing processes. And (2) allow the definition of the minimum feature size by the available state-of-the-art fabrication techniques. Table 3.25.1 presents a summary of the common materials used for fabricating micro features for hot embossing.

Normally mold inserters are presented in wafers or discs of 76 to 152 mm (3 to 6 in.) diameter [typically for prototypes a medium size around 100 mm (~4 in.) in diameter]. But molds can be found in many different shapes, since the overall shape is not a limitation for this process.

Thermoplastic Materials for Hot Embossing

Every application requires its own characteristics. Table 3.25.2 compiles most commonly used thermoplastics for the vast majority of applications. Also, it is important to note that the demolding temperature should be at least 20°C (36°F) below the T_g, but again this is something that should be optimized for each feature size and application.

It can be noted that Tables 3.25.1 and 3.25.2 gather information about the thermal expansion coefficient as it is a critical value to keep in mind for demolding the plastic sheets.

Production Quantities

As already mentioned, hot embossing techniques can cover middle-range quantities, that is, about 100 to 10,000 pieces/batch. However, lower quantities for prototyping can also be addressed by these methods. In fact, these methods shorten the gap between development and commercialization. For higher ranges like $>10^6$ pieces, injection molding would be a better choice. Typical applications mentioned earlier in this chapter are not yet commercialized in very large quantities, making hot embossing an excellent fabrication choice.

For production efficiency hot embossing is not as good as injection molding but still is fair enough. Cycle times range from about 2 to 5 min per piece. So in the case of one piece per mold, users can produce about 100 pieces per day, in an 8-h work day (multiply that by the number of devices that can be produced in a single mold).

On the one hand, cost per piece depends on three main factors: the number of pieces of that series, the feature sizes, and the embossed material. It is important to notice that one mold could be used for different materials but it will only last for a number of cycles. Manufacturing prices are typically on the order of a dollar per embossed piece (or less for larger series). Finished micro fluidic devices (drilled and bonded) are commercially sold for tens of dollars per piece, depending on the complexity of the system (number of layers, inserts, etc.).

On the other hand, mold fabrication costs are highly variable depending on the material and fabrication technique used. The choice of the mold type depends on different factors including the application of the device (resolution, surface roughness, etc.), the size of the fabrication series (durability and materials). In the case of rapid prototyping, it is possible to find molds with an estimated cost of a few hundreds of dollars (taking into account only materials and labor). However, there are some limitations,

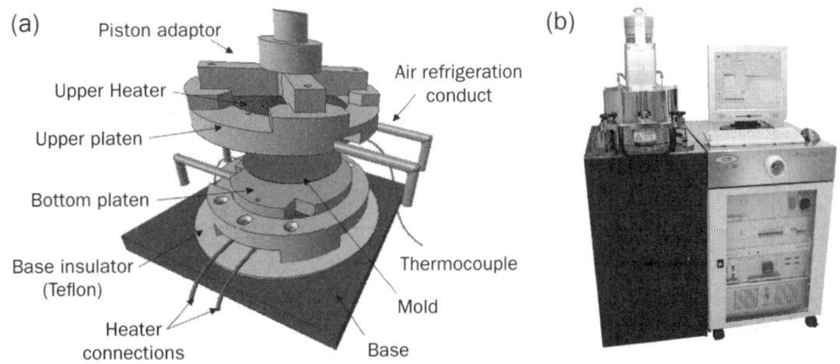

FIGURE 3.25.4 Pictures of embossing equipment (*a*) schematic of a home-designed embosser without vacuum system and (*b*) commercial system EVG® 510HE Semi-automated Hot Embossing System. (Source: EV Group.)

for example, minimum feature sizes of 200 to 300 μm fabricated by nickel electroplating on steel wafers. PDMS molds are also suitable as well for trials with a similar cost estimated to be about a few hundreds of dollars. On the contrary, molds for large series either fabricated in silicon in a clean room or in metal by ultraprecision machining are expensive. Numbers might vary, but we can expect several hundreds to thousands of dollars.

Equipment

Figure 3.25.4 presents a CAD image of a general schematic of a hot press and a commercial machine from the EVG group (Germany). This company presents a product brochure of more automated systems for industrial scale fabrication. Among other similar instruments from other brands, the general specifications are comparable in terms of embossing force, heating and cooling time, cost, etc.

Design Recommendations

Perhaps the great potential of hot embossing is precisely its ability to cover the gap between research and development and large-scale fabrication. It is a very flexible technique, allowing users to cover a great number of applications. Overall there are some general design recommendations that are important to follow for a successful fabrication.

Layout Design

This is the starting point, the presentation of the design of the micro fluidic device or optical micro component. Unless the systems are equipped with vacuum system, it is critical to avoid close features that will definitely trap air between the plastic and the mold preventing the structure replica. Also whenever possible it is preferable to keep all the features at the same size approximately for mold fabrication processes as well as for embossing and demolding.

It is also important to note the differences between positive and negative structures (see Fig. 3.25.5) and to decide which one is more appropriate for the final application. Usually micro fluidic devices will be better achieved using a positive mold and

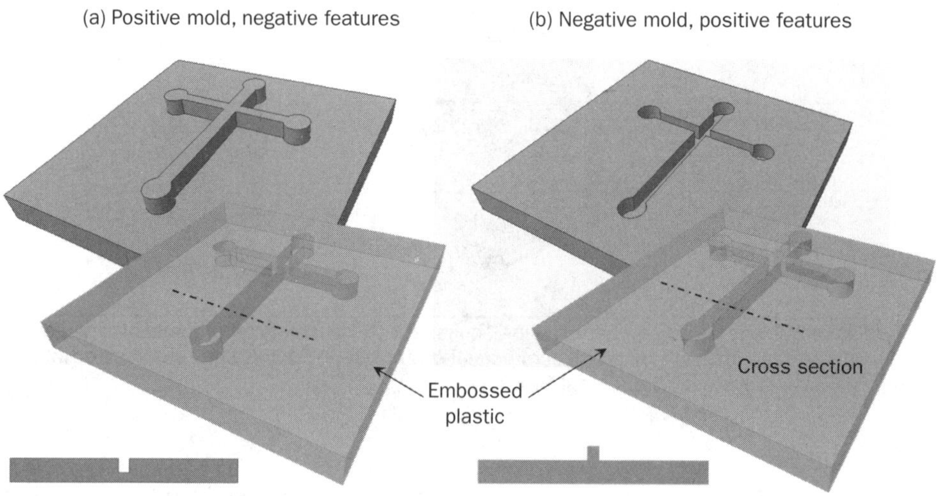

Figure 3.25.5 Differences between positive and negative structures and its consequences on the plastic outcome.

Fabrication Technique	Feature Sizes (μm)*	Height (μm)*	Surface Finishing	Comments
Micro machining	10^2–10^4	10^2–10^3	Roughness of tool	Difficult to process
Dry etching (RIE, plasma)	1–500	10	Silicon grade	High costs
Deep RIE	1–500	600	Silicon grade	High costs
Wet etching	10–500	100	Silicon grade	Isotropic or anisotropic
Electroplating	10–10^3	1–500	Substrate selected	Very well control of height
Casting	10–500	1–200	Uniformity of the resist	Usually performed over Si

*1 μm = 39.73 × 10^{-6} in.

Table 3.25.3 General Characteristics of the More Common Mold Fabrication Methods

presenting the micro channels, trenches, or holes embedded in the plastic substrate. A negative mold requires vacuum in the system and could be more appropriate for freestanding structures, walls, support features, etc.

Process Design

Once the layout is complete it is necessary to select the right material and fabrication parameters, considering the true dimension of the stamp and final piece (according to shrinkage due to α_L).

Table 3.25.3 presents some of the mold fabrication techniques with some rough numbers about precision and capabilities. There are also some other possibilities such as including a draft angle for easy demolding that are not reflected in this table.

It is obvious that increasing the aspect ratio will increase complexity and risk of the process. Also not all the ratios will be appropriate for all materials and there are some limitations caused by demolding risks. Due to the thermal expansion coefficient, there is an increase in the demolding forces that could damage either the piece or the mold (potentially causing elevated costs and delays).

Heat conductivity should not be a problem since embossing is normally processed on sheets of plastics and therefore heat transfer is quite efficient in this configuration.

Process Recommendations

There are some parameters that work nicely. The embossing temperature is probably the most critical value and should be set at least 10°C (18°F) above the T_g. Take into account that the value of T_g will increase when applying pressure and that the temperature control can have 2°C to 3°C (4°F to 5°F) of variation. Embossing force is normally kept between 1 and 2 MPa (145 and 290 psi) but usually below 4 MPa (580 psi). For a disc of 100 mm (~4 in.) these values are achieved with around 10^4 N (2250 lb) of force with an acceptable variation of 2%, which is about 0.1 MPa (15 psi) of uncertainty.

It is recommended that fabricators create a written optimization routine for each thermoplastic and mold and a troubleshooting guide to pass to coworkers. Designers need to have access to the same information and both sets of users need timely awareness of the minute and controllable-uncontrollable variations that affect the product.

CHAPTER 3.26

Roll-to-Roll Micro Embossing

LaRoux K. Gillespie

Manufacturing Consultant
Andover, Kansas

Roll-to-roll (R2R) embossing involves pulling a flexible film or sheet of polymer, paper, or metal through two hard rollers, one of which has a mold that is pressed into the film (Fig. 3.26.1). Under the right conditions, hundreds or thousands of embossed features or parts can be produced each hour. It is one of the lowest cost production processes for micro features and micro parts. With the newly developed flexible ceramics, it may be possible to do the same with that material. While one roller usually does the pressing into the film and the other provides a rolling hard resisting surface into the film, the process can emboss features into both sides by putting molds on both rollers.

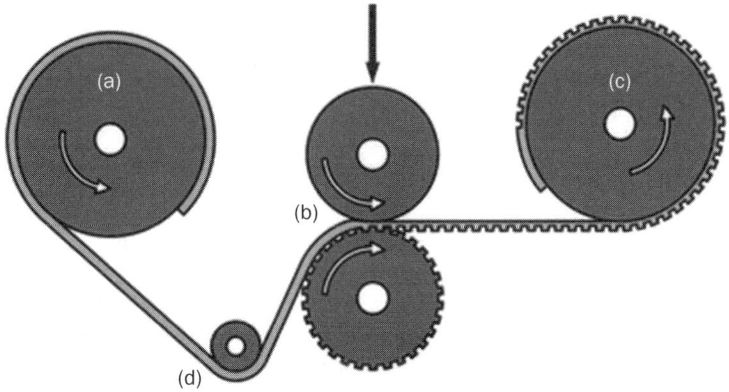

FIGURE 3.26.1 Roll-to-roll hot embossing schematic. (Courtesy VTT Technical Research Centre of Finland Ltd.)

Current applications include the manufacture of micro fluidic chips (individual chip layers), micro mechanical functions, electronic circuits, and potentially liquid crystal displays, solar cells, Fresnel lens, and data-storage devices.

Embossing has been in existence since Roman times if not prior to that. It has been used to make wax seals for official documents as well as for decorations in wallpaper, writing paper, greeting cards, and in its form as coining to make coins, tokens, and medals. Embossing comes in several forms: (1) discontinuous one sheet at a time pressings, (2) reel-to-reel room temperature pressings, (3) reel-to-reel warm pressings, (4) reel-to-reel light cured patterns, and (5) reel-to-reel resist cured metal-plated features. There are benefits to each approach since each approach targets a particular part material.

Just like single-sheet printing in the 1800s micro patterns are being made by pressing a 3D master into a softer material. R2R embossing provides much faster results yielding hundreds of parts and features per hour. Micro fluidic devices in the life sciences are typically thrown away after a single use to prevent cross contamination and as such are ideal candidates for low-cost embossing. These devices are made of silicon, glasses, and polymer foils; R2R embossing is ideal for the polymer versions. The fidelity of mold-to-product transfer depends on several process variables: product preheat temperature, embossing roller temperature, embossing pressure, and directionality of the pattern compared to the material movement direction. While the concept is simple, reaching desired depths of impression and quality of features requires experimentation. Statistical design of experiments (DOE), Taguchi experiments, and response surface methodology (RSM) are some of the fastest and least expensive means for optimizing results.

Replication quality is highly dependent on the pattern direction. As an example if a person takes a 2 × 4 piece of wood and pushes it straight down into soft mud and pulls it straight back out, the depression is a very good impression of the board. If, however, one takes the same board and rolls it into the mud as if it were a paddle on a paddle wheel steamboat and continues the motion until the board is out of the mud, the top of the resulting cavity would be rounded and wider than the bottom, somewhat like the shape of the space between miniature gear teeth (but no micro patterns are as deep as the gear engagement).

By taking a rectangular beam and pressing one end in the mud, then slowly rotating the beam in a longitudinal manner into the mud then out again, the top edges of the mud would be sharp and the channel made would be a very good impression of the board. With any material the vertical rather than parallel orientation will reduce tearing, sliding, and improve edge sharpness. To improve surface qualities, fluorinated coatings on the mold can be used.

Thermal Processes

Polyvinyl chloride (PVC), polymethyl methacrylate (PMMA), polyethylene terephthalate (PET), polypropylene (PP), cyclic olefin polymers (COP), and polycarbonate (PC) films are used in the thermal micro embossing processes. PMMA is transparent, making it a good choice in place of glass, and it is the most used polymer for thermal micro embossing. Generally, the temperatures involved are 50°C to 100°C (122-244°F) above the glass transition temperature T_g. The pressure roller will be the warmest one, and the embossing roller will only be 20°C to 120°C (68 to 248°F). The material thickness is generally several times that of the deformed features, for example, a 50-μm-deep (0.002 in.) channel in 325-μm (0.013 in.) polymer foil.

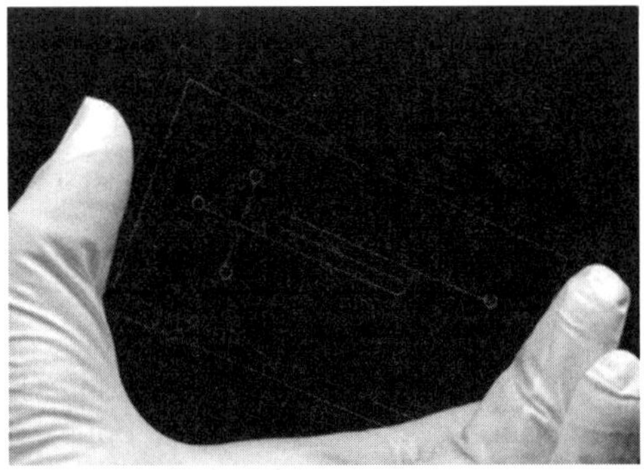

FIGURE 3.26.2 Hot embossed micro fluidic product. (Courtesy VTT Technical Research Centre of Finland Ltd.)

Increasing the pressure, increasing the temperature, and holding the mold (also called the stamp) longer on the part increases the percentage of filling into mold cavities.

Pulling the stamp out of the material can cause several surface defects, some of them based on contraction of the material after leaving the hot stamp. Putting stress-absorbing protrusions on the stamp away from the critical areas, and letting the protrusions take most of the contraction forces reduces impact on the critical features. Keeping demolding temperatures close to those of the embossing process also helps reduce damage to surfaces and features. The temperature choices need to be made in a careful relationship that depends on material and features involved.[1]

For thermally embossed polymers, it is not uncommon that the achieved channel or pattern depths are only 85% of the mold master protrusion. Channel widths as a ration of desired depth are critical with aspect ratios (depth/width) of 1 being normally close to the maximum that are achieved. Typical channel depths produced in this process range from 10 to 200 μm (0.0004 to 0.0079 in.). For example, 125- to 130-μm-thick (0.0049-in. to 0.0051-in.) PMMA and COC foils are used to make micro reactors with channels of 100 μm wide by 30 μm deep (0.0039 in. by 0.0012 in.). Figure 3.26.2 illustrates a typical product from this process.

Cold Embossing

Adding thermal energy to the process improves performance for many materials, but embossing cold eliminates the heating times and energy as well as the expansion and contraction of the film used. The lack of thermal effects makes cold embossing of micro structures more accurate. Cold embossing of photosensitive organic-inorganic sol-gel materials gives results similar to glass.

UV Resist-Based Fabrication

Most R2R embossing produces micro features, that is, features whose depth is in the order of 10 to 200 μm (0.0004 to 0.0079 in.). UV R2R embossing has achieved holes only

[1]Hayden Taylor's PhD thesis provides a wealth of information on both experimental and simulation process effects.

50 nm (2 µin.) in diameter over areas up to 120 cm^2 (18.6 $in.^2$) at rates of 10 m/min (400 in./min.) in PMMA. These are formed on silicon substrates with minor shrinkage. Features with the latter nanometer depths are typically referred to as nano imprinting or nano embossing rather than micro embossing.

The advantages of UV embossing include using lower temperatures, low embossing pressure, and high process speeds.

In UV R2R embossing, a carrier film is coated with a liquid resin resist material. The master is impressed into the resist coating, which then flows up into the mold cavities (since it is still liquid); the tensioning forces pull the film away from the roller as it rotates away, then UV radiation cures the resist, making a hard product. Resin viscosity, surface tension, UV curing speed, mechanical and optical properties after curing define the resin to be used. Viscosity affects the coating thickness and uniformity and surface tension affects adhesion to the polymer film. Epoxy resins are one choice for the resist. The coating thickness is normally in the range of 20 to 200 µm (0.00079 to 0.0079 in.). In addition to the resin choices, film speed, and pressing pressure, key issues are slot widths or feature dimensions. While the coating may be only 75 µm (0.003 in.) thick, the pressure on the protruding portions of the mold force the liquid up into the cavities to depths greater than the film thickness [can be as much as 300 µm (0.0118 in.) deep, but normally less].

Curing energies are an exponential function of coating thickness. Doubling the thickness of a coating requires 10 times the UV intensity. By choosing a resin with the appropriate chemical structure, the mechanical properties of the UV crosslinked resin (brittleness and hardness, for example) can be modified. Resins can be formulated to give specific optical properties as well. UV polymers shrink by a few percent as part of the curing cycle.

Polymethyl methacrylate and PET are the two most commonly used films for R2R UV fabrication. Both are optically transparent; PET is impact resistant while PMMA is softer, easily scratched, and will swell in organic solvents. Since they are transparent, the UV light can be focused from the bottom, non-imprinted side, while the film is still in contact with the mold. These films can be obtained in widths of 500 mm (20 in.) allowing several parts to be made side by side.

The roller molds can be made of metal (copper-based rolls are used) or polymers. Flexibility of the mold is a desirable property for very thin molds. A UV reactive polymer such as polyurethane (PU) provides good properties for a flexible mold. A commercial fluorocopolymer, ethylene-tetrafluoroethylene (ETFE) has also been used to make the master roller pattern. Etching a copper foil that is prelaminated on a liquid crystal polymer film has also been used. Photolithography is used to provide the pattern followed by etching similar to printed circuit fabrication. To fold around the roller uniformly the copper foil may only be 18 µm (0.0007 in.) thick. Double-backed adhesive tape has been used to adhere the mold to the roller since the mold is so flexible. More positive means are used as well.

When features on the final product need to have metal surfaces, metal can be coated on the hardened imprinted web of material.

For one application, a fast thermal curing PDMS that cross links in a few seconds at 120°C (248°F) works well. For another, a UV curable low-viscosity liquid epoxy silicone is desirable since it can be used at room temperature and low pressures. Feed rates of the film are in the order of 2 m/min (79 in./min).

The submicrometer mold masters require state-of-the-art lithography and dry etching techniques which limit their availability for many applications, but the deeper and larger patterns on molds require more conventional manufacturing.

The Fraunhofer-Institut fur Biomedizinische Technik (IBMT) has a discontinuous R2R hot embossing process for daily volume production of large area foils at rates of up to 10km/day. The structures produced have features ranging from 50 nm to 200 μm (0.000002 to 0.0079 in.) deep. This process uses 0.3-m-wide (118–in.) foils of cyclic olefin copolymer (COC), PC, polystyrene (PS), or thermoplastic elastomers such as PU.

For all variations of the process, production is more efficient when the aspect ratio of the features is less than 1.0. Most of the issues explained in more detail in Chap. 3.25 apply to R2R micro embossing.

Equipment

Commercial machines exist for micrometer scale features, and at least one company claims the ability for nanoscale R2R embossing. Several facilities have performed nanoscale embossing with equipment of their own design and an international organization exists to further develop capabilities. Aspect ratios for the nano imprinting are currently limited in R2R to about 1.

Several references are cited below for additional details of this process.

References

Dumond, J. J., K. A. Mahabadi, Y. S. Yee, C. Tan, J. Y. H. Fuh, H. P. Lee, and H. Y. Low, High Resolution UV Roll-to-Roll Nanoimprinting of Resin Moulds and Subsequent Replication via Thermal Nanoimprinting Lithography, *Nanotechnology*, vol. 23, no. 48, 2012.

Shift, H., Roll Embossing and Roll Imprint, Chapter 5 in *Science and New Technology in Nanoimprint*, Volume editor Y. Hirai. Frontier Publishing Co., Ltd., Japan, 2006, ISBN4-902410-09-5, English edition pp. 74–89.

Taylor, H. K., Modeling and Controlling Topographical Nonuniformity in Thermoplastic Micro- and Nano-Embossing, PhD Dissertation, Massachusetts Institute of Technology, 2009.

Zhang, J., M. Sahli, J-C Gelin, and T. Barrière, Roll Manufacturing of Polymer Microfluidic Devices Using a Roll Embossing Process, *Sensors and Actuators A: Physical*, vol. 230, July 1, 2015, pp. 156–169.

CHAPTER 3.27

Laser-Assisted Micro Fabrication

LaRoux K. Gillespie

Manufacturing Consultant
Andover, Kansas

Several of the micro fabrication processes can be enhanced with lasers. The laser does some of the work, while facilitating the other basic micro process. Thus while some processes cannot provide the desired size or depth or performance, laser actions can make that process a viable solution. Several of the basic processes are described in earlier chapters of this handbook and readers are referred to those chapters for more detail. This chapter is designed to provide an overview of what can be done at the micro level. With the exception of the laser CVD processes, use of these processes requires the user to add the laser to the exiting process equipment. Lasers are readily available for the CVD processes. Among the laser-assisted processes are:

- Laser-assisted milling
- Laser-assisted turning
- Laser-assisted micro grinding
- Super plastic forming (heating)
- Super plastic forming (plasma pulse)
- Laser-assisted deep drawing
- Laser-assisted hot embossing
- Laser chemical vapor deposition
- Laser chemical liquid deposition
- Pulsed laser deposition
- Laser-enhanced electroplating
- Laser-based combined annealing and texturing

- Laser fusion 3D printing
- Laser chemical etching
- Laser-assisted ablation + printing (glass + polymer)

Laser-Assisted Cutting and Grinding

Focusing the laser directly ahead of milling cutters softens the metal reducing the cutting forces and increasing the tool life. It has been effective on hard metals (HRC 62 for example). It also has been effective on some ceramics similarly reducing forces that break micro cutters, as well as reducing the cracks that occur in parts when machining without heating. The fact that the material is more ductile after irradiation means that if the laser spot size falls outside the cutter diameter path, burrs raised by the cutter will be larger than if the material had not been softened. In addition to a tempered region, a heat-affected zone (HAZ) exists around the laser path.

While the process is typically applied to micro milling and turning with conventional cutters, it has also been combined with ion beam machining for ultraprecision lithium, niobate micro resonators (Xu, 2015). The latter, as noted in Chap. 3.48, is for features measured in nanometers or low micrometer sizes and precision.

The process is also used to assist grinding. The laser heats the area ahead of the grinding wheel, reducing the temperature gradient and surface layer damage. This provides higher stock removal rates. The laser can also be used to condition resin-bonded CBN grinding wheels while in work.

Laser-Assisted Forming

As noted in the previous list, there are two forms of laser forming. The use of lasers for heating portions of a part is the most used approach. Chapter 1.11 provides more detail on the simple aspects of this process but its applications are more amenable to micro work than larger work. Incremental laser forming is described there.

For more challenging micro applications, super plastic forming is possible when the part is heated to high temperatures. It is even viable on materials such as metallic glass. By elevating the temperature of a product to half its melting temperature, the material can be twisted, bent, turned, or similarly modified up to 1000%. It requires materials having small grains [less than 10 μm (0.0004 in.)] and slow strain rates (10^{-4} to 10^{-2}). Micro size parts are ideal for this process since the laser can provide the heat required quickly and focused exactly where required (the laser beam spot comes close to exactly matching the micro feature size). Macro size parts, in contrast, require much more energy which is dissipated to areas where it is not needed.

Hemispheres of bulk metallic glass (BMG) have been formed by blow molding using the laser to heat the material to its plastic temperature. In this application, 6-μm-thick (0.0002-in.) strips of BMG have been formed into watch-like coiled springs (Schroers, 2007) and 4-μm-wide (0.00016-in.) gear teeth. Silicon tooling used to form the BMG strip is totally destroyed in the process, however. Tooling for other applications may not necessarily be destroyed with each use.

At the nano and micro levels, the process can make 1-μm-wide (39-μin.) grooves and half micrometer (20 μin.) high-raised lines in BMG. The effect is to make a fast shaping technique for elastic and ductile materials (polymers and metals) as well as such brittle materials as silicon, silicon dioxide, and ZnO. While it is an emerging micro field, the metallurgy behind it is known.

The plasma pulse variation of laser super plastic forming relies on a plasma pulse to rapidly move the sheet of material just as a punch would do. It does not use any active tooling—just the sudden pulse.

Laser-Assisted Deep Drawing

Deep drawing converts a flat sheet blank into a hollow shape by forcing it down into a die. The thickness of the blank stays essentially the same as it was in the blank. The bottom of the punch provides the force on the blank. This force causes a tensile stress in the sheet metal. There is however a portion of the material where a compressive stress exists. Heating the tensile-stressed area does not help the process. Local heating of the compressive stressed area does help, by influencing smoother material flow. Focusing the laser at the right spots increases the amount of drawing that can be accommodated.

The limits of deep drawing at micro levels of dimensions and foil thickness are still being studied, but as noted in Chap. 3.24 micro cups only 0.4 mm (0.016 in.) in diameter have been drawn without use of lasers.

Laser-Assisted Hot Embossing

Hot embossing is a common process for micro polymer parts. It suffers, however, with rather lengthy cool-down times. The high power density and beam focusing of the laser provide quick *localized* surface heating. This, in turn, provides faster cool-down periods. In the work done to date, parts and mold have to be transparent or opaque to the laser's wavelength. Special additives can be added to the thermoplastic so that, although visually transparent, it becomes opaque to a laser. It is also possible to make a black thermoplastic transparent to a laser beam.

With appropriate process settings a 3-s operation can produce 400-μm-deep (0.0157-in.) channels in 3 to 20 s depending on the degree of opaqueness in the sample and mold. These channels have been made in semicrystalline polymer, polyethylene (PE), and an amorphous polymer, PMMA; finite-element simulations can readily predict results.

Laser-assisted hot embossing can produce arrays of micro pillars of 20 μm (0.0008 in.) or larger diameter in PMMA. It provides a different solution for hot embossing when faster production rates are necessary. Chapter 3.25 provides a more complete summary of hot embossing.

Laser Chemical Vapor Deposition

Chemical vapor deposition is discussed in Chap. 3.29. The use of a laser to heat the part focuses the energy locally which pinpoints deposition better than whole part heating. There are two variations of the process. The first is *thermally activated CVD*, and the second *laser photochemical CVD* in which the laser concentrates a large quantity of reactive species at the substrate (part to be coated). In the latter case the laser excites the reacting molecules which then settle on the part. In the first approach the laser can write lines which become gold or other micro metallic electronic circuits.

By repeatedly writing over previous depositions, three-dimensional structures can be made. In that sense the process becomes an additive process like other rapid prototyping approaches. Fibers can be made by moving the substrate away parallel to the laser beam axis at a rate equal to the deposition rate of the fiber. Coiled springs can be made by this latter approach as well as fibrous scaffolds, or fiber-reinforced laminar composites. Such composites have not yet been made but are a logical product from the process.

The process to build refractory metal structures this way is in work today. The output is components made from rhenium, tantalum, niobium, tungsten, and their alloys with complex internal features provided by other additive processes.

Pulsed Laser Deposition

Pulsed laser deposition (PLD) is a physical vapor deposition (PVD) process in which the laser is focused on a target of the material that is to be deposited. The material is vaporized in a plasma plume and deposits as a thin film on the part which faces the target. The process works in a vacuum, or in a gas such as oxygen. Oxygen helps deposit oxides on the part.

Laser Chemical Etching

Femtosecond lasers are being combined with chemical etching to accelerate the fabrication of micro tunnels in photoetchable glass. The primary goal is to make channels of any configuration for lab-on-a-chip diagnostic systems. The laser pattern developed on the silica-based chips accelerates the etching rate. Tunnel patterns 5 mm (0.197 μin.) long are made by scanning at a rate of 100 μm/s (0.0039 in./s). Ten lines spaced 3 μm (0.0001 in.) apart are scanned in this time.

Lasers have also been used to produce micrometer size lines in metals and sapphire. Each material type requires a different etch chemical. Hydrogen fluoride is used on the metallic glasses while potassium hydroxide and phosphoric acid have been used on Co, Cr, Cu, and Pt. In all cases the etch rate (as opposed to the laser scan rate) is measured in micrometers/hour.

Titanium micro channels having aspect ratios up to 16 have been etched to a width of 10 to 30 μm (0.00039 to 0.0012 in.) via this process. The cross-sectional profile ranges from V shape to U shape depending on laser energy.

Laser-Enhanced Electroplating

Very high-plating rates occur when electrodes are illuminated by a laser beam. Experimental electroplating, electroless plating, and etching techniques that use a focused laser to define the localized plating indicate that the plating rates can be 1000 or 10,000 times faster than conventional electroplating. Gold, nickel, and copper electroplating have been studied for micrometer-sized circuit personalization and repair.

Laser-Based Combined Annealing and Texturing

Surface texturing is important in both metal and bio applications. Laser texturing is a relatively rapid way to add a variety of textures to surfaces (see Chap. 1.3 for a discussion of this process). For micro parts the time required can be miniscule and if needed the laser can provide some annealing heat to the same part as part of a single operation.

3D Printing Laser Finishing

Fused filament fabrication (FFF) 3D printing has been combined with femtosecond direct laser writing (DLW) to make biodegradable scaffolds. In this instance the process is actually a two-step process rather than the combined processes working at the same time. While not directly the target of this chapter, this combination does hint at the impact of these hybrid processes. In this instance, the scaffolding was begun with a $1000 printer. (The femtosecond laser, however, is much more expensive.)

Laser-Assisted Ablation + Printing

Femtosecond lasers have been used to scribe glass. This was then followed with an etch and then polymers were added to the etched area. Again while not performed at the same time as the other processes cited here, the laser becomes a part of the process. It is focused as a hybrid method for making functional glass/polymer composite 3D biochips.

Chapter 1.11, while directed to conventional size parts, provides more details on some additional laser-assisted processes that apply to micro features. The following references provide additional details for laser assisted micro processes.

References

Alemohammad, H., O. Aminfar, and E. Toyserkani, Morphology and microstructure analysis of nano-silver thin films deposited by laser-assisted maskless microdeposition, *Journal of Micromechanics and Microengineering*, 2008, p. 115015.

Bellouard, Y., A. Said, M. Dugan, and P. Bado, Fabrication of high-aspect ratio, microfluidic channels and tunnels using femtosecond laser pulses and chemical etching, *Optics Express*, 2004, vol. 12, no. 10, pp. 2120–2129 (doi: 10.1364/OPEX.12.002120).

Erween R., N. Warap, and Z. Mohid, Thermal-assisted machining of nickel-based alloy, in *Superalloys*, Dr. Mahmood Aliofkhazraei (Ed.), InTech, 2015 (doi:10.5772/61101).

Feinäugle, M., B. Mills, D. J. Heath, C. L. Sones, and R. W. Eason, Laser-assisted transfer for rapid additive micro-fabrication of electronic devices, in *EPSRC: Manufacturing the Future*, September 2014, Glasgow, UK.

Gelchinski, M. H., R. J. von Gutfield, and L. T. Romankiw, Laser enhanced electroplating, in *Interfaces Under Laser Irradiation*, L. D. Laude, Dieter W. Bäuerle, and M. Wautelet (Eds.), Springer, 2013, pp. 349–358.

Krassenstein, E., Ultimaker 3D printer + laser microfabrication + stem cells = Biodegradable hard tissue implants, *3D Print.com*, October 1, 2014.

Lee, C.-M., W.-S. Woo, D.-H. Kim, W.-J. Oh, and N.-S. Oh, Laser-assisted hybrid processes: A review, *International Journal of Precision Engineering and Manufacturing*, 2016, vol. 17, no. 2, pp. 257–267.

Lu, C., Development of novel micro-embossing methods and microfluidic designs for biomedical applications, PhD Dissertation, Ohio State University, 2006.

Lu, C., Yi-Je Juang, L. James Lee, D. Grewell, A. Benatar, Analysis of laser/IR-assisted microembossing, *Polymer Engineering & Science*, 2005, vol. 45, no. 5, pp. 661–668.

Melkote, S., M. Kumar, F. Hashimoto, and G. Lahoti, Laser assisted micro-milling of hard-to-machine materials, *CIRP Annals—Manufacturing Technology*, 2009, vol. 58, no. 1, pp. 45–48.

Nowak, R., S. Metev, and G. Sepold, Laser-assisted wet chemical etching of metals for microfabrication, *Proceedings of SPIE, 2207, Laser Materials Processing: Industrial and Microelectronics Applications,* 1994, 633 (doi:10.1117/12.184772).

Schroers, J., Q. Pham, and A. Desai, Thermoplastic forming of bulk metallic glass—A technology for MEMS and microstructure fabrication, *Journal of MEMS*, 2007, vol. 16, no. 2.

Shin, Y. S., and S. H. Jeong, Laser-assisted etching of titanium foil in phosphoric acid for direct fabrication of microstructures, *Journal of Laser Applications*, 2003, vol. 15, pp. 240–244.

Son, S. W., M. K. Lee, K. H. Oh, and S. H. Jeong, Fabrication of titanium microchannels using laser-assisted thermochemical wet etching, *Journal of Laser Applications*, 2006, vol. 18, no. 2, pp. 131–137 (doi:http://dx.doi.org/10.2351/1.2164473).

Terrill, R. E., K. H. Church, and M. Moon, Laser chemical vapor deposition for microelectronics production, *Aerospace Conference, 1998 IEEE*, Snowmass at Aspen, CO, 1998, vol. 1, pp. 377–382 (doi:10.1109/AERO.1998.686934).

von Gutfeld, R. J., L. T. Romankiw, and R. E. Acosta, Laser-enhanced plating and etching: Mechanisms and applications, *IBM Journal of Research and Development*, March 1982, vol. 26, no. 2, pp. 136–144 (doi:10.1147/rd.262.0136).

Wu, D., S.-Z. Wu, J. Xu, L.-G. Niu, K. Midorikawa, and K. Sugioka, Hybrid femtosecond laser microfabrication to achieve true 3D glass/polymer composite biochips with multiscale features and high performance: The concept of ship-in-a-bottle biochip, *Laser & Photonics Reviews*, 2014, vol. 8, no. 3, pp. 458–467.

Xiong W., Y. S. Zhou, X. N. Xiangnan He, Y. Gao, M. Mahjouri-Samani, T. Baldacchini, and Y. F. Lu, Three-dimensional micro/nano-fabrication by integration of additive and subtractive femtosecond-laser direct writing processes, *Proceedings of Nanomanufacturing Conference, 31st International Congress on Applications of Lasers and Electro-Optics* (ICALEO 2012), September 2012, pp. 23–27, Anaheim, CA.

Xu, Y., J. Lin, Z. Fang, M. Wang, J. Song, N. Wang, L. Qiao, W. Fang, and Ya Cheng, Fabrication of high-Q lithium niobate microresonators using femtosecond laser micromachining, *Scientific Reports*, 2015, vol. 5, no. 8072 (doi:10.1038/srep08072).

CHAPTER 3.28

Micro Extrusion

Simone Maccagnan

Gimac Microextruders
Castronno (VA), Lombardia, Italy

Process

The extrusion process forces the raw material through a die, and the die opening design will have a size proportional to the finished desired one, while the shape will derive from the desired final shape, corrected on the base of shrinkages/deformations. It is normally a continuous process producing long lengths of the item with constant cross section. While parts are extruded and cooled down continuously, parts can also be cut at final length or subject to post-processing. Post-processing includes being recut to desired length, etched, coated, chamfered, flared, skived, or otherwise improved before cutting to length. Up to thousands of small parts can be produced in minutes in this manner and thus it is one of the lowest cost processes where high production or unique shapes are required.

Even if frequently underestimated, the complexity of a continuous process of extrusion is higher than injection molding; this is because a change in viscosity of raw material would influence dimensions, something that would not have happened in a closed and pressurized volume like a mold.

Micro extrusion is frequently used for manufacturing small parts, but the original purpose of micro extrusion process is to deliver, independently from the size and shape, products in which micro structural properties are preserved and conditioned and it is these properties that allow the production of small parts where walls are thin and flows are low.

In order to select properly the processing equipment or the processing partner is of the essence to verify that it does not fall short on the below listed aspects.

Processing quality and traceability depend on equipment, controls, cleaning, and people, but proper size of extruder, optimal design of flow channels, cleanness are also critical for producing well-performing small parts. For micro parts in particular, edges in the flow channels, dirty tools, oversized extruder, gear-pumps, etc. will always produce unstable, untraceable, and random quality.

When parts made out of soft polymers like TPU, soft PEBA, EVA, etc. are collected on spools or in bundles, adhesion between spires or parts can be a critical issue, so special care has to be taken. When parts are small, in fact, stickiness, tackiness, and interaction between them constitute a critical issue, as during handling the separation force could be higher than can be structurally tolerated by parts. Handling could lead parts to be stretched, torn apart, or damaged on the surface, especially if parts are multilayer, thin walled, and multi-lumen (cross section has several openings). While preventing melt overheating and finding proper process temperatures in general are manufacturing issues, designers need to be aware of their impact on part quality and if necessary control them with design. Critical material may need to be placed on specifically treated surfaces and controlled further in packaging and shipment.

Processing Equipment

A micro extruder system is composed of the elements shown in Table 3.28.1.

Gear-pumps should always be avoided as they introduce edges (in which melt slows down and eventually burn), residence time, pulsatility, difficult cleaning, extra unjustified costs, and other issues. Design of flow channels is key to limit thermal and mechanical stress into the melt. Excessive stress during extrusion can lead to polymer degradation and loss of performances; each polymer has a different viscoelastic behavior not only in solid, but also in molten phase. Optimal results result from a specifically designed set of screw, die-head, and sizing tools.

Machine	Function	Role in Performance
Dehumidification system	Remove moisture from polymer	Minimize polymer hydrolysis and bubbling
De-dusting	Remove powders and particles from pellets	Reduce gels and impurities
Extruder	To melt polymer/pump molten polymer	Flow precision/molecular weight preservation
Die-head	Convey molten polymer	Melt orientation/molecular weight preservation/avoid flow-lines/ form cross section/control surface roughness
Melt sizing tools	Size melt before solidification	Compensate shrinkage/size cross section/mechanical properties
Vacuum sizing	Form precisely the melt while cooling	Size precision
Cooling	Solidify polymer/control cooling gradient	Physical properties (mechanical, optical, etc.)
Online measurement	Assess measure/feedback to process	Dimensional precision/quality assessment and reporting
Pulling system	Stabilize speed/control twist	Dimensional stability/control of twist
Cutter	Determine correct length	Length precision/avoid debris
Blow-off	Select parts/collect parts	Lot quality/segregate defective parts
Off-line measurement	Select parts/collect parts	Lot quality/understand defect's root cause

Table 3.28.1 Elements of an Extrusion System

Material	Common Smaller End of Diameters	Comments
Plastics		
	0.010 mm (0.0004 in.) OD	Tubing[1]
	0.007 mm (0.00028 in.) ID	Tubing[1]
	0.100 mm (0.004 in.) OD	Multi-lumen[1]
Metals		
Aluminum	0.46 mm (0.018 in.) walls; 1.2 mm (0.060 in.) ID holes	
Ceramics[2]		
YTPZ Zirconia	0.100 mm (0.004 in.) rods	1 μm (0.000040 in.) grains
Alumina (96–99.9%)	0.100 mm (0.004 in.) rods	6 μm (0.000240 in.) grains

[1]Wall thicknesses thinner than 0.05 mm (0.002 in.) and tolerances of less than ±0.01 mm (±0.0004 in.) can be held.

[2]One manufacturer notes that with straight rods or cylinders of these materials the company can maintain ±0.090 mm (0.0035 in.) tolerances on these diameters in the as fired condition and with secondary machining it is possible to hold ±0.001 mm (0.00004 in.) tolerances. For hollow tubes in these materials, internal diameters can be held as small as 0.080 mm (0.003 in.) with ±0.001 mm (0.00004 in.) tolerance depending on geometry. Wall thicknesses can be held to 0.125 mm (0.005 in.).

Table 3.28.2 Production Part Sizes Made by Extrusions

Micro Extruded Sizes

While researchers continue to refine the minuteness that extrusion can produce, Table 3.28.2 highlights some of the commercial sizes that have been produced.

The large grain size of aluminum (and some other metals) is a major impediment to its extrusion miniaturization. New advances are providing smaller grains, but in general it is not possible to approach the smallness of plastic and ceramic extrusions with the same process.

Shapes

Extrusions, whether micro or macro are often round, but square, rectangular, oval, and other shapes are also readily produced. Miniature gears are also extruded so the shapes do not have to be "normal" or simple geometries. Multi-lumen shapes have two or more shaped holes in them and may have unusual but symmetrical outer configurations. Nonsymmetrical shapes could be more complex to produce because frictional and shrinkage forces can warp, tear, or add undesirable subsurface residual stresses and internal shear bands.

Materials

In principle micro extrusion has no limitations in terms of material, but this does not mean that every material has reached the same level of technology development. Today micro parts are extruded from metals, thermoplastic and non-thermoplastic polymers, and ceramics. That makes it one of the most versatile processes for simple shapes.

While technology may not allow normal direct micro extrusion of very high melting temperature materials (e.g., steel, titanium, or copper), it is possible to mix the desired material with a thermoplastic sacrificial carrier, where the carrier will allow the metal particles to flow even though not molten. After being extruded, the part can be exposed to a thermal or thermo-chemical treatment designed to limit deformation and return a fully metallic or ceramic part. This process has been successfully used for the micro extrusion in the form of complex cross sections of stainless steel and ceramic material.

In principle it is also possible to perform direct extrusion of metals and ceramics. Making complex cross sections though, especially if co-extruded, can be prohibitively costly, because of the need to use a much higher softening temperature material than the ones commonly used for building tools. Co-extrusion can be limited also by the different shrinkage rate and volume of involved materials when exposed to extremely high temperature as well as during cooling phase.

While extrusion generally is thought of for manufacturing piece parts at some stage, the process also can produce fibers from molten basalt, tungsten, glass, carbon, and even composites. Long micro fibers for medical applications are also being produced by extrusion. While electrospinning (see Chap. 3.35) and related processes are widely used for nanometric fibers, extrusion allows manufacturers to co-extrude in the same cross section a variety of structural and medical needs (drug delivery for example) into a single process and product.

In principle, most of plastic materials or compound can be processed through micro extrusion and in general any thermoplastic material having a processing temperature lower than degradation temperature is relatively easy to process through micro extrusion. However different forms of raw material (pastes, powders, pellets, flocks, etc.) may need different equipment. Pastes, for example, obviously do not feed in the same way pellets do, but powders also feed in a very different way than pellets. Some pilot scale or experimental grade materials may be delivered in critical forms (like flocks, sticky powders, or irregular chunks). In these last mentioned situations properly customized equipment is required so that process is reproducible and stable. Pellets are the more known and accessible of the available plastic forms.

When parts are small and the target is not a polymeric part, but a metallic or ceramic part, several additional factors become extremely important. The manufacturer must control impurities, humidity, melting temperature, gas generation, as well as cross linking, shear, and time at temperature.

Because the "perfect material" doesn't exist, and there is not a polymer for every application, the search of a polymer can be a challenging task. But co-extrusion allows a combination of different polymers for a functionalized cross section. For micro co-extrusion, the influence of the above parameters is particularly critical, but it will enable the possibility to produce parts that could not be made in any other way.

Table 3.28.3 defines the more commonly utilized polymers.

The word "polymer" is frequently associated with thermoplastic polymers, but some *non-thermoplastic polymers*, sometimes having a degradation temperature lower than processing temperature, offer a lot though they need special equipment and processes that cannot be shared with any other material. PTFE, HFP, PMVE, and silicone, for example, even though they need special processing technologies, can offer unique features; the low friction of PTFE does not have competitors yet; and the elastic properties of

Family	Material	Common Raw Material Form	Positive Aspects	Negative Aspects
Common thermoplastic polymers	PP, PE, PVC, PS, etc.	Pellets, powder, precursors	Cost, availability literature	Performance
Technical polymers	PA, PU, PEBA, PET, PBT, ABS, PMMA, PC, PVDF, etc.	Pellets, precursors	Performance/ cost ease of processing	Pretreatment (Dehumidification)
Performance polymers	PPS, PSU, PEEK, PEKK, PI, PEI, PES, LCP, etc.	Pellets	Heat resistance and mechanical properties	Cost, cleanability
Fluoropolymers	PVDF, FEP, PFA, ETFE, THV, etc.	Pellets	Chemical and heat stability	Cost, special metallurgy
Silicone	Silicones	Liquid, paste	Elastic flexibility at high temperatures	Preservation, special process
Thermosetting polymers	PTFE, UHMWPE	Powder	Wetting (contact angle), low friction	Preparation, special process

Table 3.28.3 Commonly Utilized Polymers

silicone at high temperatures is also unique. These two materials need special processing and preservation techniques. Preparation, processing, post-treatments can be particularly critical, and several companies specialize in processing just these materials.

Several industry online tools such as the following or polymer vendors' websites can help in identifying the polymer with the right properties:

- http://www.campusplastics.com
- http://www.matweb.com
- http://polymer.nims.go.jp
- http://www.polymersdatabase.com

In choosing the best polymer, it is always important to involve process engineers, material engineers, and operators in order to avoid unpleasant surprises during prototyping or manufacturing, like discovering too late that a polymer does not have enough melt strength to withstand thin walls and complex sections or that a polymer theoretically ideal is practically manufactured with broad lot-to-lot variability.

Raw material quality is of the essence on final product quality, but when items are produced in small amounts, it can be difficult to get consistent quality and quick delivery of the raw materials. This requires good planning and discussions with raw material vendors. Different batches of raw material can vary in quality, so it may be important to buy big enough lots of polymers from the same batch. Some polymers are subject to aging, especially those with significant lot-to-lot-variability so storage can be an issue. Establishing accurate prescreening and verification of raw material at delivery may be a

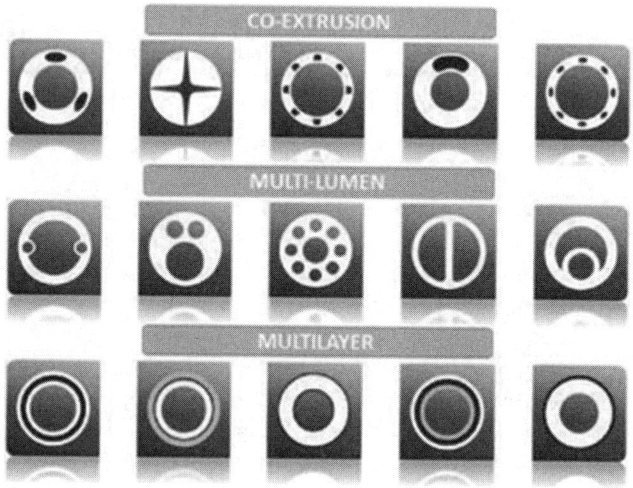

FIGURE 3.28.1 Cross-sectional configurations in extrusion.

more applicable approach for conditioning features. Molecular weight measurement by GPC (gel permeation chromatography is an analytical process that separates analytes on the basis of size or in other words gives information about molecule sizes), viscosity (by capillary rheometer), and fluidity distribution by filter tests are examples. Some of those tests may need expensive and complex to use equipment, so it is often preferable to have an agreement with raw material vendor or with an independent laboratory to deliver prescreened material.

Product Cross Sections

Cross sections (sections perpendicular to the length of the extruded part) can be divided into three main families as shown in Fig. 3.28.1 and described here.

1. Co-extrusion: the whole cross section is divided in portions, each of which is made of a different material that may have specific properties, such as color, radiopacity, flexibility, electrical properties, etc.
2. Multi-lumen: multiple channels (lumens) running parallel to each other in a straight or twisted configuration, characterize the cross section allowing the passage of fluids or devices.
3. Multilayer: a specific kind of co-extrusion in which multiple layers have different properties on each surface and through the wall thickness.

The above cross sections can be then combined with other subsequent or post-processing to produce a composite or a more complex part. Examples include post-co-extrusion, and jacketing of the first extrusion. Cross sections of a semifinished part can be introduced in the melt flow, resulting in a final part composed of different materials that cannot be extruded in a single operation because they need very different transforming process.

By developing composite material cross sections, one can achieve extreme performance and compensate for deficiencies of a specific material. The integration of properties fulfills the demanded properties while if taken alone the materials couldn't do the whole job expected from the part. The composite approach can also take advantage of changing longitudinal sections.

Longitudinal Sections

Possible longitudinal sections (sections parallel to the length of the extruded part) are divided into four main families:

- Tapered: The dimensions of the part changes along the length in a controlled and independent way, for example, a tube could have an external diameter that changes while the internal diameter does not or vice versa.
- Bump: A specific type of tapered in which dimension changes in a natural way as an indirect result of a process change, for example, a tube that changes its outer and inner diameters as a consequence of a change on pulling speed or inflation pressure.
- Totally intermittent: Also called interrupted extrusion is a process in which constituent material switches totally between a first and a second, generating a change in stiffness, radiopacity, hydrophilicity, color, or other features.
- Variable thickness extrusion: Also called intermittent layers extrusion (ILE) is a multilayer structure, the external and the internal materials remain the same, but the percentage occupied over the whole wall thickness changes along the tube (Fig. 3.28.2).

Multi-lumen parts which change stiffness along their length are typically only economically possible by extrusion. Dynamic extrusion allows users to have alternate hardness or other properties along the length of the tube. Unfortunately, only few OEM companies are equipped for this process and because of the unique approaches used, the processes are often proprietary.

Surfaces

Surfaces are often an extremely important aspect of the product beyond appearance. Surfaces can increase or reduce friction as well as chemical, UV, scratch resistance, and mechanical properties such as cracking points.

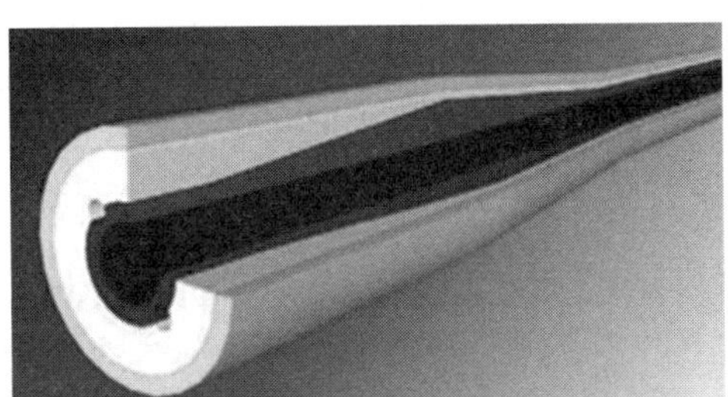

Figure 3.28.2 Variable longitudinal configuration possible in micro extrusion.

The material constituting the surface often determines its wettability and intrinsic properties, but recent research indicates that minute surface morphology can also be very impactful. Extruded surfaces can range from shiny smooth to rough or even a silky feel. Virtually every material can generate a rough as well as a smooth surface. Defining it by means of the parameter Ra is a good indicator, but it could also be misleading as the perception is driven by several factors and different materials could give the same feeling or opacity by different Ra. It must also be said that in some cases it could be difficult to get a specific Ra roughness value out of a certain material without the need to put additives into the material. Specific criteria could be defined on the base of each application and on the base of which effect the surface should be triggering.

Economics

Micro extrusion is almost synonymous with high-added value and small parts, especially if the micro structural properties of the same are important. The risk of a defective part is magnified on micro parts by the difficulty of measuring and inspecting them. Inspection considerations for micro parts and features are highlighted in Chapter 3.43, but must be considered in the design stage as well as in manufacturing.

Because there are only a small number of specialty houses extruding high-quality micro size products, it is important to realize that success may depend on a small series of design of experiments (DOE) and other characterizations as well as proven vendors. Advances occur every month in the micro world, but continued success on some micro products can be exceptionally challenging, and even less stringent parts may require exceptional knowledge.

With this said anyway, micro extrusion is also used for increasing competitiveness of existing processes; in some low marginality industrial applications, micro extrusion is able to generate parts with the same properties as others, but using less material, for example, a pressurized-air tube may reach the same burst pressure with a thinner wall, while produced at higher speed.

CHAPTER 3.29

Chemical Vapor Deposition

LaRoux K. Gillespie

Manufacturing Consultant
Andover, Kansas

Chemical vapor deposition (CVD) uses gases or vapors which settle on a substrate to produce a desired surface film. CVD coatings deposit uniformly on surfaces so they conform to all the surface topology equally. They deposit in an atomic layer deposition such that extremely thin layers of material can be produced. The process is widely used for the production of micro electronic devices, for photovoltaic element fabrication, and to provide chemical-resistant coatings for metal surfaces.

The CVD process is activated by thermal, plasma-enhanced, and photolysis or laser-induced methods. There are 13 different approaches to producing CVD coatings, but the basic approach is to heat the coating material, or reduce the pressure around it until the material vaporizes. The suspended vapor material settles on the target surface forming a uniform coating. The thickness of the coating is controlled by adjusting the temperature and duration of the process.

It is widely used for semiconductors and electronics of all kinds, but it is also used to provide polymer coatings, not just metal or silicon coatings. For metals, the heated metal vapor deposits on a cooler substrate. For coating with polymers, two or more different precursor compounds (monomers) are fed into the chamber where the parts are and they react to form polymers that deposit. By using very high-temperatures, sheets of graphene can be produced, and CVD has been used to produce large arrays of carbon nanotubes. These nanotubes can be CVD coated to improve their strength.

Low-pressure CVD (LPCVD) reactors operate at 0.1 to 2 Torr (0.002 to 0.039 psi). Atmospheric reactors do not need vacuum work.

The process works inside vacuum chambers and to indicate size scale, hundreds of silicon wafers can be run at the same time. Parts can be any size and shape but the surface on which deposition is desired has to be available to the gas, so the bottom surface of parts laying on a flat surface will not be coated. Parts can dangle from overhead conveyors to coat all sides (Fig. 3.29.1).

FIGURE 3.29.1 Any shape can be coated with the many CVD materials as shown by the micro structure of high-emittance black rhenium coating showing pyramidal dendrites formed by CVD. (Courtesy Ultramet.)

The advantages of using CVD (as defined by Ultramet) are:

- Can be used for a wide range of metals and ceramics
- Can be used for coatings or freestanding structures
- Fabricates net or near-net complex shapes
- Is self-cleaning—extremely high-purity deposits (>99.995% purity)
- Conforms homogeneously to contours of substrate surface
- Has near-theoretical as-deposited density
- Has controllable thickness and morphology
- Forms alloys
- Infiltrates fiber preforms and foam structures
- Coats internal passages with high length-to-diameter ratios
- Can simultaneously coat multiple components
- Coats powders

Ultraviolet-activated CVD (UVCVD) uses ultraviolet energy to create the reaction rather than thermal energy. It operates at room temperature to 300°C (572°F) rather than from the higher temperatures used in the other process variations. In contrast, for diamond coating the target part is heated to 700°C to 900°C. Diamond can be deposited at rates up to 1 mm/h (0.040 in./h). High-growth rates like this normally only apply to very small deposition areas [«1 cm^2 («0.156 in^2)]. Deposited film quality and growth rate are inversely related. Optically transparent high thermal conductivity diamond films are deposited at less than 10 μm/h (0.0004 in./h).

Materials Deposited

The materials commonly deposited by CVD include polysilicon, silicon nitride, silicon oxide, tungsten, aluminum, copper, molybdenum, tantalum, titanium, nickel, graphene, and synthetic diamond. In addition to the pure elements, the process can provide coatings of metals and their alloys, carbides, nitrides, borides, oxides, and intermetallic compounds (Table 3.29.1).

Since it is a gaseous process, composite materials can be produced not only on flat or normal external surfaces, but also in meshes. Stated differently, initial shapes can be infiltrated using CVD materials (Fig. 3.29.2). The results then can become ceramic matrix composites. Carbon-carbon, carbon-silicon carbide, and silicon carbide-silicon carbide composites are produced with this chemical vapor infiltration (CVI) version of CVD.

Metals	
Rhenium (Re)	Iridium (Ir)
Tantalum (Ta)	Platinum (Pt)
Tungsten (W)	Boron (B)
Niobium (Nb)	Silicon (Si)
Molybdenum (Mo)	Metal alloys: W-Re, Ta-W, Mo-Re, Ir-Pt
Nonoxide ceramics	
Tantalum carbide (TaC)	Niobium carbide (NbC)
Tantalum nitride (TaN)	Niobium nitride (NbN)
Tantalum diboride (TaB_2)	Niobium diboride (NbB_2)
Tungsten carbide (WC)	Titanium carbide (TiC)
Tungsten diboride (WB_2)	Titanium nitride (TiN)
Hafnium carbide (HfC)	Titanium diboride (TiB_2)
Hafnium nitride (HfN)	Vanadium carbide (VC)
Hafnium diboride (HfB_2)	Vanadium nitride (VN)
Zirconium carbide (ZrC)	Boron carbide (B_4C)
Zirconium nitride (ZrN)	Boron nitride (BN)
Zirconium diboride (ZrB_2)	
Silicon carbide (SiC)	
Silicon nitride (Si_3Ni_4)	
Oxide ceramics	
Alumina (Al_2O_3)	Titania (TiO_2)
Hafnia (HfO_2)	Yttria (Y_2O_3)
Silica (SiO_2)	Zirconia (ZrO_2)
Tantala (Ta_2O_5)	

TABLE 3.29.1 Refractory Materials Deposited by CVD (Courtesy Ultramet)

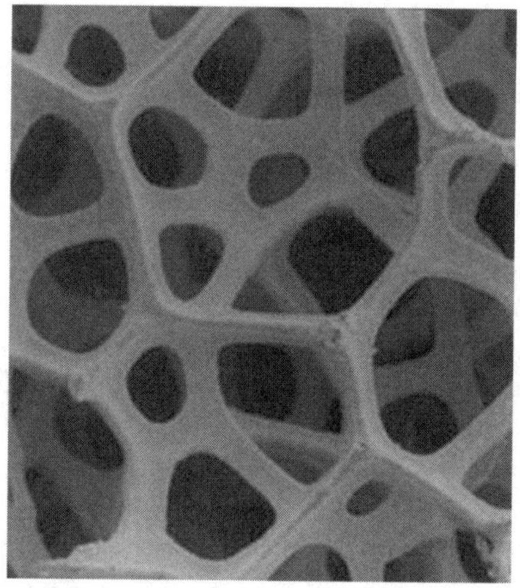

FIGURE 3.29.2 Rhenium foam fabricated by CVI, showing uniform coating and open-cell structure. (Courtesy Ultramet.)

Designers may take advantage of the fact that micro parts may be made by coating a mandrel of conventional materials that is later dissolved to leave a desired shape of the CVD material. Layers of different CVD materials may be deposited as necessary. One company calls this "inside-out" CVD processing since it begins coating what will be the inside material and then continues to overcoat that until the final size and material is on the outside. Since the deposited layers can be only a few atomic layers thick, the resulting parts are still microscopic (or if desired the part may become a very large final outer part with a microscopic interior).

Carbon nanotubes have been spun into fibers right in the reaction chamber that did the CVD deposition, so the CVD process can be part of a continuous multi-process operation. Continuous flow systems are common, so this can be a high production rate process for micro parts.

From a micro part and micro feature standpoint, the biggest challenge is determining how to handle micro size parts in a gaseous and heated environment.

CHAPTER 3.30

Magnetorheological Finishing

LaRoux K. Gillespie

Manufacturing Consultant
Andover, Kansas

Magnetorheological finishing (MRF) uses an abrasive-filled magnetic fluid to polish surfaces. When the fluid is in the presence of a strong magnetic field, it switches from a thin water-like fluid to a stiff clay-like consistency which is four orders of magnitude stiffer than before it was in the field. Despite using abrasives that are as fine as 1 to 5 μm (39 to 197 μin.), the magnetic field allows the abrasive to both aggressively polish as well as to accurately finish surfaces to precision tolerances and finishes as fine as 1 nm (0.039 μin.). Surface accuracies of better than 0.1 μm (3.9 μin.) are also achieved in automated machines. The part is rotated through the slurry in a manner to polish its surface uniformly.

The process is largely used on glasses for lenses, but it works on metals, ceramics, and hard materials of all kinds. Single crystal silicon blanks, for example, can be improved from 1300 to 8 nm (51 to 0.3 μin.) with this process.

One of its principal applications is the precision finishing of optical lenses (Fig. 3.30.1), but it can finish a wide variety of materials and equipment exists for finishing parts only a few millimeters in size to those that are over 1 m (39 in.). It is used for concave products, convex and flat surfaces. Prisms, square, and rectangular products are also finished, and it has been used to polish micro wire EDMed rectangular pockets as well. It provides pit-free and scratch-free surfaces.

Several magnetorheological (MR) fluids are used. Cerium oxide carried in a water-based suspension of carbonyl iron (CI) powder is a common one. Recently, iron powders sintered with carbon nanotubes have been shown to increase abrasive life. Nanodiamond powder is used for calcium fluoride, IR glasses, silicon, sapphire, and hard polycrystalline ceramics. Silicon carbide and aluminum oxide are used for metal finishing as well as the other materials mentioned. Water, silicone oil, mineral oil, glycerol, and paraffin oil are used as the carriers and stabilizers are added in many applications.

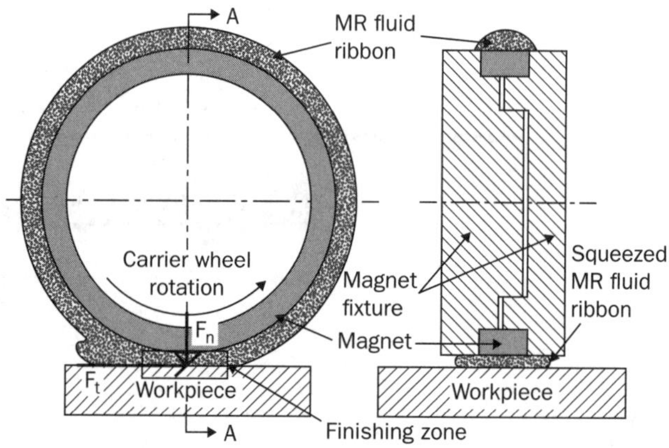

FIGURE 3.30.1 Schematic of magnetorheological finishing. (Reproduced courtesy V. K. Jain et al., copyright IJAMT, 2011.)

Material removal rates are in the order of 0.01 to 0.05 mm^3/min (0.0000006 to 0.000003 $in.^3$/min).

For metal finishing, abrasives of 11 to 22 μm (0.0004 to 0.0009 in.) have been used to obtain finishes of 44 nm (1.7 μin.) in 30 min from a starting lapped finish of 290 nm (11.4 μin.) for 52100 steel.

Several variations of the process exist and the variations are known as magnetorheological fluid abrasive flow finishing process (MRAFF), magnetorheological abrasive honing (MRAH), ball end magnetorheological finishing (BEMRF), magnetorheological jet (MR Jet) finishing, magnetic float polishing, and magnetic abrasive finishing (MAF) (the latter discussed in Chap. 3.19)—all have similar actions and characteristics. MAF, however, does not require a carrier solution.

References

Jain, V. K., S. Kalia, and A. M. Sidpara, "Some aspects of fabrication of micro devices by electrochemical micromachining (ECMM) and its finishing by magnetorheological fluid," *International Journal of Advanced Manufacturing Technology*, vol. 59, no. 9, 2011, pp. 987–996 (DOI: 10.1007/s00170-011-3563-4).

Menapace, J. A., P. R. Ehrmann, and R. C. Bickel, *Magnetorheological finishing (MRF) of potassium dihydrogen phosphate (KDP) crystals: nonaqueous fluids development, optical finish, and laser damage performance at 1064 nm and 532 nm*, Lawrence Livermore National Laboratory report LLNL-PROC-420246, November 13, 2009.

Saraswathamma, K., "Magnetorheological finishing: A review," *International Journal of Current Engineering and Technology*, Special Issue 2, February 2014, pp. 168–173.

CHAPTER 3.31

Micro Wire Products

LaRoux K. Gillespie

Manufacturing Consultant
Andover, Kansas

There is no definition of what constitutes micro wire, but clearly wire in the range of 0.25 mm (0.010 in.) or smaller would fall into this category. Wire sizes down to 25 μm (0.001 in.) have been in place for decades, with this size being a Number 50 American Wire gauge size. Wires as small as 17.8 μm (0.0007 in.) are formed today and coiled springs with outer diameters of only 127 μm (0.005 in.) are produced commercially. One of the smallest wires made is a 5-μm-diameter (0.0002-in.) wire. Its counterpart a 7.6-μm (0.0003-in.) wire is later coated to provide a 17.8-μm (0.0007-in.) wire. The micro size wires are used as springs, bent wire restraints, electrical connectors, medical coils for intravenous surgery and wound closing, probes, puncture tools, and medical guides.

The term "micro wire" while indicating minute size, and generally in the past referring to standard wire coiled onto a spool, has another meaning today. Micro wires can refer to short "whiskers," whiskers only a few micrometers long or even just a few nanometers long. The term can refer to any pillar-like item having a small diameter and relatively long length. Some of the many processes listed in this handbook produce such pillar products. While most industrial users would refer to them as pillars, they are wire-like and thus are referred to as such by many emerging technologies. Interestingly, some of these nanowires are grown on the ends of micro wire recording electrodes. So the micro wire is the beginning for the nano size wire.

Processes

The basic processes for forming wire include coiling it into springs, bending it into a variety of bent shapes, pointing the ends, and welding or bonding it to other products.

For coils, the wire is wrapped tightly around a round cylinder with standard spring making equipment. Special computer numerical controlled (CNC) wire bending machines provide an ability to bend wire into any shape and length. Manufacturers also utilize a number of self-made tools for the micro parts. While bending wires accurately into precision 3D configurations used to be a challenge, today's CNC equipment provides a unique ability to do that almost automatically (some final adjustments are needed to programmed motions for the micro wire).

Materials

Micro wire comes in round, flat, and rectangular shapes and can be procured in other special shapes as well. The load bearing wire materials include the 300 series stainless steels (302, 304, 316L, and LVM), the precipitation hardening stainless steels (17-5 and 17-7 PH), cobalt and Molly alloys (MP35N, 35NLT, Elgiloy, and others), precious metals such as platinum, palladium, gold, silver, nitinol, beryllium copper and phosphorus bronze, tungsten, titanium, and others including coated wires.

The design considerations for micro wire are most often related to diameter, stiffness, and strength, but the full list also includes dielectric strength, elongation (stress-strain diagram), conductivity (resistance, in ohms per length), corrosion resistance, hardness, and biocompatibility. Much of the softer micro wires are used in semiconductor manufacturing or bonding semiconductors to their next assemblies.

So there are two main uses of micro wires: (1) those that need strength to hold or move things and (2) those that are merely circuit connections. A third, not widely used application is as thermal dissipaters.

Wire systems address the shortcoming for strength of the copper metals by using various nickel alloys such as MP35N (or Constantan), or bimetallic materials such as copper-clad steel and silver clad with stainless steel. Unfortunately these usually exhibit unallowable high resistance, and have high costs. A series of specialized copper alloys have been developed that address the strength limitations of pure copper without sacrificing much in terms of conductivity. These alloys are much less expensive than nickel alloys or bimetallic materials. Biocompatibility of copper, however, is an issue since copper in certain concentrations can be toxic.

Wire sizes of 25 and 50 μm (0.001 and 0.002 in.) are common for life science use. Glass-coated micro wire can be obtained in sizes 1 to 30 μm (39 μin. to 0.0012 in.). For catheter insertion products most applications use wires of 0.064 to 0.192 mm (0.0025 to 0.0075 in.), but work is going on to reduce these sizes. The insulation coating on these applications is around 0.0127 mm (0.0005 in.) thick.

Flat shapes can be obtained in sizes 0.00508 to 8.509 mm (0.0002 to 0.335 in.), and widths of 0.0508 to 38.46 mm (0.002 to 1.50 in.). Square wire is available in 0.75 μm (0.030 in.).

Table 3.31.1 provides some commercial sizes and tolerances for conventional wire and for magnet wire in the micro sizes. Larger sizes and other materials are readily available but not listed here. In addition square magnet wire is also available in some micro sizes. For example, 30-gauge aluminum 0.254 mm (0.010 in.) on a side and 35-gauge copper only 0.142 mm (0.0056 in.) on a side.

As in the instance of all micro materials, grain size and its uniformity are important, so designers need to clearly articulate the precise metallurgical conditions required before forming and after if heat treating is needed. The most minute wire sizes are almost the same size as single grains in some of the common thicker materials and that greatly affects springback.

Figures 3.31.1 and 3.31.2 illustrate some of the minute commercially made forms of springs and wire forms. In Fig. 3.31.1 the coiled spring and its welded base all fit within the penny's "0" in the date "2008." In Fig. 3.31.2 some of the springs are intertwined with each other (a common problem in micro springs not adequately spaced apart). The smallest spring lying on the dime is about 0.43 mm (0.017 in.) in diameter. A formed wire is also among the many springs of different sizes, shapes, and materials.

Material	Size	Tensile Range
304V	0.001 × 0.003 in.	150–180 kpsi
304V	0.001 × 0.003 in.	300–355 kpsi
304V	0.001 × 0.003 in.	200–250 kpsi
304V	0.001 × 0.005 in.	150–200 kpsi
304V	0.001 × 0.005 in.	200–250 kpsi
304V	0.001 × 0.005 in.	300 kpsi min
304V	0.001 in. Rd	235–265 kpsi
304V	0.001 in. Rd	300 kpsi min
304V	0.002 in. Rd	151–200 kpsi
304V	0.002 in. Rd	235–265 kpsi
304V	0.002 in. Rd	300 kpsi min
304V	0.0005 × 0.003 in.	300–355 kpsi
304V	0.0005 × 0.0025 in.	300–355 kpsi
304V	0.0015 in. Rd	235–265 kpsi
304V	0.0007 × 0.003 in.	300 kpsi min
304V	0.002 × 0.006 in.	300 kpsi min
304V	0.002 × 0.005 in.	300 kpsi min
304V	0.003 in. Rd	300 kpsi min
304V	0.00075 in. Rd	300 kpsi min
Nitinol	0.001 in.	As drawn, oxide finish
Nitinol	0.002 in.	As drawn, oxide finish
Nitinol	0.003 in.	As drawn, oxide finish

TABLE 3.31.1 A Sample of Some Micro Size Wire (Courtesy Ulbrich Specialty Wire Products, A Division of Ulbrich Stainless Steels & Special Metals, Inc.)

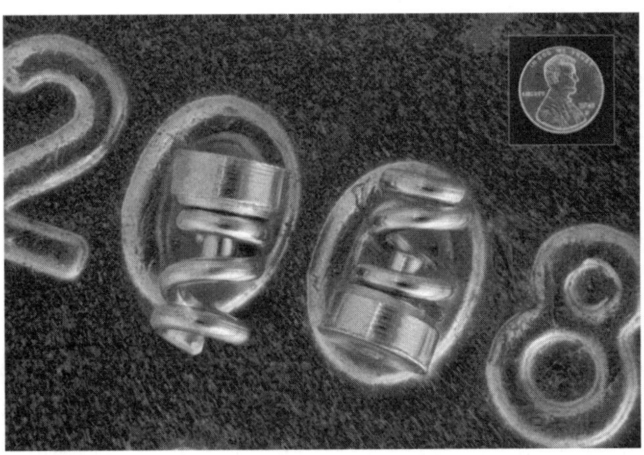

FIGURE 3.31.1 Coiled and welded micro part. (Courtesy Motion Dynamics Corporation.)

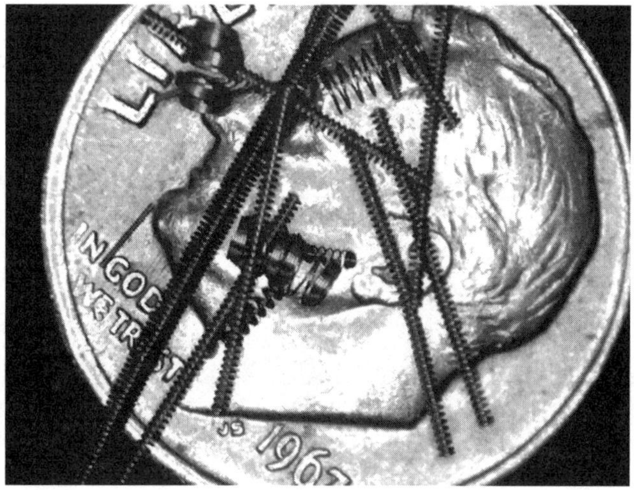

FIGURE 3.31.2 Micro springs and formed wire component. (Courtesy Motion Dynamics Corporation.)

Very small micro wires are sold by several companies and several service companies are world-class makers of the micro products from this wire.

Applications

Micro wires are used in cameras, automotive components, medical devices, consumer electronics, and even aerospace devices. They are found in cochlear hearing implants, heart probes, telemetry devices, microsurgical catheters and the tools attached to them. Micro electrode arrays of insulated 25 to 50 μm (0.001 to 0.002 in.) wires are used for neural sensors (particularly for brain activity). As an example of this application a 96-channel, 6 × 16 array of 35-μm (0.0014-in.) electrodes are spaced at 250 μm (0.0098 in.) for such use. This application requires metallic conductors that are corrosion resistant and stiff. Tungsten or stainless steel alloys are typical choices here with coatings of polyimide, Formvar, Isonel, or Teflon. These are used to penetrate living tissue and the ends must fit the intended application. Blunt, rounded, and pointed ends are all used for various applications, and the ends are sheared or ground to a variety of angles.

Cardiac ablation products have been drivers for micro wire product development. In this case the wires transmit radiofrequency (RF) energy to do the ablation and they also sense tissue temperatures. These wires are minute but also can be very long since they may have to go from the groin to the heart.

Copper and gold micro wires are used in semiconductor chips to carry signals from lead frame to the desired chip area.

For MEMs applications, shape memory alloy (SMA) wires (Dynalloy Flexinol HT) having a nominal diameter of 37.5 μm (0.0015 in.) have been used to thermally cause beam deflection. The significance here is that it does not take special shapes to cause MEMs size motions—simple (but very fine) wire can do that.

Micro wires are used as electrodes to make even smaller micro tools for EDM use. As mentioned above, they become micro probes of a number of life science measuring and recording purposes. They are being used in composites as meshes to provide mechanical strength for micro wave field tunable composites. They have superior GMI and GSI effects, soft magnetic properties, and good conductive properties. Composite makers also place these tiny wires to measure the temperatures when the fiber turns from solid to liquid to solid inside the composite part.

Nano wires are being developed to carry electronic signals for nano size systems. This is beyond the scope of this handbook, but indicates some of the future for minuteness of wire sizes. There is a similar future role for magnetic nano and micro wires, and carbon nanotubes are being used to carry micro currents.

Solar cells have been made of vertically aligned micro wires with diameters of 1 μm up to 30 μm (39 μin. to 0.0012 in.). Unlike the normal drawn wires, these are made using vapor-liquid-solid growth, deep reactive ion etching, high-pressure chemical vapor deposition, and molten core fiber drawing. These are more expensive processes than conventional wire drawing, but they provide a much smaller wire size than conventional techniques can or have. By laying several of these microscopic wires horizontally close together they trap more of the sunlight and that results in higher efficiency.

Design Considerations

- Micro wires can be laser welded with spots as small as 25 μm (0.001 in.).
- Commercial manufacturers can readily form wire down to 152 μm (0.006 in.) on their equipment and special tools, but smaller size wire is not handled well with commercial machines.
- Medical guide wires average about 0.457 mm (0.018 in.) in diameter and this diameter is typical of many medical tools. These would be considered "micro" for many companies although the wire diameters available from several manufacturers go far smaller than this.
- Sizes below 50 μm (0.002 in.) require tight control on grain size to get repeatable results.

Several references below proof additional insight into capabilities.

References

Clausi, D., *et al.*, Design and wafer-level fabrication of SMA wire microactuators on silicon, *Journal of Microelectromechanical Systems*, vol. 19, no. 4, pp. 982–991, August 2010 (doi:10.1109/JMEMS.2010.2049474).

Peng, H.-X., F. Qin, M.-H. Phan, *Magnetic Microwire Composites: From Sensors to Microwave Applications*, Springer, 2016.

Martinsen, F. A., B. K. Smeltzer, J. Ballato, T. Hawkins, M. Jones, and U. J. Gibson, Light trapping in horizontally aligned silicon microwire solar cells, *Optics Express*, vol. 23, no. 24, p. A1464, November 2015 (doi:10.1364/OE.23.0A1463).

Shaner, M. R., J. R. McKone, H. B. Gray, and N. S. Lewis, Functional integration of Ni–Mo electrocatalysts with microwire array photocathodes to simultaneously achieve high fill factors and light-limited photocurrent densities for solar-driven hydrogen evolution. *Energy & Environmental Science*, 2015 (doi:10.1039/C5EE01076D).

CHAPTER 3.32

Micro Electroforming

LaRoux K. Gillespie

Manufacturing Consultant
Andover, Kansas

Electroforming deposits metal onto a sacrificial mandrel or substrate. The mandrel is later etched away leaving only the plated metal as the finished product. The finished part is an exact reproduction of the mandrel where it was in contact with the mandrel. If the plating layer is thin, the outer surface is also a close reproduction varying only in dimension by the thickness. Nickel is the plated material in most instances, but it does not have to be nickel always. The mandrels are often made of aluminum which is easy to machine and to etch away. Mandrels can be made of non-conductive materials by covering them with a conductive coating, before the plating begins. The mandrels can be made very complex, so the resulting part can similarly have complex geometry. Internal cavities are not a problem for electroforming since they are formed over projecting mandrel features. If a male model can be made, electroforming will conform to it producing an inverse replication of the mandrel.

While tolerances of 3 nm (0.1 μin.) have been attained, typical micro parts will have much more variation. A more realistic commercial tolerance would be as low as 7 μm (0.0003 in.) but ±2 μm (±0.000079 in.) have been maintained on some designs. Individual features and spacing can be as small as 15 μm (0.00059 in.) and part thickness can be as low as 7 μm (0.0003 in.). Some commercial houses can produce 5 μm (0.0002 in.) or thinner features.

One commercial company notes that the practical limits for *photochemical etching* are 100-μm (0.004-in.) holes on 50-μm (0.002-in.) material and minimum 125-μm (0.005-in.) features on 125 μm (0.005 in.) or less material. Electroforming can produce foils less than 25 μm (0.001 in.) thick and holes under 10 μm (0.00039 in.). These two paragraphs provide a very good statement of the current state of the art and the fact that each company has different levels of ability, and in many instances some variation in its processes.

When the mandrel is smooth, the resulting mating surface will be similarly smooth. The surface finish on the top side depends on the material being electroformed, with harder materials giving mirror-like finishes and softer materials being duller. Unlike machining processes, electroforming does not make burrs which are difficult to remove from micro parts.

Electroforming produces production parts, but it also is widely used to make molds for polymer molding. It is widely used specifically for micro molds. Nickel molds are used for compression, injection, and other plastic-molding processes. They are also used for glass, rubber, and powder metallurgy die casting. Figures 3.32.1 through 3.32.4 illustrate electroformed micro items.

Figure 3.32.1 Electroformed micro bellows. (Courtesy Servometer.)

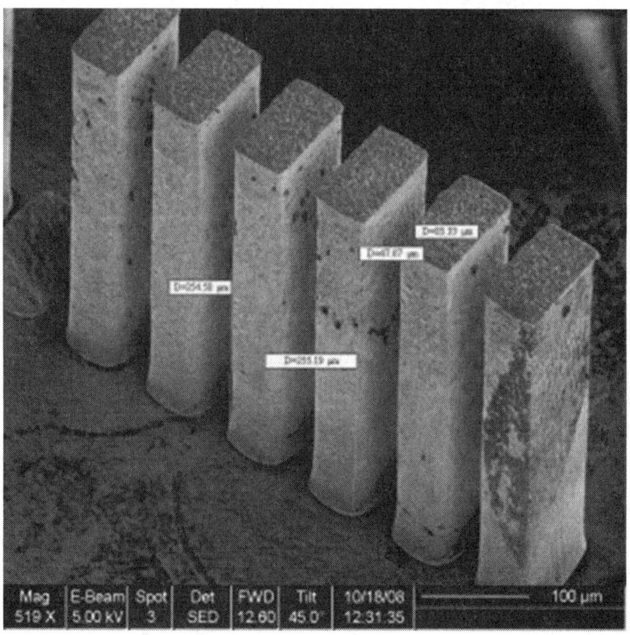

Figure 3.32.2 Micro pillars made by nickel electroforming. (Courtesy Xueyong Wei.)

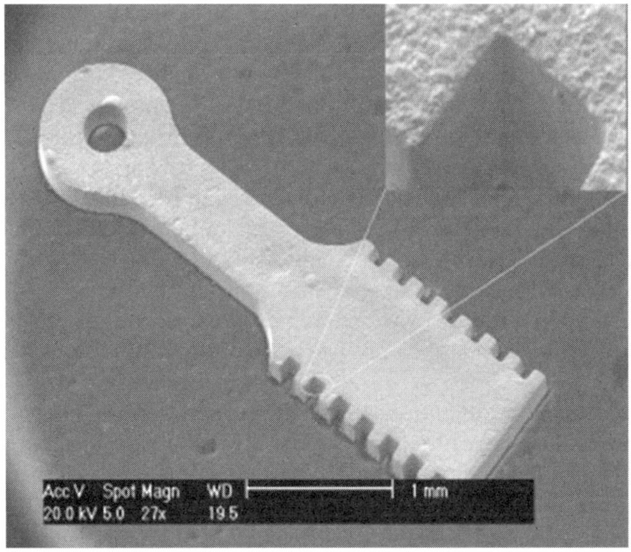

Figure 3.32.3 Micro part made by electroforming. (Courtesy Xueyong Wei.)

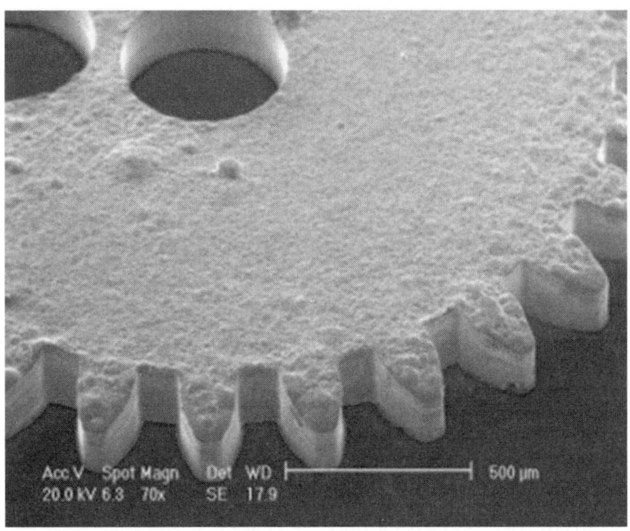

Figure 3.32.4 Micro gear made by electroforming. (Courtesy Xueyong Wei.)

LP record surfaces have spiral grooves that are 50 µm (0.002 in.) deep and 50 µm (0.002 in.) wide. These are pressed into plastic using nickel electroformed molds. The grooves are spaced typically ~100 µm (~0.004 in.) apart. These same abilities can be produced in micro-size parts. The compact discs (CDs) that are used in all computers use small pits 0.4 µm (16 µin.) wide and 0.12 µm (5 µin.) deep to store data. These pits are produced spaced about 1.6 µm (63 µin.) apart.

Metallic stamping might be the best term to use for the electroformed tools making CDs and LPs. It is not limited to these rather large parts. Such electroformed stamps are used to make micro lens or arrays of them only 36 μm (0.0014 in.) diameter in polymers. In this instance the normal description of use is "stamping of patterned media."

Encoder discs and embossing tools are other micro applications of the process, as are micro nozzles which are frequently made by this process.

Electric motor coils have been made by electroforming. Since the plating adheres to whatever the surface geometry is, the coils can be made planar or wrapped around a cylindrical mandrel. While not normally used to make electrical circuits, it plates copper in any place directed or allowed, and is not limited to copper.

Most small/micro bellows are made by electroforming.

Electroforming can also be used as a repair process, although that application is not typical.

In a unique emerging application EDM electrodes are being fabricated by additive processes such as stereolithography, coated with a metallic coating, and then electroformed to make the tool used to EDM parts.

In a different application a micro ball joint is made by WEDG in tungsten carbide. The ball end is then coated with a removable conductive material, and this assembly is then electroplated with nickel to make the shell of the ball joint. If needed the shell can then be further plated or otherwise backed up with more material to withstand high forces. A 500-μm (0.0197-in.) ball joint is the smallest that has been made to date, but by this additive process a complete assembly is made. The key to success is being able to remove the material coating the ball end after the plating has taken place.

The Microfabrica process discussed in Chap. 3.44 is a variation of electroforming. One of its main attractions is that it forms articulating layers that are working assemblies and does it as an integrated process.

The LIGA process (discussed in Chap. 3.33) uses electroforming as its principal process. With this process, parts only several micrometers wide can be electroformed to a height of 3 mm (0.1 in.). The finished parts have parallel, vertical walls and the fine precision that electroforming provides. The synchrotron X-ray radiation equipment required for LIGA exposure is not widely available. While it makes some of the most accurate micro and nano parts possible, the lack of equipment availability limits its widespread use on all but the most demanding parts. The use of LIGA to make micro turbine/cutting tools that are fed through diseased arteries to remove plaque buildup is one cost-effective application.

Just as LIGA and electroforming are combined as one process, or combination of processes, electroforming is being combined with other micro processes to yield new advances. It is used in combination with micro milling, laser cutting, WEDM, and stereolithography to produce finished products.

Laser-Evolved Electroforming (LEEF)

Traditional electroforming uses a mandrel and plates the entire surface. By coating the surface with a photoresist then exposing the photoresist to a pattern of light and then subjecting the surface to a chemical process that dissolves the exposed photoresist, a plated pattern is produced. The electroforming process will plate only the exposed areas. Thus while traditional electroforming requires a hard expendable mandrel, the use of photoresist lets light define the pattern to be plated. The process can make multiple components on sheets of expendable material, and the components can be grown three dimensionally and if needed can consist of different layers of plated material. Thus electroforming is another version of an additive process, although it is not usually defined as an additive process.

The pattern is exposed by either a mask or by a moving beam laser, with the laser gaining popularity today since it saves build time and eliminates the need to store tooling.

The process allows high aspect ratios to be achieved (greater than 1:1) with tight tolerances, micro features, and whatever complexity is desired. It can make features as small as 15 μm (0.0006 in.) while providing excellent edge detail. Since it is normally applied to sheets of metal as the substrate, it can make many parts per sheet in the same time as individual parts are produced. While most walls made by this process are straight up and down, the process can provide conical and stepped apertures walls. It can produce micro holes, micro slots, and micro cavities of almost any shape. It is widely used to make micro meshes and micro screens.

The use of lasers to provide the desired photoresist exposure removes the need for multiple masks to be used and provides quick change of paths as needed. Laser direct imaging (LDI) is a straightforward process. It accommodates production needs of tens or millions of parts and geometrical design changes normally do not incur significant costs. The turnaround time for parts from this process variation can be as short as three days, while for conventional electroforming the turnaround time depends on mandrel availability and mandrel removal times.

Materials

Nickel is the major material used for electroforming. NiCo alloys are used as well since some of them have electromagnetic, thermal, strength, and corrosion resistance properties similar to stainless steel. In addition, at least one of these alloys can be electroformed in a stress-free state which is important for some applications. Most pure metals can be electroformed. Precious and refractory metals are electroformed. Pure gold, hard gold, and copper are popular electroformed metals. Alloys of iridium and rhodium, tungsten, and molybdenum can be produced as can niobium and tantalum.

Small-diameter, seamless tubing has been made by the EL-Form™ variation of standard electroforming. That variation can produce iridium and rhenium tubing having excellent ductility. That ductility permits cold forming. For conventional metal tubes, nickel is plated onto a small diameter wire which has electrical current in it. The wire becomes both mandrel and electrode. After removing the wire, the resulting product is a tube whose inner diameter is the exact size of the wire OD. Tubes of 10-μm (0.00039-in.) ID and 5-μm (0.0002-in.) wall thickness are made this way. The tubes can be nickel or gold ID and overplated with nickel for strength. Other companies have produced 80-μm (0.003-in.) ID tubes from gold, platinum, and nickel. Tubes with tapered ID and ODs can also be made.

One commercial company makes micro bore tubing by electroplating nickel over a diamond-drawn mandrel in a continuous process. This facility provides tube IDs down to 25 μm (0.001 in.). Being able to do this in a continuous mode is a major advantage.

Emerging Aspects

Nanoparticles like alumina and carbon nanotubes have been used to enhance mechanical properties of electroformed nickel structures (Wei, 2016).

In a unique application Sandia engineers have grown 40-nm-diameter (1.6-μin.) "wires" 50 μm (0.002 in.) tall using a commercial anodized aluminum oxide mold having numerous 40-nm-diameter (1.6-μin.) pores in it. This is the smallest known feature produced by electroforming. In this instance the target was to produce hundreds of the wires in close proximity to each other to dissipate heat rapidly. This is not a commercial process or product, but it does illustrate the potential of the process.

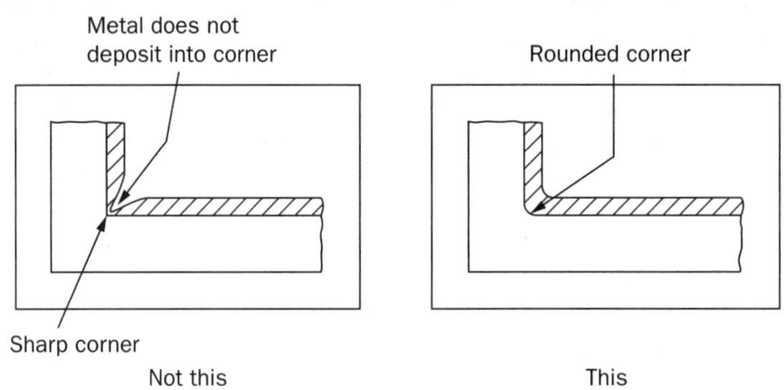

FIGURE 3.32.5 Use rounded internal corners for uniform plating (Bralla, 1999).

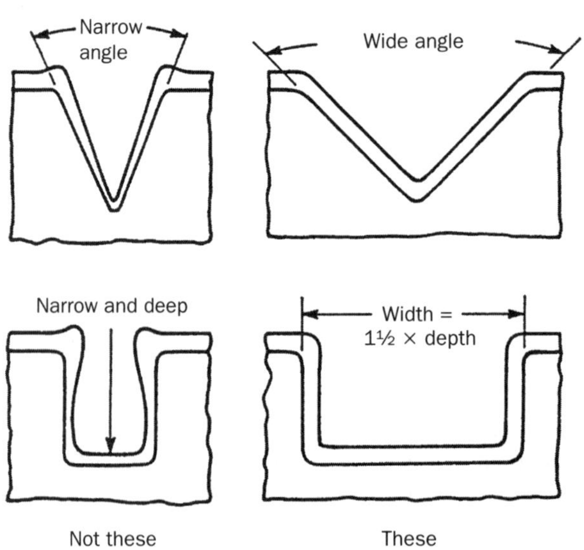

FIGURE 3.32.6 Stay away from narrow internal feature designs for uniform plating (Bralla, 1999).

Design Considerations

Electroformed parts are usually complete when removed from the mandrel or sheet they are on. Nickel can be machined, particularly by laser and wire EDM, so if the entire geometry cannot be conveniently produced by electroforming it can be finished by other processes.

One aspect that is less of an issue on micro parts than it is on larger parts, but which still needs to be considered is that internal corners do not plate as well as external corners do. External edges act as antenna for the energy and internal corners act as semi dead zones. Figure 3.32.5 illustrates good design practice for this issue. Figure 3.32.6 illustrates another aspect if parts are to be heavily plated. Most micro parts are not heavily plated and these practices may not be as significant as for larger parts and heavy deposits.

The following references provide some extensive additional details that designers might wish to review.

References

Alper, S. E., I. E. Ocak, and T. Akin, Ultrathick and High-Aspect-Ratio Nickel Microgyroscope Using EFAB Multilayer Additive Electroforming, *Journal of Microelectromechanical Systems*, vol. 16, no. 5, October 2007, pp. 1025–1035.

Bralla, James G., *Design for Manufacturability*, 2nd ed., McGraw-Hill, New York, 1999, pp. 201–202.

Chatzipirpiridis, G., O. Ergeneman, J. Pokki, F. Ullrich, S. Fusco, J. A. Ortega, K. M. Sivaraman, B. J. Nelson, and S. Pané, Electroforming of Implantable Tubular Magnetic Microrobots for Wireless Ophthalmologic Applications, *Advanced Healthcare Materials*, 2014, pp. 1–6 (DOI: 10.1002/adhm.201400256).

Chen, S.-T., A. t W.-J. Hsue, and T.-S. Luo, Fabrication of Mass Micro-Holes Using the Hybrid Precision Extruded-Wax Electroforming and Micro-EDM Processes, *Journal of the Chinese Society of Mechanical Engineers*, vol. 33, no. 4, 2012, pp. 349–356.

Dasari, P., Investigation of Micro Channel Fabrication by Electroforming, Electronic Thesis or Dissertation, University of Toledo, 2010.

Kang, S., *Micro/Nano Replication: Processes and Applications*, John Wiley & Sons, 2012, pp. 51–56.

Lin, C.-S., Y.-S. Liao, Yi-T Cheng, Y.-C. Lai, Fabrication of Micro Ball Joint by Using Micro-EDM and Electroforming, *Microelectronic Engineering*, vol. 87, no. 5–8, 2009, pp. 1475–1478.

Siegal, M. P., D. L. Overmyer, W. G. Yelton, and E. B. Webb III, *Electroforming of Bi(1-x) Sb(x) Nanowires for High-Efficiency Micro Thermoelectric Cooling Devices on a Chip*, Sandia Report Sand2006-6940 (Final LDRD Report), November 2006.

Soni, S. Ho, S. C. Park, W. Lee, and H.-K. Lee, Manufacture of μ-PIM Gear Mold by Electroforming of Fe–Ni and Fe–Ni–W Alloys, *Transactions of Nonferrous Metals Society of China*, vol. 23, 2013, pp. 366–371.

Tang, P. T., J. Fugl, L. Uriate, G. Bissacco, and H. N. Hansen, Indirect Tooling Based on Micromilling, Electroforming and Selective Etching, in *4M 2006 Second International Conference on Multi Material Micro Manufacture*, S. Dimov, W. Menz, and B. Fillon (eds.), Elsevier, 2006.

Wei, X, Z. Zhu, P. D. Prewett, and K. Jiang, Fabrication of Ni–Al_2O_3 Composite Microcomponent by Electroforming, *Microelectronic Engineering*, vol. 84, 2007, pp. 1256–1259.

Wei, X., H Lee, Z. Jiang, and K. Jiang, Thick Photoresists for Electroforming Metallic Microcomponents, *Proceedings of the IMechE Part C: Journal of Mechanical Engineering Science*, vol. 222, 2008, pp. 37–42.

Wei, X., Electroforming of Metallic and Nano Composite Micro Structures, http://www.birmingham.ac.uk/Documents/college-eps/mechanical/staff/jiang/jiang-electroforming.pdf.

Wei, X. and K. Jiang, Synthesis and Characterization of Nanoparticulates Strengthened Nickel Microcomponents. *Advances in Science and Technology*, vol. 54, 2008, pp. 299–304.

CHAPTER 3.33

Manufacturing with LIGA

LaRoux K. Gillespie

Manufacturing Consultant
Andover, Kansas

LIGA is a German acronym for the words lithography, electroforming, and molding. It is a unique process for producing high-aspect-ratio metal micro parts and features—for example, walls and slots only a few micrometers thick but millimeters tall (if that much height is needed). LIGA begins by using high-energy X-ray lithography to produce a deep nonconducting mold which is subsequently filled by electrodepositing metal. Figure 3.33.1 illustrates the process, and it is significant to note that the photoresist that is used is always put onto a solid metalized substrate of some nature. Silicon disks are a common substrate, but beryllium and titanium have also been used. Copper is often a metalized coating when the parts are to be made from nickel.

Since the LIGA parts are all typically microscopic in size, many parts can be made on a single substrate.

The final result is a miniature high-precision metal part with high-aspect-ratio (height-to-width) features. Aspect ratios of 100 with feature depths as much as 1 mm (0.039 in.) have been produced. LIGA fabrication involves several steps—mask layout, chrome mask generation, LIGA mask fabrication, substrate preparation, X-ray exposure, chemical development, electroplating, lapping (planarization), and release. Since the plating buildup vertically is somewhat uneven, planarization (lapping) levels off the top surface of the part.

If the part is to become a master from which molds are made, then after planarization a heavy layer of metal is plated onto the planarized metal.

The most commonly used photoresist is high-molecular-weight polymethylmethacrylate (PMMA). A solvent washes away the material that will become the cavities to be plated. Once plating makes the part, the remaining PMMA is dissolved (often in acetone), and the metal part is released from the metallic substrate by chemically dissolving the original metallization layer.

The processes are well established at several facilities in the United States, but the biggest restraint is the availability of a synchrotron which can provide the exposure. LIGA utilizes short-wavelength X rays generated by a synchrotron to expose the thick (hundreds of micrometers) PMMA resist. There are few synchrotrons in the United

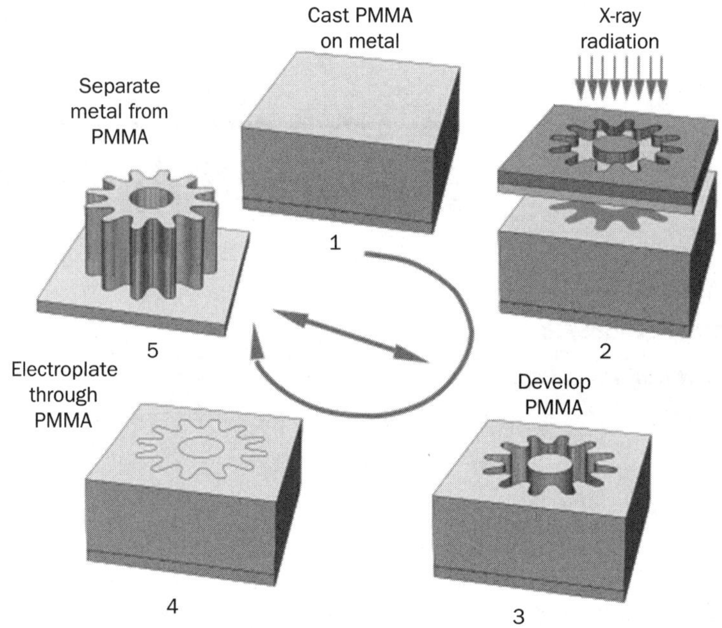

FIGURE 3.33.1 LIGA process steps. [Material provided courtesy of Dr. Michael Huff of the MEMS and Nanotechnology Exchange (www.mems-exchange.org) at the Corporation for National Research Initiatives.]

States. They are required to give the exposure for almost absolutely perpendicular walls which allow the high-aspect features.

LIGA is used to make parts in two ways. "Direct LIGA" implies the metal part is an actual part to be used; "indirect LIGA" uses the metal part to make a mold from which parts can then be molded. The latter approach is much faster and more economical.

LIGA Materials

LIGA parts can be made of metals, alloys, and even some metal-matrix composites. While nickel parts are the most common, copper, gold, and nickel-iron alloy have been used. Nickel-iron, Be-Te, NiCo, and NiMn are other metals that have been used. Electroplated nickel has good corrosion resistance and relatively strong mechanical properties. While nickel-based metals are the most commonly used LIGA material, copper has been used to make 60-μm (0.0024 in.) motor coils. Al_2O_3 particulate-strengthened nickel plating results in a metal-matrix product with higher strength than conventional nickel plating.

The elastic modulus of deposits can be influenced by the process and the crystal or grain orientation can vary from a uniform to a somewhat structured direction depending on a number of parameters (Kelly & Goods, 2005). While the features produced are not submicrometer size, the electroplating process does produce nanometer-size grains.

LIGA Products

There is literally no limit to the types of products LIGA can fabricate. Micro gears, actuators, springs, electrical contacts, rotors, and drivers are becoming common products for new generations of micro motors. Sensors of several kinds (motion, magnetic, pressure) have been made, some having fingers only 25 μm (0.00098 in.) wide and 250 μm (0.0098 in.) long. Accelerometers are another example. LIGA tools are used to emboss features in micro-assay lab-on-a-chip and micro-fluidic devices. Micro-fiber spinnerets nozzles of many cross sections produce synthetic fibers. Micro membranes and micro sieves are made via LIGA as are micro-hole structures used for filtration, separation, metering, and vaporization. Micro heat exchangers are also good applications.

Figures 3.33.2, 3.33.3, and 3.33.4 illustrate the general configurations and sizes of features that are made by LIGA. There are few limitations in configuration, but these figures are illustrative of the dimensions for widths and depths that can be produced. Figure 3.33.2 illustrates a wave guide with representative dimensions of LIGA parts. Figure 3.33.3 provides a more complex configuration representing smaller features.

A number of production houses in Europe and the United States manufacture LIGA products for companies. Some U.S. national laboratories are world leaders in the research and manufacture of such devices, Sandia National Laboratories in particular.

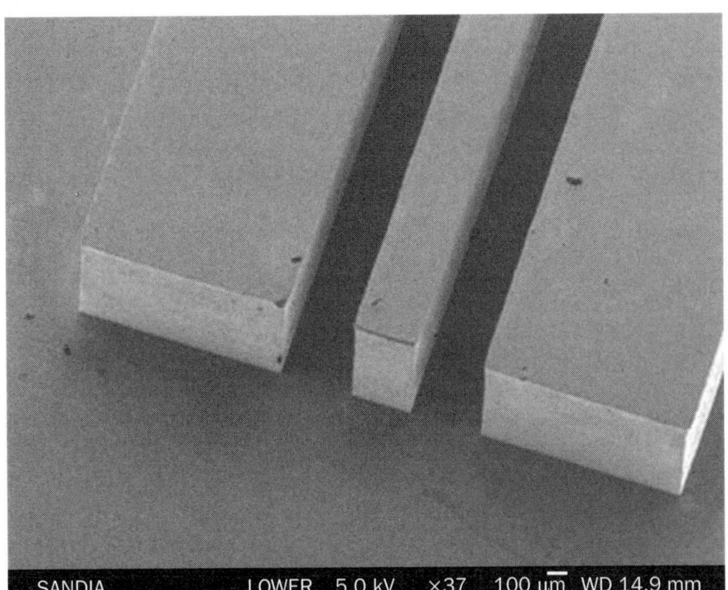

Figure 3.33.2 Wave guide fingers made via LIGA. (Michael.Forman CC BY 3.0; https://commons.wikimedia.org/w/index.php?curid=7155729.)

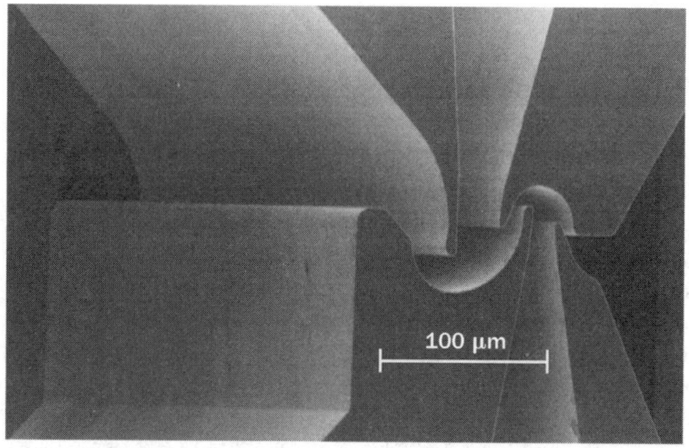

Figure 3.33.3 Uranium-enrichment nozzle made via LIGA. (Captain Bligh at the German-language Wikipedia; see Ehrfield et al., 1987.)

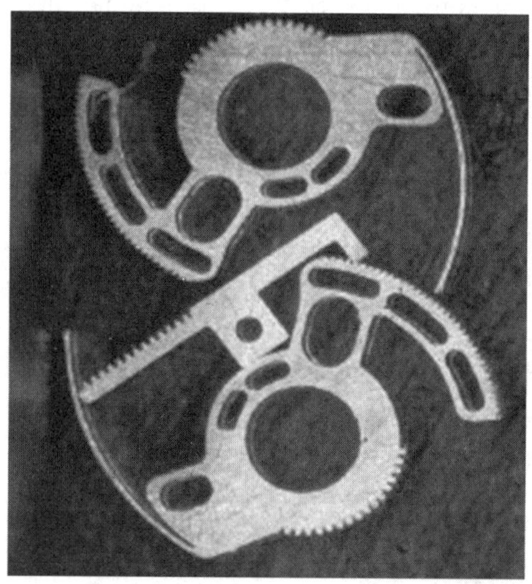

Figure 3.33.4 Long, slender features produced via LIGA (Kelly & Goods, 2005).

Alternative LIGA Approaches

Because of limited synchrotron availability, lasers have been developed to provide a lower-cost approach to pattern generation. In this instance, an excimer-based micromachining center is used to write the pattern on the photoresist. One system uses a diode laser with 18-mW average power and a 375-nm wavelength (UV). The laser wavelength is chosen to work with SU-8 photoresist, allowing a layer thickness of hundreds of micrometers. The 4-mm writing head on that system provides a precision of 1 μm (0.0039 μin.) and a writing speed of 5.7 mm^2/min (0.0088 in./min).

Inexpensive ultraviolet light from a mercury lamp can expose polymer photoresists such as SU-8, without needing synchrotron X rays. This makes UV LIGA less expensive and more accessible than traditional exposure. UV LIGA, however, cannot produce the same precision for high-aspect-ratio features, so it is used for less-demanding LIGA components.

Design Restraints

There are a number of design considerations required of LIGA components. Honeywell's FMT plant in Kansas City, Missouri, has issued a design for manufacturability guide for LIGA part fabrication. Gengenbach et al. (2008) also provides many pages on the LIGA design issues. The issues that must be considered include

- Design considerations for the lithography process
- Design considerations for the electroplating process
- Design considerations for inspecting LIGA features
- Design considerations for mechanical properties
- Design considerations for electrical properties
- Design considerations for handling
- Design considerations for assembly

As implied by this list, designers need to work with the LIGA fabrication houses to get maximum yields at the lowest cost.

Sometimes, though, a company will be required to make slight alterations to the design of a part that the part designer will not have knowledge of. For example, small cavities can entrap air bubbles that prevent the metal from depositing at the bottom of the cavity. The details will not be obvious to a designer until the design is reviewed by the manufacturing engineers to ensure that their process will not trap bubbles.

The inspection, handling, and assembly issues are largely the same for all micro parts and micro features and these are addressed in Chaps. 3.43 (inspection), 3.51 (handling), and 3.52 (assembly).

LIGA can make parts of a few millimeters in size or parts which are only a few micrometers in size. It is not a process which currently can make submicrometer-size parts or features. Many of its applications today are for parts which are roughly 1 mm in size, but with features which are a few micrometers in thickness, gap, or diameter, but which may be 10 or 20 times deeper than their width or diameter.

Some *typical capabilities*, but not absolute minimum size limits, are:

- Minimum feature size such as line widths and spaces: 20 μm (0.00078 in.)
- Aspect ratios: 70:1
- Sidewall straightness/perpendicularity: 1 μm/mm (0.001 in./in.) (~1 degree)
- Slight tapers can be produced
- Surface finish on side walls: 50 nm (~2 μin.) Ra
- Surface finish on top and bottom surfaces: mirror finish; 100 nm (~4 μin.) Ra[1]

[1]Provided by lapping before removal as individual parts.

At the *most extreme*, the following dimensions are some of the most exacting that have been produced; these are not possible by most manufacturers:

- Feature heights: up to 2500 μm (0.098 in.)
- Minimum feature thickness: 1 μm (39 μin.)
- Lateral feature tolerance: ±0.5 μm (±20 μin.)

The photoresist can expand or contract during the process and that affects dimensions and can often cause cracking of the photoresist which also shows up in metal after plating. We do not present this here because it requires a fuller treatment than space allows, but that in itself is another reason designers need to work with LIGA manufacturers to get the design right before releasing it.

Since the corners of a rectangular region will always have a local current density higher than the rest of the design, metal will deposit faster at the corners. Sharp bends holes, and irregular spacing, also result in nonuniform deposition rates near these features. Since cracks start at sharp internal corners, these corners should have radii of at least 2 to 5 μm (79 to 197 μin.) and preferably larger.

Long, thin lines in photoresist break down easily so they should be avoided where possible. From a part standpoint that means it will be difficult for very thin part walls [those approaching 1 or 2 μm (39 or 79 μin.)] to have acceptable yields.

Because of potential grain orientation issues, designers need to clearly articulate not only the material composition to be used, but also, for many applications, some definition of directional uniformity of properties such as modulus of elasticity, strength, or deflection under specified loading conditions.

As mentioned in earlier paragraphs, the materials that can be electroplated in these micro features are limited.

Because LIGA involves a lithography process, it is not possible to incorporate hidden pockets or have undercuts in the part. In almost all instances the parts must be two-dimensional.

Parts have been made that allow other materials to be inserted into pockets. Specifically, micro magnets have been inserted into small pockets in some nickel parts to allow magnetic fields to move the piece. With proper forethought, grooves, cavities, bumps, and other features can be made by LIGA that will trap other materials and micro parts for use in a final assembly, but this is not a normal LIGA ability. Extra steps are required to make anything other than a 2D part, but the part can be of almost any design. In one approach, multilayer LIGA parts were made by diffusion-bonding LIGA parts together after they were planarized. Again, this is not a normal part of LIGA fabrication, but it has been used successfully (Christenson, 1998).

Several references are provided below with additional details of the process and design issues.

References

Christenson, T. R., Advances in LIGA-Based Post-Mold Fabrication, Sandia National Laboratories Report SAND98-2374C, 1998.

Ehrfeld, W. W., P. Bley, F. Götz, P. Hagmann, D. Münchmeyer, J. Mohr, O. H. Moser, D. Münchmeyer, W. Schelb, D. Schmidt, E. W. Becker, Fabrication of Microstructures Using the LIGA Process, *Proc. Micro Robots and Teleoperators Workshop*, IEEE, November 1987.

Gengenbach, U., I. Sieber, U. Wallrabe, Design for LIGA and Safe Manufacturing, in *LIGA and Its Applications* (eds V. Saile, U. Wallrabe, O. Tabata, J. G. Korvink), Wiley-VCH Verlag GmbH & Co. KGaA, Weinheim, Germany, 2008. doi: 10.1002/9783527622573.ch6, 2009.

Guo, Y., Y. Wang, T. C. Kong, R. Du, NPTFE Electroformed Parts by SU8-Based UV-LIGA, *Proc. SPIE 7269, Micro and Nanotechnology: Materials, Processes, Packaging, and Systems IV*, 726900, December 2008. doi:10.1117/12.810665.

Holmes, A. S., Laser Processes for MEMS Manufacture, *RIKEN Review* No. 43, January 2002, pp. 63–69.

Ilie, C., C. D. Comeagă, O. G. Donţu, M. Popa, Micro Parts Errors to Precision Manufacturing using UV-LIGA Technology, *Advanced Materials Research*, Vols. 816–817, 2013, pp. 237–241.

Ilie, C., M. Popa, P. Prioteasa, I. Chiriţă, N. Tănase, Applications of LIGA Technology for the Development of Micromechanical Systems, *UPB Scientific Bulletin*, Series D, Vol. 73, Iss. 2, 2011.

Kelly, S. S., S. J. Goods, LIGA-Based Microsystem Manufacturing: The Electrochemistry of Through-Mold Deposition and Material Properties, Sandia National Laboratories Report SAND2005-2537, 2005.

Polosky, M. A., E. J. Garcia, Microsystem Product Development, *TIMA Editions/DTIP 2006*, 2006, pp. 1–7.

CHAPTER 3.34

Deburring Micro Parts

LaRoux K. Gillespie

Manufacturing Consultant
Andover, Kansas

Basic Issues

Because of the minute size and tolerances of micro parts (defined here as parts or features less than 1 mm), removing burrs is much more challenging and expensive than from larger parts. Deburring may cost 15% to 30% of the total cost of micro parts because of size and the precision required, so the most effective steps in generating desired edges include *combinations* of burr prevention, minimization, appropriate selection of deburring processes, and innovations in feature design, production, and finishing.

There is no universal cure for burrs. Because of the wide variety of shapes, tolerances, finishes, materials, and deburring side effects, it takes a careful analysis to determine the most cost-effective approach for burr prevention minimization and removal. There is no magic bullet other than burr prevention.

It is the *combinations* of edge tolerances, product and feature size, tolerances, proximity to other edges and surfaces, complexity of the basic configuration, machinability, burr properties, and accessibility of features that determine potential success of any deburring process. The burr on the 1st part is different than the burr on the 1000th part because of tool wear. Variations in burr size result in variations in deburring and finished edge dimensions, and many deburring processes change sizes and finishes on adjacent surfaces or create damage to them. External edges in general are easier to deburr than internal features like slots less than 0.5 mm (0.020 in.) wide. Holes smaller than that also become problems to deburr.

Design Issues

Designers can design parts so that (1) the part geometry and material minimizes burr formation, (2) part geometry and materials allow processes that do not make burrs, (3) part geometry allows burrs to be made where they are easy or easier to remove, (4) burrs are captured or encapsulated in assembly and do not have to be taken off, and (5) actual edge requirements are explicitly defined and not overly stated.

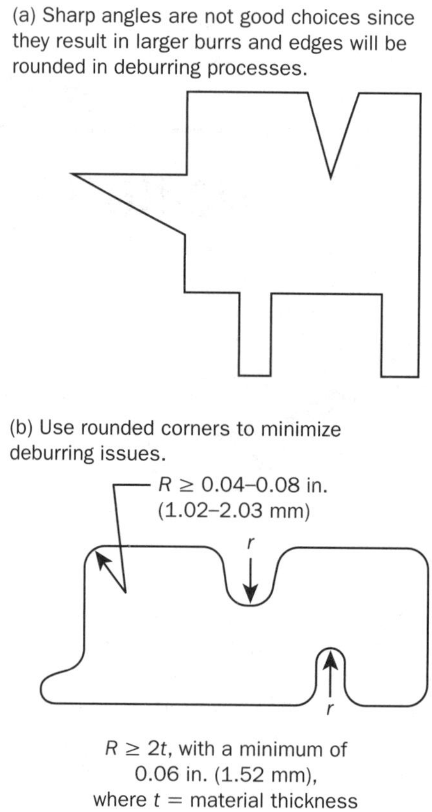

FIGURE 3.34.1 Use large radii where possible for blanked parts since they allow longer die life and hence minimize burr size (Strasser 1960).

Figures 3.34.1, 3.34.2, and 3.34.3 illustrate the first two of these concepts and several other examples are presented elsewhere (Gillespie, 1999).

Part design is critical in edge finishing. For *edge requirements* designers have to answer: What does the product actually need at edges? Why does it require that? When in the manufacturing sequence must the edges meet final requirements? Where exactly must the edge meet the most challenging requirement? Where among all the edges on the part does the most stringent requirement occur? How will the edges be verified? Using 30× magnification, manual inspection requires much more inspection time than 10× magnification, and 400× magnification required on some micro parts presents unique challenges.

The word "burr" refers to machined or stamped edges, but designers also must consider flash from molding, and dross left from laser cutting and EDM. "Edge finishing" is a more appropriate term than "deburring" for many parts. In the world of micro manufacturing, burrs typically will range in thickness from 2.5 to 76 μm (0.0001 to 0.003 in.). Thicker burrs are harder to remove and take longer to remove, so everything that minimizes burr thickness is an advantage.

Ductile materials make larger and tougher burrs than less ductile materials. At the micro level, grains and grain boundaries affect burr size as does crystalline orientation.

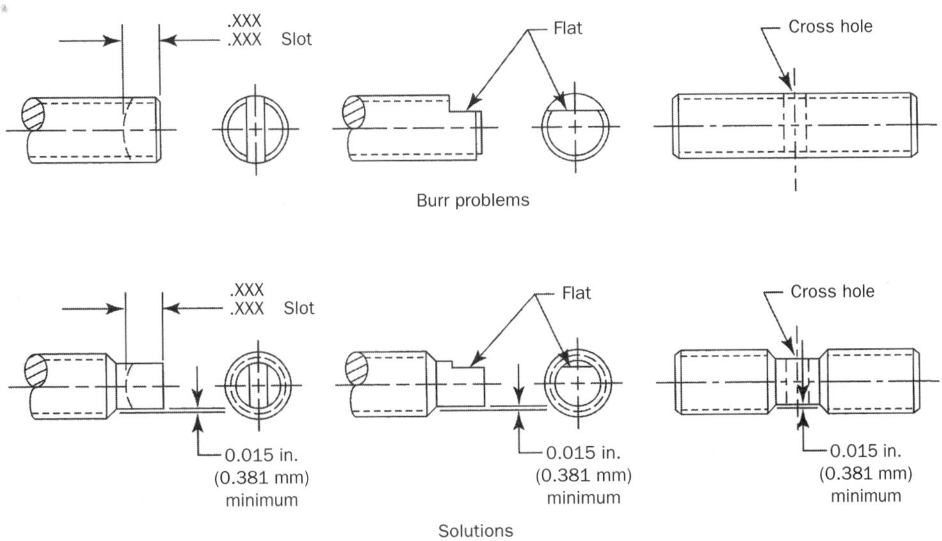

FIGURE 3.34.2 Machining through threads must be avoided wherever possible. The burrs produced are exceptionally difficult to remove without affecting the rest of the part (PMPA 1983).

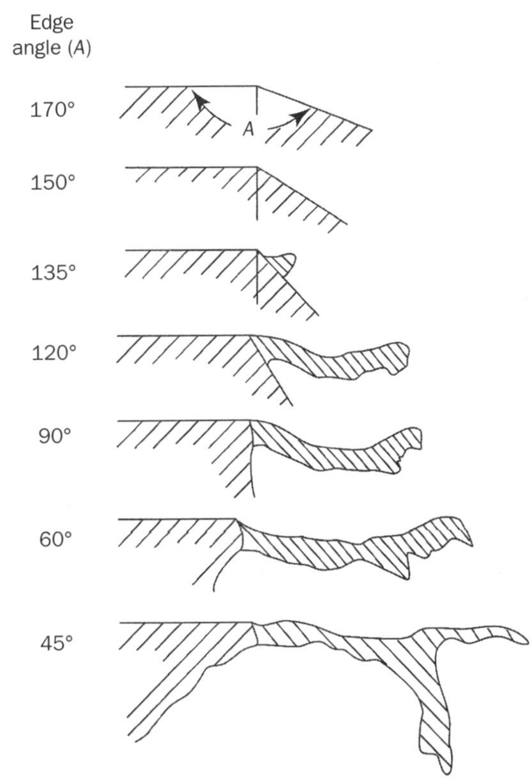

FIGURE 3.34.3 When a cutting tool is fed over a 170-degree edge, no burr is produced. When it is fed over a 30-degree edge, a thick and long burr is produced (Shafer, 1975).

Sharp tools are essential to minimizing burr size. For large grain materials such as AISI 1045, the minimum burr size is obtained while machining an undeformed chip at least equal to the average grain size. Large depths of cut for roughing followed with low depths of cut and frequent tool change for finishing result in the smallest burrs. Burrs that are 0.25 mm (0.010 in.) thick can only be removed via machining/grinding unless large changes in part dimensions are allowed. In contrast 0.01-mm-thick (0.0004-in.) burrs can be removed with conventional processes with small or no changes to size.

Preventing Burrs

It is difficult to prevent burrs with conventional machining, but some can be prevented by just altering path planning for the cutter. That may eliminate as much as 80% of some milled part problems. While manufacturers wish to eliminate all the burr, saving 80% of the costs by prevention is still a major benefit for all companies.

The burr prevention processes include:

- Electroforming
- Ion beam machining (IBM)
- Reactive ion etching (RIE)
- Electron beam machining (EBM)
- Chemical machining (CHM)
- Silicon micro machining
- LIGA
- Deep etching, electroforming, micro replication (DEM)
- Femtosecond laser machining
- Electrojet machining
- Electrochemical machining (ECM)
- Jet ECM
- Electrochemical grinding (ECG)
- Ultrasonic machining (both sinking version and a contouring version)
- Molding to shape and size
- Coining
- Hot embossing
- Abrasive jet machining (AJM)
- Micro water jet machining (μWJM)
- Extrusion
- Variants of wire EDM
- Some electrical discharge grinding (EDG) approaches
- Form tool conventional cutters
- Cutter path and force control

Precision micro impellers (400 to 600 μm thick) have been forged without flash by altering grain size of the material being forced into the die, and preventing gaps at the splits in the die. Large grains generally result in underfilling of the die and subsequent loss of dimensions.

Micro molding is a common process for making micro parts. Control systems, processing knowledge, and materials exist that allow molding with no flash at edges or parting lines, but it requires exacting precision.

Coining and vibration-assisted coining produces 5- to 10-μm deep (0.0002- to 0.0004-in.) grooves in 303 and 304L stainless steel.

Femtosecond lasers produce near-perfect edges in many cutting applications. Lasers cutting at nanosecond pulses leave obvious debris at edges. Picosecond lasers can produce debris-free edges in some materials and conditions.

Ion beam machining produces high edge quality, but it is extremely slow [typically 1 to 3 μm (40 to 120 μin.) per hour over a 1-cm (0.4-in.) area] but it can be used on glass, silicon, and a variety of metals and polymers. Some researchers claim it can produce features as small as 100 nm and hold finishes of 0.1 to 1.5 nm (0.004 to 0.06 μin.).

Electron beam machining can cut at a rate of 1 to 4 cubic millimeters per second (0.00006 to 0.000024 in.^{3}/s) and is a more typical process to be used in micro machining than IBM. A heat-affected zone is created by this process—from 2 μm to 25 μm (0.000078 in. to 0.001 in.) deep. It is used to produce holes in the order of 20 μm to 10 mm (0.0008 to 0.4000 in.). Metal foils are drilled at rates of one hole every 10 microseconds. Rates of 10,000 to 100,000 holes per second are also reported. It is normally used for thin materials.

Minimizing Burrs

Adding ultrasonic motion to micro machining often helps reduce burr thickness. It does not prevent them, but can reduce their size by 80%. That is true for mechanical micro drilling and end milling. Nearly burr-free edges have been made using elliptical vibration-assisted or orbital machining.

The manufacturing engineer controls several variables. The direction the cutter feeds into the cut makes a large difference. Up-milling produces smaller burrs than down-milling; minimizing the runout of micro tools reduces burr size. Burr height is proportional to feed rate unless the material is work hardened, like some stainless steels. Too slow a feed rate increases burr size then. Burr thickness is also affected by feed rate. Using helix angles that force chips down rather than up reduces burr size, but the cutter still has to remove chips effectively. Smaller cutting edge radii produce smaller burrs; burr size is larger in ductile materials than in nonductile materials; bigger wedge angles produce smaller burrs.

Deburring Processes for Micro Features

The most applicable deburring processes for micro features and micro parts include those listed below. In all cases they work best when the burrs are very small.

- Magnetic abrasive finishing (MAF)
- Ultrasonic deburring (USM)
- Electrochemical deburring (ECD)
- Electropolish deburring

- Electrical discharge deburring (EDM)
- Flat lapping
- Abrasive jet deburring (AJM)
- Centrifugal barrel finishing
- Centrifugal disc finishing
- Coining
- Manual deburring with dental tools

Magnetic Abrasive Finishing

Magnetic abrasive finishing (MAF) (see Chap. 3.19 for this process) is effective when precision surface finishes are required and burrs are short and thin. It leaves a burr-free, but nearly sharp edge while imparting a finish under 25 nm (1 μin.) Ra. Simple equipment is used, but some applications require equipment designed for a specific geometry. It can accommodate any size hole or other feature. It is used on external surfaces, and inside tubular parts. It is used to polish spherical bearings and a host of unusual shapes. It can also easily accommodate rectangular or box-like parts. MAF deburrs and polishes hole intersections as well as complex outer shapes in 1 to 15 min. It is particularly effective in stainless steel but it has been successfully used on ceramics and thermosetting plastics as well as a variety of metals.

MAF can remove 5- to 10-μm-tall (0.0002- to 0.0004-in.) burrs from 100-μm-diameter (0.004-in.) holes, round the edges, and smooth adjoining surfaces in 0.5 to 4.0 min per part. Other materials, features, and burr sizes may require 10 min per part. Vibration-assisted magnetic brush deburring improves the finish of planar, cylindrical, spherical, or odd-shaped parts and removes burrs at the edges of the 2-mm (0.080-in.) holes removing a 50-μm (0.002-in.) high burr in 100 s using 330 μm (0.0132 in.) iron particles mixed with an 80 μm (0.0032 in.) magnetic abrasive. A new variation of the process titled magnetic abrasive jet finishing works on the inside of hollow cylindrical parts. The magnetic media is carried by a pressurized fluid. Compressed air moves the fluid past the surfaces needing treatment, while the magnetic field outside the tube draws the media against the tube to scrub the surface and remove any slivers, burrs, or roughness there.

A similar magnetic approach and setup adds a magnetorheological fluid (see Chap. 3.30). The part is immersed in the tank with the suspensions of micrometer-sized magnetizable particles dispersed in a carrier medium with small concentrations of stabilizers. In milliseconds after turning on the magnetic field, this fluid turns into a plastic-like solid. The actual properties of the fluid are controlled by varying the strength of the field. This is typically a process used for brittle materials—optical parts for example. On BK7 glass, the surface roughness was reduced to 1 nm. It is used to polish 10 to 100 mm (0.4 to 4.0 in.) convex and concave parts. The driving vertical electrode is a needle-like magnetic pole and like the previously mentioned process the particles become a pseudo flexible brush. It is used to attain dimensional accuracies finer than 1 μm on these materials.

Ultrasonic Deburring

Ultrasonic deburring which ultrasonically agitates a mixture of fluid and fine abrasives, removes 120-μin.-tall (0.0048-in.) burrs from aluminum holes with a 750-W ultrasonic transducer in cycles of up to 180 s. It can be effective when the burrs are thin and exposed on external surfaces. It is a process which has been successful in some applications and unsuccessful in others. It is a low-cost process but some experimentation is required to select the optimum operating conditions.

Electrochemical Deburring

Electrochemical deburring typically uses a salt solution and a low-voltage current to electrochemically dissolve material from edges (see Chap. 3.9 for this process). In addition to being rapid and precise, it also leaves a very smooth surface. An electrode that is typically the mirror image of the edges to be deburred is required. Cycle times are typically 30 to 45 s and several parts or dozens of holes can be deburred in that same time. Knife blades which had some rather severe burrs have been finished in 2 s, but most parts will require 10 to 40 s. Titanium micro parts are a challenge in traditional ECD, but the use of low frequency (40 to 50 Hz) with a long forward modulation and short reverse modulation to control the oxide film re-healing during reverse modulation has worked on titanium. Titanium parts have been finished in less than 3 s to 0.005 μm (0.2 μin.) surface finishes. Stainless steel, aluminum, and copper-based alloys are the best materials for this process. Stainless-steel features having 0.45 μm (18 μin) initial Ra finish have been reduced to 0.125 μm (5 μin.) during deburring.

Electrochemical deburring can deburr dozens of intersecting holes as rapidly as it does one. When stainless steel is processed, hexavalent chrome is released in the solution, which must be discharged as hazardous waste after filtering. In addition, ECD creates a small radius, normally from 125 to 250 μm (0.005 to 0.010 in.) but larger radii can be produced. Cutting action can be tailored to specific portions of intersecting features if desired. A carbon deposit adjacent to the intersection requires brushing or cleaning for many applications. Machines can be manually operated or full automation is also available.

Electropolishing

Electropolishing uses a weak acid and electricity to polish, remove extra thickness, and to deburr parts. The advantage of electropolishing is that the electric current is concentrated at the burrs and removes them much more rapidly than taking material from the planar surfaces. In addition, this action removes the surface cracks and some particles at edges that reduce fatigue life. It is a process that has been used on miniature or micro features successfully for over 40 years. It leaves a very shiny surface, but the fact that it does remove metal from all over the part limits its application. It is used to remove laser recast from human implant stents on the majority of stents. Because the removal depends on current densities across surfaces, precision applications need to experiment some to obtain uniform removal from all surfaces. Deburring times range from a few seconds to several seconds and the process can be automated with overhead "clothes lines" dipping parts into solution continuously. Desktop units are available as are large-scale lines and most large cities have vendors specializing in the process. For very narrow slots and small holes, special electrodes may be necessary to provide best operation.

Electropolishing works on most metals and has been used for micro features for several decades.

Electrical Discharge Deburring

Micro EDM is used for deburring micro features as well as for making features. The advantage of this process for deburring is that it only attacks the edges at which it is directed. In this process a mirror image formed electrode of the area to be deburred is fed down to the edge or a small diameter EDM tool is traversed along the burr-laden edge. The latter approach is not widely used, but it eliminates the need for a special shape electrode. The simpler process is to feed down to the edges. But those who have treated the tool as an end mill note that three passes along the edges may be required to totally remove the burr and any material at the root of the burr. A rough deburring pass can be programmed to remove the top 60% or so of the burr. The semifinishing pass removes everything down to under 1 μm. To finish the edge a shallow pass is generated to assure no material is left above the nominal surface. If feeding the tool like a cutter across the material, it is slow, for example, 7 to 14 μm/s (0.0003 to 0.0006 in./s). The discharge energy for the finish pass can be programmed and might be less than a hundredth of that for rough deburring. Semifinish operation may take 20% longer. Initial burrs 200 to 1000 μm (0.008 to 0.080 in.) tall and from 80 to 200 μm (0.0032 to 0.0080 in.) thick have been fully removed after three passes. Electrodes with the mirror image of the feature edges can be fed down rather than across edges to greatly speed the process.

Flat Lapping

Flat lapping is not considered to be an edge-rounding process, but by adding polymer particles and a soap film to carry the traditional fine lapping abrasives some edge finishing can be accomplished. Urethane particles of 10 μm (0.0004 in.) diameter mixed with 1.3 μm (0.00003 in.) silicon carbide abrasive and a zinc-based soap provide good finishes and an edge rounding of 0.1 to 0.3 μm (4 to 12 μin.) on flat parts not obtained with conventional lapping.

Micro Blasting (Abrasive Micro Jet Machining)

Micro blasters have been used for several decades to remove burrs from tiny parts. They are commonly used today with up to four axes of coordinated movement (and even a 13-axis nozzle head) to deburr, and to remove oxides and dross from medical stents (see Chap. 3.42). Some use the minute sizes of blasting grains while many now rely on miniature glass beads. A hand size valve with 14 miniature intersecting holes is deburred in 1 min in one application. A 5-s blast removes burrs from a toothpick-size hole.

Manual and automated systems are available. Finished surfaces have a fine, blasted texture around the intersections. While typically used for external edges, micro blasters can be inserted into holes as small as 1.4 mm (0.056 in.) diameter to deburr hole intersections and reach under very small ledges.[1] This process is successful with nozzles as small as 0.2 mm (0.007 in.). Deburring stents, or parts having slots of 125 to 250 μm (0.005 to 0.010 in.) wide can be done with this process in most materials. It works on many plastics as well as ceramics and metals. In addition to deburring, these miniature tools can bevel hard brittle materials such as silicon wafers.

Centrifugal Barrel Deburring

Centrifugal barrel deburring is one of the most widely used processes for treating hundreds of parts at one time. Small abrasive "pebbles" or miniature-designed shapes are loaded in a rotating barrel with a fine abrasive powder, water, and a lot of parts

[1]The point here is that the nozzle can fit down into a very small hole to allow the nozzle to shoot sideways into hole intersections.

(20, 1000, or 10,000 at a time). The centrifugal action of the machine greatly increases the force between pebbles and parts and causes a slow grinding action that passes along surfaces and over edges. This is a good process for micro size parts produced on automatic screw machines since all the edges are readily accessible to the pebbles.

It is not a good choice, however, for slots, holes, and features in the range of 250 μm (0.010 in.) or less since the pebble size that can work over such edges is extremely small and the smaller the pebbles the slower any action occurs. Similarly it will not work well in internal features. It works best for external edges and features typically above 250 μm (0.010 in.). Thousands of parts can be deburred in a day in cycles of 20 to 30 min. Burrs are removed, the surface polished, and edge radii of 0.076 to 0.250 mm (0.003 to 0.010 in.) can be put on micro parts. Centrifugal disc deburring which operates similar to centrifugal barrels can be used on larger micro parts, but the 250-μm (0.010-in.) gap near the bottom of the machine will allow smaller parts to escape or be damaged.[5]

Coining

As discussed above coining by mashing a die over the edges needing burr removal is a fast and effective way to eliminate burrs from edges. The burrs are either broken loose or more often physically mashed back into the parent material, leaving the mirror image of whatever radius the die has. Part structures must be strong enough to support the coining forces. The coining can produce any radius desired and parts can be finished in rates of 50 to 100 per minute.

Hot Embossing

Embossing is normally applied to polymers or thin foils to produce a shape (see Chap. 3.26). This may be an effective means, however, to produce the desired edge quality on the light metals since it is in some sense a coining operation and the heat will soften slightly the metals or plastics being addressed.

Plasma Glow Deburring

Plasmas reduce burrs to nothing instantly. Plasma glow is a deburring/deflashing process that has been used for this purpose effectively on polymers (Gillespie, 1999). The known applications are proprietary so no open literature is available, but it is effective. It is used primarily for cleaning aerospace and electronic parts and deburring is a secondary application. It uses low-temperature plasma of oxygen or argon that surrounds all the part edges so deburring/deflashing occurs everywhere on the part.

Laser Deburring

Laser deburring can be a feasible process for micro parts since the beam can be moved at whatever rate needed. Unfortunately, there is no published data on their use although their use has been described. While nano, pico, and other lasers can be used, femtosecond lasers leave no material at the edges and no heat-affected zone so they are the best candidate for high-quality edges. Picosecond lasers can also be effective on some materials. The very slow cutting speed of femtosecond lasers is a major restraint that the other lasers do not have.

Manual Deburring

Physical cutting off and polishing edges with dental-like tools and miniature knives has been used for decades on parts with minute configurations.[6] It is an economical process when limited quantities are to be run or only small areas have to be deburred. It requires almost negligible equipment investment compared to some of the other processes, but it requires highly skilled workers. As hole sizes or slots get smaller than 0.25 mm (0.010 in.), manual tools are typically ineffective. Manual abrasive jet deburring, however, can finish features smaller than this.

Chlorine Gas Deburring

If you heat an iron-based part up to about 316°C (600°F) and immerse it in chlorine gas, the iron will react to produce iron chloride which will evaporate immediately. Since the burrs are much thinner normally than the rest of the part features, they disappear quickly. The process was developed in the 1960s and has apparently never been used commercially. Its life-safety concerns have apparently overruled its commercial use. Its potential appears unaffected by part or feature size but is limited to steel parts.

Processes Not Usually Considered as Deburring Processes

As micro parts continue to fall to finer levels of finish and precision, some of the polishing processes may have some utility for removing true micro burrs. These are not likely to be successful on traditional burrs which are thicker and taller than 76 μm (0.003 in.). They are for the very fine burrs, for example, when surface grinding steels' microscopic examination of the surface reveals long slivers of burr-like material in the recesses of the ground surface (as opposed to the edges). These can truly be called micro burrs because they tend to have thicknesses of only 2.5 μm (0.0001 in.).

- Polymer particle abrasive processing (POPAP)
- Magnetorheological finishing (MRF)
- Magnetorheological abrasive flow finishing (MRAFF)
- Chemical-assisted mechanical polishing (CMP)
- Mechanical electrochemical polishing
- Swirling flow finishing (SFF)

Rheological fluids combined with MAF can conceptually accommodate any external geometry and is capable of removing micro burrs. It can accommodate single features or the entire part in one setting. It produces finishes as fine as 0.8 nm (0.03 μin.) in glass (see Chap. 3.30).

Combining abrasive flow deburring with magnetorheological fluids has demonstrated that it can remove grinding marks and this ability also implies that it can remove micro burrs from grinding from surfaces and edges.

Chemical-assisted mechanical polishing (CMP) in which chemicals are fed over a rotating part while an abrasive-coated nonwoven nylon scrubber wipes away the soft oxide film created when the chemicals react with the part can provide smooth surfaces and would remove micro burrs. The challenge is finding the appropriate geometries that it can be effective on.

An electrochemical version of the CMP may provide more of an abrasive force against the part to accelerate the CMP process. Its use to totally remove a 100-μm-tall (0.004-in.) burr on titanium is impressive. Surfaces of 0.2 to 0.5 μm (8 to 20 μin.) roughness Ra have been produced and maintained by the process.

Measuring Micro Burrs

There are over 50 methods available for detecting and inspecting burrs, but only 12 are applicable for micro parts. Conventional stereo microscopes are widely used. Triangulated conoscopic holography with interferometry provides a 3D display of the thickness and height of the burrs. 2D and 3D vision systems can detect burrs as small as 5 µm (0.0002 in.) tall to as high as 1 mm. White-light interferometry and multi-focus 3D microscopy can provide 5-µm-height (0.0002 in.) discrimination, but doing it at assembly line rates might be an issue. Taking a part having 200 mm (~8 in.) of total potentially burr-laden edges, a white-light system would need to acquire approximately 100 images. If each image is generated in 1 s that would require 100 s which may be too long for in-line conveyor processes.

The Optimum Approach

For many applications the optimum approach is a combination of all the ideas presented: that is (1) design the parts for the most judicious burr removal; (2) prevent burrs on the hard-to-reach areas; (3) minimize the burrs where possible by better choices of cutting conditions or processes; (4) put the burrs where you can take them off; (5) remove the burrs in the operation that produced them so you do not have to handle the parts again. (In some instances this means machine and deburr while the parts are still attached to their base material. In other situations it is better to have them individually handled.) Brushing can remove small burrs easily, but it will not remove heavy burrs on micro parts. (6) If conventional commercial processes are not adequate combine two processes into one operation, and lastly (7) innovate on design, manufacturing method, material, and deburring process. To do any of this, the reader has to understand the basics of these principles which apply regardless of size. They are all described in the literature.

References

Gillespie, L., *Deburring and Edge Finishing Handbook*, 1999, Society of Manufacturing Engineers, Dearborn, MI.

Strasser, F, How Control of Burrs Aids Sheet Metal Stamping, 1960, *The Iron Age*, 185 (3):90–92.

Designer's Guide, 1983, National Screw Machine Products Association (now PMPA), Brecksville, OH.

Shafer, F., *Entgraten*, 1975, Krausskopf-Verlag, Mainz, Germany.

Gillespie, L., *Mass Finishing Handbook*, 2007, Industrial Press, New York, NY.

Gillespie, L., *Hand Deburring: Increasing Shop Productivity*, 2003, Society of Manufacturing Engineers, Dearborn, MI.

CHAPTER 3.35

Electrospinning[1]

LaRoux K. Gillespie

Manufacturing Consultant
Andover, Kansas

Electrostatic spinning can create a wide array of micro and nano-sized fibers. And because the process, patented in 1902 and usually called electrospinning, is straightforward, it's a relatively inexpensive way to produce fibers.

There are at least 16 variations of electrospinning performed today, with more expected to come online as the technology evolves. The most common is "solution electrospinning." It involves applying a sufficiently high-voltage charge (15,000 to 25,000 V and up to 60 kV for some applications) to a solution-filled syringe (or capillary tube). A pump pushes the solution through the syringe tip at a steady rate, directing the stream toward a collector plate that has an opposing charge. As the liquid falls in a circular motion, the solvent evaporates, leaving a charged, solid fiber on the collector.

Conventional fiber-spinning processes, which involve extruding a thick, viscous liquid through holes in a plate, cannot produce usable fibers smaller than 2 μm (79 μin.). The electrospinning process, conversely, can produce robust fibers from 10 nm to 10 μm (0.4 μin. to 0.00039 in.) in diameter. While typically used for the production of polymer fibers, electrospinning can also produce metal, ceramic, and glass fibers, or combinations of those materials. The process has been used to make fibers from more than 100 different materials.

Electrospinning can be used to produce solid fibers or tubes. In certain applications, the product is a mesh-like tube or a combination of materials. Liquids and gases also can be encapsulated in the middle of a fiber, and gradient structures can be produced that have properties that vary along the length of the fiber or across the thickness of finished mats made from the fibers.

Raw material costs span a broad range. Polymers range from a few cents per kilogram for gelatin to $43,000/kg for solutions of gold or other rare-earth materials. Platinum is used for some battery applications.

[1]This chapter is drawn from the author's article "Spinning Tale" that appeared in the January/February 2011 issue of *MICROmanufacturing* magazine, which is no longer published.

Many electrospun fibers can be coated by chemical vapor deposition (CVD) or other process to enhance their properties. Some fibers have been coated with silver, which kills disease-causing germs, molds, and about 650 types of bacteria.

Electrostatic-spun fibers have application in the filtration, textile, photonics, medical, energy, and aerospace industries.

Fiber Characteristics

Electrospinning is well-suited to micro manufacturing. One reason is the fibers' molecules or crystals are often oriented such that they have high tensile strength. Some boast a Young's modulus of 50 GPa (7.25×10^6 psi) or higher. For comparison purposes, 303Se stainless steel has a Young's modulus of 193 GPa (28×10^6 psi).

Because of their cylindrical structure and controllable porosity, electrospun fibers can have surface-area-to-volume (or mass) ratios up to 100,000 times higher than conventionally spun fibers.

The surface-area-to-volume ratio for a cylinder is expressed as

$$(2\pi r \times l)/(\pi r^2 \times l) = 2/r \qquad (3.35.1)$$

where r is the radius and l is the length.

So, for a 25.4 mm long by 25.4 diameter (1 in. long × 1 in. diameter) cylindrical rod, the ratio is 2. For a 25.4 mm long × 0.254 mm (1 in. long × 0.010 in. diameter) fiber, the ratio is 200. An electrospun fiber of 30 nm (1 μin.) will have an exposed surface-area-to-volume ratio that is 100,000 times that of a conventional 76 μm (0.003 in.) fiber.

Fiber diameter can be increased by concentrating the solution and be decreased by reducing the electrical conductivity of the solution—typically by varying salt content. Reducing viscosity and lowering flow rates also reduces fiber diameter.

Porous fibers are produced by mixing solvents that cause phase separation or by producing them in a high-humidity environment, which creates "water islands" in the mixture that evaporate as the fibers cool. Salts with small ionic radii produce fine fibers, while those with larger radii produce coarse fibers. Solvent volatility greatly affects fiber porosity.

Fibers can be laid parallel to each other by using parallel collector plates. Precise fiber placement is made possible by spacing the plates a few millimeters apart instead of the typical 10 to 20 cm (3.94 to 7.87 in.).

Process variables include solution viscosity; polymer concentration; conductivity of the solution, polymer, and solvents; solution surface tension; polymer molecular weight; and volatility of solvents used (i.e., vapor pressure). Other factors influence process results, too, including electrical field strength; flow rate; distance from nozzle tip to the collector; tip design and placement; collector design; take-up velocity on the collector plate; temperature; humidity; and atmospheric pressure.

Co-electrospinning

Metals, glasses, and ceramics can be added to fibers via a process called "co-electrospinning." With co-electrospinning, different materials are pumped through two coaxially arranged chambers. The process allows spinning of hollow fibers and fibers having a core and a shell.

To incorporate these nonpolymer materials into fibers, low-molar-mass precursor materials, such as metal salts, must be added to the outer polymer. Subsequent treatment of the precursor removes it. (This can be achieved by chemical means, evaporation, or sintering.)

Several layers of materials can be deposited, allowing the creation of complex fibers. The outer shell could be designed to change over time, for example, or to modify itself as the environment it is used in changes.

Because multiple materials can be encapsulated, electrospun fibers can also incorporate nanoparticles, including ferromagnetic or superparamagnetic particles. Use of these particles permits manipulation of fibers by an external magnetic field. And because the magnetic field can be turned on and off, handling of micro sized components is simplified.

As mentioned earlier, one property of micro and nano fibers is that their surface-to-volume ratio is high. This can greatly enhance their ability to absorb ions in capacitor applications, and, hence, allows them to produce powerful miniature capacitors. Carbon nanotubes have already been made into miniature field-effect transistors that, compared to semiconductor-based devices, exhibit a twenty-fold improvement in transconductance. The high electrical conductivity of carbon nanofibers (490 S/m) opens up many potential electrical applications. For comparison, the electrical conductivity of Nylon 6 is 1×10^{-15} while annealed copper is 5.8×10^{7} S/m.

Nanofiber Applications

Among the current applications for nanofibers are filtration and separation devices and medical products for treating wounds. Electrospun fibers also are used to remove bacteria and viruses during the manufacture of common drugs, such as Tylenol, as well as high-end heart medicines. The superfine fibers ensure that molecules of unwanted, dangerous life forms do not pass through the solutions into the final product.

Fabrics made from electrospun fibers are commercially available and electrospinning is key to certain medical applications, including the production of biomaterial scaffolds used in human-tissue engineering and drug-delivery systems.

Two potential areas of application for electrospun fibers are polymer batteries and fuel cells. Polymer batteries already power some notebook computers and cell phones. Lithium-ion cells built with polyvinylidene fluoride (PVDF) have better wettability and thermal stability than existing battery materials. Anything that shrinks the distance between a battery's electrodes improves performance and reduces its size. Nanofibers conform to the "smaller is better" mindset of today's designers.

Because polymers can be converted to carbon rods, it may be possible to electrospin 20-nm (0.8-µin.) electrodes for EDMing operations.

Fiber diameters of 20 nm (0.8 µin.) are available commercially today and TiO_2, MgO, and Al_2O_3 fibers as small as 5 nm (0.2 µin.) in diameter have reportedly been made in the laboratory, but not mass-produced. Mass-production is currently limited to producing fibers with diameters of about 50 nm (2 µin.) in most materials.

Equipment

Electrostatic spinning machines, like the fibers they produce, are available in a wide range of models from several manufacturers. Single-nozzle electrospinning machines like those typically used in a laboratory are readily available and machines able to electrospin fibers can be easily constructed from simple lab supplies for much less, assuming a high-voltage power source is available. Equipment for mass-producing electrospun fibers are also readily available.

Organic fiber production lines can make 80- to 700-nm-diameter (3- to 28-μin.) fibers at speeds of 0.2 to 5 m/min (7.9 to 197 in./min). Setup times range from 5 to 20 min. Machine maintenance takes about 15 h per month for systems with four spinning electrodes.

Some of the fastest machines, which spin 400-nm-diameter PVA fibers can produce 11 million square meters (118 million square feet) of such products annually from a single machine. Machines for inorganics—such as TiO_2, $Li_4Ti_5O_{12}$, Al_2O_3, SiO_2, ZrO_2, and other oxides—can produce 300-nm (11.8-μin.) fibers at a rate of 1370 g/h (44 oz/h) using six heads.

CHAPTER 3.36

Designing for Resistance Welding Micro Parts

Marty Mewborne

Amada Miyachi America
Monrovia, California

Resistance Welding Basics

Resistance welding (RW) is a fairly simple thermoelectric process for joining metal parts. Heat is generated at the interface of two or more parts by passing an electrical current through them. The parts are pressed together and held in contact by electrodes that apply force and provide a path for the current to flow. The name "resistance welding" derives from the fact that the heat is created by the material's resistance to the flow of electrical current. The amount of heat generated is affected by the resistivity of the materials. Everything else being equal, highly resistive parts will heat up more quickly than conductive parts. In addition to the resistance of the parts to be joined, the resistance of the electrodes also comes into play. Resistive electrodes can add heat to the parts while conductive electrodes can sink heat away.

Advantages of the RW process include very short processing times, and—due to the low voltage at the workpieces—safety. It is also a clean and environmentally friendly process. And of course, its key advantage is that it forms an extremely reliable electromechanical joint.

When the current is passed through the parts, the heat that is generated decomposes dirt and grease, breaks up any oxide film, and then softens or melts the metals. The electrode force holds the parts together as this occurs. Atoms on either side migrate to form a diffusion bond, or the metals melt and mix to form a fusion weld. The force is maintained after the current flows and the strength develops as the joint cools, during the hold time. Figure 3.36.1 shows the relationship between force, current, and time in a basic welding schedule, which can be used for the bulk of all RW applications.

Figure 3.36.2 shows the types of bonds that can be achieved with RW, along with application examples.

Figure 3.36.3 shows the components of an RW system. A basic system is depicted in the top graphic, while the bottom shows an advanced system.

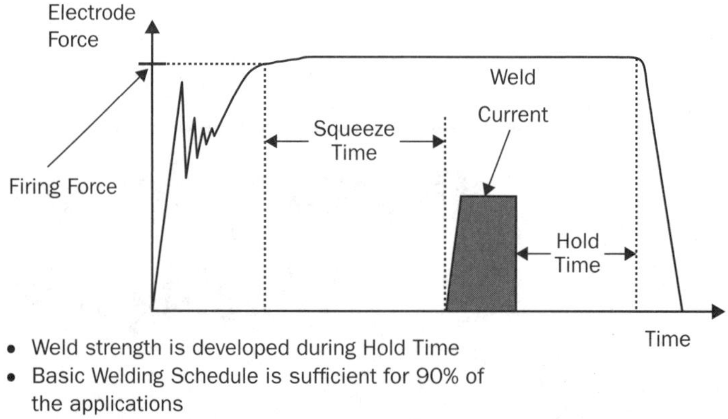

FIGURE 3.36.1 Typical force current and time chart for resistance welding.

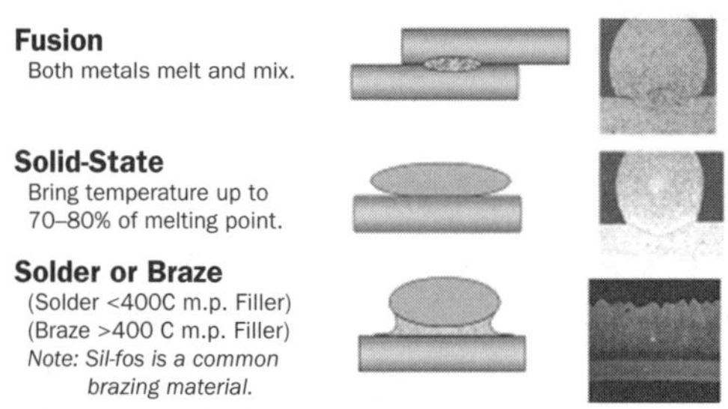

FIGURE 3.36.2 Types of resistance welding bonds.

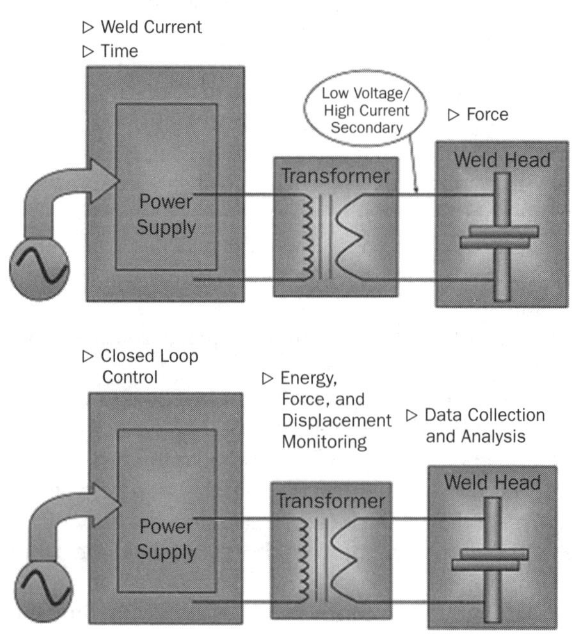

FIGURE 3.36.3 Resistance welding systems components.

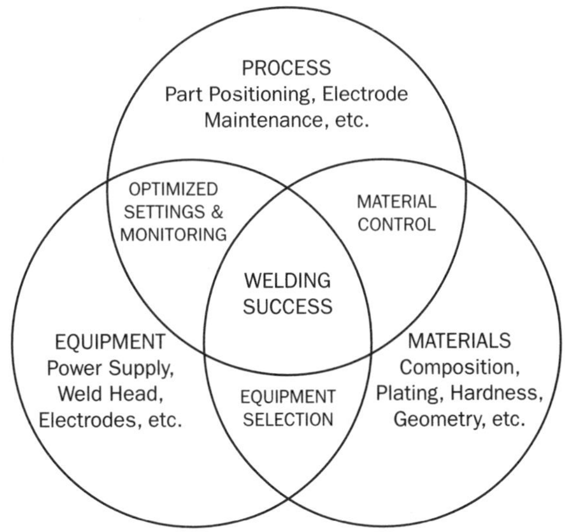

FIGURE 3.36.4 Resistance welding process, equipment, and materials overlap.

Welding engineers have traditionally focused on the three main variables of current, time, and force when developing and optimizing an RW process. Although these three factors are key, there are many other factors that cannot be ignored if welding success is to be achieved. The diagram shown in Fig. 3.36.4 lists some of the important variables associated with materials, equipment, and process and shows how these three areas overlap in a successful RW environment.

Resistance Welding for Micro Joining Small Parts

A micro part is a small part. When RW is used for joining such small parts together, it is commonly referred to as micro joining. But how small is small? The term means different things to different people. The definition of *micro parts* adopted for this handbook is "parts in which a wire thickness (or diameter) is on the order of 0.25 mm (0.010 in.) or smaller, or where sheet or foil thickness is 0.076 mm (0.003 in.) or smaller."

RW of micro parts is done in many industry sectors, but is most widespread in today's medical device manufacturing industry. A wide variety of medical devices use micro welding, including catheters, hearing aids, small surgical tools, and implantable pacemaker leads. The longstanding trend of making devices smaller and smaller to be less intrusive for patients is definitely driving the use of micro joining.

In addition, micro welding is used in automotive safety applications, namely, squibs for airbag detonators. Other than the tiniest of medical devices, the smallest wires utilizing RW are squib wires, which can be less than 0.025 mm (0.001 in.) in diameter, typically on the order of 20 μm (0.00079 in.).

Using RW for micro joining offers several advantages. First of all, welding electrodes hold the parts together, unlike laser welding, which requires a tool to hold parts together in intimate contact. Initial capital investment is also much lower for RW compared to laser welding.

Another benefit is that with RW, the heat is generated from inside out, not from the outside in as it is in other welding techniques. This means energy or heat can be focused at the interface of the parts where the weld should occur. Other methods require applying heat through the outside surfaces of the parts to get to the part-to-part interface. This additional heat has the potential to damage the small parts involved in micro joining.

RW does have a couple of drawbacks for micro joining. Small parts require the use of small electrodes, which require frequent maintenance, cleaning, and replacement, adding to the overall cycle time and ultimately affecting the production rate. In addition, the welding process may result in part deformation and marking (small but visible impressions) from the electrodes. When metals heat up, they get soft, so when the electrodes press on them, they can deform, especially wires. Similarly, since there is contact resistance at the electrode-to-part interface, this area heats up, which can cause marking (indentations on the part). Excessive wire deformation can affect weld strength; marking is normally just a cosmetic issue.

Micro Joining Design Challenges

There are several challenges faced by designers using RW for micro parts. The difficulty of designing parts for weldability increases as part sizes decrease. First of all, parts must be designed in such a way that one can manipulate them into a welding machine, and control placement and alignment between welding electrodes during a manufacturing process, which is no easy feat, considering how tiny they are.

RW relies heavily on consistency of materials and process. Part size must also be controlled to meet tight tolerances. Therefore, whether machined or produced using other methods, part size must be consistent from one to the next.

Heat balance is another concern; both parts must get hot at the same time. Heat balance is critical even when welding normal sized parts, but the challenge is definitely magnified with micro parts. If the thermal mass of one part is large in relation to the other, heat balance techniques must be used to overcome the disparity.

Many medical devices use MP35N, a high-strength, corrosion-resistant, nickel-cobalt-based alloy in their design. Other common materials used are stainless steel, titanium, and platinum-iridium. Welding each material must take into account specific material properties such as melting point, resistivity, hardness, etc.

A final challenge with micro welding is the issue of working with insulated wires, such as enamel-coated magnet wire. Since insulation, by definition, will not allow the current to pass, it must be stripped off before the wires are presented to the welder. In some cases, the insulation can be soft and easy to remove, but other hard, high temperature insulation is difficult to remove. Chemical etches, solder dipping, abrasive processes (sandpaper), or other abrasion methods can be used, but it is difficult to completely remove the insulation without nicking the wire. Residual insulation left on the wire will cause localized overheating, and can result in blown welds.

Precise Control Is the Key to Meeting Micro Joining Challenges

Designing the parts for weldability is just the first challenge when aiming for welding success. The next step is to choose the appropriate equipment for the application. Equipment must be matched to the application in terms of capacity and features. Current, time, and force ranges must be in line with the size of the parts. Power supplies must accurately control the electrical parameters—current, voltage, or power, and weld time. The weld head must be capable of maintaining precise control over the welding force throughout the process. Simply put, precise control over the electrical and mechanical variables is essential to ensuring consistent welding results.

Advances in closed-loop power supply technology have led to new solutions for micro miniature RW. These micro welders control down to as low as 5 A of current, with fast response times and precision control, which works especially well for the smallest of micro joining applications.

Force control for micro joining has also drastically improved. Equipment operators welding very small parts have been accustomed to applying the welding force by pressing on a cable pedal with their foot. The operator typically holds the parts by hand while looking through a microscope, and then presses on the cable pedal to control the downward motion of the upper electrode. They hover to get the alignment just right, then continue pressing with their foot to complete the welding sequence. The operators have historically preferred cable pedals over pneumatic weld heads, since pneumatic heads do not allow them to hover over the parts before welding. Unfortunately, if the operator presses too quickly on the cable pedal, there can be too much impact force on the small parts. Or, if they press too hard on the pedal, they may apply too much welding force.

To solve the challenge, new electronic weld heads have been developed that can accurately control the welding force down to 200 g. Equipped with a linear foot switch, these new heads allow the operator to control the downward motion like a cable pedal, but once the force is reached, the electronics takes over to control the force and timing during the weld. The operators feel comfortable with the new linear foot switches since they mimic the manual cable pedals they are accustomed to using. The operator still looks at the parts, hovers, and when ready to weld, presses down on the pedal. The head comes down and makes the weld, with force and timing being controlled electronically.

Electrode Design and Tooling

When parts are small, electrode size, shape, and material are even more critical than with larger parts. If welding a 50-µm-diameter (0.002-in.) wire to a terminal, the electrode contacting that wire needs to be flat on its face—flatter than 50 µm (0.002 in.). If it is not, it will not only touch the wire, it will short down to the terminal below. Flatness of the electrode face becomes critical. Electrodes must be frequently maintained to keep them flat during welding at all times. Likewise, any tooling that keeps parts and electrodes in place during welding must be maintained to ensure proper alignment.

Three Areas to Consider When Designing Micro Parts for Resistance Welding

Proper designing of parts for weldability requires attention to three main areas: material properties, surface conditions, and physical part design.

Material Properties

Material properties that must be considered include electrical and thermal conductivity, melting point, hardness, and welding compatibility. Metals can be classified into

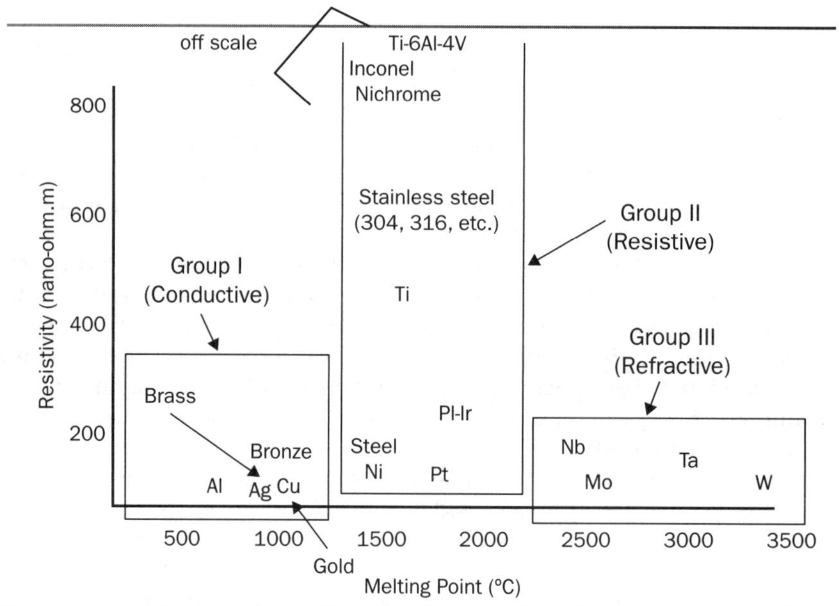

Figure 3.36.5 Metal groupings.

	Group I (Conductive)	Group II (Resistive)	Group III (Refractive)
Group I (Conductive)	• Solid-State • Braze or Solder	• Solid-State • Projection on Group I	• Solid-State • Fine projections on Group III
Group II (Resistive)		• Solid-State or Fusion • Easies to Weld	• Solid-State • Braze • Projection on III
Group III (Refractive)			• Solid-State • Braze

Figure 3.36.6 Welding dissimilar metals.

three categories based on melting points and resistivity: Group I (conductive); Group II (resistive); and Group III (refractive). Figure 3.36.5 shows which groups the metals used most for micro joining applications fall under.

Many times, part designers are locked into using specific materials, for example, a Group II metal like MP35N to itself. Or, they may need to weld copper, a Group I metal to platinum, which is in Group II. In some cases they may be able to select plating that makes a combination better for welding.

Figure 3.36.6 provides some guidance on welding dissimilar materials. Welding materials in Group II to Group II is the easiest. This includes steel, nickel, and platinum. Welding Group I to Group III can be the most challenging.

Soft, conductive materials in Group I, such as copper, can be difficult to weld since the bulk and contact resistances are low. Heat generation can be difficult. It is not uncommon to add tin plating to copper parts to allow diffusion bonding to occur. On the other hand, harder parts in Group II require higher welding forces to control the contact resistance at the part-to-part interface. Nickel is typically specified at quarter-hard to make the parts easier to weld. Group III metals have high melting points and can become brittle when welded. It is not uncommon for a platinum interface to be used between tungsten and molybdenum when welding these two refractive metals to each other. Specifying two metals that are compatible with each other (with similar melting points and properties) will make welding easier. If at all possible, choose wisely and choose two materials that are compatible.

Figure 3.36.7 is a weldability chart for common metals. For each combination of materials, it includes codes for weldability, preferred electrode materials, and comments to provide more information. There is also a section with some notes on designing parts for better weldability.

To use the chart, refer to the Key, which represents a quadrant at the intersection of each row and column in the chart. For example, the combination of titanium and MP35N has the following results:

Weldability:	2: Good
Electrode Material Choice (Titanium Side):	A: Glidcop (see chart for definition)
Electrode Material Choice (MP35N Side):	A: Glidcop
Comments:	a: High joint strength is possible

As shown in the chart, not all combinations of materials lend themselves to welding. Also, note that the electrodes may be different from each other, depending on the materials being welded. In the end, this chart should only be used as a rule of thumb reference tool. There are many other factors that can affect welding results and they all should be considered when choosing materials and developing a welding process.

Surface Conditions

Surface conditions are important to RW because they affect contact resistance. Surface conditions to be considered include plating, oxidation, roughness, insulation, and contamination. Plating is normally added to protect a material from oxidation. If welding a copper terminal for example, thin tin plating or nickel plating is advisable. However, it should be noted that welding copper parts with nickel plating is often more difficult than those with tin plating. With thick nickel plating, a copper-based electrode must be used to avoid sticking. Molybdenum or tungsten electrodes will stick to nickel plating. If a thin layer of tin is used on copper parts, tungsten or molybdenum electrodes are preferred as they will help to generate heat and will not be prone to sticking. If possible, ensure clean, smooth parts—free of insulation and oxidation.

Physical Part Design

Consider a part's size, shape, and access to the welding area with the electrodes when designing for weldability. Figure 3.36.8 shows just few application examples with a variety of part sizes and shapes.

Figure 3.36.9 shows possible electrode configurations. An opposed weld, in which the electrodes face each other on top and bottom, is by far the best option, as it offers the easiest control over the current path.

Materials to be Welded (Melting Point °C)	Aluminum		Brass		Cold Rolled Steel		Copper		Dumet		Galvanized Steel		Gold		Inconel, Kanthal, Kovar, and Nichrome		Molybdenum and Tungsten		MP35N		Nickel		Niobium		Phosphor Bronze		Platinum		Stainless Steel		Titanium	
Titanium (1670°)	5	A,E	4	A	2	A,B	4	E	3	A,B	3	A	4	A,E	1	A,B	2	A,B	2	A	2	A	2	A,B	5	A	2	A	2	A,B	1	A,B
	A,B	c	A,B	c	A,B	a	A,B	c	A,B	d	A,B	d	A,B	d	A,B	a	A	e	A	a	A,B		A,B		A,B	c	A,B		A,B		A,B	a
Stainless Steel (1450°)	4	A,E	4	A	1	A,B	4	E	3	A,B	2	A	4	A,E	1	A,B	3	A,B	2	A,B	1	A	2	A,B	5	A	2	A,B	1	A,B		
	A,B	c	A,B	c	A,B	a	A,B	c	A,B	d	A,B	d	A,B	d	A,B	a	A,B	e	A,B	a	A,B	a	A,B		A,B	c	A,B		A,B	a		
Platinum (1770°)	4	A,E	4	A	2	B	4	E	3	A,B	3	A	3	A,E	2	A,B	2	A	2	A,B	1	A	2	A,B	4	A	1	A				
	A,B		A,B		A,B		A,B	c	A,B	d	A,B	d	A,B		A,B	a	A	b,e	A,B	a	A,B	a	A		A,B	c	A	a				
Phosphor Bronze (900°)	4	A,E	3	A	4	B	3	E	3	A,B	4	A	3	A,E	4	A,B	5	A,B	5	A,B	4	A	5	A,B	2	A						
	A	c	A	c	A	c	A	c	A	d	A	d	A	c	A	c	A	c,e	A	c,d	A	c	A	c	A	c						
Niobium (2470°)	4	A,E	4	A	3	B	4	E	3	A,B	3	A	4	A,E	2	A,B	2	A,B	2	A,B	3	A	2	A,B								
	A,B	c	A,B	c	A,B		A,B	c	A,B	d	A,B	d	A,B	c	A,B	a	A,B	e	A,B	a	A,B		A,B	a								
Nickel (1450°)	4	A,E	3	A	2	B	3	E	2	A,B	2	A	3	A,E	1	A,B	2	A,B	2	A,B	1	A										
	A	c	A		A	a	A	c	A	d	A	d	A		A	a	A	e	A	a	A	a										
MP35N (1400°)	4	A,E	4	A	3	B	4	E	3	A,B	3	A	4	A,E	2	A,B	2	A,B	1	A,B												
	A	c	A	c	A		A	c	A	d	A	d	A	c	A	a	A	e	A	a												
Molybdenum (2000°) & Tungsten (3400°)	4	A,E	4	A	4	B	4	E	4	A,B	4	A	4	A,E	3	A,B	2	A,B														
	A,B	e	A,B	e	A,B	e	A,B	c,e	A,B	d,e	A,B	d,e	A,B	e	A,B	e	A,B	b,e														
Inconel, Kanthal, Kovar, and Nichrome (1400°–1500°)	4	A,E	4	A	2	B	4	E	3	A,B	3	A	4	A,E	1	A,B																
	A,B	c	A,B	c	A,B	a	A,B	c	A,B	d	A,B	d	A,B	c	A,B	a																
Gold (1060°)	4	A,E	2	A	4	B	2	E	3	A,B	4	A	2	A,E																		
	A,E	c	A,E		A,E	c	A,E	c	A,E	d	A,E	d	A,E																			
Galvanized Steel (1450°)	4	A,E	4	A	2	B	4	E	3	A,B	2	A																				
	A	c,d	A	c	A	d	A	c	A	d	A	d																				
Dumet (N/A)	4	A,E	3	A	2	B	3	E	2	A,B																						
	A,B	d	A,B	d	A,B	d	A,B	d	A,B	d																						
Copper (1080°)	4	A,E	2	A	3	B	2	E																								
	E	c,d	E		E	d	E																									
Cold Rolled Steel (1450°)	4	A,E	4	A	1	B																										
	B	c,d	B		B	a																										
Brass (900°)	4	A,E	2	A																												
	A	c	A																													
Aluminum (660°)	3	A																														
	A	d																														

Key:

Weldability	Electrode Material Choice ↑
Electrode Material Choice ←	Comments

Weldability Codes:

1 Excellent
2 Good
3 Fair
4 Difficult
5 Very Difficult

Designing Parts for Weldability:

- In general, resistive parts are easier to weld than conductive parts.
- Consider the melting point, hardness, and electrical and thermal conductivity of the materials.
- Balance the thermal mass of the parts to be welded.
- If the thermal imbalance is too great, add projections to the part with the greater thermal mass or to the most conductive part.
- Design parts for easy electrode access. An opposed electrode configuration is always preferred.

Electrode Materials:

Code	Description and Applications
A.	Glidcop—0.15% Al Oxide Dispersion Strengthened Copper. Long life, high strength electrodes primarily for welding resistive parts.
B.	RWMA2—Copper Chromium Alloy. Used for welding steels, nickel alloys, and other resistive parts.
C.	RWMA3—Copper Cobalt Beryllium Alloy. Used for welding resistive parts requiring high welding forces.
D.	RWMA11—Copper Tungsten Alloy. Used for welding cuprous and precious metals.
E.	RWMA13—Tungsten. Usually inserted into RWMA2 shank. Very hard. Used to weld non-ferrous metals such as copper and brass.
F.	RWMA14—Molybdenum. Usually inserted into RWMA2 shank. Used for welding copper, silver, gold, and their alloys.

Comments:

a. High joint strength is possible.
b. Use power supply with closed loop feedback.
c. Low joint strength is possible.
d. Electrode sticking may occur.
e. Short weld times may be necessary.

FIGURE 3.36.7 Weldability chart.

FIGURE 3.36. 8 Various part sizes and shapes.

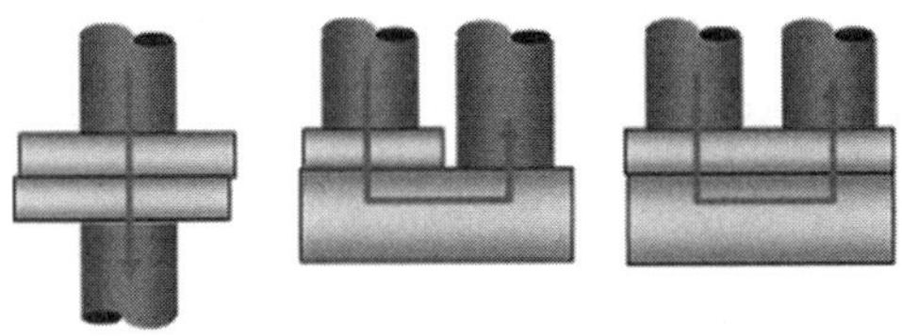

FIGURE 3.36.9 Three different electrode configurations.

Sometimes it is not possible to get to the bottom of a part and both electrodes must come down from the top—the step weld or parallel gap choice. If this is necessary, operators must be able to control the spacing (gap) between the electrodes and must control the force on both electrodes instead of just one. This adds variables compared to the use of an opposed weld.

Small parts require small electrodes. Such electrodes simply cannot handle the amount of current a larger electrode can handle. Electrodes must be designed to handle the amount of current or energy required for the weld to avoid overheating, bending, sticking, or wearing out too quickly. The smallest electrodes used are 2.54 mm (0.010 in.) and they are only used for the most minute welds.

Cycle Times

Typical production quantities for many micro parts, including the smallest medical devices, can be substantial, but far lower than those found in such industrial applications as automotive components, where parts may be produced every 2 s or faster. This means that minimizing cycle times for micro joining applications tends not to be as important as it may be in other industries where high volumes of parts are manufactured.

More important than minimizing cycle times is ensuring that parts are aligned properly between the electrodes and aligned to each other. The actual welding time, when the current is flowing, is extremely short (less than 50 ms), with a total time of

3 to 5 s, including indexing parts, aligning them, actuating the head, and welding. So in actuality, for micro joining, there is much more time involved in getting the parts into place and lined up than actual "welding," or time the current is flowing.

Heat Balance and Specific Design Recommendations

Heat balance is a key concern with any type of RW, but it is especially important with very small parts. When designing for micro welding, a shape that accommodates opposed welding is desirable, because this allows access to the bottom of the bottom part and the top of the top part, making heat balance easier. If there are two parts that are similar in size and thermal mass, it is easier to balance the heat between the two using an opposed weld.

It is also easier to weld flat parts than wires or round parts. If there is a big enough difference in the thermal mass of the parts, it may be necessary to insert features in the bigger parts, for example, a projection, or dimple. Weld projections promote heat balance by reducing the thermal mass of the thicker piece, increasing current density, and increasing part interface resistance. Projections also extend electrode life because a larger electrode face can be used. The projection also ensures current path, minimizing the effects of shunting. Figure 3.36.10 is an example of a projection used for heat balance.

Advances in Micro Resistance Welding Technology on the Horizon

Since micro parts are getting smaller and smaller, there is a real need for better control of the welding parameters and accurate monitoring of the welding process. This has led in recent years to controls that can work very accurately at low current levels and to electronic weld heads that precisely control the welding force. Monitoring processes are now being used that can provide accurate measurement of current, voltage, power, and resistance, in addition to welding force, and in some cases, displacement—if the parts have enough displacement to measure.

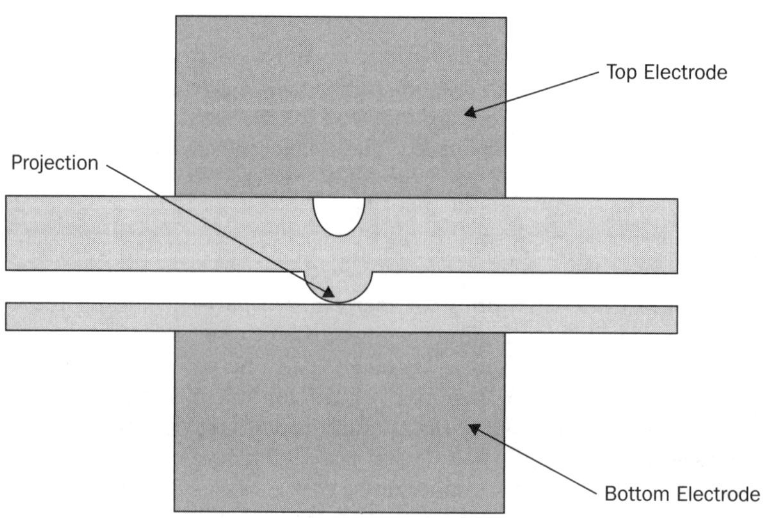

Figure 3.36.10 Cross-sectional view of projection weld.

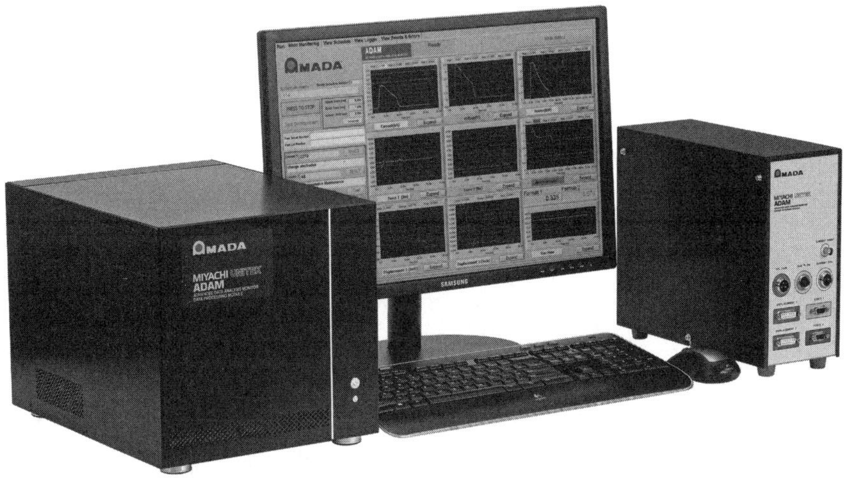

FIGURE 3.36.11 Resistance welding monitor.

The availability of this high level of monitoring data for electrical and mechanical data provides the whole picture of the weld. This is critical for both medical device and automobile safety applications. Operators must be able to select the optimum parameters for their welder to optimize their process. They must then be able to monitor the optimized process and store the monitored data, so if necessary they can go back to a specific weld for a failed part and have all the data on welding parameters used. Figure 3.36.11 shows an example of an external monitoring system that can provide this level of data.

Summary

Many factors need to be considered when designing micro parts for RW. The challenge of designing for weldability increases as the parts get smaller. The effects on heat balance should always be considered when choosing materials. Surface conditions and physical part design can greatly affect the welding results positively or negatively depending on choices made during the design phase. Equipment selection, electrode design, and methodical process development can lead to successful welding. Advanced welding technologies offer considerable process control and monitoring tools to help with difficult micro joining applications.

CHAPTER 3.37

Practical Guide to Laser Micro Welding

Geoff Shannon

Amada Miyachi America
Monrovia, California

Introduction

Laser micro welding has been in the industrial marketplace since the mid-1980s. Suitable for working on a wide range of joint geometries and materials, laser micro welding features low heat input, tailored weld dimensions, flexible joint geometries, and a process geared to both manual and automated production. This makes it an ideal noncontact joining process for spot and seam welding of the small, and in many cases mechanically delicate, components within medical devices, automotive sensors and safety equipment, batteries, and consumer electronics.

While there has always been some debate on an actual definition of micro welding, it typically refers to applications with less than 0.5-mm (0.02-in.) penetration, and average power of less than 500 W. It's worth noting that parts can be significantly smaller than this, for instance welding a 10-μm (0.0005-in.) foil, 100-μm-diameter (0.004-in.) wires or parts that can only be seen with optical magnification. With laser-focused spot diameters down to 25 μm (0.001 in.) very small parts can be welded—as long as you can tool them.

Laser Micro Welding Basics

Laser micro welding systems are comprised of the laser, a flexible fiber cable to deliver the laser to the workstation, a focus head to focus the laser onto the part, and motion (as-needed) for the part, the focus head, or both. Figure 3.37.1 shows a typical laser micro welding system.

The weld is formed as the intense laser light rapidly heats the material in fractions of milliseconds with power densities in excess of 10^6 W/cm^2. This high energy density maximizes weld penetration with minimal heat input.

FIGURE 3.37.1 4-axis laser micro welding system. The laser sits under the workstation for space efficiency, with typical dimensions of 762 to 914 mm (30 to 36 in.) wide × 762 to 914 mm (30 to 36 in.) deep. The workstation is on casters so can be moved as needed, then locked in place with leveling feet.

There are three types of welds, which are defined by their power density: conduction mode, transition or partial keyhole mode, and penetration or keyhole mode. Figure 3.37.2 shows a cross section of the three weld types.

Conduction mode welding is performed at low energy density, which forms a shallow and wide weld nugget. The heat creating the weld occurs by conduction from the surface.

Transition or partial mode occurs at medium energy density, and typically provides welds with a very similar width and depth. The keyhole is present but has shallow penetration.

Penetration or *keyhole mode* welding is characterized by deep narrow welds. In this mode the laser light forms a filament of vaporized material known as a "keyhole" that extends into the material and provides a conduit for the efficient delivery of laser light. This direct delivery of energy into the material maximizes weld depth and minimizes the heat into the material, reducing the heat-affected zone—and part distortion. The laser is one of the few welding sources capable of sufficient power density to enable keyhole welding, which significantly increases *weld penetration and joining efficiency, therefore minimizing heat input.*

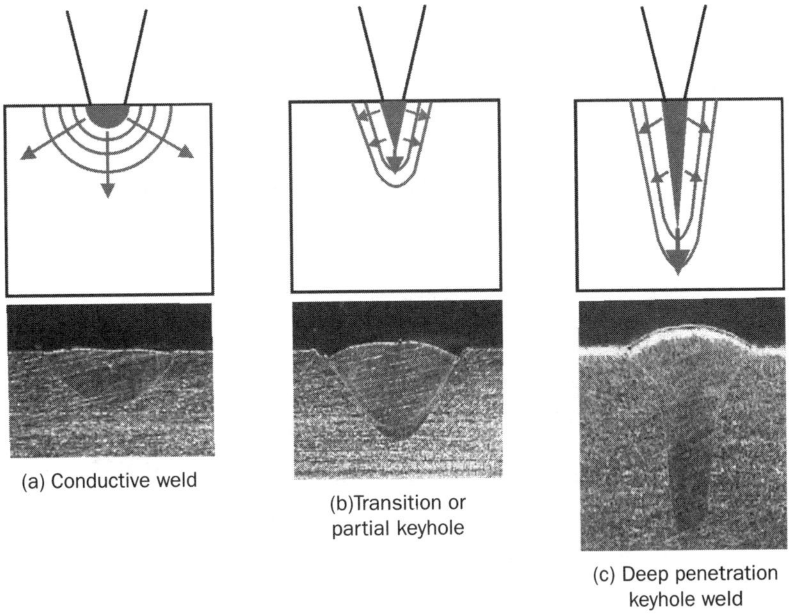

FIGURE 3.37.2 Laser weld types.

Laser Types for Micro Welding

Lasers suitable for micro welding include pulsed neodymium-doped yttrium aluminum garnet (Nd:YAG), Yb:fiber, and diode, each of which offers unique features aligning to specific applications.

The pulsed Nd:YAG laser has by far the largest installed base, with peak powers and pulse widths designed for micro welding. As an example 50-W pulsed Nd:YAG lasers are routinely used to seam weld 0.015-in.-thick titanium cases for implantable devices.

More recently developed fiber lasers offer excellent flexibility in tailoring weld dimensions and the best penetration per watt performance, which enables high-speed seam welding. A 300-W fiber lasers can seam weld 0.010-in.-thick airbag detonator casings at 2 in. per second, while a 20-W pulsed fiber laser can produce 0.001-in.-diameter spot welds in 0.001-in.-thick foil.

The diode laser is a well-established laser technology that has been used for mainly plastic welding applications, notably in the automotive industry for welding the rear light housing. It is likely these lasers will also expand into metals joining. Figure 3.37.3 provides several examples of micro welds.

Selecting the Correct Material for Weldability and Functionality—*Metals*

A successful laser micro welding process depends on selecting the correct material. The most common metal by far for micro welding is 300 series stainless steel, followed by aluminum alloys, titanium, and a range of other materials. Table 3.37.1 lists the most

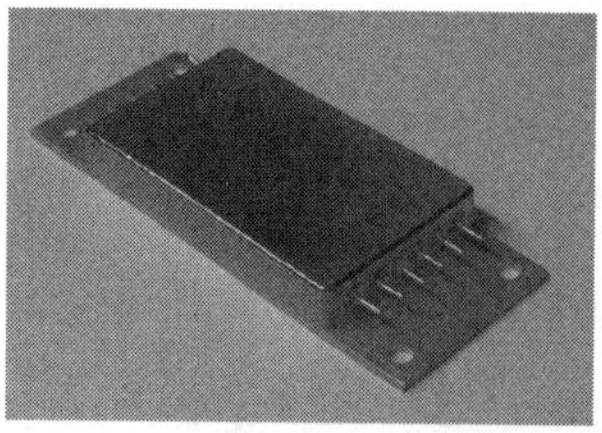		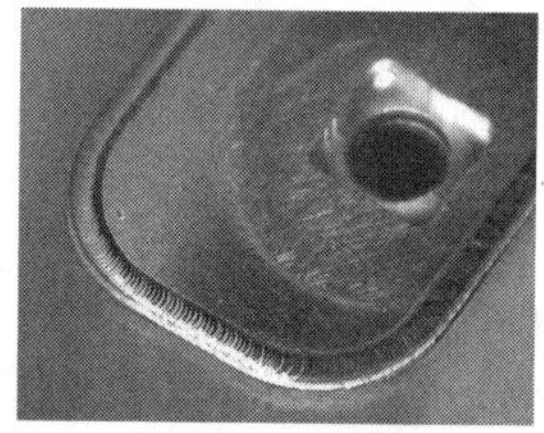
Seam sealing RF packages	Seam sealing airbag initiators	Battery can seam seal and tab welding
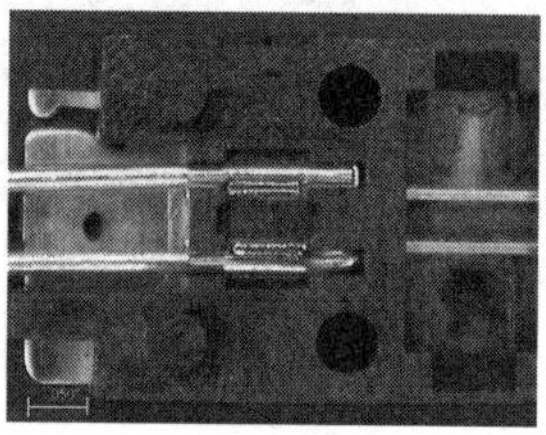		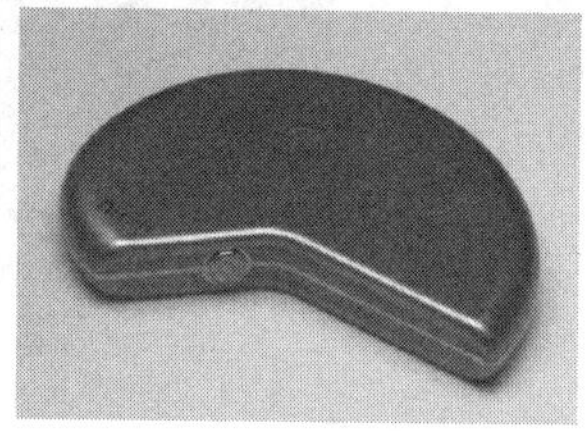
Electrical connections	Seam sealing medical implantable devices	Implantable medical device seam sealing

Figure 3.37.3 Examples of micro welding; from left to right and top to bottom; RF package with gold-coated 6061 aluminum body (uncoated or brushed clean of coating at weld interface) to uncoated 4047 aluminum lid, airbag initiator with 304 stainless steel, small prismatic can seal with 3003 aluminum, electrical contacts gold-coated copper, pressure sensor with 316 stainless steel, pacemaker with titanium.

common metals used, notes which provide good welds, and provides comments on practices that can be used to improve weldability.

Welding Dissimilar Metals

The need to weld dissimilar materials continues to increase to enable optimized product performance and reduce costs. For example, the battery manufacturing industry requires welding of copper, steel, and aluminum in various combinations.

It is important to assess the joining metallurgy of welds when considering using of more than one metal, even if the alloys are in the same material family. Many material combinations can create *intermetallics,* which can cause brittleness. This makes it even more critical to ensure that the weld will survive the component life cycle and to confirm this judgment with appropriate testing. Table 3.37.2 provides general guidelines on the more common dissimilar material selection for battery, medical, and aerospace industries.

Welding different alloys within the same family of metal should also be considered as dissimilar welding—and approached with the same caution. The common families are copper, stainless steel, and aluminum. Welding stainless steels within the 3XX series is generally successful, but it is worth noting that 303 and 316 are problematic materials.

Material	Comments
Aluminum	1050, 3003, and 6061 to 4047 are good for micro welding Continuous wave welding increases weldability of alloys such as 5052 and 5082 Aluminum alloys should be tested thoroughly for crack sensitivity
Beryllium copper	Good welds Potential safety hazard exists from the beryllium oxide fumes
Carbon steel	Good welds Carbon content should be less than 0.12% for pulsed welding, up to 0.3% for continuous wave welding
Copper	Good welds High power density is required to overcome surface reflectivity unless 532 nm welding laser is used
Kovar	Good welds
Nickel	Good welds
Nickel alloys	Good welds Good for alloys such as Hastelloy-X, Inconel 600 and 718
Phosphor bronze	Good welds
Stainless steel	304 and 304L produce excellent welds 316 and 316L produce good welds provided Cr/Ni ratio is greater than 1.7 303 is not recommended due to cracking tendencies but can be matched with friendlier materials, such as 304; a CW laser can be used to increase weldability 400 series require testing for crack sensitivity
Titanium	Good welds
Tungsten	Brittle welds

Table 3.37.1 General Guidelines on Material Selection—Metals

Because 303 is free-machining steel containing sulfur, it is not good for welding, causing cracking and porosity. However, pairing it with a 304L can produce a weldable combination—providing the mixing ratio that favors the 304L. This can be further mitigated by using a CW rather than pulsed laser, as the CW laser reduces the thermal cycling of the parts.

For the 316, the final chromium/nickel ratio of the weld material must be greater than 1.7 to ensure reliable welds devoid of cracking. Again, the use of a CW laser helps in welding of 316.

Welding the 4XX series can be problematic, due to its carbon content and the other alloying elements for ferrite stabilization. However, welding is generally helped by welding to 3XX steels.

Welding alloys within the copper family is good, but brass must be avoided due to its zinc content. Zinc has a low boiling point and will create porosity at best—and weld explosions at worse. Aluminum alloys have limited joining across the many alloy

Material 1	Material 2	Comments
Aluminum	Copper	Can be bonded Brittle intermetallics are created at the interface Fitness for purpose testing essential
Stainless steel	Nitinol	Can be bonded Brittle intermetallics are created at the interface Fitness for purpose testing essential
Stainless steel	Inconel	OK with certain alloys (304 with 600/700) Need to watch for cracking When welding offset into the steel to promote high Cr/Ni ratio in weld metal
Stainless steel	Titanium	Can be bonded Brittle intermetallics are created at the interface Fitness for purpose testing essential
Stainless steel	Copper	OK
Copper	Phosphor bronze	OK
Titanium	Aluminum	OK with certain aluminum alloys (1xxx and Ti-6Al-4V)

TABLE 3.37.2 Guidelines for Dissimilar Material Selection

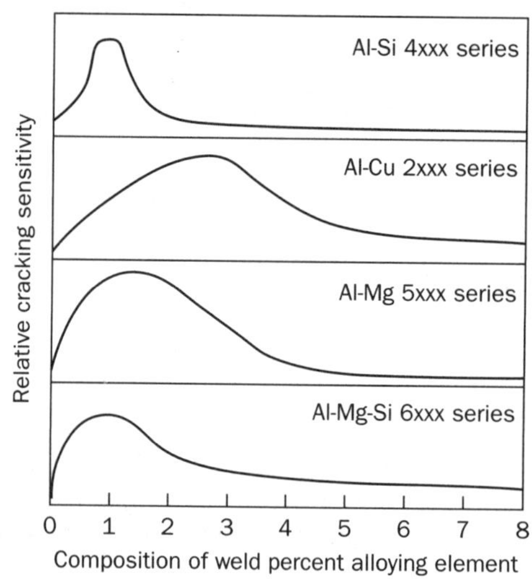

FIGURE 3.37.4 Joining aluminum alloys.

combinations. The key factor is to ensure that the correct percentage of the alloying element is in the weld, as shown in Fig. 3.37.4.

For example, 6061 aluminum cannot be welded to itself (without filler material) as the resulting Si content is around 0.6%. However, when paired with 4047, the resultant weld is good. With 10% to 12% Si content in 4047, the final weld Si content is pushed to around 6%, well above the percentage for cracking.

Plating Material	Comment
Tin	Low melting point tends to create spatter When lap welding there is a potential for weld porosity
Nickel	Preferred plating with high melting point that creates stable welding
Gold	Can be welded up to 1.27 μm (50 μin.), beyond this thickness cracking tendency increases

Table 3.37.3 Plating Material Guidelines

In other welding technologies, mismatches in material can be overcome by using filler materials. This is not the case for laser micro welding, because achieving reliable small diameter wire feeding into a very small melt pool (that may even be pulsing) is simply not production viable.

The figure also highlights how the alloy content percentage relates to crack sensitivity. For different alloys in the 2xxx, 4xxx, 5xxx, and 6xxx series, a certain percentage promotes cracking. Therefore, matching alloys that move the content percentage higher enables crack-free welding.

Metal Plating Affects Welding Process

Plating material, thickness, and plating method can have a significant effect on the welding process. For micro welding, the electrolytic plating method is recommended. Electroless plating can create welding problems due to the inclusion of phosphor and other contaminants such as brighteners during the plating process.

When hermetic seam sealing aluminum packages, no plating is allowed in the welding area; however, Kovar and stainless steel can be sealed with electrolytic nickel plating. Table 3.37.3 provides general guidelines about the use of a variety of plating materials.

Selecting the Correct Material for Weldability and Functionality—*Plastics*

As with metal welding, plastic welding success is heavily based on material selection. There are two methods for welding plastic based on the laser wavelength that determine both the materials and also the type of weld joint that can be welded. Figure 3.37.5 shows the two different plastic welding mechanisms according to the laser wavelength being used.

The top graphic shows the *transmission/absorption method* used with 1-μm-wavelength sources. The transmission/absorption method relies on the top layer being laser transmissive and the bottom layer being laser absorptive. The laser passes through the top layer and heats the bottom layer. With time and some pressure, typically under 3.73 kg (10 lb), the bottom layer heats the top layer by conduction and creates the joint. These welds can be made using both fiber and diode lasers of around 1-μm wavelength. Due to the transmission/absorption mechanism and the need to apply force, this method works only in the lap weld configuration.

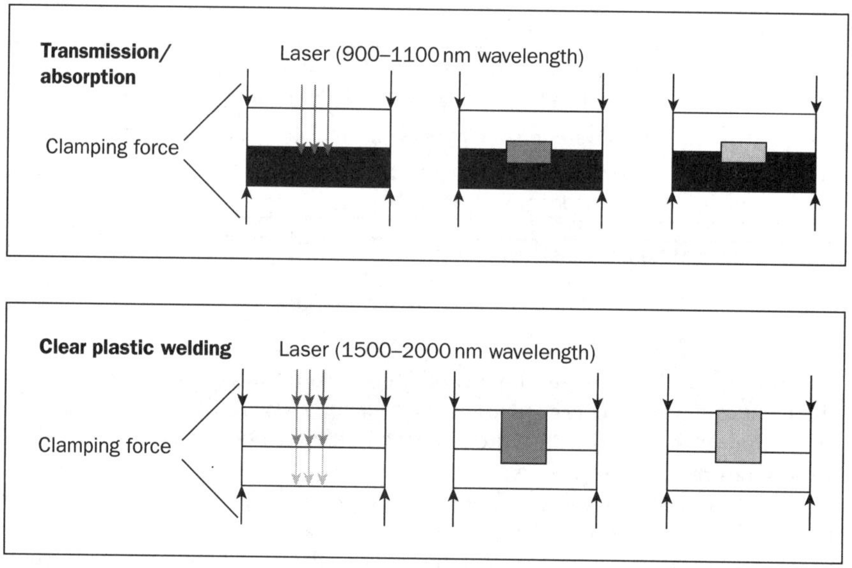

FIGURE 3.37.5 Plastics welding schematics.

The *clear plastic welding method* uses 1.5- to 2-μm wavelength sources and works by a balance of local absorption and transmission to enable the weld to be created from the top of the surface through to the welding interface. This method expands the joint configuration to include butt and fillet weld geometry and materials with no or little color or pigmentation. As at this wavelength, there are absorption peaks in a number of thermoplastics somewhat independent of colors or pigmentation such that suitable heating in the material can occur for welding.

For either method used, thermoplastics are preferred for laser welding. For the transmission/absorption method, the top layer should generally have less than 20% glass fill or other types of fillers to avoid scattering of the laser such that insufficient power reaches the bottom layer.

Certain colors tend to work better, but it is important to understand how the color is created. For instance, a black material using carbon black for pigmentation will be absorbed by the laser, so it is good for the bottom material but not for the top. Pigmentation and color of the materials can be tailored, and there are a number of plastic suppliers that offer laser weld friendly plastics or material additives. Typically, black and white are the most common for either top and bottom materials.

Because weldability of different plastics can differ greatly, the material should be evaluated for welding at an early development stage by actually trying to weld the materials using a laser and verifying its transmission/absorption characteristics.

Joint Design, Part Tolerances, and Fit-up

Laser welding is a noncontact process that requires only single-sided access, so many joint geometries can be welded. The most common joint geometry types are butt, fillet, and lap. Each type results in manufacturability tradeoffs in terms of fit-up tolerances,

Characteristic	Butt	Fillet	Lap
	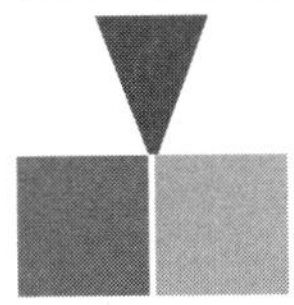	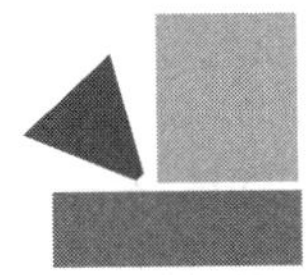	
Allowable gap between faces (t = thickness of part)	<0.05 × t Or up to 0.10 mm (0.004 in.) max*	<0.7 × t Or up to 0.127 mm (0.005 in.) max*	<0.1 × t Or up to 0.152 mm (0.006 in.) max*
Welding efficiency	High	Medium	Low
Laser to joint alignment	<20% beam dia	<30% beam dia	No alignment
Vertical tolerance of focus position	±0.127–0.500 mm (±0.005–0.02 in.)		
Part edges	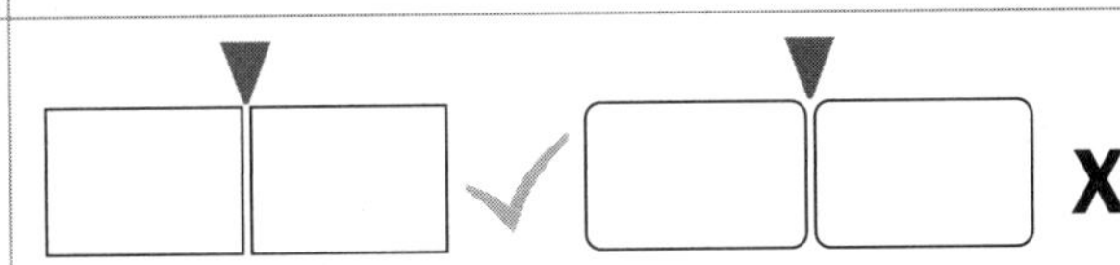 To maximize weld tolerances, part edges should be square, not rounded		

*Highly dependent on spot size and pulse energy—should always be minimized rather than pushed to limits.

TABLE 3.37.4 Comparison of Pertinent Characteristics for Weld Geometry Types

weld efficiency, and beam alignment to the joint line. Table 3.37.4 is an overview comparing pertinent characteristics for each weld geometry type.

The most important requirement for reliable laser micro welding is to ensure that the parts to be welded are in intimate contact. The level of contact is known as fit-up, and each joint geometry has a specific range of tolerances. These tolerances can be pushed by using excess power/pulse energy and large optical spot sizes; however, the process may become unstable. As with any process, knowing the process window—and where you are in the process window—is key for production. Using excess power or energy to produce the weld also affects heat input to the part and weld efficiency. Each joint geometry also has an impact on weld efficiency. Butt welding is the most efficient joint geometry because the laser aligns directly down the joint line, but it has the smallest fit-up and beam-alignment tolerances.

In contrast, lap welding is most inefficient because the laser must penetrate the top layer to reach the joint line, in which case weld strength is defined by the width of the weld, not penetration. However, there are no beam-alignment requirements and part clamping is also much easier. The merits of each joint geometry need to be carefully considered to match the part and, more specifically, how well part tolerances can be held.

Steps for Ensuring an Optimal Laser Micro Welding Process

Laser micro welding offers many advantages as a joining technology. To ensure an optimal process with high yield, consider the following factors in the design for a manufacturing process using laser welding:

- Select the correct material for like-to-like alloys or dissimilar materials or alloys.
- Clearly define the weld's requirements, for example, weld strength, electrical conductivity, or hermeticity to define the level of weld quality needed.
- Select a joint geometry that best enables the weld to be made under production constraints.
- Ensure the part-to-part fit-up of production parts aligns well with the use of laser welding and the joint geometry selected.
- When moving to laser welding from another welding technology, make sure that the previous points have been addressed *and* there is a viable tooling solution *and* an understanding that laser welding does not use a filler material.
- Test production-representative parts for laser welding to assess the best laser to use and ensure materials and joint fit-up are OK.

If these steps are followed and the correct decisions are made with suitable testing of parts, success in production is ensured.

CHAPTER 3.38

Micro Electron Beam Welding

LaRoux K. Gillespie

Manufacturing Consultant
Andover, Kansas

Electron beam (EB) welding of micro features has been a staple of the industry since 1962 when it was first used to seal thin components. EB welds are very precise and narrow unless width is needed; they minimize the heat input into the assembly, and that also minimizes the heat-affected zone surrounding the weld. Unlike most other welding systems, EB welding requires vacuum chambers to operate (to generate the beam); a vacuum is also necessary when reactive metals are joined, and also for hermitic seals. The vacuum also prevents oxidation from the weld heat. EB welding is fast—in one early application 96 welds were completed in 2.5 s. The process is used on most metals and can accommodate materials as thin as 25 to 50 μm (0.001 to 0.002 in.). Titanium and stainless steel foils of 20 μm (0.00078 in.) thickness have been welded together, but thinner materials are not generally used in this application. The power used for micro welds is only a few watts while conventional EB welding uses a hundred or many kilowatts of energy. Even the low-power scanning electron microscopes have been used to weld micro features, but that requires modification of the equipment and is not a production practice.

One manufacturer provides a 2-kW beam machine that welds, solders, serves as an SEM to photograph welds and measure widths, drills, and even hardens surfaces—all within a single machine. That machine can focus the beam down to 30 μm (0.0012 in.) for low currents. For micro welding, a range as low as 10 to 500 W may be adequate, but most machines are designed to allow a wide range of weld conditions, and thus provide more power than this. Figure 3.38.1 illustrates watch components some of which are EB welded, and Fig. 3.38.2 illustrates two thin membranes EB welded. Expandable bellows, made of thin metals like this, are a typical EB-welded product because of the speed and ability to assure leak-proof joints at high-production rates.

The beam can be focused to only a few micrometers in diameter and since its position is controlled by electromagnetic coils, it can be positioned exceptionally quickly and accurately and in addition to a continuous welding mode, EB welding can also

FIGURE 3.38.1 EB weld connects a shaft to an EDM cut piece. Note the nail-like structure of the weld. Smaller welds are also made. The assembly is used in pressure sensors for nuclear power plants. (Courtesy EB Industries.)

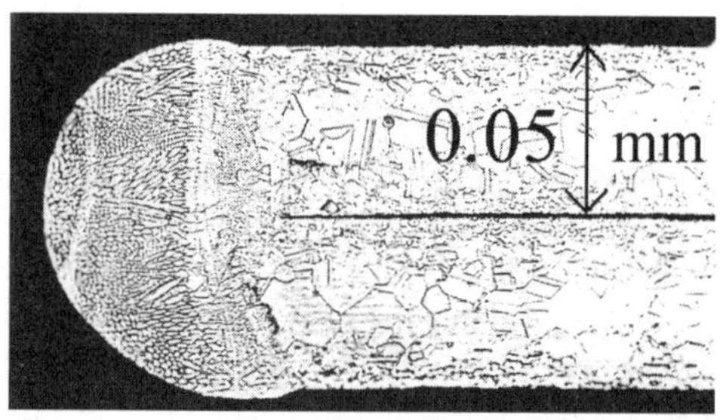

FIGURE 3.38.2 Two 0.05-mm (0.002-in.) membranes which have been micro EB welded. (Public domain image.)

operate in a pulsed mode that further reduces energy input and allows much wider applications. In a rather unique capability when making circular closures if the weld is started at one point and continuously moved around the circle, the starting point will have cooled slightly by the time the weld is finished and that can cause distortion or unacceptable buckling. The EB is fast enough that the beam can be switched back and forth to points around the weld to balance the temperature effects. Several parts can even be welded at the same time with this capability. Today's machines are fully automated such that operators do not need to provide any manual controlling actions.

Heating unbalances occur when materials having dissimilar conductivity or thicknesses are welded, but the quick joining via EB welding eliminates much of that concern. Welding copper to stainless steels for example faces both thermal expansion and melting point differences, but that can be accommodated by micro EB welding.

From a design aspect intimate contact is required throughout the weld. That means that mating parts must match their opposed surfaces exactly. The depth of weld penetration, joint preparation, cleaning, inert gas or vacuum environment, weld joint accessibility, proximity to heat-sensitive materials, productivity, and cost are all issues that must be considered in the design of joints, but other than weld depth they are common to EB welding of any size product. While micro parts do not often require deep spike

Bimetallic Joints: Weldability

	Ag	Al	Au	Be	Cd	Co	Cr	Cu	Fe	Mg	Mn	Mo	Nb	Ni	Pb	Pt	Re	Sn	Ta	Ti	V	W	Zr
Silver (Ag)		C	S		C	D	C	C	D		C	D	N	C	C	S	D	C		C	D	D	
Aluminum (Al)	C			C				C		C					C		N	C					
Gold (Au)	S					C	D	S	C			C	N	S		S	N		N		D	N	
Beryllium (Be)		C			N										N			D	D				
Cadmium (Cd)	C			N		D	D		D	S	D	N	N	D	C		N	C	N		N	N	D
Cobalt (Co)	D		C		D		C	C	C		C			S	C	S	S						
Chromium (Cr)	C		D		D	C		C	C		C	S		C	C	C	S	C		S	D	S	
Copper (Cu)	C	C	S			C	C		C		S	D	C	S	C	S	D	C	D		D	D	
Iron (Fe)	D		C		D	C	C	C		D	C	C		C	C	S					S		
Magnesium (Mg)		C			S				D			D	N				N		N	D	N	D	D
Manganese (Mn)	C				D	C	C	S	C			D		C	C		N					D	
Molybdenum (Mo)	D		C		N		S	D	C				S		D	C		D	S	S	S	S	
Niobium (Nb)	N		N		N			C		N		S			N				S	S	S	S	S
Nickel (Ni)	C		S		D	S	C	S	C		C				C	S	D						
Lead (Pb)	C	C		N	C	C	C	C	C		C	D	N	C			N	C	N		N	D	
Platinum (Pt)	S		S			S	C	S	S			C		S			C					S	
Rhenium (Re)	D	N	N		N	S	S	D		N	N			D	N	C		D			D		
Tin (Sn)	C	C		D	C		C	C				D			C		D					D	
Tantalum (Ta)			N	D	N			D		N		S	S		N					S		S	C
Titanium (Ti)	C						S			D		S	S						S		S	C	S
Vanadium (V)	D		D		N		D	D	S	N		S	S		N		D			S		S	
Tungsten (W)	D		N		N		S	D		D	D	S	S		D	S		D	S	C	S		
Zirconium (Zr)					D					D			S						C	S			

PTR

PRECISION TECHNOLOGIES, INC.

www.ptreb.com

Note: While the data is observably consistent, it has not all been verified.

light gray fill, intermetallic compounds formed—undesirable combination; D, insufficient data for proper evaluation—use with caution! N, no data available—use with extreme caution! S, solid solubility exists in all alloy combinations—very desirable combination; C, complex structures may exist—probably acceptable combination.

Table 3.38.1 Electron Beam Weldability of Metals and Alloys Compiled from Many Sources (Courtesy PTR-Precision Technologies, Inc.)

welds, EB welding can provide them and that feature is a unique aspect of the process. Joining thin foils to thick sections is regularly performed with this process.

EB welding is a well-proven process and the uniqueness for micro parts is its ability to work successfully on thin materials, quickly and cleanly. The biggest challenge for manufacturers on micro sized parts is the handling of parts and fixturing of them to prevent contamination before welding. Chapter 3.51 discusses solutions to handling.

Typical metals which are EB welded include titanium, 300, 400, and precipitation hardening stainless steels, aluminum, cobalt, copper, tungsten, gold, and nickel (Table 3.38.1 provides a more detailed list). EB welding requires that the materials to be joined are electrically conductive. Some ceramics are electrically conductive and can be welded. Materials with high vapor pressure at the melting temperature, like zinc, cadmium, and magnesium are not typically successfully joined with this process.

Commercial equipment for micro applications are more often desktop sized than that used for automotive parts.

CHAPTER 3.39

Micro Welding for Assembly and Rapid Turnaround Changes

Scott Malkasian

MicroArc Welding
Worcester, Massachusetts

As with generic welding, but scaled down, micro welding is commonly used to join small, precise, or delicate parts together as a step in the production/assembly or repair process. However, where micro welding really shines is when it is used to add or create shapes out of weld material in three dimensions. It is here that micro welding delivers its greatest value: as an enabler during the design and prototyping phase, as well as part of the ongoing repair and maintenance process.

Welding might generally be described as the process of joining metals together by heating them to a melting point and allowing the molten portions to combine together. Most times welding requires the addition of filler material, identical or compatible with the base material in order to ensure that gaps between the parts are filled completely.

There is no standardized meaning of the term "micro welding." Micro welding is a specialized type of welding employing specially adapted versions of the same type of welding processes, equipment, and materials used in conventional welding.

Among industry professionals micro welding or micro dimensional welding is generally understood to mean the high-precision placement of weld material, many times to a specific dimension or tolerance, using very thin filler wire, carefully controlling heat and minimizing its unintended effect—more generally, the ability to put a weld no matter how small, exactly where it is needed and nowhere else.

The vast majority of micro welding is done by hand, by experienced craftspeople, with specialized equipment, under a microscope. There are many types of welding equipment and methods in use today. Micro welding as discussed in this chapter typically employs either TIG (tungsten inert gas welding), or laser welding. While micro TIG welding can be performed by any shop, in practice there are few firms capable of performing some of the minute and precisions work described here.

FIGURE 3.39.1 Four small TIG-weld beads were added to permit a correction to only the grooves at the top of this part. Note the curved body lying across the welded area is an eyelash.

Perhaps the only quantitative delineation of micro welding might be the size of the weld filler wire used: Micro welders normally use filler wire diameters between 635 down to 76 μm (0.025 to 0.003 in.). To illustrate the scale of its potential application, the diameter of an average human hair is 102 μm (0.004 in.) (Fig. 3.39.1).

Micro TIG versus Laser

Both TIG and laser welding are used regularly at the author's company. Although both are done under a microscope, micro TIG is slightly less precise than laser. This is because it uses an electric arc to melt the metal. The arc is electrical current that "jumps" from the sharpened tip of the tungsten electrode to the "grounded" workpiece when it is moved close enough, sometimes only a few thousandths from the workpiece. The view of the weld is obscured by the arc which is very bright, so it requires a darkened shield to protect the eyes and the electrode itself is very close to the weld. TIG generates much more heat to the surrounding area and because the electrode is typically used freehand it takes greatly developed skill and a steady hand to control precisely. The smallest practical weld bead using TIG would be approximately 0.25 mm (0.010 in.) across using weld wire 76 μm (0.003 in.) in diameter. The smallest laser weld would be slightly narrower, but laser welding at these dimensions is more controllable with less danger of damage or distortion from heat.

Laser welding uses a focused beam of light energy (laser beam) instead of an electric arc to melt the material, making it much more precise. The beam is adjustable in both the width and power. The operator uses a joystick to manipulate laser beam over the workpiece. The welders look through the microscope and use crosshairs to position the head over the target. When the intended weld location is in focus, those crosshairs represent the laser beam's focal point when it is turned on. Although the lens filters out the harmful light, the laser is much less bright and the details of the weld area are fully visible.

The laser energy is highly controllable and generates very low residual heat which quickly dissipates and has almost no effect on the area outside the weld bead. Material just a few thousandths thick can be welded without distortion or even producing a heat signature on the back side.

Because laser welding is typically seen as the latest and most advanced of the two, it is assumed by many as the "best." While this is true in many applications, it is not always the case.

When you work with TIG, the heat keeps the weld "pool" melted longer, providing more time for material imperfections to work their way out. A laser weld solidifies extremely quickly, so if imperfections are there, they may become trapped.

For some applications, micro TIG remains the preferred welding method, especially if the job requires a lot of weld buildup. With micro TIG, you can build shallow, flat beads or tall, thin beads, more quickly depending on the amount of current (amperes) used. This is all possible with laser welding, but it's more difficult.

Laser welding requires an unobstructed "line of sight" to the weld area and a substantially perpendicular angle to the weld surface. Because of this TIG may be able to reach weld areas the laser cannot. One frequent example is a very narrow, deep slot with a side wall that needs to be welded. The angle of attack from the laser may deflect the beam off the sidewall losing its energy. In this case a micro TIG's tungsten tip may be able to reach into the slot to apply the weld, since the arc will naturally be drawn to the grounded sidewall. However, if that same narrow, deep slot requires a weld on the bottom, the laser may be the best choice. For some projects, a micro welder may use micro TIG for most of the welding, then call upon the laser to finish more delicate areas as necessary.

As is many times the case, innovation is born out of economic necessity. Micro welding as a specialty grew from serious challenges facing the plastic injection mold making industry. Molds used to produce plastic parts are precision machined, highly complex, one of a kind parts. These molds are made from hardened tool steel requiring multiple, sometimes dozens of precision machining operations with tolerances specified down as small as micrometers (tenths of thousandths of an inch), requiring weeks or even months to build.

Although the makers of this type of component are expert tool makers, even under the best of conditions parts can be damaged as a result of machining errors from tool breakage, shipment or handling, material defects, mid-process design changes/corrections, engineering specification errors, or miscommunications to name just a few. When this happens, with thousands of dollars and time already invested, scrapping the part and starting over is simply unthinkable.

Shown in Fig. 3.39.2 is a small but complex, expensive injection mold half that was damaged after completion. This part took weeks of work and a multitude of precise machining processes to complete. What is also shown in Fig. 3.39.2 is what a skilled micro welder can do when called upon to remedy this type of situation. After precisely replacing lost or damaged metal in just the damaged area, they can be remachined, preserving all the work already done with the final results virtually undetectable from the original. The original piece was not welded back, but instead base material was added using weld wire.

Prototyping and Iterative Design

These very same challenges and the micro welding solutions taken from the mold building industry are directly transferable to the prototyping or iterative design processes

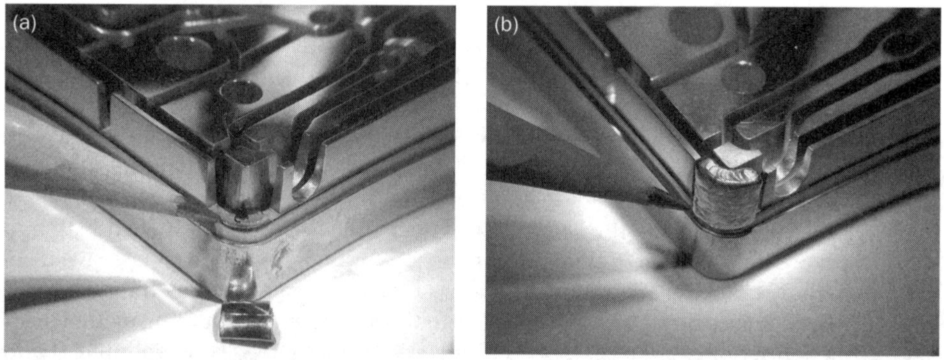

Figure 3.39.2 (*a*) A high-cost injection mold with a delicate corner piece broken off accidentally. A TIG micro welder rebuilt the corner by adding layer upon layer of weld wire (*b*). The corner was then remachined restoring the part to its original condition.

Figure 3.39.3 A design change facilitated by TIG micro welding. (*a*) Before, (*b*) after adding weld material, and (*c*) the completed new design after machining.

widely used today. As we will see, this process of adding new material can be used not only to make repairs, but to facilitate design changes late in the process.

A prototype is not unlike the case of a complex injection mold. When making a prototype it is not uncommon to have to wait for this "one of a kind" job to fit within a production machine shop schedule. If, after completion a small design change is desired, or error made, remaking a multistep part could take weeks.

Figure 3.39.3 illustrates one such example. It shows a before, during, and after image of the tip of a longer more complex and expensive part (not shown). Much work had been done to create a precision tapered and grooved shaft, but the tip shape required an engineering change. Without micro welding it would have required a complete remake.

Skills developed by micro welders go well beyond what can be done with normal welding to help clients solve manufacturing problems and have changed the way they do business. Micro welding techniques can be used when making parts to not only save money but to uniquely enhance function. One method for applying micro welding is to reduce machining. Although this is changing, most machining today involves removal or cutting away material. If a part has a large open surface with a standing detail, it can be cheaper to add the detail with weld than to remove everything around it. Another novel and not so obvious idea can be to mix alloys on the same component. For example, a part can be made of low-carbon steel but may require that a specific area or feature be wear resistant or very hard or corrosion resistant. A micro welder can add

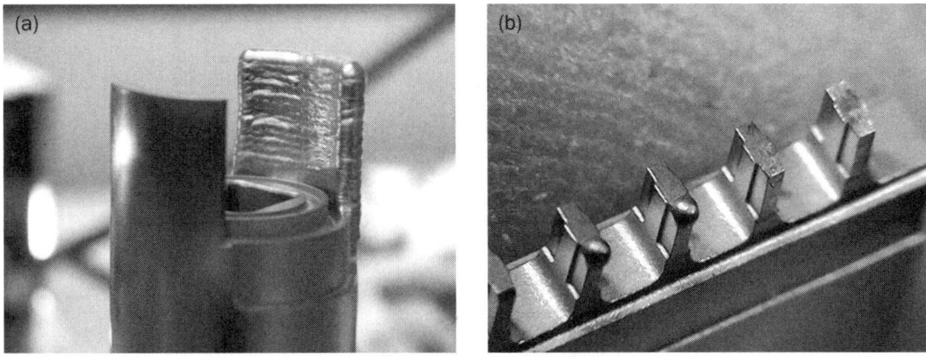

Figure 3.39.4 (*a*) Rebuilt flange. (*b*) Tips of two chipped teeth were built up via TIG micro welding, permitting complete restoration.

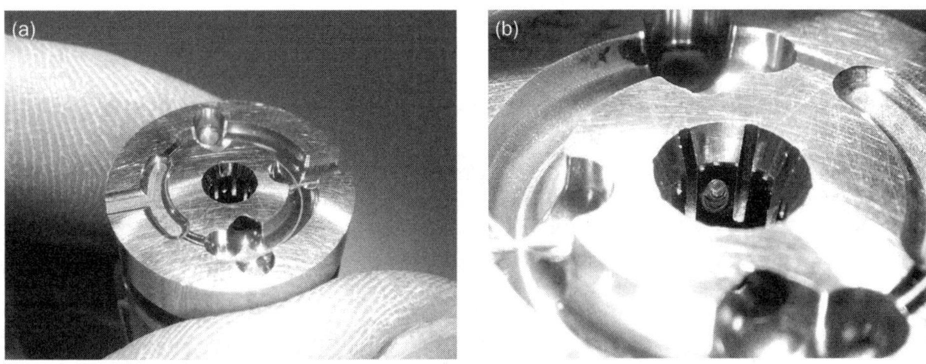

Figure 3.39.5 (*a*) A small hole can be seen deep below the center hole. Seen in (*b*) is an enlargement of the center interior showing a TIG closed hole.

material or shapes using weld filler wire of specific alloys making them part of the base block before machining or after the fact.

While not at all exhaustive, the following examples offer designers ideas and inspiration on how micro welding may be used to assist with their projects.

Figure 3.39.4 depicts two complex prototype parts that were damaged during the machining process. Rebuilding damaged areas with weld permits quick restoration without starting over. One side of the sleeve was lost, but restored with precisely built-up layers of TIG-weld material. In Fig. 3.39.4*b*, the tips of two teeth were slightly chipped. TIG micro welding restored the damaged metal, permitting complete restoration.

Micro welders are called upon to place weld in some very challenging places without damaging surrounding areas as shown in Fig. 3.39.5. In this instance, a design change required that a small passageway (small circle) be laser welded closed.

Over and above the challenge of precise weld placement, the micro welder must also understand the dynamics and control of the inevitable and potentially destructive by-product of the welding process—*heat*. Thermal expansion is a function of temperature. By definition, welding is the application of sufficient heat to the weld area to raise the temperature of the material above its melting point. In the case of steel, this is

Figure 3.39.6 Precision part, machined from a single block of stainless steel, which required a design change that required a larger height. A TIG micro welder added material avoiding an extremely expensive remake. The tall thin sections are extremely vulnerable to distortion from heat stresses, but were completed with no unwanted distortion and no unwanted weld into adjacent surfaces. (typically, there is no weld splatter with TIG welding).

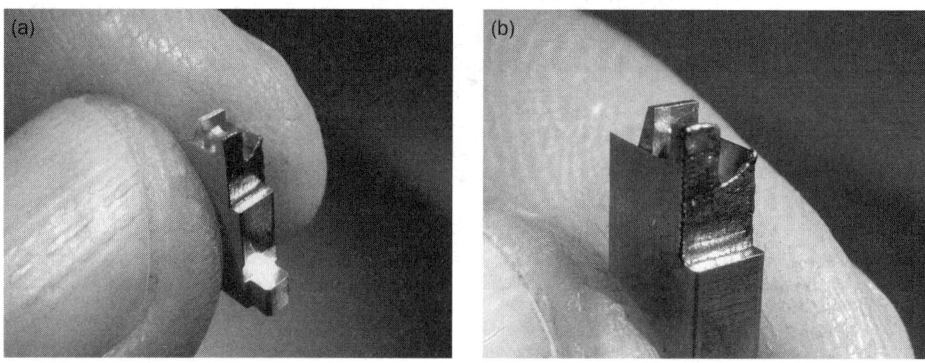

Figure 3.39.7 (*a*) This part needed an additional 178-μm-thickness (0.007-in.) laser added on the face (*b*) and required fully square surfaces.

approximately 1510°C (2750°F). This creates a massive temperature differential between the weld area and the much cooler surrounding material. As the molten weld cools, its rate of contraction is much greater than that of the surrounding metal, creating "tons" of pulling force toward the cooling weld area. The greater the amount of weld applied, the greater the heat. Unless this effect is minimized or carefully controlled, it can easily distort and/or destroy a valuable part in an instant. It should be noted that while heat is potentially destructive, once mastered a weld can be used in a carefully controlled manner to manipulate or "pull" parts into alignment that have been bent or distorted.

Figures 3.39.6 and 3.39.7 depict some of the challenges and usefulness for micro welding, commonly encountered. The precision part depicted in Fig. 3.39.6 was machined from a single block of stainless steel. A design change called for material where it had already been machined away. Instead of scrapping the part, a TIG micro welder added material avoiding an extremely expensive remake. The tall thin sections

FIGURE 3.39.8 A precision assembly involving laser welding on a small stainless steel valve.

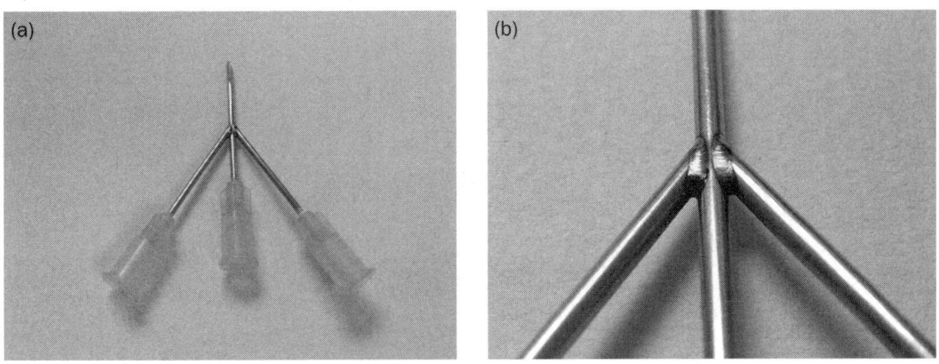

FIGURE 3.39.9 (*a*) A prototype three-into-one needle, made by laser joining 3- to 0.813-mm (0.032-in.) stainless steel needles (*b*). Note that the needle walls are 152 μm (0.006 in.) thick and all three must provide full flow.

are extremely vulnerable to distortion from heat stresses, but were completed with no unwanted distortion and no weld splatter onto adjacent surfaces.

The trick to micro TIG welding is keeping the electrode to workpiece gap as small as possible for maximum control and to concentrate the heat. Micro arc welders hold the pin-sharp tip of the tungsten electrode 76 μm (0.003 in.) above the workpiece [recall that the average human hair is 102 μm (0.004 in.)]. This requires a very steady hand since when adding wire the molten puddle rises and falls constantly. Figures 3.39.8 to 3.39.10 showcase the capability and level of skill in weld placement and material control achievable by one who has years of practice and experience in this precision craft. These are examples of micro welding as an integral part of the production assembly process.

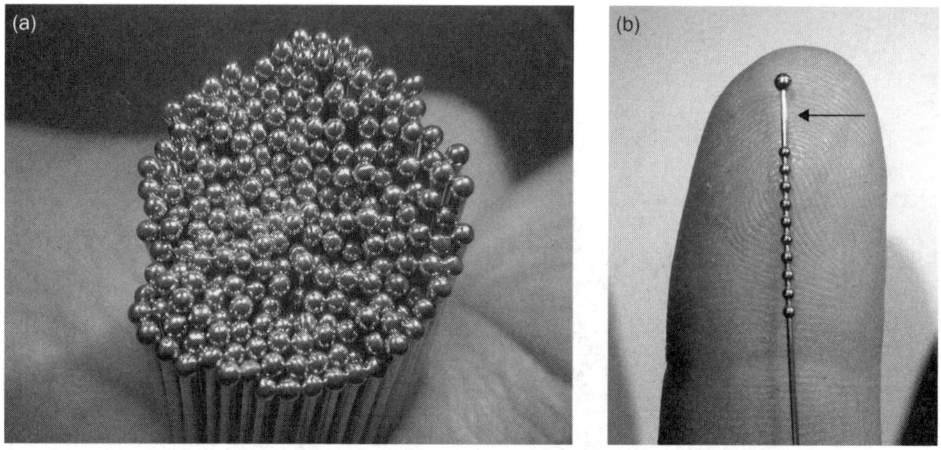

FIGURE 3.39.10 (*a*) Part made from laser welding bored out ball bearings onto a straight 0.762-mm (0.030-in.) wire. The ball at the top (*b*) is made from balling up melted weld wire.

FIGURE 3.39.11 0.762-mm (0.030-in.) stainless steel hollow tubes joined to body with laser weld (shown by arrow). The finished welds require unimpeded flow through the tubes and pencil tip is used to show size of part.

The three uprights in Fig. 3.39.11 are hollow tubes 0.762 mm (0.030 in.) in diameter. The finished welds must permit unimpeded flow through the tubes.

The tapered flange surrounding the outer lip of the part shown in Fig. 3.39.12 was added using a laser welder. Note the near finished symmetry and concentricity of the shape created using the weld material.

A high heat finned part is welded into a cylinder during the assembly process, see Fig. 3.39.13. Specifications often call for precision welds with exotic materials and tight seals without damage to parts. In addition to the initial manufacturing, micro welding can be used during the maintenance cycle, where small or expensive parts become

Figure 3.39.12 A bidirectional tapered flange was added with a laser welder (*a* and *b*). Note the near finished symmetry and concentricity of the shape added with laser weld material.

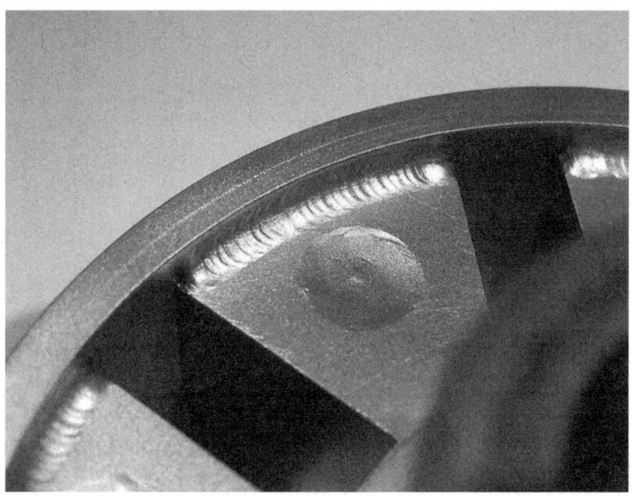

Figure 3.39.13 Finned Hastelloy part, TIG welded into cylinder with 5 mm inside diameter, used in high heat, corrosive environment.

worn (Fig. 3.39.14). In Fig. 3.39.15 a precision shaft had worn down beyond specification. Instead of purchasing a new part, a layer of laser weld material less than 0.25 mm (0.010 in.) thick was added along its length. This shaft was then recut to the correct diameter using a lathe.

A steady, practiced hand can create or fill precise shapes without damage or material spatter onto existing surfaces as depicted in Figs. 3.39.16 to 3.39.19.

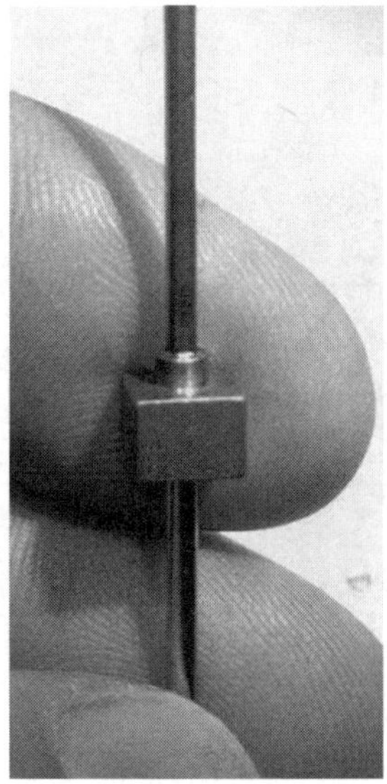

Figure 3.39.14 A small hollow tube laser sealed as it pierces the block, permitting unimpeded flow after completion of the weld.

Figure 3.39.15 A shaft worn down beyond specification repaired by adding a layer of laser weld material less than 0.25 mm (0.010 in.) thick along its length, permitting turning down to the correct diameter using a lathe.

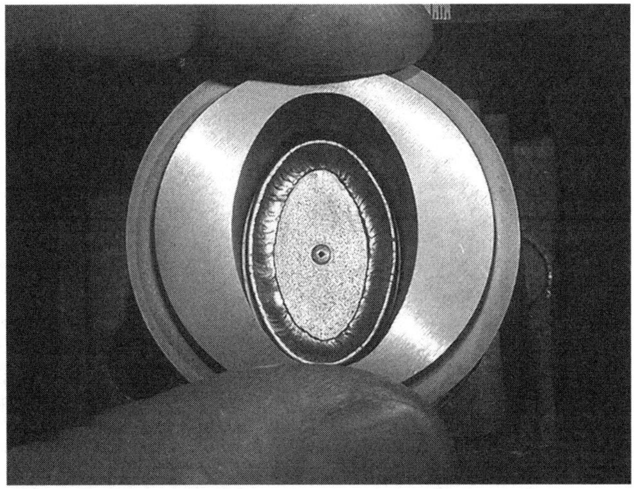

FIGURE 3.39.16 A very difficult TIG weld at the bottom of elliptic hole. Note that weld is spaced evenly from the sidewall without causing any ancillary damage.

FIGURE 3.39.17 (*Right*) A design change specified reducing the depth of this 0.762-mm (0.030-in.) slot (laser weld).

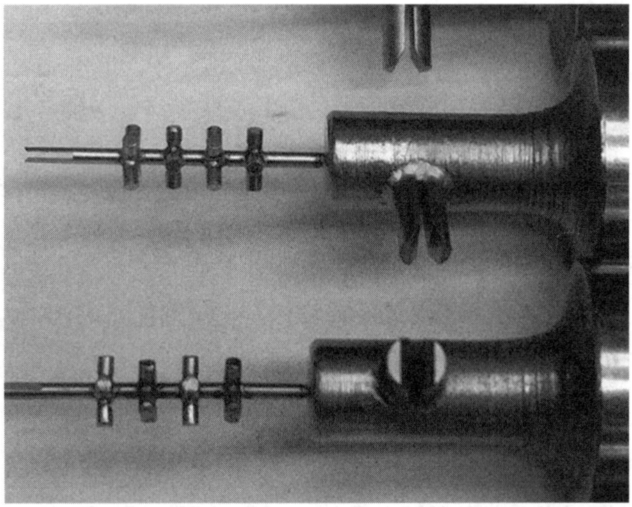

FIGURE 3.39.18 (*Right*) A prototype part made by laser welding this "tree-like" structure from 0.635-mm-diameter (0.025-in.) material. Note the "fork" cut at the left.

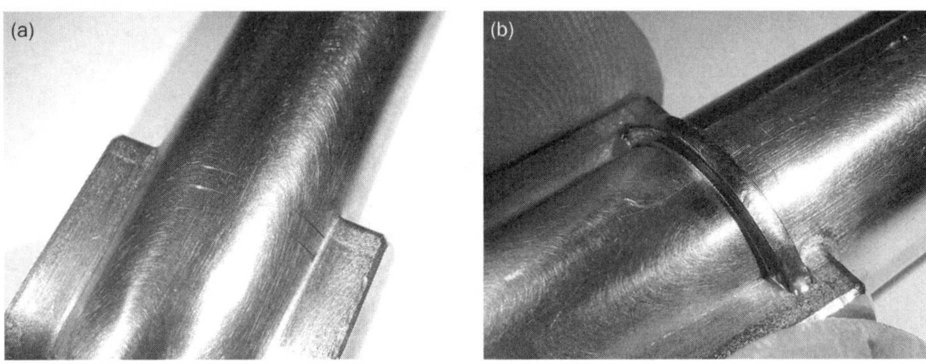

FIGURE 3.39.19 This part required the addition of a feature. The score lines (*a*) designate the location of the desired feature. Shown at the right is the completed weld ready for finishing. Note that this detail was repeated on the opposite side as well. Also note that the part is hollow in the center.

Invented to solve real problems in the mold-making industry, micro welding is now widely used in all industries including electronics, medical devices, aerospace, defense, oil and gas, and firearms to name just a few. It should be seen as an option for use in the manufacturing process when very small or delicate parts cannot be made or machined as a single part, or as part of the assembly process. Micro welding can join parts together structurally or provide an ultrahigh vacuum seal. Any material that can be welded is a candidate.

Extremely large quantities involving a precise repetitive process may warrant investment in automation where the cost can be justified. The majority of micro welding is performed by hand or using semiautomated techniques. When called upon to solve unique problems as those shown above, the utility and cost effectiveness of micro welding is without question and unmatched.

CHAPTER 3.40
Ultrasonic Micro Welding

LaRoux K. Gillespie

Manufacturing Consultant
Andover, Kansas

Process

Ultrasonic welding (USW) relies upon a metal tool (called a horn) pressing against two stacked materials lying on an anvil. The horn is oscillated at ultrasonic speeds (16 to 40 kHz for example). Each vibration of the horn involves a motion of 13 to 102 μm (0.0005 to 0.004 in.), with the smaller being more applicable to micro features. This generates compressive stresses in the materials, and when the right frequency is chosen, the energy is focused at the interface of the materials and liquefies that junction until the operation is turned off (Fig. 3.40.1). Force from the anvil, amplitude of vibration, energy, power, time, and weld depth are variables that can be set and controlled.

Ultrasonic welding is widely used for joining plastic components, but it is also a process used in metal joining, and can be a good choice for both micro applications. It can be fully automated or manual systems are also available for short runs. Welding cycles are typically in the range of 1 s or less. Loading and unloading parts will take more time than processing time. Because a force is involved, the setup has to be able to hold the two parts securely and allow a compressing vertical force—which can be an issue on micro parts. The fact that no external heat is applied is a major benefit for assuring that properties of the part materials are not adversely affected.

The process is widely used in medical and electronic applications involving some of the truly micro applications, although it was developed for larger parts originally and widely used in aerospace and automotive applications also.

While three weld joints are used in USW, on micro parts there may not be adequate surface to use anything other than butt or lap joints (Fig. 3.40.2). Additionally while for larger parts an energy director saves processing time, the production of such a feature on micro parts may be more difficult than advantageous. O-rings can be enclosed in the joint as well as other micro items.

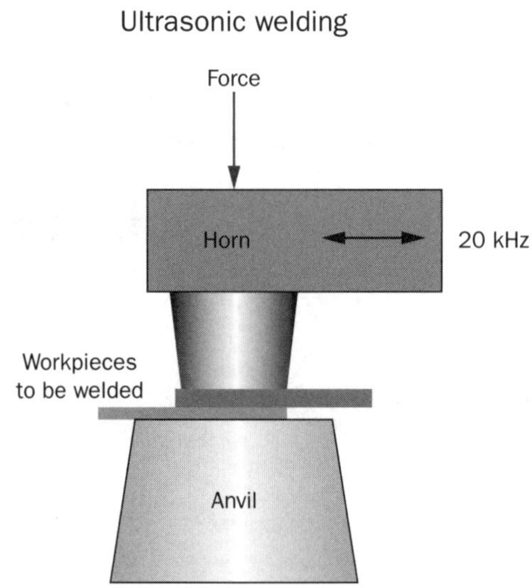

FIGURE 3.40.1 Schematic of ultrasonic welding process. (Courtesy Dr. Dmitri Kopeliovich, www.substech.com.)

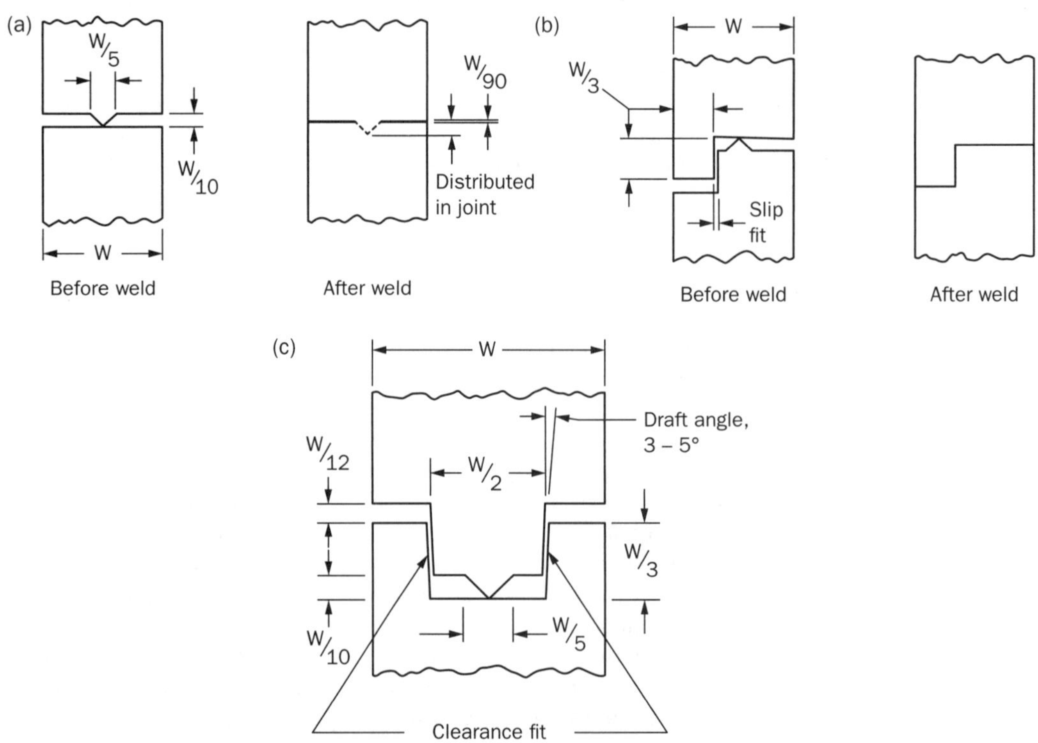

FIGURE 3.40.2 USW joint and energy director designs. Many of these designs will not be suitable for thin areas and irregular micro contours. In all cases the final assembly joint surfaces will be tightly flushed. Lap and butt welds will be more common for micro joints (Bralla, 1999).

Polymer Parts

Some plastics cannot be joined by solvent welding, so ultrasonics is a particularly good choice. Polyether ether ketone (PEEK), polymethyl methacrylate (PMMA), polystyrene (PS), amorphous polycarbonate (PC), and poly vinyl chloride (PVC) are common polymers bonded with USW. The manufacture of micro fluidics often relies upon bonding layers of thin flat sheets having channels, cavities, and even working devices in them. Polystyrene parts are joined to polystyrene film in a micro fluidics application [1.5-mm-thick part to 0.15-mm (0.059- to 0.0059-in.) film].

While welding parameters vary with conditions, one typical USW of a micro PVC assembly had parameters of 35 kHz, 340 N (79 lb), 350 ms ultrasound with 4-μm (0.00016-in.) amplitude, and 0.8 s cooling time.

The equipment for micro parts utilizes tabletop machines. Some machine controls display the weld process as a graph for which tight process limits can be defined. The curve of the joining velocity is a unique "fingerprint" and should be a linear function since constant joining velocity assures a stable weld process. Such visible operation provides immediate understanding of issues.

In an unusual potential it has been shown that the micro USW equipment can also be used to plasticize polymer sheets into micro parts. Since the process liquefies the polymer with minor adaption to the USW equipment, it can force the liquid polymer into a shaped cavity to "mold" micro features and parts.

Ultrasonic hot embossing can be matched with micro USW to make a setup capable of both, and with appropriate tooling the tooling can also cut wires of other features in the assembly to provide electrical isolation.

Metal Joining

In contrast to polymer joining, in metal welding the USW process is a solid-state joining process. The ultrasonics scrub oxides and other contaminants for the interface and allow the surfaces to bond as a single piece of the same material, although they may be of two different metals (Fig. 3.40.3). Aluminum, copper, gold, and silver are common

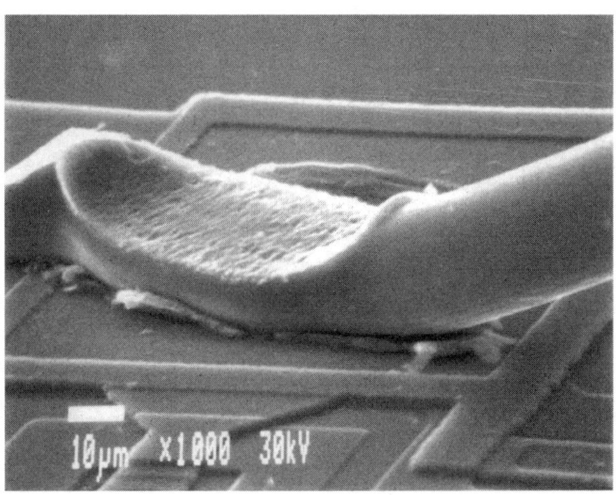

FIGURE 3.40.3 Ultrasonically joined 25-μm (0.001-in.) aluminum wire on an integrated circuit chip. (Courtesy Precision Joining Technologies, Miamisburg, Ohio.)

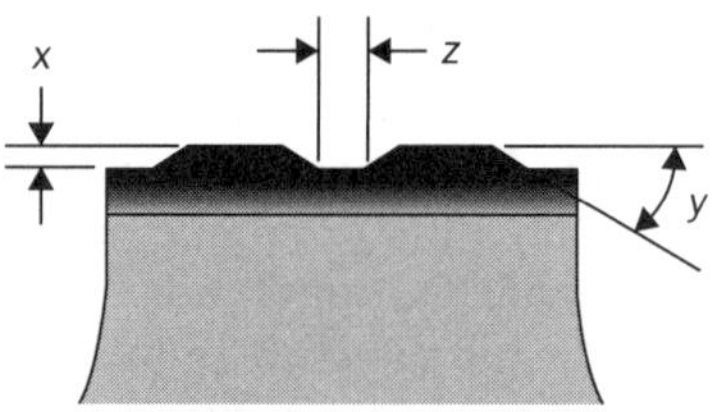

Figure 3.40.4 Horn knurl design variables. (Courtesy EWI, Columbus, Ohio.)

USW materials. Nichrome and tin-plated parts are other metals that are bonded by this process. Welding foils to metal tabs in lithium ion battery cells is one high-production application. In the instance of welding thin foils, particularly more than one at a time, the anvil touching the foils needs to have a patterned surface, in some ways similar to a knurl (Fig. 3.40.4). Appropriate parameters for the knurl dimensions yield stronger welds and significantly reduce foil damage and required energy. As many as 30 10-μm-thick (0.0004-in.) foils at a time have been welded to a solid base. Most metal weld configurations are confined to lap welds.

Joining Metals to Nonmetals

USW is used to embed electronic wires in polymer assemblies with little issue. The liquidification melts the polymer and the forces involved keep the polymer tightly bound to the wires so a seal is made. Similarly, it can seal 3D-printed wire traces inside polymer assemblies.

As a result of recent research, it has been shown that light metals can be bonded via USW to carbon-fiber-reinforced polymer (CFRP) sheets but that is not necessarily a combination used in micro joining. Ultrasonic nanowelding is being developed to join carbon nano tubes (CNTs) to Ti electrodes for the development of carbon nanotube-based photovoltaic cells.

USW has been used successfully to bond gold bumps to gold-plated surfaces where the gold bumps were on silicon wafers and the plated gold was on aluminum nitride substrate. The ultrasonics caused no fractures of the wafers or substrate—they passed right through.

Key Design Considerations

The design considerations are common to large and micro parts, but harder to physically do on micro parts.

- Parts must be carefully aligned
- The ultrasonic horn has to be directly over the weld area
- The type of joint (butt, shear, lap, etc.) will make assembly easier or harder (see equipment manufacturer's design recommendations)
- More surface area makes a stronger joint and increases hermeticity seals
- Sharp edges increase stresses locally

- Cutouts under the weld area will adversely affect weld in those areas
- Venting paths may be needed in some designs to prevent weld bubbles in seals
- Thin fragile projections near the joint may break from the vibrations

References

Bralla, J. G., *Design for Manufacturability*, McGraw-Hill, New York, NY, 1999, pp. 6.138–6.139.

Khan, M. I. and A. K. Pandey, Micro/Nano Welding—Current Status and Future Trends, *International Journal of Technical Research and Applications* (IJTRA), Volume 1, Issue 4, Sept–Oct 2013, pp. 1–5.

Michaelia, W., E. Haberstrohb, and W.-M. Hoffmann, Ultrasonic Welding of Micro Plastic Parts, *Multi-material Micro Manufacturing*, Whittles Publishing Ltd., Cardiff, UK, 2008.

Schomburg, W. K., K. Burlage, and C. Gerhardy, Ultrasonic Hot Embossing, *Micromachines*, Volume 2, 2011, pp. 157–166.

CHAPTER 3.41

Micro Adhesive Bonding

LaRoux K. Gillespie

Manufacturing Consultant
Andover, Kansas

Process

There are several means for bonding micro products:

- Soldering
- Welding
- Direct and indirect bonding via high temperatures [200°C to 500°C (392°F to 932°F)]
- Adhesive bonding

Soldering and welding involve heating portions of the assembly to temperatures that are harmful to many assemblies or at least undesirable. Direct bonding refers to taking two surfaces, often the same material, and heating them such that the atoms move across the interface in both directions so that it becomes a single material. That can be done with two chemically clean surfaces of many metals for example. Heat and pressure speeds up the process. It also works with silicon wafer materials. Indirect bonding uses a different material between the two to help accomplish the same atom or molecular bonding.

While adhesive bonding can use, but often does not, heating, it is generally at a much lower temperature than the processes listed above. Thus adhesive bonding has fewer impacts upon the final product.

Almost all materials can be bonded with some adhesive. Glass to any other material is no problem. While several adhesives are multipurpose, many are designed for specific material combinations and cure times.

Adhesive Bonding

Adhesive bonding offers several advantages over other forms of joining. It can bond a variety of different materials (glass, metals, ceramics, and polymers) to each other in a manner that is essentially totally stress-free. It uses no or very low temperature so that the joined parts are not affected or distorted by temperatures—a very important factor in many bio applications. It can be a very fast process with many parts per second processed. Adhesives can include electrically conductive metal powders or additives to provide electrical connections. The adhesives can be optically clear for transmission of light and later viewing of processes. They can assure perfect sealing.

Micro applications require careful considerations of several factors in bonding:

- Material properties at bonding and in required use (mechanical, electrical, bio, chemical, etc.)
- Surface properties where parts are to be joined
- Accuracy and repeatability of part alignment and adhesive positioning
- Actual placement of the second piece to be bonded and the timing of it
- Stability of the final product over time (aging effects) and effects of their installed environment (sterilization, temperature shock, chemicals)
- Volume and repeatability of required adhesive
- Production rate or application rate required
- Gap and pressure between the two parts to be bonded

Adhesive bonding utilizes several mechanisms to accomplish the bond. Room temperature curing, hot melt systems, and ultra violet (UV) light curing are the three most common mechanisms.

Solid film adhesives (polyurethane adhesive foil and hot melt glue) have also been used in micro bonding and changed to a liquid via a hot gas stream. Solvents can assist some of the adhesives, but generally are driven off by heat in some manner.

For any of the application choices, it is imperative for designers to define all the properties required of the bonding agent—specifically, optical, strength, chemical resistance, hardness, stability over time, outgassing, extent of expected coverage, and thermal properties (both during application and in use).

Equipment

Micro bonding requires high-precision movement of either the product or the dispensing head. It also requires that the user have a means to assure absolute alignment between the parts to be bonded. It is incumbent upon the designer to provide features to align the parts—pins and mating holes, repeatable edges, uniform locating platforms (parts may not be flat). It is also important that the positioning features not fall into the path of adhesives unless they are intended to be carried along throughout all assembly.

Vision is also required for most micro bonding. It is required for setup, and in-process monitoring. Vision is required to assure droplet size as well as accurate placement of the drops or adhesive path (adhesives may be dispensed as a continuous path or as droplets). One system uses a 12× and a 120× microscope camera to observe operation. Very simple semi-manual machines are available as are fully automated high-volume ones.

Micro dispensing is a field in itself driven by the need for inkjet printing, micro assemblies, and medical needs. While inkjet printing, also called just "jetting," is the principal method of applying micro quantities of adhesive, contact dispensing is also used. When contact dispensing is used, a drop is formed at the end of a minute

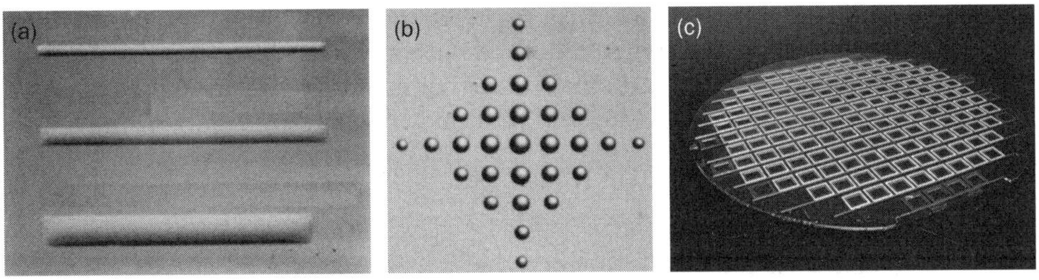

FIGURE 3.41.1 (*a*) Adhesive lines of varying thickness and width can be deposited with a single dispenser by using different print patterns; (*b*) adhesive dots of varying sizes can be printed with the same device; (*c*) wafers and similar products can be easily bonded by jetting. (Courtesy MicroFab Technologies Inc.)

nozzle and that drop contacts the part to be adhered before the drop falls from the nozzle. Inkjet printing uses a piezoelectric system to shoot micro size drops from the nozzle (Fig. 3.41.1). The direct contact approach can use stiffer (higher viscosity) adhesives that can be injected from inkjet heads. Direct contact is a slower process and it may be difficult to reproduce exactly the microscopic amounts needed. The drop is deposited by touching the part surface then moving the head laterally in a wiping off motion. Jetting is not significantly affected if the part surface is not flat, but that can be an issue with the direct contact approach. Jetting does not require the lateral wiping motion which makes the jetting operation faster.

The jetting processes can eject drops as small as 8 pL (488 pico in.^{3}, i.e., 488×10^{-12} in.^{3}). Droplet diameters of 30 to 100 μm (0.0012 to 0.0039 in.) are typical, but at the 8-pL size they will be smaller. The rate of droplet ejection can be as high as 100 kHz (100,000/s), but a more common rate would be 2000/s. The accuracy of the drops can be controlled as precise as <1% and viscosities up to 200 Pa s (0.029 pound force seconds/in.^{2}). The xyz platforms that move the head or the product can be positioned within 1-μm (39-μin.) accuracy and heated nozzle heads are available for temperatures up to 160°C (320°F). Inkjet printer technology can now deliver adhesives through 75-μm (0.003-in.) nozzles that include 4-μm (0.00016-in.) silver particles representing 70% by weight of the adhesive.

Cyanoacrylates, epoxy, UC cured, silicones, acrylics, and urethane adhesives can be applied by jetting. Even some hot melts can be applied in this process. At least seven different suppliers provide jettable adhesives and at least one equipment supplier provides a heated reservoir that lowers adhesive viscosity to make it more suitable to jetting.

Hot Melt Approaches

Hot melt adhesives are single-component thermoplastics involving no solvents. They melt between 60°C and 180°C (140°F to 356°F). They can be put on the micro parts via spheres, melt liquid form, powder, adhesive foils, and stencil or screen printing. The adhesive can set on the product for long times without affecting future bonding performance. Once heated the adhesive softens, bonds and when cooled is cured hard. Hot melts set very rapidly—full strength can be reached in less than 1 s. The cycle time for a complete assembly however may be as long as 2 min because of the need for heat up and cool down times. Heat would normally be provided through the two bonded parts. The adhesive can be pigmented if desired but most are clear or straw colored.

Some typical minimum dimensions of hot melt volumes are shown below, but the technology is rapidly advancing and these may not stay lower limits long.

- Balls—500 μm (0.020 in.) diameter
- Balls—nonuniform picoliter sizes
- Films—25 μm (0.001 in.) thick
- Spin-coated soluble—1 μm (0.000039 in.) thick
- Powder particles—15 μm (0.0006 in.)
- Stencil-printed dispersions—150 μm (0.006 in.)

Some hot melt adhesives can also be presented by noncontact equipment. Drops as small as 0.4 nL (0.000000024 $in.^3$) have been made by commercial equipment.

Hot melt adhesives are based on ethylene-vinyl acetate (EVA), polyolefins, polyethylene, atactic polypropylene (PP or APP), polybutene-1, oxidized polyethylene, polyamides, polyurethanes, styrene block copolymers, polycarbonates, fluoropolymers, silicone rubbers, thermoplastic elastomers, or polypyrrole (PPY), and a few other variations.

UV Curable Adhesives

If the parts being bonded allow UV light to reach the adhesive then UV curable adhesives can be used. Hydroxyethyl methacrylate (HEMA) is one of the UV curable adhesives; it is often used in dental labs for dental appliances. It is a liquid when applied but UV light quickly cures it and it dries smooth and shiny. Several acrylate formulations are used, particularly the urethane-based ones. UV adhesives are widely used for mounting optical lenses.

The UV adhesives come in a variety of properties. Some offer excellent optical and superb physical properties. Some are biocompatible per USP Class VI testing. Some can be spin-coated with high thermal stability, and some have NASA levels of low outgassing. High chemical resistance and strong insulation properties characterize others. Shrinkage, adhesion, modulus of elasticity, and elongation are the factors that the designer must consider when selecting the appropriate adhesive composition. So are shock, vibration, and moisture resistance. For some applications, the refractive index and hardness are also important. Each manufacturer provides data sheets for their adhesives containing these and other application insights. Since it is light rays that set the bond, light must be able to reach all areas of the bond. Any masking will not provide the expected bond.

Like jettable adhesives, micro parts will typically be joined using a single dot of UV curable adhesive, although spin coating may also be used for larger areas.

Unfilled nylon, Delrin®, Teflon®, polyethylene, and polypropylene are not good choices for bonding with these adhesives.

Full curing depends on both the time of cure and the intensity of the UV rays reaching the bond area. Cure time may be as high as 12 min for some adhesives and 0.1 s for others, but all depend on intensity of the light source. Most adhesives cure in 2 to 10 s.

Some adhesive manufacturers offer 25 different adhesive combinations, each meant for a specific application.

Additional Design Thoughts

Some designs require that the glue line be clearly distinguishable after hardening. Designers in this instance can choose to install a micro channel around the area to be bounded or sealed and fill the channels with the sealing adhesive. Others have a fluorescing ingredient that shows where bond lines are.

A hot gas stream of nitrogen or other gas can be passed over adhesives to cause some of them to cure. This may be preferable to other heat sources.

Silicon-to-silicon bonding has been accomplished with a variety of the direct and indirect bonding approaches. The indirect approaches include anodic bonding, glass-frit bonding using low-melt glasses, eutectic, foil bonding, and resist bonding.

References

Böhm, S. K. Dilger, J. Hesselbach, J. Wrege, S. Rathmann, W. Ma, E. Stammen, and G. Hemken, Micro Bonding with Non-viscous Adhesives, *Microsystem Technologies*, Vol., 12, No. 7, June 12, 2006, pp. 676–679.

Dziuban, Jan A., *Bonding in Microsystems Technology*, Springer Netherlands, Netherlands, 2006.

Böhm, S., K. Dilger, and E. Stammen, Micro Bonding Using Hot Melt Adhesives, *Journal of Adhesion and Interface*, Vol. 7, No. 4, 2006, pp. 28–33.

Rathmann, S., A. Raatz, and J. Hesselbach, Concepts for Hybrid Micro Assembly using Hot Melt Joining, *Micro-Assembly Technologies and Applications*, Vol. 260, 2008, pp. 161–169.

Rathmann, S., A. Ratz, and J. Hesselbach, Strategies for the Use of Hot Melt Adhesives for Highly Accurate Micro Assembly, *Proceedings 2nd CIRP Conference on Assembly Technologies & Systems*, Toronto, Canada, September 2008.

CHAPTER 3.42

Micro Blasting

Colin Weightman and Jason McNary

Comco Inc.
Burbank, California

Micro abrasive blasting, also called "micro blasting," is precision micro scale sandblasting. In this method, a fine abrasive media mixed with compressed air propels through a miniature nozzle. The nozzle increases the velocity of the mixture as it exits the nozzle opening and allows an operator to precisely direct the abrasive stream to a target.

This versatile technology is used by many industries to texture, deburr, clean, and erode a wide variety of materials—even the most delicate. It provides a quick, cost-effective, and environmentally friendly solution. The hallmark of micro blasting is precision, and its micro size makes it ideal for many micro parts and features. Micro blasting is capable of processing parts as small as a grain of rice to as large as a basketball. It is precise enough to deburr the ID of a capillary tip 25 μm (0.001 in.) in diameter, and efficient enough to texture selected sections of a basketball-sized part without masking.

Micro blasting equipment varies from simple desktop-size machines for manual applications to larger fully CNC-controlled automated systems (Figs. 3.42.1 and 3.42.2). This system was originally designed to remove oxide layers from vascular implants with motions similar to that of a lathe. Other systems have additional capabilities and axes of motion. This machine is designed for micro fluidics to drill precise holes in ceramic substrates like silicon or glass.

Abrasives

The abrasive media (Table 3.42.1) is the most significant factor in the blasting process. One abrasive may clean a part, while another will erode it. Micro blasting requires the removal of fine, coarse, and foreign particles from the media to closely control particle distribution. Particle size variance and impurities impact consistency and performance. A bottle of carefully sized sodium bicarbonate is superior to supermarket baking soda or the industrial abrasive supply equivalent.

Table 3.42.1 lists common abrasives used in micro blasting and their properties.

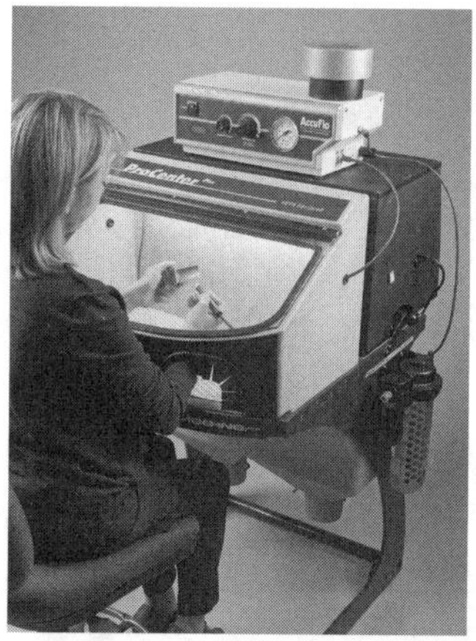

Figure 3.42.1 A typical manual micro blasting system.

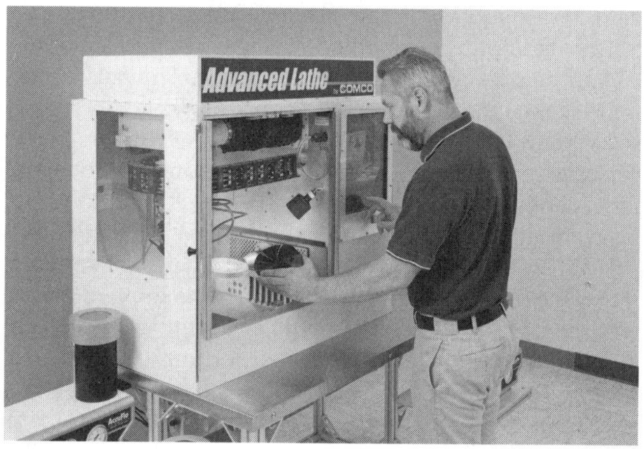

Figure 3.42.2 Sample automated micro blasting system.

Nozzles

Nozzles provide focus and acceleration to the abrasive stream generated by the blaster. Abrasive particles must travel fast and straight in order to be effective. Proper focus increases the efficiency of the abrasive, ensuring every particle strikes the targeted area. The interior geometry of a nozzle must limit turbulence. The application requirements dictate the nozzle shape and size (Fig. 3.42.3). Nozzle designs include:

- Round: 0.254- to 3.175-mm-diameter (0.010- to 0.125-in.) nozzle opening
- Rectangular: 0.203 × 0.508 mm to 0.381 × 4.826 mm (0.008 × 0.020 in. to 0.015 × 0.190 in.)
- Extended: Up to 254 mm (10 in.) in length
- Angled nozzles: 45 or 90 degrees; will fit into a bore as small as 2.032 mm (0.080 in.) and up to 254 mm (10 in.) in length

Abrasive		Particle Size	Shape	Hardness (Mohs)
	Aluminum oxide	10–150 μm (0.0004–0.006 in.)	Blocky & sharp	9
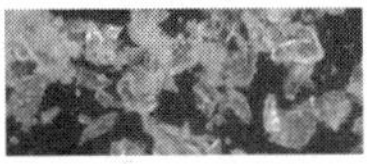	Crushed glass	50 μm (0.002 in.)	Blocky & sharp	Between 5 and 6
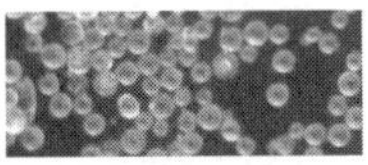	Glass beads	35–100 μm (0.001–0.004 in.)	Spherical	6
	Plastic media	50–200 μm (0.002–0.008 in.)	Blocky	Between 2 and 4
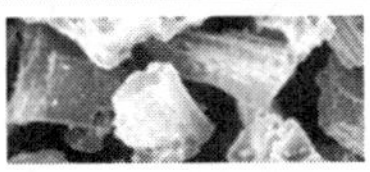	Pumice	50 μm (0.002 in.)	Blocky & sharp	6
	Silicon carbide	20–50 μm (0.002–0.008 in.)	Blocky & sharp	9+
	Sodium bicarbonate	50–100 μm (0.002–0.004 in.)	Monoclinic	Between 3 and 4
	Walnut shell	250 μm (0.010 in.)	Blocky	Between 3 and 4

Table 3.42.1 Common Abrasives Used in Micro Blasting and Their Properties

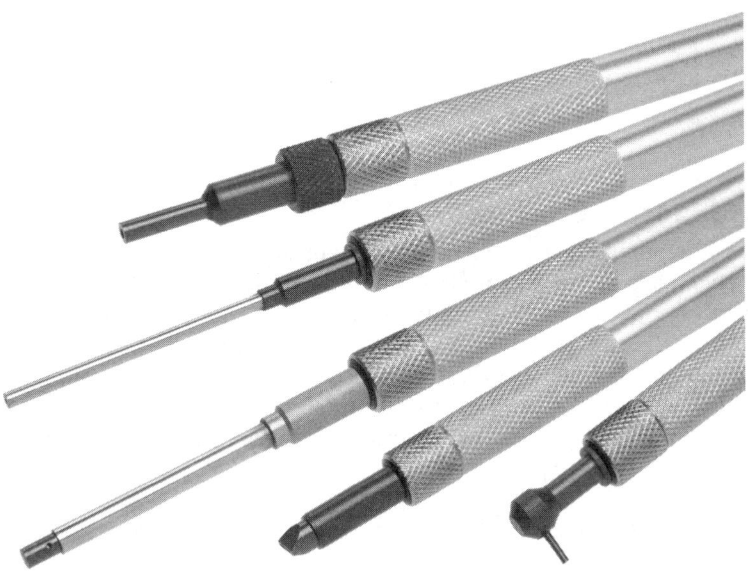

Figure 3.42.3 Micro blasting nozzle options.

Key Variables

The four key variables in micro blasting—*type, size, velocity,* and *quantity*—greatly influence results. Surface texturing requires focus on three particular variables to produce the desired result, while controlled erosion requires focus on a different combination of variables. Type and size are intrinsic to abrasive media, while blasting technology exclusively controls velocity and quantity.

- *Type: The composition and shape of the abrasive particle.* Abrasives run the gamut in hardness: from soft compounds like sodium bicarbonate, to hard ceramics like silicon carbide. Abrasive particles range from spherical to blocky to pointy. Some are water soluble, some organic.
- *Size: The size of the abrasive particle.* Micro blasting is optimized for very fine particle media in the 10 to 50 μm (0.0004 to 0.0020 in.) range and effective with media up to 300 μm (0.012 in.).
- *Velocity: The speed of the particle as it strikes the target.* Air pressure primarily determines velocity, but nozzle size, hose length, abrasive feed rate, and nozzle-to-part distance play important roles.
- *Quantity: The number of particles striking the target per unit time.* Both blast duration and abrasive stream density control quantity.

Important Notes

- Media feed rate into the system needs to be constant. Velocity drops if abrasive pushes through in pulses.
- The distance between the nozzle and the target should remain constant. The abrasive stream continues to accelerate after it leaves the nozzle. Particle velocity stops increasing at about an inch away from the nozzle; stays constant for the next 50 to 75 mm (2 to 3 in.), then slows down.
- Nozzle design is critical. A properly designed nozzle increases particle velocity by as much as 30%.

Controlled Erosion Overview

Material removal by controlled erosion with micro blasting is uniform and so precise that part integrity remains intact. The focused impact of millions of very fine abrasive particles removes a small quantity of material at a time without forming micro cracks and stress concentrations. Micro blasting does not generate heat or shock, and it provides tight control over the removal process down to 70 nm (2.8 μin.).

The distinction of using micro blasting is the ability to feed fine abrasives [10 to 100 μm (0.0004 to 0.0040 in.) of material] and to control abrasive output [1 to 50 g (0.035 to 1.76 oz) of abrasive per minute].

Materials Suitable to the Controlled Erosion Process

The shockless nature of micro blasting is well suited for machining very hard, brittle materials, including but not limited to silicon, quartz, sapphire, glass, ceramics, graphite, composites, hardened steel, and titanium. A high percentage of energy transfers from blasted abrasive media into a target's surface causing small chunks of surface to shatter away. Ductile parts have a natural elasticity, causing the abrasive energy to absorb and deflect abrasive. Micro blasting can erode ductile parts, but slowly and inefficiently. The target material should be homogenous to maintain surface uniformity. The likelihood of pitting increases on sintered parts, but controlled erosion is still viable on them.

Abrasive Characteristics

The hard, sharp edges of aluminum oxide make it the primary abrasive used in controlled erosion applications. Abrasive should have a narrow distribution of particle sizes to create uniformity during this process. The abrasive particle size ideally suited to controlled erosion is 17.5 to 25 μm (0.0007 to 0.001 in.), which yields an Ra surface finish of 0.3 to 0.6 μm (12 to 24 μin.). Coarse media creates a rough surface finish, and fine media erodes a surface too slowly.

The characteristics of the substrate material also impact surface roughness. The finer the grain structure and the fracture mechanics of the substrate, the lower the Ra after blasting.

The rate of controlled erosion has a linear relationship with abrasive quantity and velocity. A precise feed system lays the foundation and does most of the work, so that multiple passes simply need to even out slight variations. Without an effective feed technology, peaks and valleys appear on a surface.

Typical particle velocities during controlled erosion are 150 to 225 m/s (492 to 739 ft/s) based on air pressure, nozzle geometry, and blast distance from the nozzle to the part.

Appropriate Applications: Masking versus Direct Machining

There are two methods for performing controlled erosion using micro blasting: direct machining and masking. Both methods accurately remove a few micrometers to a millimeter of material with multiple passes. Direct machining relies on the focus of the abrasive stream and the accuracy of an automated CNC platform. Masking protects small or delicate part features and limits a part's exposure to the erosion process.

Masking is necessary if the application requires creating a large number of features; creating features that are 127 to 178 μm (0.005 to 0.007 in.); or creating features with tolerances that are ±51 μm (±0.002 in.). For example, masking must be used when cutting holes or vias into silicon wafers for micro fluidics. Each wafer may need several hundred holes drilled, and the drilling process slows as a hole grows deeper. Due to this drop in efficiency, the practical limit on depth is approximately two times the hole diameter.

Pocket creation in a part benefits from masking. The mask is used to create the profile, and the abrasive stream, to control depth. Tolerances of ±0.5 μm (±20 μin.) are possible through automation.

Direct machining relies on the precision of micro blasting to remove large, simple sections of material. A small nozzle and accurate parts-handling capabilities create a consistent and focused abrasive jet stream that cuts a pattern into a surface, traces a pattern of the cut geometry, or cuts a sharp edge. The transition from blasted to unblasted surface can be as sharp as 178 μm (0.007 in.). Direct machining is the more preferable and efficient method, because it eliminates the need for masking and consumes less abrasive.

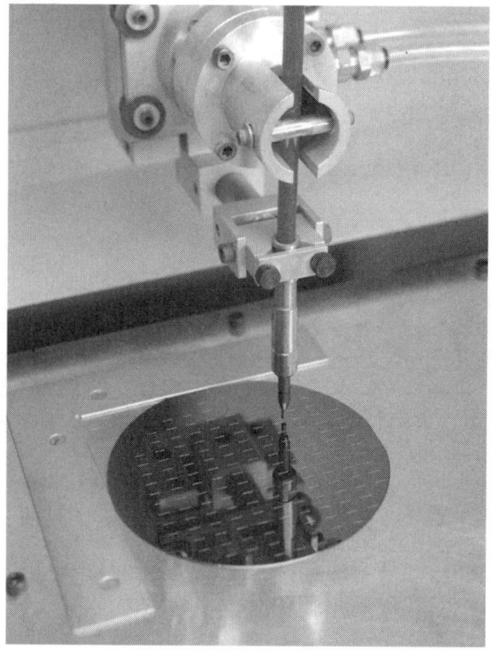

FIGURE 3.42.4 Cutting slots in silicon wafer using micro blasting.

As an example of precision erosion, every inkjet printer cartridge contains a silicon wafer that connects the ink reservoir to the nozzles. The silicon manages both the electronic signals and the ink flow. Ink must pass uninhibited through the center of the wafer. Micro blasting erodes an organic flow path, free of turbulence (Fig. 3.42.4).

Selective Cleaning

Selective cleaning leverages abrasive qualities to carefully erode layers of unwanted material from a base substrate. The process takes advantage of the different physical properties of these two layers. Brittle materials can be removed from soft ductile layers, or soft ductile layers can be removed from brittle layers. Selective cleaning differs from controlled erosion because selective cleaning works with dissimilar materials, whereas controlled erosion removes homogenous material.

Materials Suitable to Cleaning

Abrasive selection is everything to this application. Abrasive must be aggressive enough to consistently remove the top layer, but lack the energy to cut into the base material.

Higher particle velocities clean a part faster, but risk damaging or pitting a part's exposed surface. The similarity of the coating layer to the part surface determines the upper limit of velocity. If these two materials are very different, high pressure cleans the coating layer away quickly. Parts with similar material properties require low pressures and slow removal rates.

	Remove a Brittle Material from a Ductile One	Remove a Ductile Material from a Brittle One	N/A
Aluminum oxide	X		
Glass beads	X		
Plastic		X	
Pumice		X	
Silicon carbide			X
Sodium bicarbonate		X	
Walnut shell		X	

TABLE 3.42.2 Abrasive Cleaning Suggestions

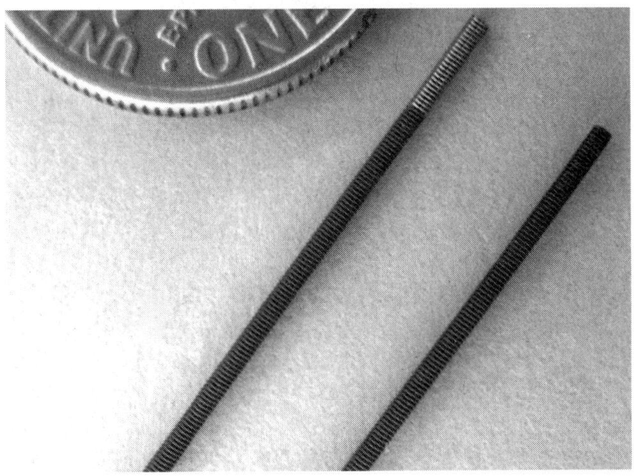

FIGURE 3.42.5 Coating removal from guidewire.

Selective cleaning uses abrasive sized 25 to 50 μm (0.001 to 0.002 in.). Small particles increase the number of strikes, which, in turn, speeds the removal process (Table 3.42.2).

Coating guidewires with PTFE improves lubricity. It is far more efficient to coat guidewires in long reels than in individual lengths. After the wire is coated and cut to length, sections of the coating need to be removed (cleaning process). There are three critical aspects to this application:

- The delineation from coated to uncoated must be sharp.
- The coating must be completely removed.
- The integrity of the wire must be maintained.

Thanks to its physical properties and water solubility, sodium bicarbonate is an ideal abrasive media for selectively cleaning guidewires (Fig. 3.42.5). These abrasive particles are sharp enough to cut through the PTFE's long fibrous strands, but soft enough (3 to 3.5 Mohs hardness) to leave the guidewire unharmed. It is ideal for

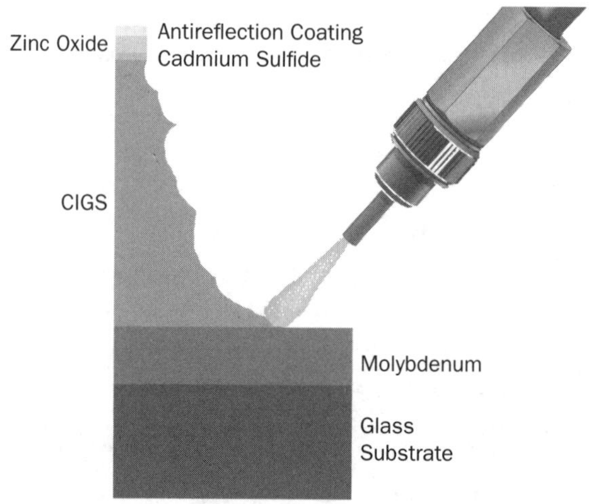

FIGURE 3.42.6 CIGS removal on solar cell.

interventional use, because it leaves no residue following a standard cleaning process using deionized water with or without an ultrasonic cleaner.

Brittle Coating Removal: CIGS from Molybdenum

A solar cell has an absorber layer (CIGS), a conductive layer (molybdenum), and a base substrate (glass). The cell's CIGS layer absorbs the energy of the sun and converts it into electricity. This electricity is transferred to the "moly" layer, where it can be drawn out of the cell.

The entire cell is coated in CIGS during manufacturing. Creating access points to the moly requires careful selective removal. CIGS is a brittle material, and moly is ductile. There are three critical aspects to cleaning in this application:

- The delineation from coated to uncoated must be sharp.
- The CIGS coating must be completely removed.
- The integrity of the moly (the conductor) over the glass must be maintained.

Glass beads is an ideal abrasive for this application. The hard spherical shape will fracture the CIGS coating, shattering it away from the moly (Fig. 3.42.6). The elastic properties of molybdenum absorb and displace the energy of the spherical bead without eroding.

Precision Deburring

Precision deburring is the removal of sharp particles or edges often left on parts as a result of machining operations. Micro blasting is more consistent, safer, and less dependent on operator skill than traditional deburring methods particularly for micro features. The pinpoint accuracy of a blasted abrasive stream removes burrs without damaging or causing dimensional changes to the part.

Micro blasting is effective at removing thin and brittle burrs without damage to the base material. Thick burrs or ductile metal left at edges absorb the energy of the abrasive stream and causes erosion in the surrounding region rather than fast burr removal.

Part Material and Abrasive Selection

Part material determines the proper attack method on a burr. Hard components made from stainless, titanium, or tool steel require a different approach than much softer polymers like PEEK and nylon. Hard metal burrs break free with an abrasive stream focused at a 45-degree angle to the burr root. Burrs on soft materials require multiple passes from different angles.

- *Hard metals:* Depending on the size of the burr, aluminum oxide is the most common abrasive used to deburr hard metals. Aluminum oxide tends to deburr much faster with less risk of corner damage but leaves a matte finish. Glass beads and sodium bicarbonate are slow and soft. These two abrasives are used against very fine burrs only, but are beneficial when preserving surface finish.
- *Polymers:* Sodium bicarbonate easily cuts through the root of burrs found on this ductile material.
- *Ground edges:* Ground edges have brittle regions. Brittle region removal significantly extends the life of a tool, particularly a cutting instrument. Maintaining a sharp edge is a challenge during deburring, honing, or cutting off the brittle region. Sodium bicarbonate is the best choice for this application. This abrasive is hard enough to knock off the brittle layer, but lacks enough energy to dig into the ductile base material.

Micro blasting is generally used to target a brittle burr on a more ductile part. The difference in material properties provides a buffer that protects the underlying part and prevents overblasting (Table 3.42.3).

Manual versus Automation

The random nature of burr location on a part requires manual micro blasting. The natural dexterity of the human hand, unencumbered by a heavy glove, allows an operator to specifically target problem areas and determine the best method of attack.

When volume absolutely requires automation, all targeted parts must be treated as worst-case scenarios or sorted into different levels or locations of burrs.

Thanks to its biocompatibility, PEEK is a staple in the medical device industry. Though easily machined, PEEK often develops fine feather burrs (Fig. 3.42.7). All burrs must be removed during the manufacturing process for the part to function properly and to meet FDA requirements.

Since PEEK is an expensive raw material, any slip of the hand during knife or pick deburring results in a costly mistake. A rich abrasive stream flowing at a low pressure allows an operator to follow the contours of the implant, covering several features in a single pass. Multiple passes remove the burr without causing damage to the part. Sodium bicarbonate's soft nature gently deburrs without damaging this ductile material.

Material	Abrasive	Effectiveness	Considerations
PEEK	Sodium bicarbonate	Highly effective	Removes some of the shine from the surface but does not texture.
Soft aluminum	Any abrasive	Ineffective	Soft aluminum tends to smear. It lacks the brittle root required for micro blasting.
Titanium or stainless steel	Sodium bicarbonate	Effective at removing feather burrs	Does not alter surface finish.
	Plastic media	Effective at removing a range of burrs	Has some impact on surface finish.
	Walnut shell	Effective at removing a range of burrs	Has some impact on surface finish.
Nylon	Sodium bicarbonate	Moderately effective	Removes burrs effectively, but the process is slow.
Torlon	Glass beads	Effective	Removes burrs quickly without impacting surface finish.
Hardened steel	Glass beads	Effective	Fine burrs on precision machined or ground edges.

Table 3.42.3 Abrasive Selection Guide for Precision Deburring

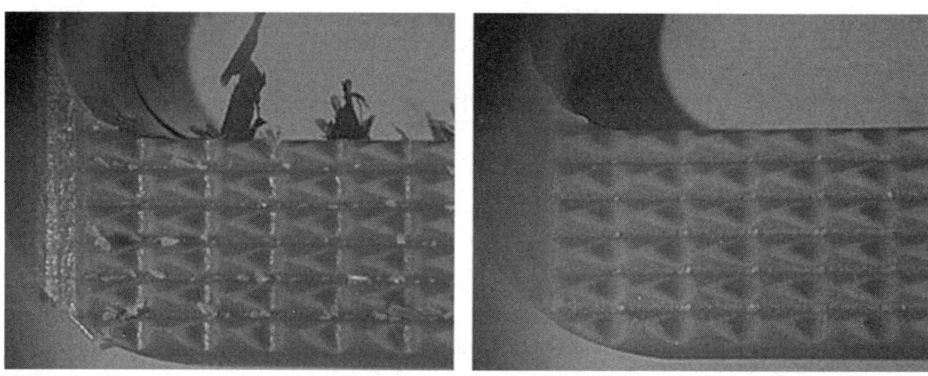

Figure 3.42.7 PEEK device before (*left*) and after (*right*) precision deburring.

Surface Texturing

Surface texturing is the creation of specific peaks and valleys on a part surface to enhance adhesion characteristics, control chemical reactions, or create a cosmetic finish. Micro blasting produces accurate and uniform surface finishes on a variety of base materials—from soft polymers to hard ceramics.

Ra or Sa

The most common goal of texturing is to increase surface area, but surface roughness is difficult to measure and difficult to control. Specifying a surface finish can be tricky. Repeatability requires close attention to all setup details. Ra is the most common method for a surface finish callout. Ra is the arithmetic mean of the absolute value of the deviations of the profile from the mean height. It is a measurement tool developed for machining, but adopted by micro blasting to quantify the effect of blasting a surface. Ra is sometimes represented in three dimensions as Sa. Combining Ra, coverage, surface area ratio (Sdr), and shape data creates a complete picture of the surface topography.

Coverage

All blasted surfaces should be specified to have a minimum of 100% coverage. Ra increases linearly with percentage coverage until 100% coverage is reached on a blasted surface (Fig. 3.42.8).

If an entire surface is not blasted, the resulting Ra value is low. Inspection at 10× magnification can identify any smooth or shiny areas. The entire surface of the part must appear altered. Blasting beyond 100% lowers the Ra value by breaking off peaks instead of digging new valleys. Variations in Ra indicate incomplete coverage, so multiple Ra readings provide a more comprehensive understanding of surface coverage (Fig. 3.42.9).

Surface Area Ratio

Developed surface (Sdr) is a measurement of the surface area ratio, comparing an idealized smooth, flat surface and the newly textured one. Sdr is usually represented as a percentage of the amount of additional surface created. Two surfaces with the same Ra may have different developed surface ratios due to different blasting parameters. Sdr rises to peak at 100% coverage and then drops significantly if blasting continues.

Shape

Shape refers to the size and spacing of blasted craters. Shape impacts the functional properties of a textured surface. Ra cannot distinguish between a satin finish and a matte finish.

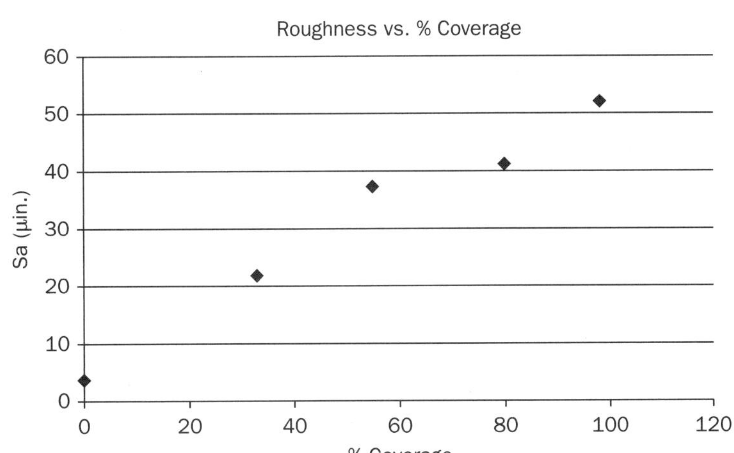

FIGURE 3.42.8 Measurements of surface roughness showing linear trend of roughness versus % coverage for 100 µm (0.004 in.) aluminum oxide.

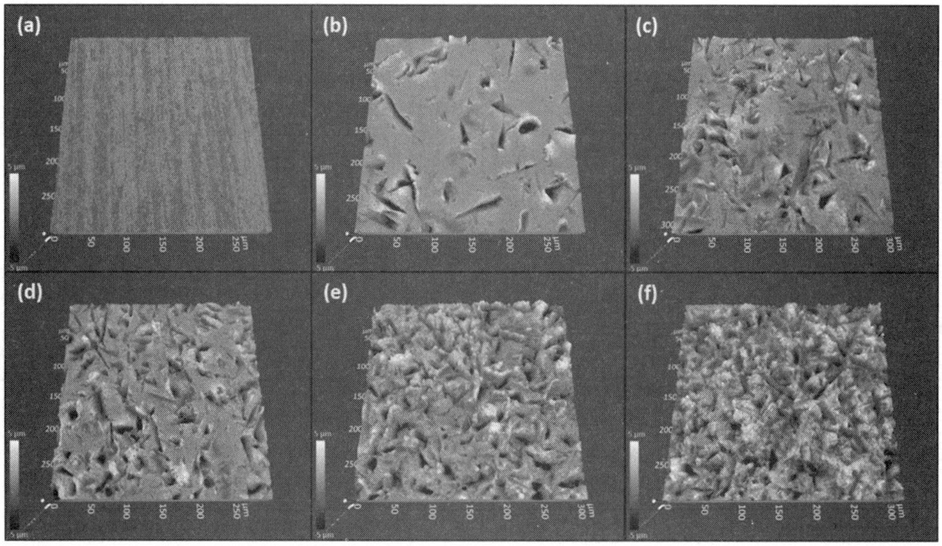

FIGURE 3.42.9 Progression of coverage from unblasted (*a*) to 33% (*b*), 55% (*c*), 80% (*d*), 100% (*e*), and 400% (*f*). Blasting media is 100 μm (0.004 in.) aluminum oxide.

Materials Suitable to the Process

Micro blasting is a flexible process, adaptable for use across industries to texture tool-steel molds, ceramics, polymers, medical-grade stainless steel and titanium, glass, Kapton, silicon, sapphire, and the list goes on. Part materials fall into one of two categories based on physical properties: ductile or brittle.

On ductile materials, the resulting finish is a combination of the crater depth (the impact of each abrasive particle) and the displaced ring around the edge of the crater. Micro blasting creates either a matte or a satin finish depending on the abrasive media.

On brittle materials, the fracture mechanics and grain size of the substrate determine the surface finish. Brittle materials tend to break along grain boundaries and lack a displacement ring. Micro blasting only produces a matte finish on brittle materials, regardless of abrasive.

Abrasive Characteristics

Angular particles cut jaggedly and leave a matte finish on a ductile surface, whereas spherical particles peen craters to leave a satin finish. All brittle surfaces fracture to create a matte finish, regardless of abrasive used. The more energy an abrasive can store, the rougher it leaves a surface. Figure 3.42.10 illustrates the roughness produced with different media size at different velocities on stainless steel.

Abrasive particle size significantly impacts the resulting Ra value. Large particles yield large dents, which become more pronounced at higher velocities. The delta between the smoothest Ra surface created and roughest surface created increases as particle size increases. For instance, a 10-μm (0.0004-in.) aluminum oxide blasted at low pressure onto stainless steel creates a 0.1-μm (4-μin.) finish and a 0.2-μm (8-μin.) finish

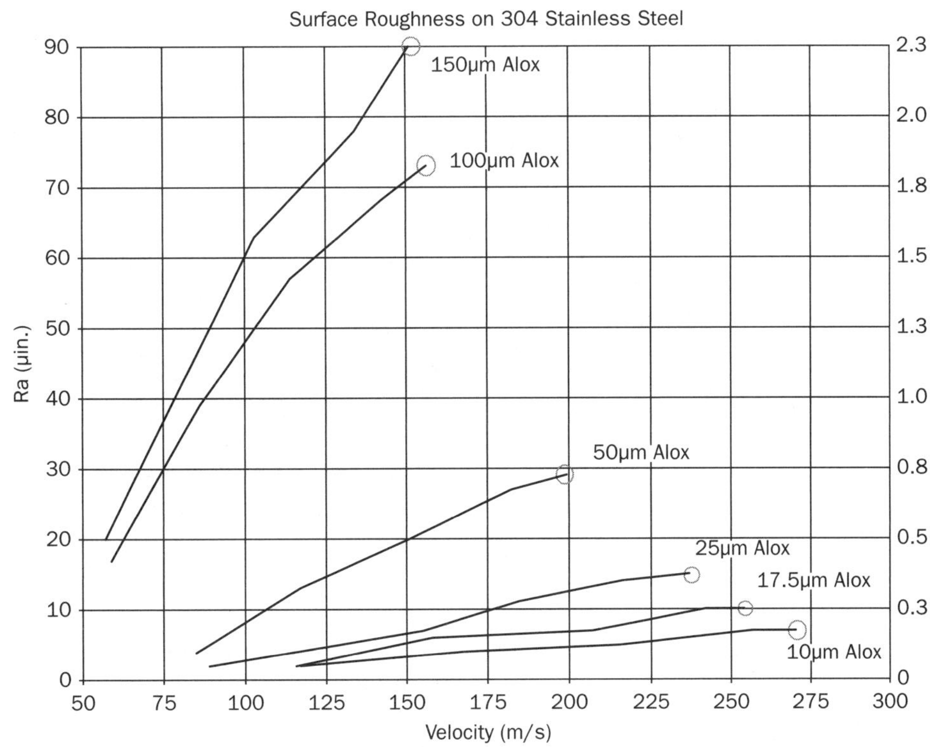

FIGURE 3.42.10 Surface roughness on 304 stainless steel.

at high pressure. A 150-µm (0.006-in.) aluminum oxide creates a 0.5-µm (20-µin.) finish blasted at low pressure and a 1.75-µm (70-µin.) finish at high pressure.

Important Notes

- Crater depth increases as the speed of the abrasive particle increases. Blast pressure primarily determines velocity, but nozzle design, propellant or gas type, and nozzle distance grow with importance depending on the surface finish requirements. Air, nitrogen, and carbon dioxide can be used as the gas propellant.
- An increase in media feed rate results in fewer interactions between air molecules and abrasive particles, which decrease particle velocity.
- The relationship between velocity and Ra is linear for ductile materials, but not for brittle materials, which tend to break along grain boundaries.

Example: Dental Implants

The surface texture of a dental implant determines how quickly the implant integrates into a jaw bone. So texture is a critical design component. Studies indicate that the

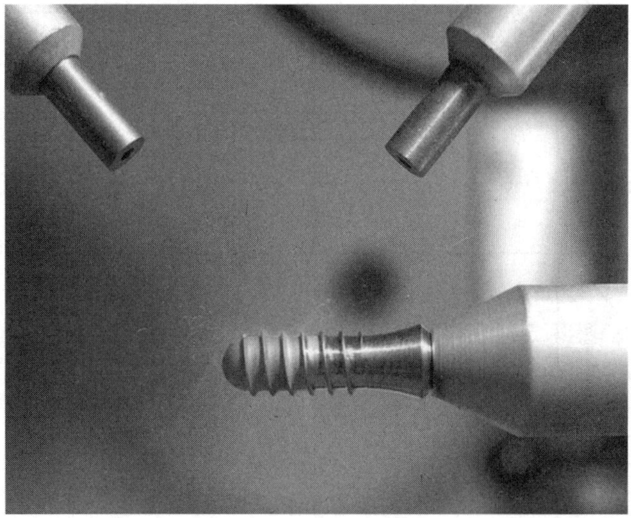

FIGURE 3.42.11 Texturing surface of dental implant using automated micro blasting.

surface of the dental implant needs to be more than just rough. Several characteristics impact the rate of integration. Micro blasting controls two of these characteristics: Ra and Sdr.

Automated micro blasting textures dental implants best. It provides careful control over particle velocity and coverage to create surfaces as smooth as 0.25 μm (10 μin.) or as rough as 5 μm (200 μin.) (Fig. 3.42.11).

CHAPTER 3.43

Micro Part Inspection

LaRoux K. Gillespie

Manufacturing Consultant
Andover, Kansas

The inspection of micro parts and micro features is an often overlooked aspect of design. Verification of features and dimensions can be made via several techniques, but defining requirements needs to be carefully considered. As an example, specifying inspection at 30× magnification may require five times the inspection time that a 10× magnification requires. Further, what magnification actually assures complete visualization of the point of interest?

Handling

Micro parts are notoriously difficult to handle, and to position. In addition, many have such thin walls that they are easily distorted by handling. For parts that are dime sized or larger this may not be an issue unless they are so thin that they stick to flat surfaces. Chapter 3.51 describes considerations for just the handling aspect of micro products.

Inspection Approaches

Micro inspection equipment includes touch probing, hard gaging, optical comparators, lasers, white light scanners, and even video systems. Digital X-ray and computed tomography (CT scanning) also are increasingly being used and displayed in three dimensions. In addition to the physical measuring, it is important to consider how lighting, translucence, and product color affect good measurement. These are the important factors to be considered particularly for translucent, shiny, and very dark-colored features.

For the very micro features, a scanning electron microscope (SEM) may be needed to clearly see them. The SEM allows measurements as well as visualization of features. The SEM can photograph at 10,000× with everything in view being in sharp focus. SEMs are widely used in research and analytical laboratories. Figures 3.43.1 and 3.43.2 illustrate two micro products that are inspected by a variety of means.

FIGURE 3.43.1 Micro screws are inspected with hard pitch gages as well as highly magnified comparators or laser systems.

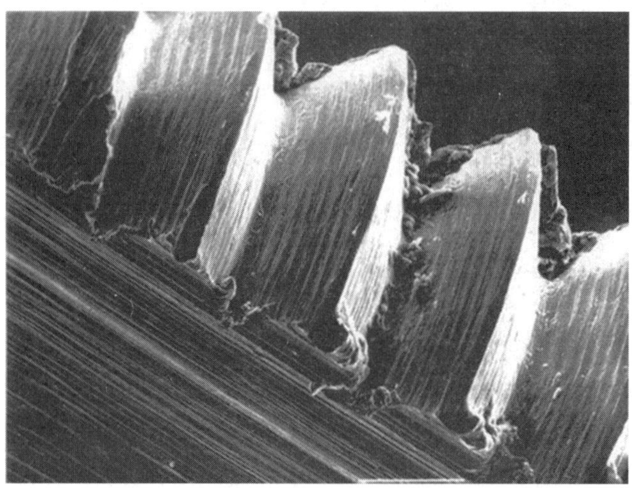

FIGURE 3.43.2 Microscopes are used by a tap user in incoming inspection to detect burrs on these "highest quality" micro thread taps. 30× magnification may be used to see them, but a scanning electron microscope (SEM) at up to 1000× was used to clearly photograph them.

Touch Probing

Conventional coordinate measuring machines (CMMs) utilize touch probes to make hundreds of measurement a minute and establish best-fit or cloud-based best definition of features. For the tiniest of parts and the tiniest of features, however, touch probing is a problem. It either moves the part on the table or deflects the part while measuring or the probe measuring it is too large for the feature to be probed (i.e., cannot fit in the hole or space).

Today probing has also gone micro. One manufacturer makes a 127-μm-diameter (0.005-in.) stylus. With some associated precision rotary indexer, it has been used on fuel injector nozzles having 229-μm-diameter (0.009-in.) holes. While it still applies pressure on the part, it is measured in only milligrams.

The traditional CMM, however, has become more than a probing device today; it can be and is in some equipment a multi-sensor-ready machine. It can use touch trigger probes for tactile measurement of features that are hard to image. It can employ laser scanners in place of the probe, or even scanning white-light sensors. Lasers also can be fit to a CMM head.

Hard Gaging

Hard gaging is still used for metal parts and rigid polymers for simple measurements. Because of the tolerances now being assigned to micro features, using hard fixed gages is becoming less and less the norm. Micro threads of 0.5 mm (0.020 in.) and larger can be measured for go-no-go in that manner, but laser systems or vision systems are able to accurately define the contours. So, the manufacturer must need to know whether dimensional data is required or simple accept/reject is sufficient.

Microscopes

Microscopes are good for verifying presence and absence of features. Some "scopes" can also superimpose a measurement grid that allows manual measurements. In general that approach is neither rapid enough nor accurate for the increasingly common tolerances of 2 to 3 μm (79 to 118 μin.).

Microscopic inspection of the presence or absence of features or issues is being performed at 10× to 400× magnification. Verification of the absence of burrs on microscopic gears or threads is one example of good use of microscopes. In the latter instance, stereoscopes are required though rather than conventional classroom microscopes. Stereo viewing adds considerably to accurate views.

Both manual and high-production systems are available that display output through lens, large computer-like screens, or simply as pass-fail computer-analyzed statistics. They may be called inspection stations as opposed to microscopes.

Some of these displays also allow users to annotate comments on the screen and save them for later use.

Optical Comparators

Optical comparators are one of the lowest cost inspection devices for micro parts, but they are used primarily to determine if the contours fit within predetermined boundaries. A magnified version of the part is focused on a glass screen. This image is then compared to a 2D image of the CAD design. It can detect contours, locations of through features, the presence of burrs or flash and some chamfers, scratches, and dents. It is relatively fast for a manual system and it is possible to add mechanized positioning and acceptance but that is not the norm. Some comparators magnify the image to 100×, but most are 30× or 50× at the maximum. Typical parts inspected with comparators are larger than 25 mm (1 in.) and often are several inches in size, but systems for the smaller parts exist.

White Light Systems

White light technology (which can actually be white, blue, green, or red) can scan with accuracies up to less than 25 μm (0.001 in.). It is widely used on larger parts and can scan a million or more points a minute. It works by shining parallel lines on the product and to the eye these straight lines curve because of the contour of the part. The computer takes photos of each shot of the lines and makes a point cloud of the resulting images which portrays the actual contour. The resulting data can then be matched with drawing requirements. While it can be used on micro products, laser systems provide more accurate data which is generally needed for the micro parts. White light systems are also called structured light systems.

For both laser and white light systems, parts may need to be coated with a light coating of powder or paint to minimize reflection or absorption of the light. Both have the advantage of noncontact, so can accurately inspect pliable products.

Laser Scanners

Laser scanners now collect up to 225,000 measurements per second. Also known as laser line scanners some can scan shiny parts and like most modern computer-based systems they can compare the results against the computer design model to display conformance and errors. Some of the most accurate machines have accuracy and a repeatability of 6 μm (0.0002 in.). These machines are placed on CMM tables for maximum accuracy, but smaller tabletop systems also exist. Measurement times per part are often in the range of 3 to 4 min depending on the number of points to be measured, and can be less. Some units are barely larger than the home coffee maker machine.

Video Systems

These systems typically employ a video camera with computer-driven magnification capabilities. They employ a flat screen using high-definition projection. The product or the imaging head is moved back and forth and can be locked in position. Major systems have built-in lighting that can be adjusted either manually or digitally to better highlight specific areas.

Most of these are desktop systems with abilities up to 250× magnification and come with measurement ability, but few are as accurate or repeatable as the laser or touch probe systems. They are quick and convenient for work that is not high volume or high pace inspection.

Video systems, however, can be automated for high production. The commercial on-the-shelf systems generally do not fit that market. Special design optical measuring systems have been married with LabView to measure 1200 parts a minute with an accuracy of ±12 μm (±0.0005 in.) and a precision of ±3 μm (±0.00012 in.).

These also may be referred to as inspection stations as opposed to video systems.

Digital X-Ray

X-ray systems today are used to inspect electronics for the presence of features and undesirable issues. Just as they do for body parts they can show the internal features and present them in a 3D manner if desired.

CT Scanning

Like conventional X-rays, CT scanning is used to display internal features or attributes. The process is essentially a fully automated scanning process that interlaces 2D X-ray images to provide 3D representations. Unlike X-rays alone, CT scanning allows users to verify the dimensions as well as the presence or absence of items. In this instance, the data set produced is fed into a separate program to provide the measurements. CT is used to inspect precision plastic parts, small castings, and complex parts and assemblies, and it can provide powerful visualization of the product as well as an analysis of the data. CT is suitable for a wide range of sample sizes and material densities. It can provide accuracies up to 10+ μm (0.0004 in.). Accuracies depend on the sample size and lessen as the sample length increases.

Because micro parts and features are so small, the X-rays can penetrate almost all materials they are made of. They are used for plastic injection mold validation of process and tooling as well as part dimensions after prove-in to detect shrinkage, deformation, and dimensional errors. They have been used to verify micro gear and micro threads as well as complete assemblies.

The equipment is expensive, but inspection services can be hired to perform that inspection. Also because they take layers of inspection data which may involve thousands of points, the process is much slower than measuring external features.

Other Approaches

Acoustic micro imaging is used to detect abnormalities at material interfaces (delamination for example). Gaps between materials as thin as 100 Å (0.4 μin.) can be detected as can internal anomalies like warped or tilted internal layers. The detection, however, requires the transducer to be coupled to the part by a column of water or other liquid.

Because of the manufacturing processes, it may be necessary to make some critical measurements before the process is complete. In-process inspection is common to some products because some features are covered up later or altered by the assembly process. Designers can assist in making this approach work by the design they use and notes that allow acceptance in process rather than just in final condition.

A new approach for micro parts that have hard-to-measure internal features is to destructively sample some of them. Since micro parts are easily distorted by forces and temperature, they are encased in a slow curing thermoplastic resin. Then thin slices are shaved off the top of the encased block until the top of the part is reached. Gradually additional thin slices are made or rather shaved from the block. A camera takes a photo of each stage of the slicing which may involve layers only 25 to 50 μm (0.001 to 0002 in.) each. One equipment maker uses a resolution of the photos of 1550 pixels/mm^2 (1 million pixels/in.^{2}). The process continues until the bottom of the part is reached. The images are then compiled to make a 3D point cloud that defines every contour of the part. That cloud is then compared to the part definition.

Environment

The environment in which measurements are taken influences the measurement and/or the product. For most applications the environmental issues to be considered are temperature change, and for polymers and fabrics humidity effects. For the biomedical products the presence of any contaminants is a major concern and inspection can create them or not detect them when they exist.

Foam parts are subject to both temperature and humidity as well as physical forces, so they need to be isolated from all of them from the time they are made until assembled. While thermal expansion or contraction on micro features will similarly be micro, they may cause some distortion as well as size change.

The product dimensions may also change in the using environment, and while that is not an inspection issue, it may mean that the designer needs to define dimensions during product manufacturing that is different than what they will be in use.

Validating the Process Rather Than the Product

For many parts it is more economical to validate that the process is making good parts than it is to measure the parts themselves. This means that every aspect of the manufacturing process has to be closely controlled. It is a common approach for some micro molded products or at least it limits the number of measurements that have to be made. The mold is measured as well as the materials, temperatures, ejection, cooling, etc.—every aspect of the process is controlled to assure the finished part will be the same for every part. Simple sample inspection or destructive sampling is used then to check for any unexpected changes.

References

Kavuri, S. H., Size Effects in Human Visual Inspection for Micro/Meso Scale Parts, master's degree thesis, University of Nebraska—Lincoln, 2015.

Ljubicic, D. M., and B. W. Anthony, 3D High Speed Profilometer for Inspection of Micro-manufactured Transparent Parts, *Proceedings SPIE 8082, Optical Measurement Systems for Industrial Inspection VII*, 808211 (May 26, 2011).

CHAPTER 3.44

Advanced Additive Manufacturing: The MICA Freeform Process

Richard T. Chen

Microfabrica Inc.
Van Nuys, California

MICA Freeform Process

MICA Freeform is a high-precision, high-volume, additive manufacturing technology targeted at devices in the millimeter and submillimeter range with features and tolerances measured in micrometers. Like 3D printing, highly complex 3D structures can be fabricated directly from a digital model. Unlike many 3D printing technologies that use powdered materials, the MICA Freeform technology produces devices composed of chemically pure, fully dense materials. Many applications that use MICA Freeform require high reliability and are either stressed for millions of cycles or exposed to high-impact environments. MICA Freeform has been commercially available since 2009 and has been used in a variety of industries including semiconductor, medical, industrial, electronics, optical, and aerospace.

While most 3D printing processes are not easily scaled to high-volume applications, the MICA Freeform process is routinely used to produce millions of parts each year. The MICA Freeform process can also fabricate functional "printed assemblies" comprising multiple moving parts for which no assembly is required. Figure 3.44.1 shows images of example parts fabricated using the MICA Freeform process.

As with other 3D printing processes, parts fabricated using the MICA Freeform process are built in layers. The MICA Freeform process involves three primary steps per layer. First, a structural metal is electrodeposited onto a substrate in selected regions corresponding to the desired cross section of the part to be fabricated. Deposition occurs through openings in a photoresist that are patterned using a submicrometer resolution photomask, in a method similar to that used in semiconductor fabrication

FIGURE 3.44.1 Example devices fabricated using the MICA Freeform process: (*a*) medical biopsy forceps; (*b*) medical-implanted electrodes; and (*c*) micro fluidic device. (Images provided courtesy of Microfabrica.)

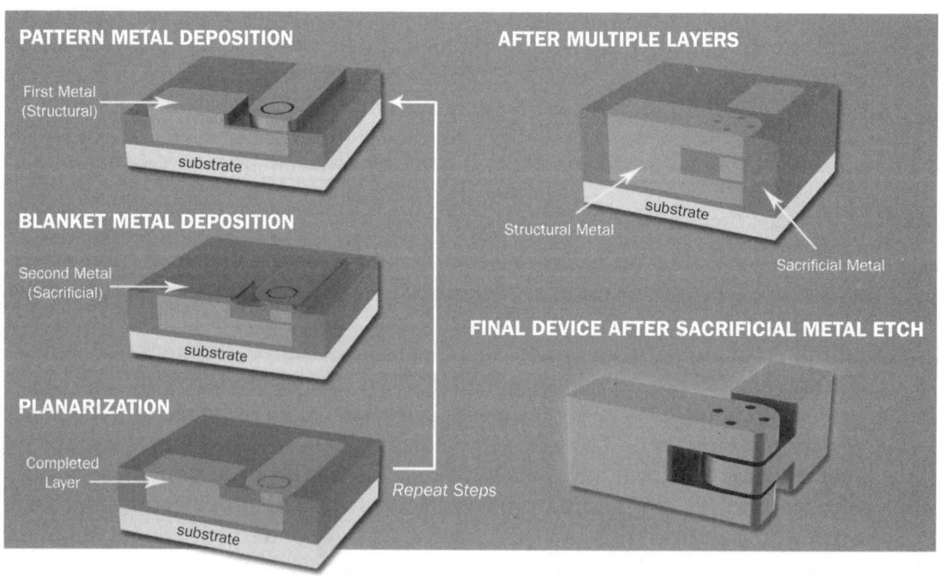

FIGURE 3.44.2 The MICA Freeform process. (Image provided courtesy of Microfabrica.)

facilities. After removal of the photoresist, a sacrificial metal is electrodeposited over the structural metal. Finally, both metals are planarized (abrasively lapped) to yield a layer that is flat, planar, and of precisely controlled thickness. The three steps are then repeated for all layers required, after which a chemical etchant is used to dissolve the sacrificial metal, releasing the parts. Figure 3.44.2 shows a diagram of the MICA Freeform process.

Capabilities of the MICA Freeform Process

The MICA Freeform process provides tolerances of approximately ±2 μm (±0.000079 in.), and minimum features as small as 10 μm (0.0004 in.) in the horizontal plane and 5 μm (0.0002 in.) along the vertical plane. Parts made by the MICA Freeform process are typically less than 1.25 mm (0.049 in.) thick and 10 mm (0.394 in.) on a side. This process produces parts that are true to design and do not require any secondary operations such as deburring or polishing.

MICA Freeform creates 3D objects using layers with thicknesses between 5 μm and 50 μm (0.0002 and 0.0020 in.). The thickness of each layer is determined by the designer. For example, a part can be made with all 25-μm-thick (0.0010-in.) layers or each layer can be of a different thickness (e.g., 5, 10, 25, 5, 10, 25). The number of layers used to create a part is also defined by the designer. Parts can consist of a single layer or over 50 layers.

Unique Features

The MICA Freeform process can produce features that are unique compared to conventional micro fabrication technologies as described in the following sections.

Precision Holes

Holes are simultaneously fabricated as the part is built and virtually any hole shape can be fabricated. The circularity of round holes, blind holes, and counterbores is maintained at sizes as small as 40 μm (0.0016 in.) in diameter. The concentricity of inner and outer diameter features can be held to ±2 μm (±0.000079 in.). Examples of holes fabricated with the MICA Freeform process are shown in Fig. 3.44.3.

Micro Channels

Channels are also simultaneously fabricated as the part is built, allowing for both open and hidden channels. Channels can be as narrow as 10 μm (0.0004 in.) and can have a depth as large as required or as small as 5 μm (0.0002 in.). Hidden channels can be fully embedded within a part and do not need to have line-of-sight access. Figure 3.44.4 shows an example of a hidden channel that is fabricated in a micro fluidic part that is 0.9 mm (0.035 in.) in diameter. The center hole is connected to radial blind holes via a hidden channel.

Micro Bosses and Ribs

Bosses and ribs can be fabricated with features as small as 10 μm (0.0004 in.). The walls of the features are vertical and form 90-degree sharp corners with the base.

FIGURE 3.44.3 Example of precision holes. (Image provided courtesy of Microfabrica.)

(a)

(b)

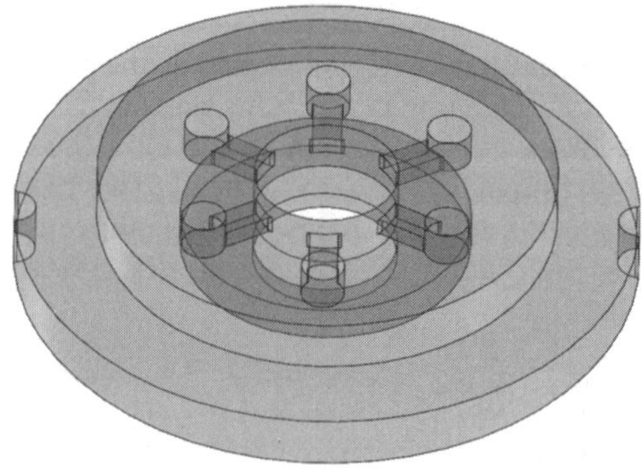

FIGURE 3.44.4 (*a*) Example of a micro fluidic part with hidden channels. (*b*) Transparent drawing showing the hidden channels connecting the center hole with the radial blind holes. (Images provided courtesy of Microfabrica.)

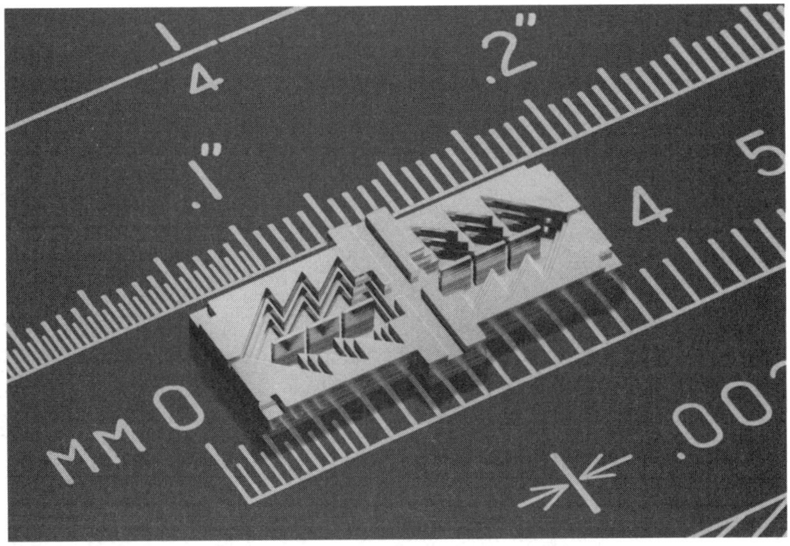

Figure 3.44.5 Micro mold insert with precision boss features. (Image provided courtesy of Microfabrica.)

The outside corners of these features can have sharp corners with 5-μm (0.0002-in.) radii. Figure 3.44.5 shows a micro mold insert which has several precision bosses.

Undercuts

Undercut features can be virtually impossible to create at the micro scale using conventional technologies. Undercuts can be used freely without restriction in the MICA Freeform process. Examples of undercut features can be seen in the medical tissue staple shown in Fig. 3.44.6.

Assemblies and Mechanisms

Perhaps the most unique feature that can be fabricated using the MICA Freeform process is an articulated assembly or mechanism. All moving parts of a full assembly or mechanism can be fabricated simultaneously in its assembled state, eliminating the need for micro assembly. This has the advantage of avoiding assembly cost and reducing yield loss. In some cases, micro assembly of complex mechanisms would be impossible. Hinges, journal bearings, and linear bearings are examples of mechanisms that can be fabricated as a single unit. The image in Fig. 3.44.7 shows a gear train that was monolithically fabricated in a fully assembled state. The gears are held in place by journal bearings.

Materials

The process uses several engineering-grade metals which include Valloy-120, a nickel-cobalt alloy, as well as pure palladium, rhodium, and copper. Valloy-120 has been optimized for strength and ductility and is similar in strength to many common grades of stainless steel. It is also 24-h biocompatible and can be used for applications such as medical tools.

(a)

(b)

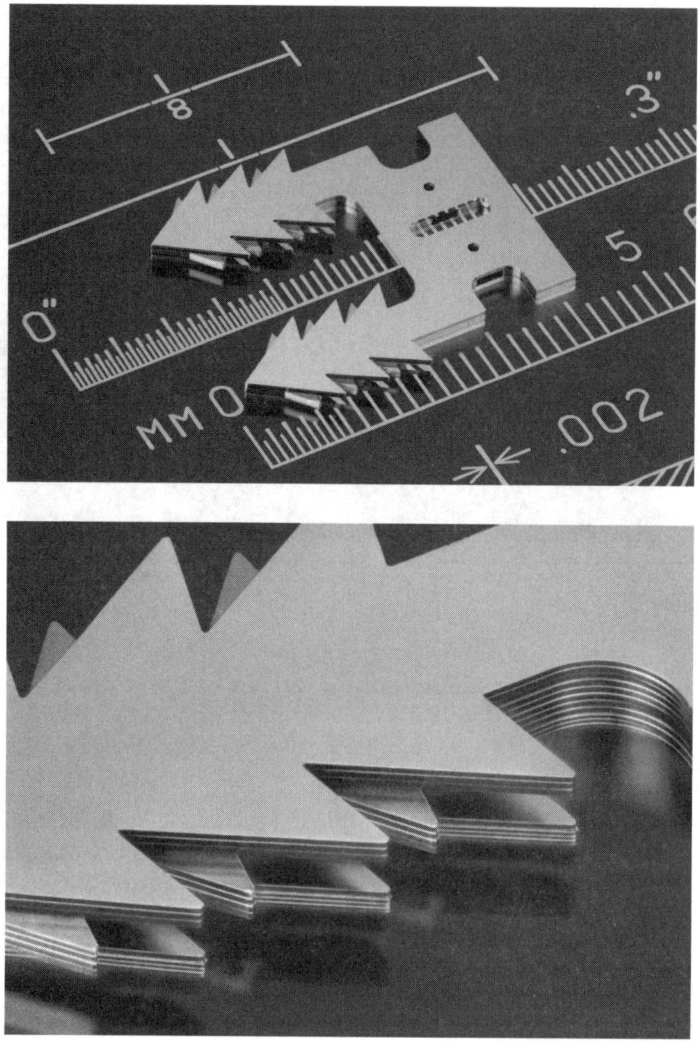

FIGURE 3.44.6 (*a*) Example of undercut features in a medical tissue staple. (*b*) Close-up of undercut feature. (Images provided courtesy of Microfabrica.)

Palladium is a platinum group metal that is biocompatible, implantable, nonmagnetic, corrosion-resistant, and radio opaque. It is commonly used for both medical and electronic applications. The palladium deposition process ensures very small grain sizes, producing strength much higher than typical stock palladium. Rhodium, another platinum group metal, has high Rockwell hardness and high wear resistance. Rhodium does not oxidize, and therefore makes a very good electrical contact material. Rhodium is typically placed strategically within a device where high wear resistance or good electrical contact is needed. For example, rhodium is often used in micro connectors. Copper can be embedded within a device to provide high current carrying capability or high thermal conductance. A brief summary of common

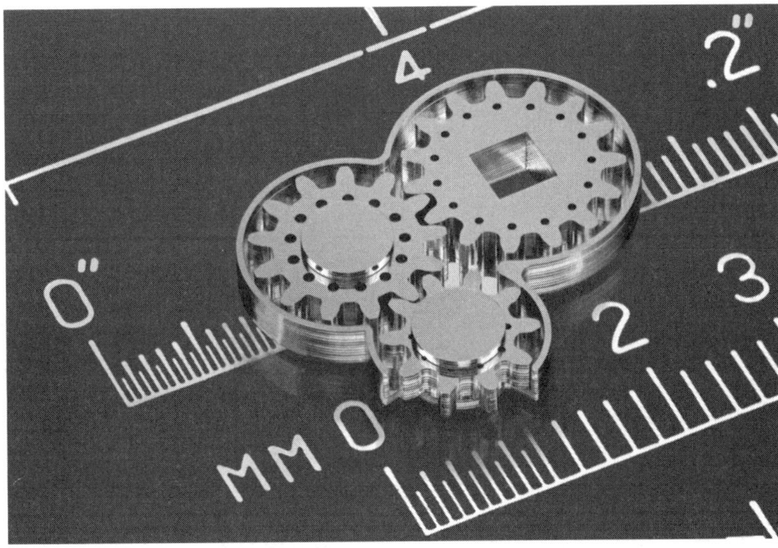

Figure 3.44.7 Gear train fabricated monolithically without assembly. (Images provided courtesy of Microfabrica.)

material properties for the structural metals can be found in Table 3.44.1. Other materials continue to be developed. A more complete discussion of the most up to date materials can be found on the Microfabrica website (www.microfabrica.com).

Multiple materials are sometimes used together to form micro composite structures. As mentioned above, rhodium and copper are often used in conjunction with Valloy-120 or palladium. This allows a micro device to have characteristics of all the metals used, therefore optimizing performance. For example, a spring connector can be made with Valloy-120, rhodium, and copper. Valloy-120 provides the mechanical strength for the spring; rhodium provides the high wear resistance and excellent electrical contact for the connector's contact point, while copper provides a low-resistance path to pass high electrical current. No single material would be able to provide this combination of properties.

Design Recommendations

It is recommended that the designer uses a computer-aided design (CAD) tool and designs the device with layers in mind. Each device is fabricated based on its own unique layer scheme. A layer scheme is the sequence of layer thicknesses used for a particular fabrication run, from layer 1 through N (where N is the last layer). The designer specifies how many layers are to be used and the thickness of each layer.

Generally, thinner layers can produce smaller feature sizes. The total layer count determines cost and cycle time. To optimize for faster lead times and lower costs, the thickest layers possible should be used to define vertical features, lowering the total layer count. For example, the sphere in Fig. 3.44.8 can be fabricated using 20 fixed thickness layers (right) or 13 variable thickness layers (left). The overall shape of the spheres

		Stainless Alloys		Microfabrica
		316L (cold worked)	17-4 (H1100)	Valloy-120
UTS	MPa	862	1000	1100
Yield 0.2%	MPa	689	910	900
Hardness	Rockwell C	11	34	36
Hardness	HV (Vickers)	199	335	425
Modulus of elasticity	GPa	200		170
Electrical resistivity	Ω - m	7.4E-8		1.05E-7
CTE (0-100C)	/K		1.08E-05	1.34E-05

		Pt-10% Ir		TI	Microfabrica	
		As cast	Annealed	F136 Ti-6Al-4V ELI (annealed)	Palladium* (as fabricated)	Palladium* (annealed)
UTS	MPa	329	350	896	1350	275
Yield 0.2%	MPa	239		827	1100	175
Hardness	Rockwell C			35	43	11
Hardness	HV (Vickers)	113	130	341	400	100
Modulus of elasticity	GPa		164.8	124	110	75
Electrical resistivity	Ω - m	2.5E-7		1.8E-7	1.5E-7	1.5E-7
CTE (0-100C)	/K	8.40E-06	8.40E-06	8.60E-06	1.11E-05	1.11E-05

		Stainless Alloys		Microfabrica
		316L (cold worked)	17-4 (H1100)	Valloy-120
UTS	ksi	125	150	160
Yield .2%	ksi	100	130	130
Hardness	Rockwell C	11	34	36
Hardness	HV (Vickers)	199	335	425
Modulus of elasticity	10^6 psi	29		25
CTE (32-212F)	/F		6.00E-06	7.49E-05

		Pt-10% Ir			Microfabrica	
		As cast	Annealed	F136 Ti-6Al-4V ELI (annealed)	Palladium* (as fabricated)	Palladium* (annealed)
UTS	ksi	47.7	50.7	130	220	40
Yield 0.2%	ksi	30		120	160	30
Hardness	Rockwell C		11	35	43	11
Hardness	HV (Vickers)	113	130	341	400	100
Modulus of elasticity	10^6 psi		24	18	16	11
CTE (32-212F)	/F	6.20E-05	6.20E-05	4.78E-06	6.20E-05	6.20E-05

*Preliminary

TABLE 3.44.1 Common Microfabrica Process Material Properties

Figure 3.44.8 Example of a sphere designed with 13 variable thickness layers (*left*) and 20 fixed thickness layers (*right*).

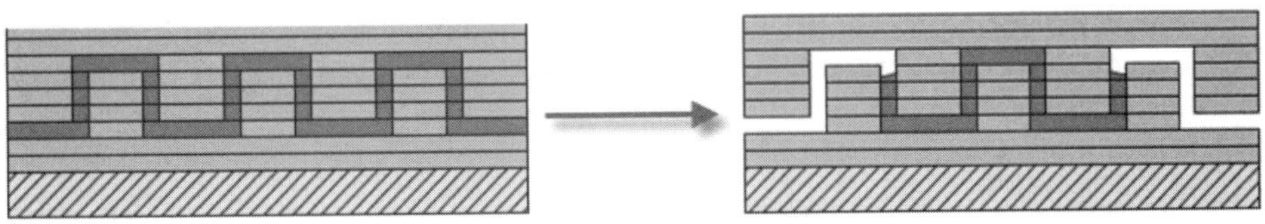

Figure 3.44.9 Example of a complex channel geometry.

in both options remains relatively similar, but the cost and lead time for sphere using a reduced number of layers will be lower. Any of these layer schemes is valid for the MICA Freeform process and the designer determines which layer scheme is the best fit for a particular application.

As shown in Fig. 3.44.2, fabricated devices are initially embedded with sacrificial metal which is removed at the end of MICA Freeform processing. Since a wet etchant is used to remove the sacrificial metal, devices must be designed such that the release etchant can access all the regions from which sacrificial metal should be removed. Due to diffusion and other effects, removal of sacrificial metal from long narrow channels can be difficult, especially if the channel is meandering (Fig. 3.44.9). The complex channel geometry (left) may not clear completely after the release etching (right). Lengthy and convoluted channel structures should be avoided in order to facilitate release etching.

To maximize the probability of complete release, devices should be designed such that the sacrificial metal is as fully exposed to the release etchant as possible. Sacrificial metal cannot be released from fully enclosed, hollow geometries. In some cases, entrapped sacrificial metal is desired. For example, if the sacrificial metal is copper, fully enclosing the copper will form a composite structure with either Valloy-120 or palladium as the outer shell, providing mechanical strength, and copper will remain in the center, providing a low-resistance path for electrical current. Areas from which it is difficult to remove the sacrificial metal include narrow, tortuous channels and narrow gaps between wide surfaces. In these cases, use of release holes is recommended. Release holes should be at least 60 μm (0.0024 in.) in diameter.

When to Use MICA Freeform

The MICA Freeform process is ideally used for millimeter-scale applications that require high precision or complex features. Also, if high-volume production of very uniform parts is desired, then the MICA Freeform process can provide the manufacturing solution. When conventional machining techniques are unable to produce the desired part, the MICA Freeform process is often a viable alternative.

If the device can easily be made using photochemical etching or stamping, then the MICA Freeform process may not be cost competitive if the lower accuracy and definition of the conventional processes are acceptable. As a given part geometry gets smaller, it becomes a better commercial fit for the MICA Freeform process. In some cases, the MICA Freeform process is enabling when an existing product must be reduced to smaller size. While conventional technologies may have been adequate for the original product, the smaller, next-generation product may require a new process to enable miniaturization while preserving product performance.

New concepts and possibilities are often enabled when the flexibility of the MICA Freeform process is leveraged during the initial development of the design. So, designers are encouraged to fully explore the current capabilities at the time they begin their design.

CHAPTER 3.45

Micro Stereolithography

LaRoux K. Gillespie

Manufacturing Consultant
Andover, Kansas

Micro stereolithography (μSL) is an adaptation of conventional stereolithography in micro scale. The basic processes are the same, but the fine scale of micro features has some additional constraints and much finer resolution. It is also a field that continues to expand beyond just polymer materials. Today it is possible to blend ceramic particles in the polymer, and after completing the SL process, burn out the polymer and sinter the resulting material to a final product shape. Micro ceramic gears have been made in this manner, although it is not a standard process. Some colors can also be blended with the monomer.

In the most basic terms, a UV laser is scanned on a liquid monomer resin. This hardens a layer of roughly 10 μm (0.0004 in.) (can be more or less) in the places the laser hits the polymer. The platform on which the part is made is lowered down in the solution (in the order of a few micrometers, e.g., 0.0002 to 0.002 in.) and another scan is made. This is continued until the full height of the part is achieved. Each layer represents what a section at that location would look like if sliced out of the part. It is a layer-upon-layer fabrication process. Like conventional SL, several parts can be made at one time on the same platform. Chapter 2.11 presents details of the SL process, while this chapter concentrates on the capabilities of the micro adaption.

Different line widths and layer depth are produced by adjusting the laser beam diameter and the exposure energy. Laser exposure energy is a function of scan speed, laser power, and spot size. The smallest spot size and hence line width to date is 1 to 2 μm (39 to 79 μin.). With this discrimination micro gears, micro channels, micro tubes, biomedical capillaries, micro bioreactors, and a variety of 3D structures are made. Layers are produced at the rates of 100 to 1000 layers per hour, depending on the equipment used. Most resins are acrylic or epoxy variations, but new resins are being developed each year that allow additional features to be incorporated.

Experimental data for polymerized hexanediol diacrylate (HDDA) indicates that the cure depth is a straight-line function of the *logarithm* of laser exposure where exposure is measured in J/cm^2. The line width is a straight-line function of *just* J/cm^2. As an example of capability, an HDDA monomer line was made with thickness of 10 μm (0.0004 in.) and its width was controlled to 1.4 μm (55 μin.) with a 2.2-μW laser. In the same equipment an 800 μm (0.031 in.) long tube with ID of 50 μm (0.002 in.) was made, with an aspect ratio of 16. With some equipment, line depths as thin as 1 μm (39 μin.) have been made.

Applications

Micro SL has several applications for the micro world.

- Quick production of a first part micro prototype (either complete part or just key features to validate design).
- Continuous production of low-quantity micro parts.
- Production of parts which after later electroforming produce molds which are used for micro molding or micro hot embossing.

Micro SL makes MEMS, micro devices of many designs, micro gears, micro channels for fluid separation and analysis, micro fluid mixers, optic mounts, micro sensors, micro connectors, and a variety of biomedical cell application parts. In the bio world, tissue scaffolds, nerve guidance conduits, and cardiovascular stents are manufactured via this process and micro needle arrays with *integral* drugs have been made. Figures 3.45.1 through 3.45.5 illustrate the details and micro features that are made by μSL.

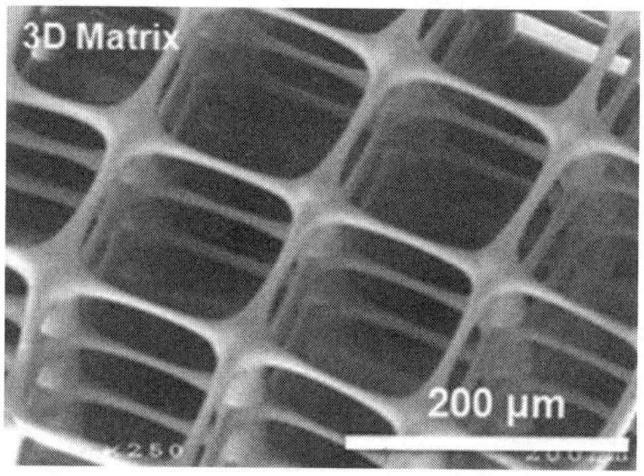

FIGURE 3.45.1 SEM photo of 3D matrix made by μSL. (Courtesy Fan Zhou, fzhou1@stanford.edu.)

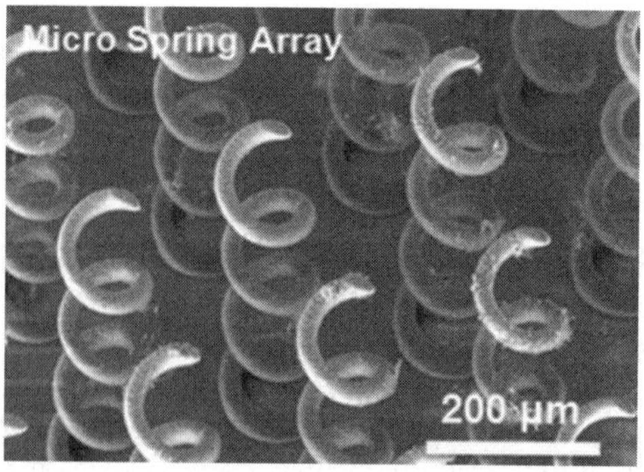

FIGURE 3.45.2 SEM photo of spring array made by μSL. (Courtesy Fan Zhou, fzhou1@stanford.edu.)

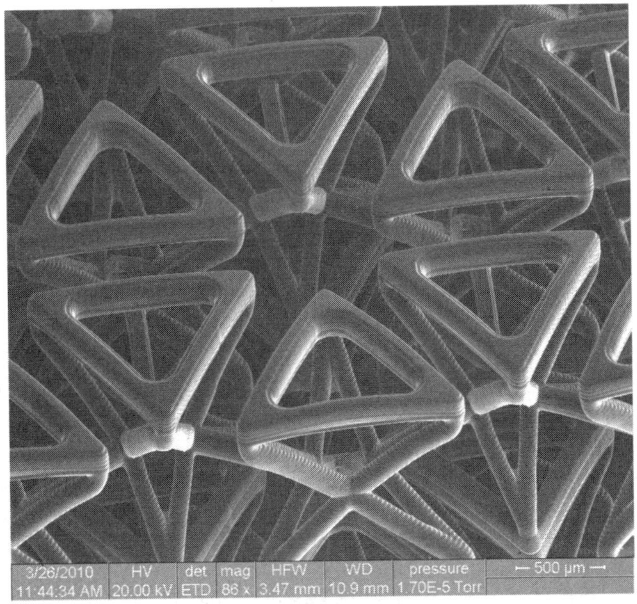

FIGURE 3.45.3 SEM photo of auxetic structure made via μSL. (Courtesy Fan Zhou, fzhou1@stanford.edu.)

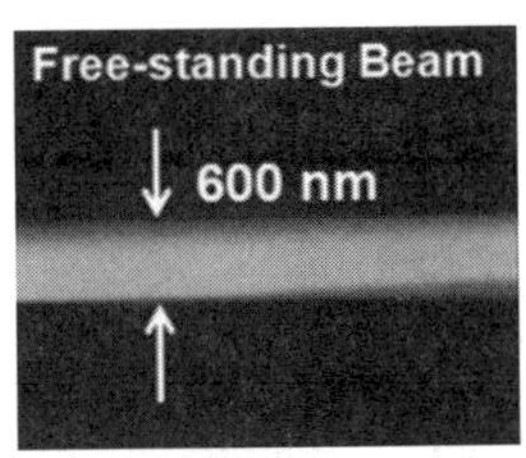

FIGURE 3.45.4 Free standing micro beam made by μSL. (Courtesy Fan Zhou, fzhou1@stanford.edu.)

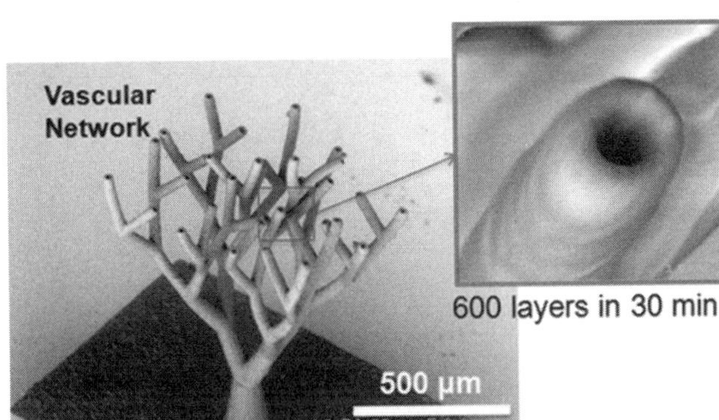

FIGURE 3.45.5 Vascular network made by μSL. (Courtesy Fan Zhou, fzhou1@stanford.edu.)

The monomers or polymers can incorporate materials such as powders or particles of ceramic. With appropriate concentrations after hardening, the product can be heated to burn out the polymer and then sintered to produce a ceramic part. While this is not a normal process, it does illustrate the potential that designers can take advantage of.

A variation of the process (called projection micro stereolithography) uses a flood of UV light passing across a dynamic mask generator. Then, the beam containing the mask image is focused on the surface of the UV curable polymer resin through a projection lens. The lens reduces the image to the desired size.

The advantage of the process is that it does not need a laser to make fine detail. A digital micro mirror device (DMD) is used as the dynamic mask generator. A DMD consists of hundreds of thousands of small mirrors,[1] each mirror corresponding to a pixel that can be independently turned on or off. This process variation has the ability to produce Z-axis discrimination as fine as 1 μm (39 μin.), which appears to be the very finest that can be achieved. Because projection micro stereography provides the entire image at one instant rather than scanning back and forth across the platform, it is logically faster for many parts and it can provide finer resolution for many as well. Both are proven commercial processes with equipment of several sizes. Several companies also provide the service of making parts. A number of universities also have equipment and expertise in both variations of the process.

The issues that designers need to be aware of are the same as defined in Chap. 2.11 and are not repeated here.

References

Lehtinen, P., Projection microstereolithography equipment, Master of Science Degree Thesis, Aalto University School of Science, Aalto, Finland, 2013.

Sun, C., N. Fang, D. M. Wu, and X. Zhang, Projection micro-stereolithography using digital micro-mirror dynamic mask, *Sensors and Actuators A: Physical*, Volume 121, Issue 1, May 31, 2005, pp. 113–120.

Zhang, X., X. N. Jiang, and C. Sun, Micro-stereolithography of polymeric and ceramic microstructures, *Sensors and Actuators*, Volume 77, 1999, pp. 149–156.

Zhou, F., W. Cao, B. Dong, T. Reissman, W. Zhang, and C. Sun, Additive Manufacturing of a 3D Terahertz Gradient-Refractive Index Lens, Advanced Optical Materials, 2016; 4:1034–1040.

[1]One system employs a million mirrors in a 2 × 2 cm (0.79 × 0.79 in) area.

CHAPTER 3.46

Micro Electromechanical Systems

LaRoux K. Gillespie

Manufacturing Consultant
Andover, Kansas

Micro electromechanical systems (MEMS) combine mechanical and electrical features in a manner to make a functioning product with feature sizes ranging from micrometers to millimeters. They are typically made using methods similar to those used for integrated circuits (ICs). Since hundreds or thousands are made on a single IC wafer, they can be made relatively inexpensively. While the products themselves are typically measured in micrometers and millimeters, feature sizes may be on the order of the wavelength of light. Thus they can be used for many optical applications. Micro sensors such as the accelerometers for automobile crash detection and pressure sensors for biomedical applications are typical MEMS products. Micro actuators for moving arrays of micro mirrors in projection systems are another example of commercial MEMS. RF switches and inkjet printheads are other MEM examples in everyday use.

MEMS combine micrometer-size features with near weightless mass in a manner that can make them exceptionally fast switches. Micro mechanical switches allow phase shifting and signal switching at speeds impossible using conventional size switches. Today the nomenclature includes nano levels so the term N/MEMS conveys the concept of both nano and micro levels of the basic capabilities. For this chapter the acronym MEMS will also include the concept of nano level.

MEMS Manufacture

Micro products of any kind face several limitations and require some unique approaches to get past issues of adhesion and friction surface forces and electrostatics that are not major issues in more conventional size products. Further MEMS must be made with processes amenable to IC-style fabrication and that limits the number of facilities that can produce them (Fig. 3.46.1).

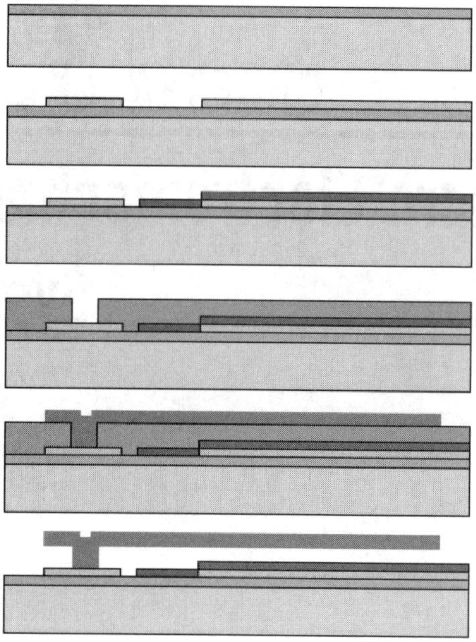

FIGURE 3.46.1 Typical MEMS manufacturing steps to produce a cantilever beam. (Courtesy Silvaco and Yong Qing Fu, Jack Luo, Stuart Milne, Andrew Flewitt, and Ahmed Nejim.)

MEMS manufacturing is not by definition limited to one specific type of manufacturing, but in general terms they are made by one or more of the following processes: surface micro machining, bulk micro machining, or micro molding.

Bulk micro machining implies machining many features at one time and can be accomplished via isotropic or anisotropic wet etching of silicon or glass substrates. Potassium hydroxide (KOH) or tetramethylammonium hydroxide (TMAH) etch silicon substrate preferentially depending on the crystalline planes in the direction of etching. The etch rate is tens to hundreds of time faster in the [100] crystalline plane than in the [111] plane.

Deep reactive ion etching (DRIE) uses a plasma etcher rather than wet chemicals. Silicon, as well as some other materials, can be etched very quickly, making possible very thick structures with small widths. Figure 3.46.2 illustrates one DRIE product.

"Surface micro machining" uses alternating layers of structural material and sacrificial layers to make MEMS devices. Films are deposited on a substrate and exposed using photolithography much like printed circuit board fabrication (but with much more precision). The structural layers become the MEMS product while the sacrificial layers support the structural components during the fabrication process. More than one layer can be produced in order to provide the necessary design. After completing the exposure, the sacrificial material is removed by wet chemical etching. The result is a product that can pivot, or turn relative to the base substrate.

Polycrystalline silicon (polysilicon) is one of the primary structural layers, but several other materials have been used, including metals (gold, nickel, aluminum, copper, chromium, titanium, tungsten, platinum, and silver), oxides and ceramics (nitrides of silicon, aluminum, and titanium as well as silicon carbide and others), as well as

FIGURE 3.46.2 Titanium mirrors bulk micro machined using deep etching of a titanium substrate. (Courtesy Brigham Young University.)

	Part Complexity	Typical Size	Fabrication Complexity	Aspect Ratio
Surface micro machining	High	Small	Low-high	Low
Bulk micro machining	Low	Large	Low	High
Molding	Medium-high	Small-large	Medium	High

TABLE 3.46.1 Characteristics of Common Processes Used for MEMS (Courtesy Brigham Young University)

polymers. As an example of the process, fingers for an electrostatic drive are made with polysilicon. The "fingers" of the device are about 35 μm (0.0014 in.) long, and each layer of the composite device is about 2 μm (78 μin.) thick with gaps between the layers (a result of the sacrificial oxide film thickness) of 2 μm (78 μin.) in this application.

Molding MEMS relies upon either LIGA or UV light photolithography technology (described in Chaps. 3.21 through 3.26, Chap. 3.33, and Chap. 3.44). Most MEMS devices will require chemical mechanical polishing (CMP) which is described in Chap. 3.14, and the recommendation made there to discuss designs with IC foundries early in design applies here to MEMS design as well.

Table 3.46.1 characterizes the complexity and capabilities of these processes.

MEMS substrates can be bonded together to make more complex parts and features. They can be bonded by adhesives if spacing and alignment between bonded parts can be closely controlled. Silicon and glass wafers are bonded more accurately by applying temperatures of about 1000°C (1830°F) to a clean interface between the surfaces. This bond can be as strong as the intrinsic strength of glass or silicon. Anodic bonding can be used at a temperature of about 400°C (750°F) with a voltage of about 1000 V placed between the wafers. The electric field generated accelerates bonding.

The full range of manufacturing processes used or that can be used to make MEMS include electron beam lithography, ion beam lithography, ion track technology, X-ray lithography, diamond patterning, wet etching, isotropic etching, anisotropic etching, HF etching, electrochemical etching, dry etching, vapor etching, xenon difluoride etching, plasma etching, sputtering, reactive ion etching (RIE), and die preparation (wafer backgrinding, and dicing).

Designing for MEMS

MEMS require some exceptionally detailed considerations and are not something that a general product designer will do. They are designed with unique software and an understanding of the detailed manufacturing process capabilities, and with libraries of basic knowledge of how to design single aspects of MEMS. As for many other products, MEMS are conceptually visualized, then modeled, then simulated and revised as required at each level of the design process. One firm defines the design process as five levels of design: (1) first principles, (2) component level, (3) device level, (4) algorithmic level, and (5) system level.

First principles predict material behavior (either directly or in environments like etching solutions). Software programs help define this aspect for specific materials. These programs are essential since they predict how growth, diffusion, and etching will affect the resulting size and geometry of the part produced (e.g., what are the necessary mask dimensions to result in the desired physical dimensions). Component-level models model a single aspect such as beam deflection or vibration. Libraries of proven feature performance are used as the starting point to get correct features. Several different levels of granularity in simulation may be required for the desired design since not all configurations are in the library, and each may require a different software program. Device-level models provide the complete design of the item on the wafer. They will model or simulate the effects of forces, temperatures, electrostatics, fluid flow, damping, deflection, and motion. Algorithmic models define the behavior of logic or control aspects of the design, and system-level models model the entire MEMS micro system. Because at the micro and nano level a system may have 100,000 to 1,000,000 degrees of freedom, modeling at this level requires significant simplifying assumptions. Etching may not provide perfectly vertical sidewalls. The resulting angles or non-perpendicularity causes other micro level changes in performance, changes which are often not captured in the models. The modeling programs and inherent experience provide the detail needed and not generally conveniently defined.

The point of the above paragraph is to illustrate that MEMS design is done by professionals that live in that environment. Product designers will normally carry a design concept to the MEMS designers and the MEMS designers actually do the detail design with interplay back to the original product designer.

Design for MEMS Actuation

While the details for accomplishing movement inside a MEMS device requires extensive explanation, an overview of the issues is straightforward. Causing one or more elements of a MEMS device to move is typically the result of employing electrostatic forces, thermal, magnetic, vibration, piezoelectric pressure, and electrical stimuli.

Some form of comb drive electrostatic actuators is one means to cause movement between two elements. If any of the interlacing fingers become slightly off-center during motion, however, or if they are bent, the entire comb will be bent sideways, potentially causing contact which can cause catastrophic melting or vaporization of the fingers.

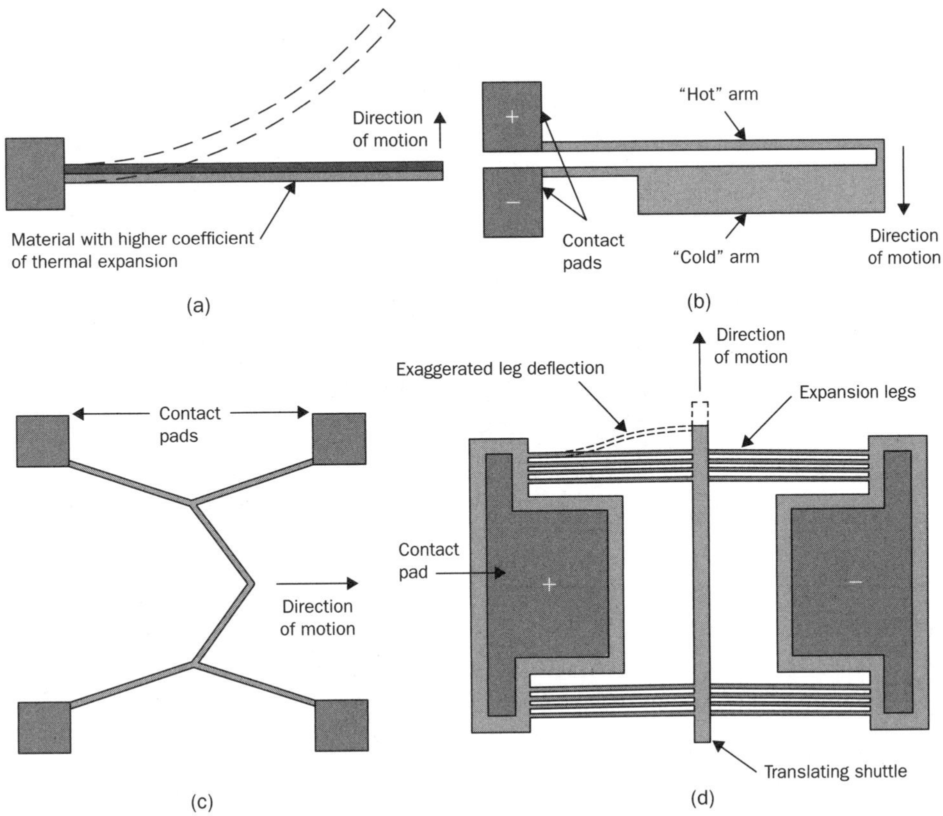

FIGURE 3.46.3 Design approaches to generate movement in a MEMS device. (Courtesy Brigham Young University.)

Because materials expand when temperature increases, this property is the simplest means to cause motion. The temperature change can be caused by the environment or heating by electrical currents. By using materials having two different expansion coefficients bonded together, the heated element can be designed to bend sharply multiplying the effect of temperature expansion. Alternately if there is more mass in one-half of the bimetallic beam it will similarly deflect more when a current is passed through it. Some simple spiderweb-like structures also force one element to move much more than a simple beam. These and another variation are shown in Fig. 3.46.3. In Fig. 3.46.3*d* the design uses legs having a small cross-sectional area which result in high electrical resistance. The induced heat from electrical currents causes the legs to heat, thus moving the shuttle. This device design concept provides high deflections and large forces; the forces induced depend on the number of legs used.

Figure 3.46.4 illustrates a design by Sandia National Labs that provides microscopic gear movement in a manner similar to a mechanical watch. The two thin "pullers" rotate the ratchet wheel which in turn rotates the gear under it and that in turn rotates the adjacent gear. For scale, the smallest center pin is slightly smaller in diameter than 10 μm (0.0004 in.). These are MEMS made in silicon.

FIGURE 3.46.4 MEMS timing device made in silicon. (Courtesy Sandia National Labs.)

Sensors

MEMS are widely employed as micro sensors. Piezoelectric effects and capacitance are two design features that are widely used for sensor design. They are also used to make RF switches that can operate at up to 40 GHz, employing simple microscopic or nano level movements to change capacitance.

What a New Designer Should Do When MEMS Design Seems Applicable

Before any attempt at detailed MEMS design, product designers need to

1. Conceptually illustrate what the device in mind looks like.
2. Define the size of features conceived for the device features.
3. Define the desired materials and options.
4. Define the expected/necessary properties of the device (i.e., how the device should/should not respond to thermal, mechanical, electrical, optical stimuli).
5. Define the expected quantity to be made and price points.

At this point the designer is ready to begin discussions with MEMS facilities to determine how realistic the concept is. At this point a number of alternatives can be explored.

Constraints

MEMS products almost universally require IC-style processes which are exceptionally expensive to install. Designing MEMS requires ingenuity and sound grounding in physical attributes, and the manufacturing of them is relegated to plants that specialize in them. MEMS are also largely a high-production item since the costs to make them can be very expensive; it is only economical to use this approach when high volumes are required (millions of units for example). At the same time, unique products such as space flight products for which only one or two are needed, but weight and size drive feasibility are good MEM targets (but expensive).

Assembly of all micro size products is an issue, and if the MEMS are married to other micro products, assembly can be a severe restraint. MEMS design houses can provide needed designs and assembly solutions. Several design and manufacturing facilities provide needed support. Many universities both research and offer design and some manufacturing capabilities. MEMS technology is documented in almost 500,000 documents of some nature.

References

Akkaraju, S., N/MEMS Design Methodologies, IntelliSense White Paper, Woburn, MA, circa 2015.

Davies, B. R., M. Steven Rodgers, and S. Montague, Applying Macro Design Tools to the Design of MEMS Accelerometers, Intelligent Micromachine Department, Sandia National Laboratories, Albuquerque, New Mexico, circa 2016 White Paper.

Fu, Y. Q., J. Luo, S. Milne, A. Flewitt, and A. Nejim, Process Flow Simulation and Manufacture of Variable RF MEMS Capacitors, TCAD Simulation Standard, 2005, Silvaco, St. Ives, Cambridgeshire and Engineering Department, University of Cambridge, Cambridge, UK.

CHAPTER 3.47

Origami Micro Fabrication

LaRoux K. Gillespie

Manufacturing Consultant
Andover, Kansas

Assembly of micro components is a major challenge as reported in Chap. 3.52. With careful thought, however, simple changes in configuration and performance can be made via origami principles. Today, with origami approaches, parts can unfold into the shape desired, and in some instances even self-change functional properties such as electrical, electrostatic properties. A miniature origami robot has been demonstrated that can fold itself, walk around, and even self-destruct. To date the origami approach has been used to design medicine delivery capsules, electronic switches, hinges, and micro mirrors which self pop up. A pigskin wrapped product has also been demonstrated that after unfolding from an ice capsule can attach and by external magnetic manipulation move button batteries from stomachs and move them to disposal through the intestine.

State of the Art

Defining the state of the art is not straightforward since several different approaches have been used to make devices via origami. MEMS processes provide the most varied opportunities since the processes already are focused on micro sizes, involve very thin materials, and can take advantage of minute differences in material properties to make large effects. In addition, the MEMS technology allows high production at reasonable unit costs. Most approaches to date involve liquids, internal strains caused by material differences, lithography of some kind, etching, and/or thermal differences. The technology is focused on micro sizes of 25 μm (0.001 in.) or larger (typically larger but still only measured in micrometers). The largest immediate market is in drug delivery through the blood stream or digestive system. But the capability to produce micro mirrors and microscopic switching devices or similar movable pieces could advance the field far beyond simple drug delivery.

Manufacturing Approaches

Several fabrication approaches can be used to make these minute features and products. The most common include:

- Surface tension forces pulling features away from their base configuration
- Developing pre-stressed crystal lattices that can be released by some mechanism
- Volumetric changes that cause deflection

These in turn are initiated by

- Basic MEMS technology
- Chemicals that dissolve or release pre-stressed features
- Mechanical micro machining of thin bend lines
- Lithography approaches to producing shapes and thin bend lines
- Laser cutting of shapes
- Laser writing of materials that subsequently cause bending or release
- Thermal or electrical heating of key areas and materials
- Other combinations of forces and chemical reactions

Simple Shape Changes

Researchers have cut geometric shapes a millimeter or so across from thin, rubbery membranes ranging from 40 to 80 μm (0.0016 to 0.0031 in.) thick. By placing a drop of water large enough to reach all corners on top of the shapes, they caused the shapes to fold inward on themselves making a semi-closed 3D ball. As the liquid evaporates, its surface tension tugs the membrane closer around the water's decreasing volume, until the remaining liquid is completely encapsulated inside a sphere, tetrahedron, or some other shape. Rather than closing up, the materials can also be programmed to open to release their contents. Plastic membranes can be made of several different materials that depend on the actions required and process used to cause that.

In a different approach, gold-plated nickel is attached to flexing legs by a three-layered joint made of a 50-nm-thick (2-μin.) layer of chromium, a 250-nm (10-μin.) layer of copper, and a 3500-nm (138-μin.) layer of polymer. The copper layer is under tension. It normally would be bent but the polymer layer holds it flat. Exposing the device to acetic acid dissolves the polymer, allowing the copper to bend. To reverse the bending the device is exposed to hydrogen peroxide which dissolves the copper and leaves the chromium layer, which folds the arms flat. The nickel plays no role in the folding or unfolding, but as part of the body allows the device to be manipulated by external magnets (Figs. 3.47.1 and 3.47.2).

Similarly, using silicon nitride wafers as the material, manufacturers carve small hinges into the wafers to make bend paths. When exposed to a liquid the flat sheets morph into microscopic cubes, pyramids, and other geometric shapes. Capillary forces cause the changes. Capillarity is particularly dominant at micro scales. This approach works best with very minute parts and features.

In a different approach, the origami idea is expanded to making micro switches that move under electrical stimuli or even thermal or optical. All that is needed in this

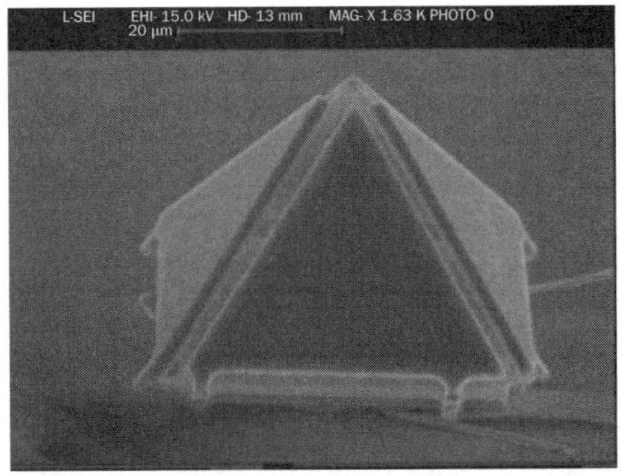

FIGURE 3.47.1 Origami-inspired micro part made via PolyMUMPS. (After Gagler, 2008.)

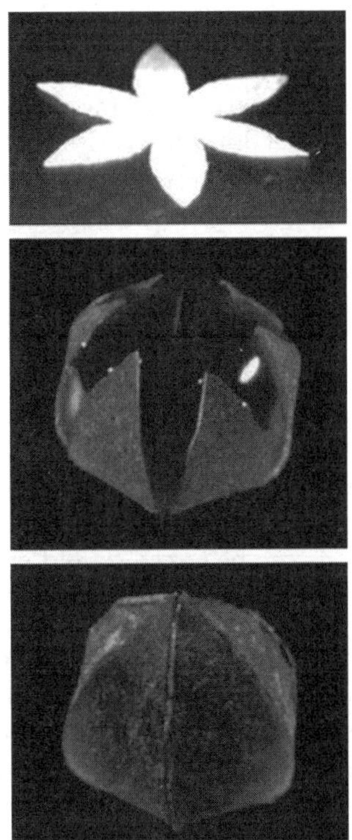

FIGURE 3.47.2 Micro origami enclosure. [Courtesy *Physical Review Letters* (Charlotte Py et al. article).]

instance is to have the object make or break electrical contact under some stimuli. An example has been demonstrated using a GaAs substrate, an Al0.52Ga0.48As/AlAs digital alloy, and, an In0.19Ga0.81As layer. SiGe/Si is another combination that has been successfully used for micro origami.

Complex Changes and Abilities

In a more complex shape change design, innovators have generated patterns in polysilicon that is placed on a thin film of gold, via the commercial silicon wafer process PolyMUMPs. Then the polysilicon is removed off the hinge areas by etching. The parts are then electrocoated with permalloy to make them magnetic. The mask used for electrocoating leaves the hinge areas uncoated. This assures that the hinge area is where the device will fold. Magnetic force is used to bend the hinges. Then water pressure, capillary forces, and drying cause the final folding.

Combining partial folding of structures by the combination of magnetic actuation and liquid closure to bring the structures to their final closed state is just one of several useful techniques for mass production of large arrays of micrometer size. Designs can take advantage of several of the means of flexing on a single product, but more environments can create more restraints for some actions.

Other innovators have developed the technology to coat a single layer of graphene (one atom thick) with a layer of silicon dioxide glass [half a nanometer thick (0.02 μin.)]. Then, using heat, electric charges, or pH solutions, they can predict how the sheet would expand and contract, resulting in folds. This is the most extreme and thinnest origami micro fabrication process developed. It is not clear how useful it is for actual products, but the unique properties of graphene imply that it has several potential applications. Although not yet done, it is possible to print a circuit on a graphene sheet, and then fold it into a structure that absorbs light, makes a calculation, or sends a signal or even sends back light at a different frequency.

Another approach to folding involves selective etching of pre-stressed multilayered structures. The etching releases internal stresses that cause bending or flexing even into rolls, spirals, and tubes. This approach works by controlling material composition, mask design, and the etching process. While somewhat an exaggeration, the statement that "Robust hinges for movable parts, can be applied to any pair of lattice mismatched epitaxial layers, in semiconductors or metals," is in principle accurate and a focal point for designers to use.

In a new approach lasers or other strongly focused light is used to cause motion of MEMS origami devices. As in the examples described earlier, depending on how they are made, the flap portions can be forced to fold either down or up. As one researcher noted, "If you have a polymer, or thermally sensitive layer, that's sufficiently thick, it will prevent the layers from folding because you have this other material on top, so it is like a spring that's compressed." Then, by shining a light (laser from up close or far away) on it to heat it up, that polymer softens allowing the residually stressed layers to fold.

The tiny robot mentioned above measures approximately 1.7 cm (0.67 in.) long and weighs about one-third of a gram (0.011 oz). The robot's primary feature is a neodymium magnet layered with polystyrene and PVC. When heated, the material allows it to take on another form similar to an origami. A multistage folding process can also be achieved by heating the robot at varying degrees. Low heat can give the shape-shifting robot one form, but placing it back on the heating pad and raising the temperature can result in a secondary design. Instead of using a set of wires to make the robot move, it uses an external magnetic field made up of four coils to guide it along its path. These magnets power its movement by cycling on and off at 15 Hz. It can walk or roll as desired at up to 3 or 4 cm/s (1.2 to 1.6 in./s) and can be disintegrated by exposing it to acetone (Fig. 3.47.3).

Figures 3.47.4 and 3.47.5 illustrate some basic fabrication approaches that are independent of processes used. Figure 3.47.6 illustrates some laser approaches. While not

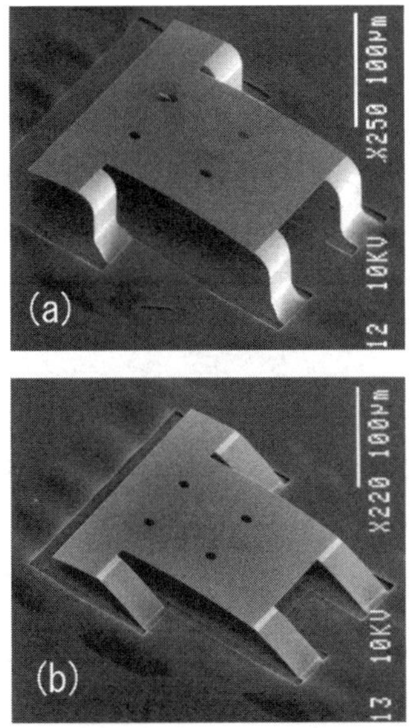

FIGURE 3.47.3 A more complex movable origami-inspired product (Vaccaro, 2003).

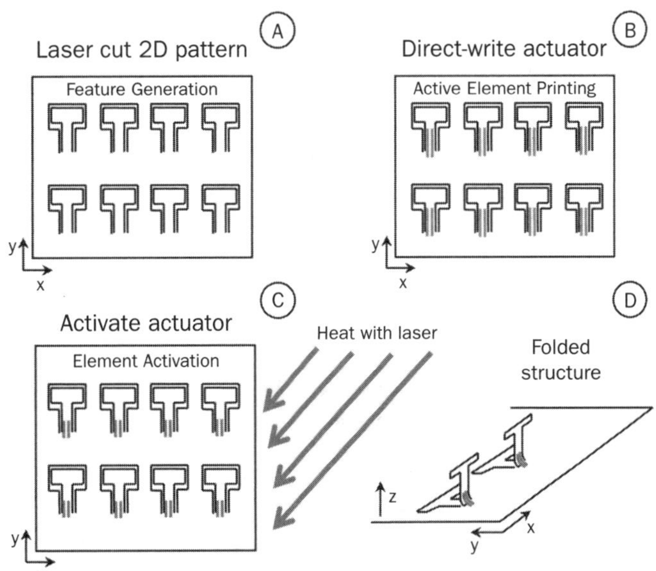

FIGURE 3.47.4 Basic steps in making closed origami container. (Courtesy Elisabeth Smela Laboratory for Micro Technologies.)

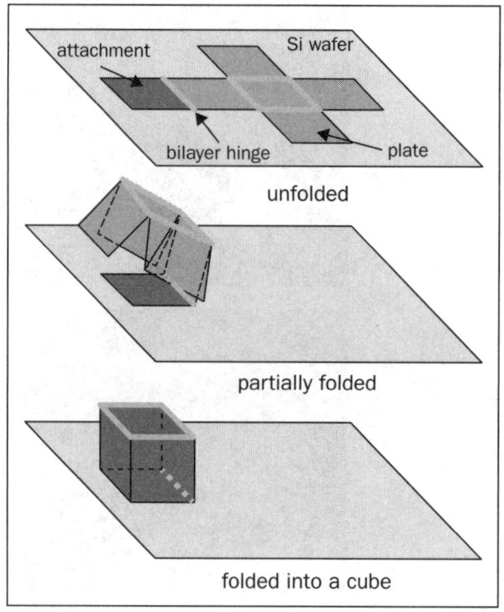

FIGURE 3.47.5 Bilayer attached to silicon substrate. Hinges 30 × 30 μm (0.0012 in.) have been fashioned attached to 150 × 150 μm (0.006 in.) plates and caused to flip over. The hinges can accommodate 1-mm-wide plates as well. (Courtesy Elisabeth Smela Laboratory for Micro Technologies.)

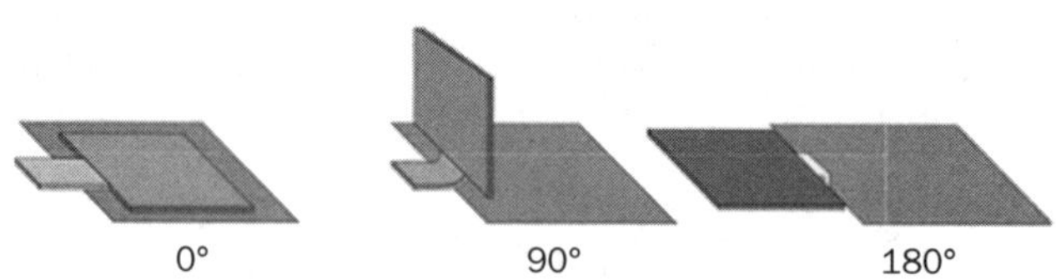

FIGURE 3.47.6 Laser-based origami fabrication provides micro machining, laser printing, laser heating (activation). (Courtesy Naval Research Laboratory.)

presented here, design details for using additive processes for origami foldable panels have also been presented by Deng and Chen.

Design to Accomplish Change

There are several considerations in designing origami-inspired configurations. They include:

1. The product material has to be affected significantly by an external force.
2. The product needs lines of weakness that can allow bending or movement.
3. The environment in which the change is designed to survive in must be compatible with that which caused the change (i.e., the use environment must not cause undesired changes).

4. Most origami devices are single use, but not all. Defining the desired complete life path is essential at the beginning of the conceptual stage.
5. While no known lower limit exists for size of product that can be changed, a common lower limit seems to be about 25 μm (0.001 in.) on a side for the assembled product. Centimeter size parts have been made by micro origami.

From an initial starting position in the technology of origami design, the choices are to either (1) review all the literature searching for material combinations that have already been reported to work, or (2) develop a database of material properties of interest searching for inherent properties that can be exploited. Designers typically know the materials they would prefer to use, but it may take a review of molecular and crystal structure properties to determine the potential for that property.

The simplest way to create flexing or bending motion in these micro devices is to bond two different materials together and let heat or cold cause bending. Polypyride (PPy) coated on gold film when exposed to an electrochemical potential causes a volume change that forces bending. Using this principle, and these materials on a silicon substrate, one can force extensive curling if the composite has been surface micro machined to leave long strips of the materials before immersing in an electrolyte of 0.1 M NaDBS. Curling by itself can cause switching, but curling effect also can be used as an active hinge.

Bimetallic strips and shape memory alloys have long been used to react in the described way, but the conventional materials used today for larger applications may not be as effective in micro sized environments as some of the materials being explored and used today.

Lasers can replace the MEMS fabrication or part of it as needed. In one approach a fused silica substrate was back irradiated with an indium-tin oxide (ITO)/copper multilayer film. The copper product on this substrate is then laser machined to generate the desired unfolded configuration. Then the laser is used to drive heat across the surface that is desired to be bent (moved out of the flat plane). The ITO releases everything above it, causing bending along the interface between laser irradiation and no laser irradiation.

This approach allows the laser to bend whatever elements it needs in whatever sequence it desires. The extent of the bending depends on the laser pulse intensity.

In yet another laser-based approach, metallic nanoinks made of suspended metallic nanoparticles in a highly viscous, organic solution are laser printed by forward transfer of high-resolution patterns. Laser micro machining cuts the desired part configurations. Then laser curing of the inks evaporates the organic material and sinters the metallic particles. This causes substantial volumetric shrinkage. This change in volume then causes bending of the underlying polyimide membranes.

A third approach involves placing pre-stressed polymer strips across the fold joints and then using the laser to heat the array causing heat-induced shrinkage to fold the material.

In these approaches the laser cuts the shape, deposits (writes) material in critical places, heats the desired portions, and the heat induces changes. The laser performs all or most of the necessary operations, eliminating much of the need for very costly fabrication facilities, and it greatly reduces experimentation times.

Liquids like water, acetone, and acetic acid cause change, but temperature, light, vibration, electrostatic fields, and magnetic fields are also key tools. When a sequence of flap opening or closing is required, the designer needs to carefully consider how to accomplish that. Like other processes, both design and fabrication are tightly interwoven. Fabricators spend a lot of time experimenting with different means of controlling movement.

Directing water from different directions can sequence movement as can drying. Varying the time of drying or wetting can produce tighter seams. If designed properly, the structure remains folded after the evaporation because the parts remain stuck together. These are not aspects that most designers intuitively recognize.

For experimentation and small-scale production, shapes affected by liquids can be changed by applying small droplets with a syringe rather than larger scale equipment.

While large-scale production for several of the approaches is envisioned, most of them require MEMS technology to accomplish. Fabrication with that technology is outsourced typically to MEMS foundries because of fabrication costs. The MEMSPRO CAD software is how some of these origami-inspired designs have been accomplished. MEMS design is discussed in Chap. 3.46.

Several world centers for advancing this technology exist and they produce a variety of shapes as small as 30 μm (0.0012 in.) on a side. To illustrate the minuteness of that, 1 in. = 25,400 μm and 833 of these micro items could be placed side by side in less than an inch.

References

Alberto P., S. A. Mathews, N. A. Charipar, A. J. Birnbaum, Laser origami: A new technique for assembling 3D microstructures. *Proceeding SPIE 8244, Laser-based Micro- and Nanopackaging and Assembly VI,* 82440B, February 9, 2012 (doi:10.1117/12.909187).

Arora, W. J., H. I. Smith, and G. Barbastathis, Membrane folding by ion implantation induced stresses to fabricate three-dimensional nanostructures, *Microelectronic Engineering*, vol. 84, 2007, pp. 1454–1458.

Birnbaum, A. J. and A. Piqué, Laser induced extraplanar propulsion for three-dimensional microfabrication, *Applied Physics Letters*, vol. 98, 2011, p. 134101.

Blees, M. K., A. W. Barnard, P. A. Rose, S. P. Roberts, K. L. McGill, P. Y. Huang, A. R. Ruyack, J. W. Kevek, B. Kobrin, D. A. Muller, and P. L. McEuen, Graphene kirigami, *Nature: Letter*, vol. 524, no. 7564, August 13, 2015, pp. 204–207.

Boyle, R., Graphene origami produced by world's thinnest folds, *New Scientist Daily News*, March 15, 2016.

Cowen, A., B. Hardy, R. Mahadevan, and S. Wilcenski, PolyMUMPs Design Handbook, MEMSCAP Inc., Durham, NC, Rev.13, 1992.

Danescu, A., C. Chevalier, G. Grenet, Ph. Regreny, X. Letartre, and J. L. Leclercq, Spherical curves design for micro-origami using intrinsic stress relaxation, *Applied Physics Letters*, vol. 102, no. 12, 2013.

Danescu, A., C. Chevalier, R. Artinyan, and J. L. Leclercq, Self-assembly 'micro-origami' photon cages as hollow micro-resonators, *Proceedings of SPIE—The International Society for Optical Engineering*, vol. 9520,·May 2015.

Deng, D., and Y. Chen, Assembled additive manufacturing—A hybrid fabrication process inspired by origami design, University of Texas Symposium manuscript, online paper, circa 2015, pp. 174–187.

Deng, D., and Y. Chen, Origami-based self-folding structure design and fabrication using projection based stereolithography, *Journal of Mechanical Design*, vol. 137, February 2015, pp. 021701-1–021701-12.

Evans, J., Micro-origami lets scientists get a grip, *New Scientist Daily News*, December 9, 2008.

Gagler, R., A. Bugacov, B. E. Koel and P. M. Will, Voxels: volume-enclosing microstructures, <http://iopscience.iop.org/journal/0960-1317> *Journal of Micromechanics and Microengineering*, <http://iopscience.iop.org/volume/0960-1317/18> vol. 18, <http://iopscience.iop.org/issue/0960-1317/18/5> no. 5, April 15, 2008, p. 055025.

Na, J.-H., A. A. Evans, J. Bae, M. C. Chiappelli, C. D. Santangelo, R. J. Lang, T. C. Hull, and R. C. Hayward, Programming reversibly self-folding origami with micropatterned photo-crosslinkable polymer trilayers, *Advanced Materials*, vol. 27, issue 1, January 7, 2015, pp. 79–85.

Pandley, S., E. Gultepe, D. H. Gracias, Origami inspired self-assembly of patterned and reconfigurable particles, *Jove*, no. 72, February 4, 2013.

Piqué, A., S. Mathews, A. Birnbaum, and N. Charipar, Microfabricating 3D structures by laser origami, *SPIE Newsroom*, November 9, 2011.

Py, C., P. Reverdy, L. Doppler, J. Bico, B. Roman, and C. N. Baroud, Capillary origami: Spontaneous wrapping of a droplet with an elastic sheet, *Physical Review Letters*, vol. 98, April 13, 2007, p. 156103.

Ranosa, T., MIT tiny origami robot can fold itself and self-destruct: Perfect profile for future micro surgeon? *Tech Times*, May 30, 2015.

Seffers, G. I., Meet laser-triggered, origami-like pixie dust, *Signal (AFCEA)*, May 1, 2013.

Smela, E., O. Inganäs, and I. Lundström, Controlled folding of micrometer-size structures, *Science*, June 23, 1995, pp. 1735–1738.

Strellman, P., T. Buchner, W. J. Arora, and G. Barbastathis, Dynamics of nanostructured origami, *Journal of Microelectromechanical Systems*, vol. 16, 2007, pp. 932–949.

University of Twente, Three-dimensional microtechnology with origami folding art, *Science Daily*, November 14, 2014.

Vaccaro, P. O., K. Kubota, T. Fleischmann, A. Vorobev, J. M. Zanardi Ocampo, S. Saravanan, and N. Saitoh, Self-positioning micromachined structures made by micro-origami, Design Guidance Document, Advanced Telecommunications Research Institute International, WEL Department of Photonics, Japan, circa 2016.

Vaccaro, P. O., K. Kubota, T. Fleischmann, S. Saravanan, and T. Aida, Valley-fold and mountain-fold in the micro-origami technique, *Microelectronics Journal*, vol. 34, 2003, pp. 447–449.

Vorob'ev, A., P. Vaccaro, K. Kubota, S. Saravanan, and T. Aida, Array of micromachined components fabricated using "micro-origami" method, *Japanese Journal of Applied Physics*, vol. 42, part 1, no. 6B, June 1, 2003.

CHAPTER 3.48

Ion Beam Machining

LaRoux K. Gillespie

Manufacturing Consultant
Andover, Kansas

Ion beam machining, also called ion milling, uses a focused ion beam (FIB) microscope which can image a spot on a part, sputter away electrons leaving a machined surface, analyze the elements coming off the surface when sputtered, and it can also deposit material onto a surface. All of that can occur within a single instrument. FIB provides the finest resolution and smallest feature size of any material removal process. It is also among the very slowest of the machining processes. It makes superb ultra micro features, but at an exceedingly slow rate [~0.5 μm^3/s (31 μin.^{3}/s)].[1] FIB machining has to be performed in a high vacuum which also slows production.

FIB machining typically uses a beam of Ga ions focused down to 5 to 8 nm (0.2 to 0.3 μin.) to physically knock off surface atoms of the target material. In an overly simplified manner, it is not unlike scattering billiard balls with the cue ball. The beam can be deflected or positioned to serve just as a numerically controlled machine tool does for metal cutting. In other words the beam can cut out most shapes that a conventional milling machine can—it just does it for much smaller features and without cutting forces (Fig. 3.48.1).

In addition to normal cutting, the FIB equipment can thin samples, and cut them loose from parent material. As shown in Fig. 3.48.2, FIB can also be used as a prototype cutting tool to make micro products while the normal MEMS or other production processes are being developed or fine-tuned for faster production. FIB, in addition to having ability for making nanometer-size features, can machine any material. That alone makes it a unique tool with many micro applications.

Design Freedom

The several product images shown in the following paragraphs provide a visual concept of the design freedom that FIB brings to micrometer-size features. In many respects, FIB machining is largely unlimited except for time to fabricate which is long compared to all other processes. The process is proven, and it applies to all materials. Masking may

[1]31 μin.^{3}/s is 1 in.^{3} every 9 h.

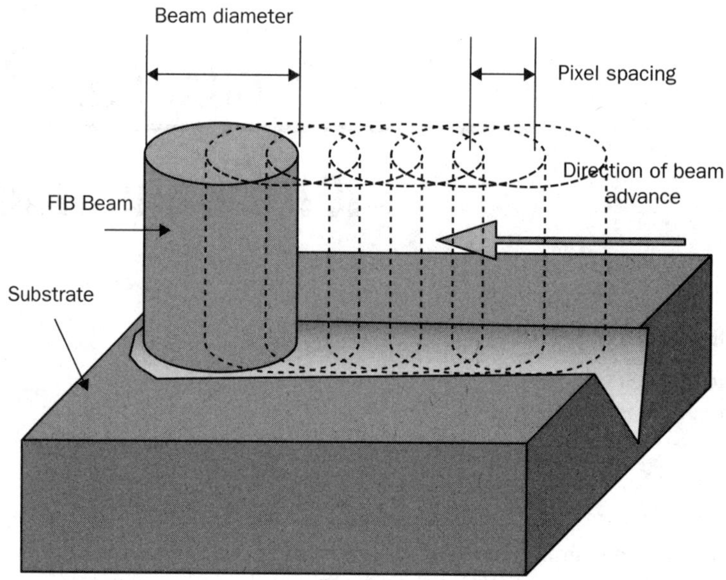

FIGURE 3.48.1 Schematic of single-pass FIB machining a channel. (Teng, 2004 ©IOP Publishing. Reproduced with permission. All rights reserved.)

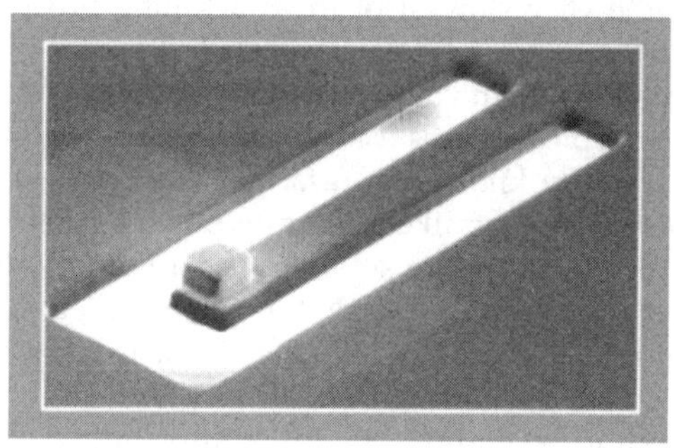

FIGURE 3.48.2 Cantilever with proof mass FIB machined in SiN to prove design principle. High-volume production would then build these devices with normal MEMS technology. (Courtesy Nanostructure Physics Section, KTH Royal Institute of Technology, Sweden.)

be required for some shapes, and some may require use of additional gas streams, or FIB deposition of additional material. Machine time is essentially the only restraint.

While the smallest feature size from FIB is about 10 nm (0.4 μin.), it is more regularly used to make features of up to 20 μm (0.0008 in.). The quantitative measure of how small a feature FIB machining can produce depends on the material, crystal orientation, ion beam incidence angle, and some operational limits due to redeposition of the material being removed.

FIB machining has been used for over 25 years to make micro channel structures, which are basic building blocks for many micro devices. FIB is used to make micro heat pipes, sensors, linear photonic crystals, optics, semiconductors, micro molds,

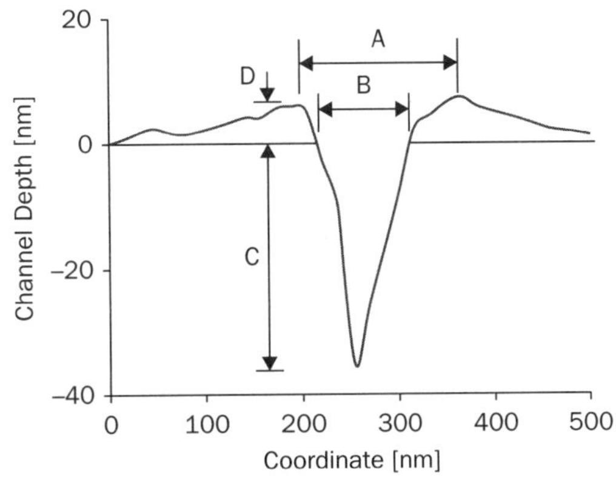

FIGURE 3.48.3 Channel milled with a single FIB pass on 125 nm (5 μin.) gold film on silicon. (Teng, 2004 ©IOP Publishing. Reproduced with permission. All rights reserved.)

microscopic cantilever beams (for biometric sensing of virus and bacteria presence), fluidic devices, and to cut and sharpen microscopic cutting tools (single-crystal diamond cutters for example). FIB's most widespread use beyond research has been its use to cut away excess material on some poorly masked semiconductor layers. It is a repair tool which can be used on an entire wafer of semiconductor products.

The beam current and diameter used for milling nanochannels are extremely small [~1 pA and 10 nm (0.4 μin.), respectively)], but this is lower than normally used for micro milling. For example, minute channels milled by a single pass used a beam current of 5 pA to mill a 125-nm-thick (5-μin.) gold film coated on a Si wafer. Figure 3.48.3 shows the V-shaped cross section of a single-pass channel milled at these conditions.

Most useful micro configurations, however, require three features: (a) flat channel bottom; (b) nearly vertical sidewalls; (c) high-aspect-ratio profile. Providing all three characteristics requires multiple passes.

The single-pass cavity shown in Fig. 3.48.4*a* has an outer diameter of 6.2 nm (0.2 μin.) and a depth of 4.9 nm (0.2 μin.). Four passes were used to achieve the configuration in Fig. 3.48.4*b*.

Figure 4.48.5 illustrates some of the ultra micro features that can be cut with ion beam.

The DBR grating shown in Fig. 3.48.6 is almost 2 μm (0.008 μin.) deep, which makes the aspect ratio higher than 10. On newer machines it is possible for hole drilling to achieve a 20:1 aspect ratio.

Figure 3.48.7 illustrates FIB's ability to both machine narrow structures and finely adjust the shape and size of narrow beams. In this instance the desire was to make a minute nanostring. As shown here, the cross section of the initial suspended structure beam was roughly rectangular, but successive FIB machining was able to produce a nearly round cross section 50-nm (2-μin.) silicon string.

Figure 3.48.8 shows an SEM of a uniform array of closely packed, high-aspect-ratio pillars obtained by single-spot irradiations of a focused proton beam. The pillars are 4.5 μm (0.0002 in.) high with a diameter of 0.6 μm (24 μin.), and a periodicity of 2 μm (79 μin.). Such a periodic array of submicrometer diameter pillars is potentially important for the fabrication of photonic crystals.

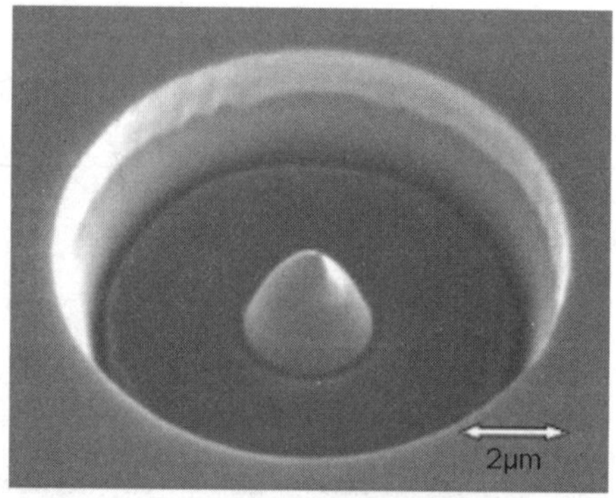

(a)

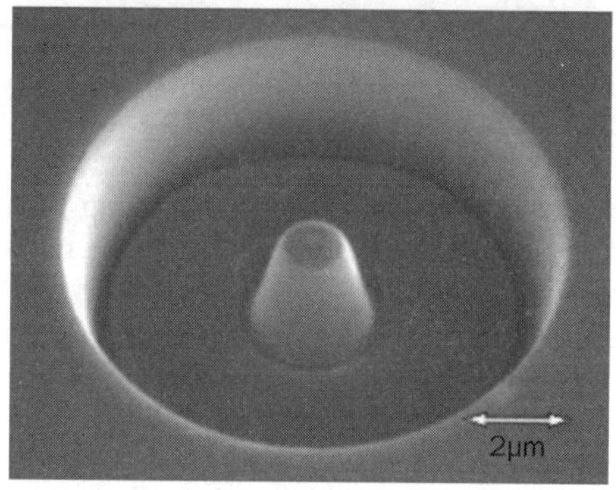

(b)

FIGURE 3.48.4 Difference between single-pass and multi-pass FIB machining. (Courtesy *Journal of Micromechanics and Microengineering*.)

FIGURE 3.48.5 Tin ball ~10 µm (0.0004 in.) in diameter is ion milled in a pattern which when viewed from above looks like a micro snowflake. The narrow slots are ~0.35 µm (14 µin.) in width. (Courtesy FEI, Hillsboro, Oregon.)

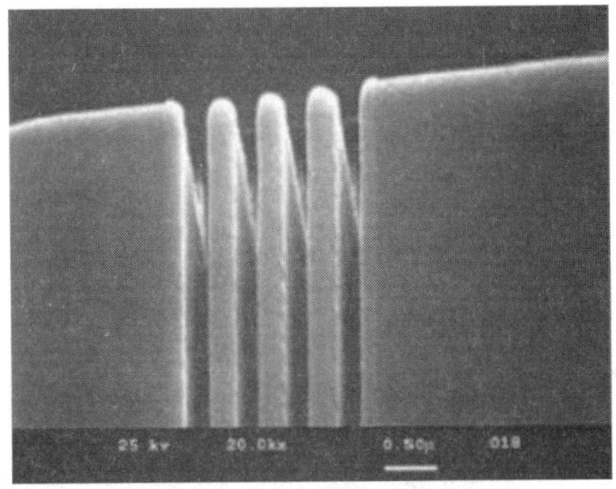

Figure 3.48.6 Thin slots (~0.20 μm) milled via FIB in GaN. (Courtesy Nanoelectronics Laboratory, University of Cincinnati.)

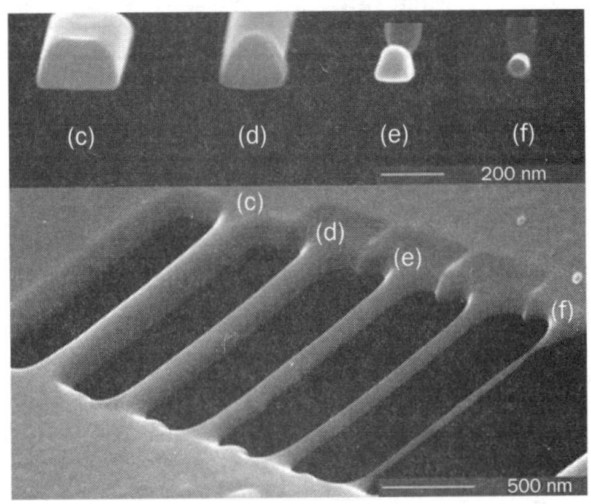

Figure 3.48.7 FIB can refine features as well as produce basic shapes. In this instance a 50-nm (2-μin.) "string" is made from an FIB-produced rectangular beam. (Courtesy Li et al., 2015, Creative Commons license.)

Figure 3.48.8 Array of high-aspect-ratio pillars obtained by single-spot irradiations; inset picture shows the profile of the pillars. (Reproduced from Teo et al., *Applied Physics Letters*, 2004, vol. 84, No. 16, with the permission of AIP Publishing.)

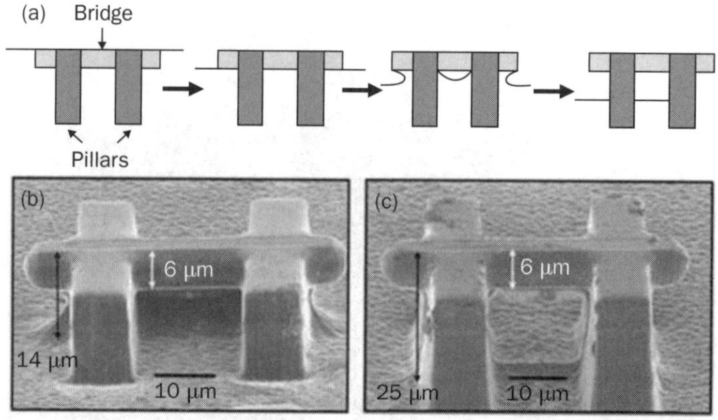

FIGURE 3.48.9 (*a*) Evolution of the double-energy irradiated structure with etching depth. (*b*) At etch depth of 14 μm (0.0006 in.), the bridge starts to separate from the substrate. (*c*) The bridge is completely free-standing at an etch depth of 25 μm (0.001 in.). (Reproduced from Teo et al., *Applied Physics Letters*, 2004, vol. 84, No. 16, with the permission of AIP Publishing.)

Multilevel structures can be created by exposing the sample to two different proton energies. Since the structure irradiated with lower energy has a shorter range, it will begin to undercut at a shallower etch depth. In this way, one can fabricate multilevel free-standing micro structures in a single etch step, rather than the multiple processing steps required in conventional lithography techniques. To demonstrate this capability, a bridge structure was irradiated with 0.5-MeV protons, and two supporting pillars with 2-MeV protons. Figure 3.48.9 shows the evolution of the etching process that led to the formation of a free-standing bridge. Initially, etching occurs in all regions except the irradiated portions. As the depth of etching goes beyond the penetration range of the 0.5-MeV protons [~6 μm (0.0004 in.)], undercutting of the bridge starts to occur. At an etch depth of 14 μm (0.0006 in.), the SEM picture in Fig. 3.48.9*b* shows that the bridge is fully undercut and separates from the substrate. It remains supported by the two pillars irradiated by higher energy. The structure of a free-standing bridge is formed after further etching to 25 μm (0.001 in.) below the surface (Fig. 3.48.9*c*). Undercutting does not occur at the pillars as the range of 2-MeV protons is 48 μm (0.002 in.).

In an application which is limited to indium, FIB can produce a mesh of spikes as shown in Fig. 3.48.10. These high-aspect-ratio nanospikes are produced when indium metal droplets formed early in the irradiation process act as self-assembled etch masks for the underlying semiconductor film. The metal droplets protect the area under them from FIB erosion and nanospikes form as the surrounding film is destroyed. Nanospikes have been created from InAs homoepitaxial films and from InAs/InP heterostructures. FIB-induced nanospikes readily form on InAs, but their formation stops once the InP is exposed. This provides a route to engineer the height of the nanospike by controlling the InAs film thickness. The designers can accurately control the locations at which nanospikes form. Templated arrays of nanospikes may be created by pre-patterning an InAs film to selectively expose the InP substrate and to limit the locations at which indium droplets can act as masks (Grossklaus, 2011). These spikes are one form of nanowires with a number of electronic applications and have different properties from those made by vapor-liquid solid buildup.

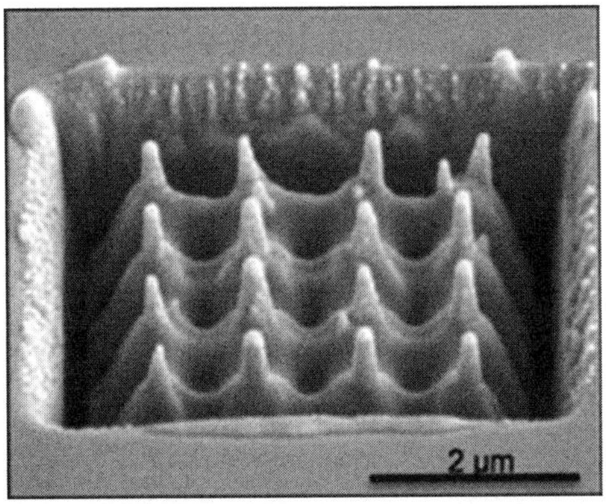

Figure 3.48.10 FIB-produced spike pattern in InAs/InP films. (Grossklaus, 2011, ©IOP Publishing. Reproduced with permission. All rights reserved.)

	Ti	Al	Cr	Au	ZnO
FIB milling rate*	100	150	200	275	150
Masking ability	Best	Good	Good	Worst	—
Typical electrical contact with Zn	Ohmic	Ohmic	Schottky	Schottky	—

*Rate relative to Ti set at 100. The cutting rate for most materials ranges between 100 and 300 on the scale used in this table. Glass, diamond, ceramics, garnet, and Permalloy, for example, have all been cut at what would be 100 here. Silver cuts at three times that rate and gold and quartz may cut at twice the lower rate here.

Table 3.48.1 FIB Milling Rates for Commonly Used Metals in Microelectronics and MEMS Technology (Courtesy Sensors CC BY 4.0)

Chemically assisted FIB [also known as reactive ion etching (RIE) or gas-assisted etching] can accelerate the machining rate by up to a factor of ten using oxygen, nitrogen, chlorine, or other gases. Water alone also hastens reactions.

FIB chemical vapor deposition (FIB-CVD) as opposed to accelerating machining (etching) uses ion energy to force CVD via a direct writing technique to a specific location. In this manner FIB can create some unusual structures.

Table 3.48.1 provides insight into the relative machining rates that can be achieved in different materials.

Other design details of note include the fact that diamond styli have been sharpened to a radius of 10 nm (0.4 μin.) via FIB and on other materials surface roughness of 0.13 nm (0.005 μin.) Ra has been reached, and in general it is possible to attain roughness <2 nm (0.08 μin.). The process can reach a resolution of ~7 nm (0.3 μin.) and a profile accuracy of <0.1 μm (4 μin.).

FIB beams can be focused to a spot size as low as 10 nm (0.4 μin.). Larger diameters are more commonly used (Table 3.48.2). Silicon nitride can be machined at rates up to 10 $\mu m^3/s$ (620 $\mu in.^3/s$).

FIB milling involves a stationary part while a beam or table deflections provide the cutting path. Research has been performed on a micro lathe that is passed through the

Machining Method	Materials That Can be Cut	Feature Size and Tolerance	Positional Tolerance	Material Removal Rate (μm³/s)
Micro machining (mechanical cutting)	Metals, polymers, composites	12 μm (2 μm)	3 μm	10,000
Micro electro discharge machining (μEDM)	Any conducting material	10 μm (3 μm)	3 μm	25,000,000
Electron beam machining (EBM)	Any conducting materials	5 μm (submicrometer)	1 μm	100,000
Femtosecond laser machining (LBM)	Any material	1 μm (submicrometer)	Submicrometer	13,000
Focused ion beam machining (FIB)	Any material	0.2 μm (0.02 μm)	0.1 μm	0.5

TABLE 3.48.2 Comparative Micro Machining Methods (Courtesy Butterworth-Heinemann)

Characteristic	Laser	Ion Beam
Environment	Atmosphere	High vacuum
Chemicals involved	No	Ga, N, H, or O; some gasses increase cut rate
Verify in process?	No	Yes
Removal rate	High	Low
Feature size	>2 μm	<0.5 μm

TABLE 3.48.3 Comparison of Two Micro Level Machining Tools

material transfer port and once inside is electrically connected to rotate the part. It is not a commercial item, but the concept has been demonstrated.

Laser beam technology is found to be useful for cutting the metal lines when the separation between adjacent metal lines is 2 μm (79 μin.) larger. For sub micro spacing of metal lines, FIB technology is a much better choice. Comparison of focused ion beam and laser beam technologies is summarized in Table 3.48.3.

FIB equipment can even be used after machining to bond the part to an internal machine probe via FIB material deposition at a contact point, then moving the part to assemble it. Once assembled the deposited material used to bond part to probe is ion machined away. With appropriate equipment the entire process can be viewed as it happens, so absolute placement is assured.

FIB equipment is produced by a multitude of equipment makers and several commercial shops provide as needed fabrication. A relatively large number of universities also have equipment that can be used by companies to validate designs or to further advance fabrication capabilities.

The following references provide much more detail on process capabilities.

References

Araneo, R., A. Rinaldi, A. Notargiacomo, F. Bini, M. Pea, S. Celozzi, F. Marinozzi, and G. Lovat, Design Concepts, Fabrication and Advanced Characterization Methods of Innovative Piezoelectric Sensors Based on ZnO Nanowires, *Sensors*, vol. 14, No. 12, 2014, 23539–23562.

Ashby, M. F., D. L. Schodek, and P. Ferreira, Chapter 8—Nanomaterials: Synthesis and Characterization, *Nanomaterials, Nanotechnologies, and Design: An Introduction for Engineers*, Butterworth-Heinemann, March 24, 2009.

Belski, A., W. Kammrath, and H. H. Gatzen, Focused Ion Beam Fiber Turning Lathe for Fabricating a Strain Gauge for Carbon Fibers, *Proceedings of the EUSPEN International Conference*, Zurich, May 2008.

Gonzalez, J. C., M. I. N. deSilva, D. P. Griffis, T. T. Miau, and P. E. Russell, Improvements in Focused Ion Beam Micro-machining of Interconnect Materials, *Microelectronics Technology and Devices, SBMicro 2002, Proceedings of the Electrochemical Society Electronics Division*, vol. 8, 2002, 404–411.

Grossklaus, K. A. and J. M. Millunchick, Focused Ion Beam Creation and Templating of InAs and InAs/InP Nanospikes, *Nanotechnology*, vol. 22, 2011, 355302.

Hodges, V. C., D. P. Adams, A. Claudet, D. Gill, W. Sweatt, M. J. Vasile, Aspherizing the Sphere: Using Focused Ion Beam Machining to Create Aspherical Diamond Mills for Micro Optics, *Proceedings of the 21st Annual ASPE Meeting*, 2006, No. 1, pp. 1–4.

Li, C., L. Zhao, Y. Wu, W. Wu, and J. Xu, Focused-Ion-Beam Induced Rayleigh Plateau Instability for Diversiform Suspended Nanostructure Fabrication, *Scientific Reports (Nature)*, vol. 5, online article No. 8236, February 4, 2015 (doi:10.1038/srep08236).

Picard, Y. N., D. P. Adams, M. J. Vasile, and M. B. Ritchey, Focused Ion Beam-Shaped Microtools for Ultra-Precision Machining of Cylindrical Components, *Precision Engineering*, vol. 27, Issue 1, 2003, 59–69.

Teng, A., Recent Developments in Micro Milling, *Journal of Micromechancis and Microengineering*, vol. 14, 2004, R15–R34 (doi:10.1088/0960-1317/14/4/R01).

Teo, E. J., M. B. H. Breese, E. P. Tavernier, A. A. Bettiol, F. Watt, M. H. Liu, and D. J. Blackwood, Three-Dimensional Microfabrication in Bulk Silicon Using High-Energy Protons, *Applied Physics Letters*, vol. 84, Issue 16, 3202–3204 (doi:10.1063/1.1723703).

Volkert, C. A. and A. M. Minor, Focused Ion Beam Microscopy and Micromachining. *MRS Bulletin*, vol. 32, May 2007, 389–399.

Yang, H. and S. Rachev, Focused Ion Beam Micro Machining and Micro Assembly, *Precision Assembly Technologies and Systems*, vol. 315 of the series IFIP Advances in Information and Communication Technology, 2010, 81–86.

CHAPTER 3.49

Dip-Pen/Polymer-Pen Technology

LaRoux K. Gillespie

Manufacturing Consultant
Andover, Kansas

Dip-pen technology allows the manufacturer to tailor the chemical composition and structure of features measuring less than 100 nm (4 μin.) via an additive manufacturing process. While it is an additive process, it is not included in this handbook's additive technology because it is very limited in applications and focused entirely on microscopic features rather than the sizes and applications used by the normal additive processes. The process writes lines, dots, and blocks.

Because of the minute size of the features that are built via this process, it will be used primarily for molecular electronics, optics, minute semiconductors, biomedical, and similar size structures rather than more commercial endeavors. Dip-pen nanolithography (DPN) or lithography uses scanning-probe microscopy to direct-write soft materials. More specifically it uses an atomic force microscope (AFM) to write with special "inks" that become the structure. Molecules of an "ink" solution are deposited on and moved by the tip of an AFM probe. The ink molecules deposit on the substrate surface via the solvent meniscus that forms between the head of the probe and the substrate (Fig. 3.49.1). Since it occurs at ambient temperatures and involves only normal meniscus formation, it is one of the simplest additive processes. At the same time because the ink and substrates have to have a mutual affinity, it is currently very limited in the materials that can be used.

The tip of an AFM is in the order of 20 nm (0.8 μin.). The process can produce resolutions as precise as 10 to 15 nm (0.4 to 0.6 μin.) and in certain applications as fine as 5 nm (0.2 μin.) widths. Tip size plays a role in deposition of some of the inks. Because the writing is done with the AFM tip, it can more technically be named "tip-substrate molecular transport."

The key to success is in finding the right ink/substrate combinations. That involves considerable science. The ink must be chosen such that it does not form globs or diffuse too much. The ink molecules are chosen so that they anchor themselves to the desired location. Stated differently, some inks have a special affinity to attach to specific

FIGURE 3.49.1 Schematic of dip-pen nanolithography (DPN). (Courtesy Nanoink1. The file was created and owned by NanoInk; CC BY 3.0.)

metals or compounds and these are the characteristics that make the process successful. For example, silver can be coated with 1-octadecanethiol (ODT). Thiols (RSH) have a particularly strong attraction to metal surfaces. Thiols, such as RSH, are organosulfur compounds which will quickly self-assemble into a monolayer to coat the entire metal surface. The sulfur attaches to the metal and the tail end of the molecule points up and away from the surface, in a manner similar to that shown in Fig. 3.49.1. With the right organic or metal molecules in the ink, the molecules then are attached to the substrate.

Organic inks can be printed on inorganic substrates, and inorganics upon organic substrates. That is significant since the process can direct-write materials or it can also in the same setup deposit photo resists that will mask features during subsequent etching and processing. Using multiprobes in unison, both organic and inorganic inks can be printed at the same time. It is important, however, to recognize that the inks must have an affinity for the substrate they are deposited on. This greatly limits the materials that can be used.

Factors which influence process success include:

- Chemical makeup and purity of the ink and surface
- Tip shape
- Distribution of ink on the tip
- Temperature at which material is printed

- Humidity of the surroundings
- Water solubility of the ink
- Substrate surface smoothness (for finest resolution)
- Grain size of the substrate

It is essential to keep a constant layer of water on the writing tip (the "water meniscus"), so a humidity-controlled box is necessary.

The AFM tip has to move at rates slow enough so that the ink transfers to the surface consistently. This is a slow process when a single tip is used, since the tip writes one line or one dot after another. While parallel tip arrays allow faster deposition, the tips today cannot be controlled individually. Parallel patterning using two-dimensional arrays of 55,000 tips has been shown, but this is not a commercial application yet.

Without going into detail, one unique aspect of the inks is that "liquid inks" deposit by a different mechanism than "molecular inks." Molecular ink's diffusion rate of the molecule differs for each molecule type involved. Feature size depends on tip/surface dwell time. Dwell time can vary from milliseconds to seconds and on the water meniscus size (water is used for most deposition, but not all). Humidity conditions determine meniscus size if the tip's radius of curvature is much smaller than the meniscus. The ink/substrate combinations that exist today include:

- Alkane thiolates imprinting onto a gold surface
- Silanes (solid phase) written to glass or silicon
- Hexamethyldisilazane (HMDS) to modify polarity of semiconductor surfaces
- Others in development

Substrates compatible with known inks include gold, silicon, and silicon oxide. Tin oxide/P-123 has been written as lines on silicon-oxide substrates. Aluminum oxide/P-123 has made nanodot structures on silicon substrates. The writing speed for these solid-state materials varies from 0.1 to 0.2 μm/s (4 to 8 μin./s). Aluminum dot structures can be built with a dwell time for each dot of 1 s. Gallium arsenide is another of the semiconductor substrates that is reportedly compatible with DPN.

BaFe magnetic blocks are built on silicon substrates with a sol-based ink and after an annealing operation provide the desired hexagonal magnetoplumbite (M-type) structure.

DPN can write one nanostructure and then fill in the surrounding area with a second type of "ink." This opens many possibilities; one of the first is the building of multilayer structures particularly with photo resist multistep processes.

The process has produced 100-nm (4-μin.) gold dots on oxidized silicon with an array of 55,000 tips. The end result was an array of 880,000,000 dots spaced 400 nm (16 μin.) apart.

The thermal dip-pen lithography (tDPL) process uses heat to deposit semiconductor, magnetic, metallic, or optically active nanoparticles. The particles are suspended in a PMMA-like polymer, and heated by the probe tip to cause flow. Resolutions of 78 to 400 nm (3 to 16 μin.) can be achieved. An oxygen plasma etch removes the PMMA matrix. When iron oxide nanoparticles are the particles, the process resolution can be as fine as 10 nm (0.4 μin.). The advantage of this process is that it does not require special solution preparation techniques. The particles used, however, must be smaller than the radius of gyration of the polymer which for PMMA is about 6 nm (0.2 μin.). Deposition rates of 200 μm/s (8 μin./s) are possible for pure polymers, but the addition of nanoparticles can slow speeds to 2 μm/s (0.008 μin.). This is still faster than the regular dip-pen lithography.

DPN is also used in a mechanical-electrochemical writing mode to write 5-nm-wide (0.2-μin.) copper lines on gold surfaces.

While the process is unique and limited in size capability, over 20 worldwide laboratories are expanding its capabilities and are there to help companies as they implement the process. At least one commercial firm supports the industry. Designers need to read the detailed technical literature and contact one of these firms to determine applicability of any material combinations not already mentioned, and to understand the production issues that would further limit production use. This is one of the youngest technology fields in existence and because of material limitations is also one of the most limited processes unless the right materials are available.

References

Fu, L., X. Liu, Yi Zhang, V. P. Dravid, and C. A. Mirkin, Nanopatterning of "Hard" Magnetic Nanostructures via Dip-Pen Nanolithography and a Sol-Based Ink, *Nano Letters*, April 2003, *3*(6), pp. 757–760 (doi:10.1021/nl034172g).

Nie, Z. and E. Kumacheva, Patterning surfaces with functional polymers, *Nature Material*, 2008, *7*, pp. 277–290 (doi:10.1038/nmat2109).

Obermair, C., A. Wagner, and T. Schimmel, The Atomic Force Microscope as a Mechano–Electrochemical Pen, *Beilstein Journal of Nanotechnology*, October 2011, *2*, pp. 659–664 (doi:10.3762/bjnano.2.70).

O'Connell, C. D., M. J. Higgins, D. Marusic, S. E. Moulton, and G. G. Wallace, Liquid Ink Deposition from an Atomic Force Microscope Tip: Deposition Monitoring and Control of Feature Size, *Langmuir*, February 2014, *30*(10), pp. 2712–2721 (doi:10.1021/la402936z).

Salaita, K., Y. Wang, and C. A. Mirkin, Applications of Dip-Pen Nanolithography, *Nature Nanotechnology*, 2007, *2*, pp. 145–155. Published online: February 25, 2007 (doi:10.1038/nnano.2007.39).

CHAPTER 3.50
Capillary Forming

LaRoux K. Gillespie

Manufacturing Consultant
Andover, Kansas

Capillary forming, as the name implies, uses capillary action to form a structure. The means by which the structure is formed and materials involved depend on which variation of the process is used.

Simple Capillary Action

In the simplest operation, a liquid rises into a mold by capillary action and then is permanently solidified by some means. Ultraviolet-curable resins respond to capillary filling and surface tension and provide an easy solidification approach, but other combinations are possible. The resulting structures have smooth surfaces and various shapes have been produced, including simple low-cost micro lens of easily varied focus points. Final shapes are a result of the resin chosen, and a proper combination of pressing pressure, pressing duration, and UV-curing dose.

The process can be used to make concave or convex surfaces as well as simple striped products by some simple variation of the process. In a process called capillary force lithography (CFL), a variety of shapes are produced (Fig. 3.50.1). This is based on the principle that the time needed for a fluid to fill up a narrow capillary cavity depends on the contact angle of the mold or the width of the cavity. The resulting product can be varied by selectively modifying the mold by UV or oxygen plasma treatment. This changes the mold's wettability in different areas. A similar effect occurs by simply varying the width of the cavity where the liquid will be forced up by capillarity. **Polydimethylsiloxane** (PDMS) molds are ideally suited for this micro capillary forming.

Polyurethane, polyacrylate, As_2S_3-propylamine, epoxy, and many other solutions and polymers have been formed in this manner to micrometer-size channels.

Carbon Nanotube Process

Working with carbon nanotubes (CNTs), the process makes pillars, twisting spires, concentric rings, and bending petals (Figs. 3.50.2 and 3.50.3). The process has made spikes as thin as 5 μm (0.0002 in.) and patterns of pillars several millimeters wide. Structures as large as 500 μm (0.020 in.) have also been fabricated.

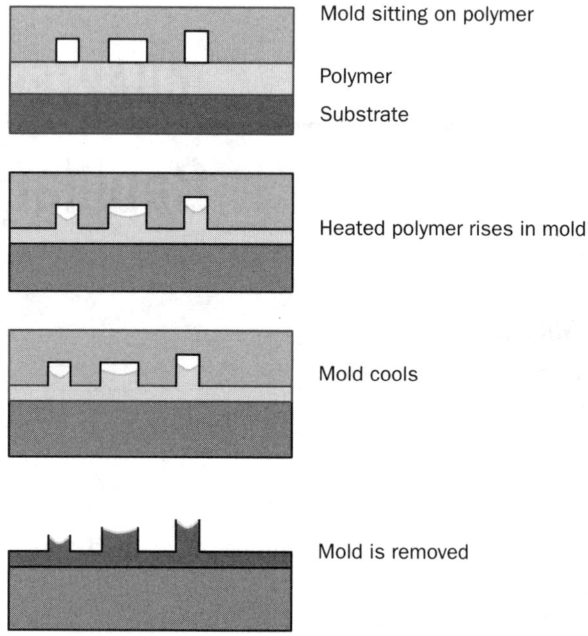

FIGURE 3.50.1 Simple capillary action moves liquids up into mold cavities. Heat, time, and liquid properties determine resulting configurations. Hydrophilic modifications to the mold channels also influence results. (Reproduced from Tsay, 2010.)

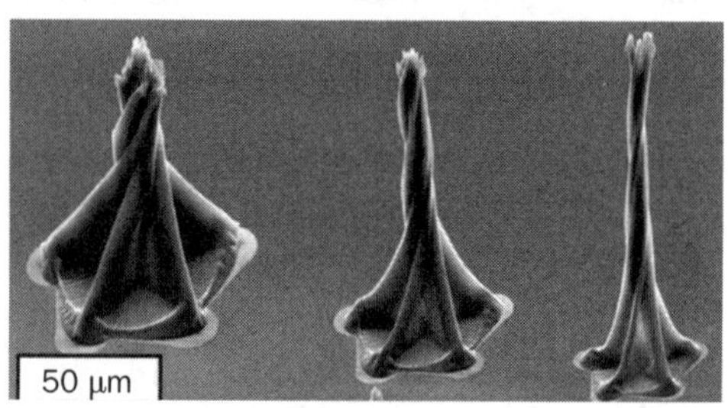

FIGURE 3.50.2 Twisted spike nanotubes from capillary forming (De Volder et al., 2010).

For the nanotube process, pieces of thin metal film (e.g., an iron catalyst) are placed on a silicon wafer by lithography. The metal film acts as a catalyst and a template. The final shape of the CNT product depends on the shape of these microscopic film bits.

The wafer and film are heated to a high temperature in a hydrocarbon gas mixture. The gas reacts with the film/catalyst, causing the carbon from the gas to form into nanotubes growing from the film. The resulting nanotube-laden wafer is suspended over boiling acetone. The acetone condenses on the nanotubes, traveling up the spaces between them via capillary action. This acetone action causes the structures to bend and twist into 3D shapes as shown in the figures.

The chemical vapor deposition (CVD) process causes the CNTs to grow vertically in a tree-like structure. The acetone causes the CNTs to alter their shapes and dimensions.

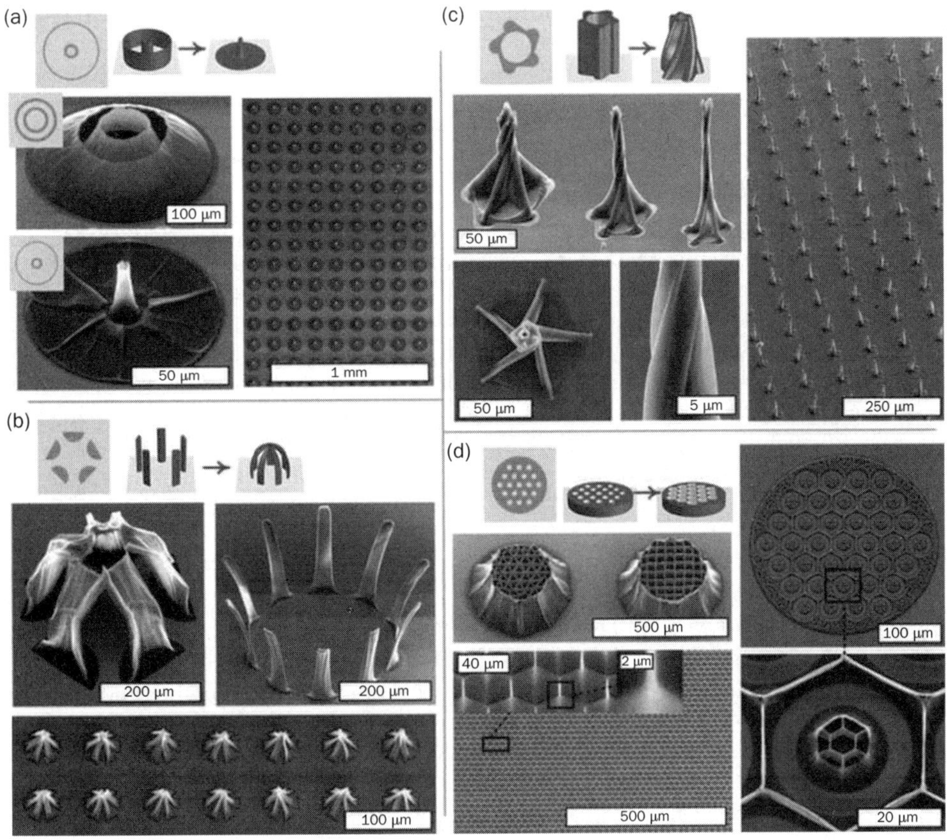

FIGURE 3.50.3 A variety of carbon nanotube shapes made via capillary forming (De Volder et al., 2010).

Because the shapes are a result of the initial 2D metal films, a variety of final configurations are possible, and large fields of the structures can be made at any one time by lithographic placement of the starting metal foils. Because CNTs have such exceptional mechanical, thermal, electrical, and chemical properties, the potential applications are extensive, although the most immediate ones are in the biomedical field. The nanotube structures are strong (up to 10 times stronger than polymers used for similar size products) so if desired they can become the model from which molds are made to rapidly produce multitudes of other materials having the same configuration.

Making vertical pillars of CNTs is a straightforward process. The fact that they can be bent or twisted by capillary forming provides geometries not possible by more conventional approaches. Producing new shapes requires an extensive knowledge of condensation and evaporation thermodynamics, the kinetics of capillary rise inside porous filament-like materials, and the mechanics of liquid-induced stresses and plastic deformations in these materials.

The CNT structures in addition to being twisted, bent, flattened, or otherwise changed can be filled or coated with polymers to provide yet different properties

(De Volder et al., 2010). In addition, the process allows thin CNT filaments to self-aggregate to make a denser pillar or product. Surface tension has long been known capable of aggregating wet hair, micro fibers, and polymer pillars. Capillary forming extends the process to CNT filaments.

Design Considerations

Simple capillary movement into molds is a relatively straightforward process, and has been used since the 1990s. It involves some experimentation. It is limited by the materials which can be hardened or solidified by some appropriate process. In some instances simple freezing may meet needs. Heat sets As_2S_3-propylamine and other polymers.

CNT capillary 3D production requires a more extensive experimental and analytical approach and there are few experts or companies to assist with this process. Despite the relative difficulty of generating new shapes, the configurations shown in Figs. 3.50.2 and 3.50.3 illustrate the potential for this process. The details for these processes are already defined for CNT materials in several of the following references.

References

Chang, T.-J., Y.-H. Kim, K.-S. Yoo, Y.-S. Kim, and S.-W. Joe, 29.3: Barrier Rib Molding Process using Capillary Action. *SID Symposium Digest of Technical Papers*, 2003, 34:1011–1013 (doi: 10.1889/1.1832459).

De Volder, M., S. H. Tawfick, S. J. Park, D. Copic, Z. Zhao, W. Lu, and A. John Hart, Diverse 3-D Microarchitectures Made by Capillary Forming of Carbon Nanotubes, *Advanced Materials*, October 2010, 4384–4389 (doi:10.1002/adma.201001893).

Hart, A. John, Fabrication of 3D Microstructures Using Capillary Forming, *Advanced Coatings & Surface Technology*, February 2011: 10+ *Academic OneFile*. Web. May 24, 2016.

Kim, E., Y. Xia, and G. M. Whitesides, Polymer Microstructures Formed by Moulding in Capillaries, *Nature*, August 17, 1995, 376:581–584 (doi:10.1038/376581a0).

Li, H., W. Yu, J. Xu, C. Yang, Y. Wang, and H. Bu, Hierarchical Structure Formation and Pattern Replication by Capillary Force Lithography, *RSC Advances*, 2014, 4(75): 39684–39690 (doi:10.1039/C4RA04797D).

Lochhead, M. J. and P. Yager, Patterned Sol-Gel Structures by Micro Molding in Capillaries, *MRS Proceedings*, 1996, 444:105 (doi:10.1557/PROC-444-105).

Tawfick, S., Z. Zhao, M. Maschmann, A. Brieland-Shoultz, M. De Volder, J. W. Baur, W. Lu, and A. John Hart, Mechanics of Capillary Forming of Aligned Carbon Nanotube Assemblies, *Langmuir*, 2013, 29(17):5190–5198 (doi:10.1021/la40022190).

Tsay, C., Y. Zha, and C. B. Arnold, Solution-Processed Chalcogenide Glass for Integrated Single-Mode Mid-Infrared Waveguides, *Optics Express*, 2010, 18:26744–26753.

Xia, Y., E. Kim, and G. M. Whitesides, Micromolding of Polymers in Capillaries: Applications in Microfabrication, *Chemistry of Materials*, 1996, 8(7):1558–1567 (doi:10.1021/cm9602196).

CHAPTER 3.51

Handling Micro Parts

LaRoux K. Gillespie

Manufacturing Consultant
Andover, Kansas

While the manufacturing engineers will determine tools and detailed process for making and assembling micro parts, it is essential for designers to consider handling issues in their design. The reason is that some simple design issues can make handling much easier.

Parts which are the size of dust specks are greatly influenced by:

- Van der Waals force (the relatively weak attraction between the electron-rich and electron-poor areas of two molecules that pass very close to each other)
- Electrostatic force, the attraction of two atoms of opposite charge
- Surface tension, caused by cohesive forces between molecules

These are not just possible effects; they are real issues for almost every micro part. When part sizes reach about 100 μm (0.004 in.), surface forces start to win out over gravity, which weakens as part volume decreases. When gripper jaws open, or tools which move the parts try to release the parts, tiny parts will often stick to the tools' surface rather than drop into place.

Room humidity or any moisture in the processes leaves a thin layer of water on surfaces, and the surface tension of the liquid sticks the parts to the surfaces. Relative humidity of 70% or higher will initiate stickiness, but if that same humidity drops below 30% electrostatic forces will cause clinging. In both situations an external force is required to break the bonds causing the attraction.

Hydrophobic (water repelling) materials such as zinc oxide and polydimethylsiloxane (PDMS) and other materials can be coated on all the tools having contact with the micro parts. Intentional rough surfaces on parts or tools also reduce the attraction because it reduces van der Waals forces.

Handling involves several physical aspects including

- Grasping
- Feeding

- Positioning
- Releasing

While a human does each of these aspects without much thinking, for micro parts there are multiple issues with each aspect and the various solutions in turn also have limitations that may not be obvious. The issue here is for designers to carefully consider the needs of handling and before releasing final designs to have discussed potential issues with the manufacturing engineers.

Handling Solutions

While the solutions to handling are the province of the manufacturing engineers, some are made easier by designer choices. Moving and handling micro parts can be implemented by

- Manual tools
- Automated mechanical approaches
- Magnetics
- Electrostatics
- Surface tension
- Vacuum and air pressure
- Adhesives
- Thermal issues
- Laser
- Bernoulli effects
- Sonic approaches
- Acoustic (levitation particularly)
- Vibratory approaches
- Fixtures
- Biological processes

Handling micro parts requires first a decision of whether contact or noncontact approaches will be used. Then the production rate and time required for handling will further refine decisions. From a design standpoint, the configuration, materials, and potential for handling damage will drive the solution. Surface damage and part release are two key issues that drive design.

Surface damage can be minimized by designing areas where contact will not harm the part, that is, specific areas for contact. Because of the forces described earlier, many parts, particularly the smooth surface thin ones, and exceptionally minute ones will tend to stick to the surfaces grasping the part.

Manual Approaches

There are literally hundreds of conventional tweezers available for micro parts. Vacuum "tweezers" are similarly available and they come in interchangeable tips for flat parts and many other configurations including hypodermic needle configurations. All tweezers can come with a variety of coatings to prevent damage and to help release. Manual electrostatic handheld units are available with integral squeeze bulb release pressure. Some even provide heat to help desolder as they move components.

The limitations to manual handling include damage to delicate parts, scratches on optical or coated products, and leaving traces of contamination on the surfaces. Releasing parts is also a major constraint for many materials. These are also considerations for most of the approaches mentioned in the following paragraphs.

Automated Mechanical Approaches

Moving parts with nanometer precision requires special designed systems. Small robots with micro grippers are commercially available and some can put glass micro fibers or arrays of fibers into plate holes. These systems can position to better than 1 nm (0.004 μin.) and can control gripping forces up to 150 mN (0.034 lb). While the tool used is generally called a gripper, it does not necessarily mean the part is squeezed between two surfaces. Special jaws can be used which open, surround the part, and then close to cradle the part to its next location.

Automation picks the parts up, moves them, and then places them in the desired location. All of that can be automated. The gripper can employ any of the approaches we have defined here.

Magnetics

Magnetic parts are easily grasped and moved but again releasing them can be an issue unless the magnetic field can be disengaged in some effective manner. Further, once magnetic approaches pick up a part those parts will retain enough magnetism to attract other parts to them, so they stick together. Separating parts from other parts via magnetism is also an issue.

Electrostatics

Both conductive and dielectric objects are attracted toward regions with a higher electric field. This has been developed with hand units for positioning, and also when combined with a small amount of vibration has been shown capable of automated alignment, positioning, transporting (moving), and releasing micro parts. Different part shapes, different part materials, and different dimensions can all be moved with the right combination of electrostatics and vibration. A variety of handling devices have been investigated and used, including simple needle-like tools which can accomplish peg-in-hole positioning. For conveyor-like noncontact movement, micro conveyors can be used which create nonuniform traveling field waves which "push" the parts down the conveyor. Typical part sizes handled via this approach include 50-μm-thick (0.002-in.) plates (polyamide), 100- to 800-μm (0.004- to 0.031-in.) glass spheres, and many other materials and geometries including those made of steel, copper, silicon, plastic, glass, and composites. Prismatic, spherical, gear-like, cylindrical, and other shapes have all been successfully handled by this process.

Releasing parts can be an issue and that is the reason vibration and acoustics are tied with electrostatics to release parts. Electrostatics also may damage delicate electrical components and can leave a residual charge on them. Releasing dielectric parts can be released by inverting the applied potential or turning it off, but for conductive parts inversion is not a solution. Picking up a single part from a group of parts can also be an issue.

Surface Tension

It is easy to pick up micro parts with only a small amount of liquid on the tip of a needle or syringe because surface tension forces provide all the needed force. Releasing them becomes the issue. Placing parts in a confining hole or pocket allows lateral motion to break the bond, but it leaves moisture on the part. Contamination from the fluid can be an issue for some parts.

Vacuum and Air Pressure

While vacuum tweezers are the most common approach to picking up delicate parts of any shape, air pressure and/or vibration are also used to move parts and by forcing them into carefully designed slots to position them for high-volume work. Compressed air can move 0.1-mm (0.004-in.) balls, electronic chips of 0.3 mm (0.012 in.) or more, and 1.2-mm (0.047-in.) screws.

Adhesives

Tacky adhesives easily hold parts but again releasing them is an issue. Several solutions have been used, but in each instance contamination and residual material from the adhesive have to be accounted for.

Thermal Approaches

Micro ice tweezers or grippers can easily and quickly move tiny and larger parts by freezing a small amount of liquid to the part. This approach is independent of electrical charge, and free of force damage. Releasing is simply a matter of heating the tweezers. It also has the advantage that it can grasp objects in liquids as well as in air. It would leave a film of moisture on the object.

In addition to using ice to move the part, ice clamping can also be a convenient holding force while assembling or performing other tasks. Cycling for both air and under liquid systems can be as short as 2 s and in some instances as short as 0.3 s.

Lasers

Lasers are very limited in their ability to handle parts. They are mainly suitable for very small components and mainly transparent ones. Heat adsorption may be an issue for some components.

Bernoulli Effect

Increasing the velocity of horizontal flow (gases or liquids) results in decreasing the static pressure above that flow (Bernoulli's law). If a vertical hole intersects a channel of such flow, parts will be sucked down from this change in pressure and released when the flow is reduced or stopped. Airplane lift is based on this effect.

Providing accurate positioning with this approach is a problem. Bernoulli grippers are easily applied to pickup movement of flat micro parts.

Sonics Approaches

Ultrasonics can create an air pressure difference between vertical plates that levitate very light parts.

Acoustic Approaches

Acoustics and sonic approaches are part of the same family. They can move and levitate light parts, but accurate placement has not been possible. The acoustics can also be used to release parts held by other approaches.

Vibratory Approaches

Vibratory feeders have been used for decades to move parts and with the aid of slots or positioning devices, they can feed a mass of parts into a continuous line of correctly oriented parts. Vibration also is used to release parts.

Fixturing

Handling parts can be simplified if parts are placed into close-fitting pockets of a tool and the tool moved to the various positions. The tool can be of any size and material. Its advantage is that parts do not have to be picked up and released multiple times. They can be designed to release parts, although most have been used for merely transporting them. Vibratory feeders can drop parts into the pockets automatically with little or no sensing.

Biological Processes

Biological handling in this instance refers to moving items via cellular effects or via animal mimicking. An example of the latter is the gecko and tree frog adhesion to vertical walls. While the underlying mechanisms may be the same as listed above, it is the combination often of two or more of these approaches that make the natural animal abilities work so well. In the former case living cells can transport nano size objects.

Specific Design Considerations

While each of the handling techniques described above was presented as a handling means, each can also become the releasing mechanism to free parts from their gripper. Designers need to consider how fragile, porous, dielectric, conductive, magnetic, diamagnetic, charge sensitive, water sensitive the parts are. Electronic surface mount devices (SMDs) are charge sensitive and would be damaged by electrostatic grippers or electrowetting techniques. Liquid grippers leave stains or foreign material on optical lenses. Fantoni's work (2003, 2004, 2011, 2014) provides an extensive summary of considerations. Most of the issues are more targeted to manufacturing engineers than designers, so we merely note his work here.

- Other design concerns are nesting together, entanglement. Slippery parts are harder to handle as are flexible or squiggly ones.
- Since adhesion effects are greatly reduced when parts are in fluids, immersing them in fluids may be a way to easily release parts.
- As previously mentioned, greater roughness reduces adhesion forces. Roughness is a result of the manufacturing process, but can also be changed by sputtering gold films on the part or electron beam or laser texturing on either

part or tools. Roughness reduces the area of contact and the high points allow self-discharge of van der Waals and electrostatic forces.

- Reducing contact area between gripper and part reduces adhesion forces.
- Conductive materials or coatings significantly reduce static charges; grounding them drains the charge.
- Use materials which do not vary significantly in EV potential.
- Hydrophobic coatings reduce surface tension. Hydrophilic coatings increase surface tension. Both can be used for specific handling applications and programmable hydrophobic surfaces are possible (Fantoni, 2011).
- As a general note, while austenitic stainless steel (303 and 304 SS) is regarded as nonmagnetic, when the metal is cold-worked by any means it gains magnetic properties. And in general one should consider that it is always at least weakly magnetic.
- A drop of liquid on the end of a hypodermic needle will pick up parts because of surface tension, but releasing it requires expelling the drop, dropping the part into a holding pocket, and moving the needle sideways to break the adhesion, or having the part placed on a bed of higher adhesion (glue, solder, thermal bonding, ultrasonic bonding, etc.).
- Manufacturing engineers can make many parts as a group of parts all held together by a strip or tab of some kind. This simplifies handling and orienting individual parts multiple times.

References

Bark, C. and T. Binneboese, Gripping with low viscosity fluid, *Proceedings of the IEEE International Workshop on MEMS*, 1998, pp. 301–305.

Fantoni, G., Assembly of mini and micro parts: development of an electrostatic feeder, *Proceedings of the 6th A.I.Te.M. International Conference*, Gaeta, Italy, September 2003, pp. 352–366.

Fantoni, G. and F. Biganzoli, Design of a novel electrostatic gripper, *International Journal for Manufacturing Science & Production*, vol. 6, no. 4, 2004, pp. 163–179.

Fantoni, G., Active surfaces, materials and tools for assembly (presentation slides), Department of Mechanical, Nuclear and Production Engineering, University of Pisa, Italy, 2011.

Fantoni, G., Grasping devices and methods in automated production processes, *CIRP Annals—Manufacturing Technologies*, vol. 63, no. 2, 2014 (slide variation of paper also found at http://docslide.us/engineering/grasping-everything.html).

Fantoni, G. and M. Ports, A critical review of releasing strategies in microparts handling, in Swetan Rachev's *Micro-Assembly Technologies and Applications: IFIP TC5 WG5.5, Fourth International Precision Assembly Seminar (IPAS 2008)*, Chamonix, France, Springer, 2008.

Gauthier, M., B. Lopez-Walle, and C. Clevy, Comparison between micro-objects manipulation in dry and liquid mediums, *Proceedings of CIRA '05*, Espoo, Finland, 2005.

Kochan, A., European project develops "ice" gripper for micro-sized components, *Assembly Automation*, vol., 17, no. 2, 1997, pp. 114–115.

Leventon, W., Coming to grips with sticky surface forces, *MICROmanufacturing*, March/April 2014, pp. 20–23.

Monkman, G. J., Electroadhesive microgrippers, *Assembly Automation*, vol. 24, no. 1, 2003, MCB University Press.

Obata, K. J., T. Motokado, S. Saito, and K. Takahushi, A scheme for micro-manipulation based on capillary force, *Journal of Fluid Mechanics*, vol. 498, 2004, pp. 113–121.

Saito, S., H. Himeno, K. Takashashi, and T. Onzawa, Electrostatic detachment of a micro-object from a probe by applied voltage, *Proceedings of the International Conference on Intelligent Robots and Systems*, vol. 2, 2002, pp. 1790–1795.

Tichem, M., D. Lang, and B. Karpuschewski, A classification scheme for quantitative scheme for quantitative analysis of micro-grip principles, *Proceedings of the 1st International Precision Assembly Seminar (IPAS 2003)*, Bad Hofgastein, Austria, March 19, 2003.

Vögeli, B. and H. von Känel, AFM-study of sticking effects for micro parts handling, *Wear*, 238, 2000, pp. 20–24.

Zesch, W., M. Brumner, and A. Weber, Vacuum tool for handling microobjects with a NanoRobot, *Proceedings of the IEEE International Conference on Robotics and Automation* (ICRA '97), 1997, pp. 761–776.

CHAPTER 3.52

Assembly of Micro Parts

LaRoux K. Gillespie

Manufacturing Consultant
Andover, Kansas

Assembly of micro parts, also called micro assembly, is the field of assembly involving parts, particles, or products having features on the order of 1 mm (0.040 in.) or less. It can also involve assembling cells, although that is not yet normal for the industrial world. It can involve items whose size is as small as 100 nm (4 μin.) and features as small as 10 nm (0.4 μin.).

Assembly consists of two major elements: (1) handling, which includes pickup, movement and positioning, and releasing, and (2) joining. For some, in-process testing or measuring is also a part of assembly operations, and for some of the newer processes, releasing the components from their substrates or other parts joined by tabs is performed in the assembly area. As such it becomes an assembly operation—freeing it so it can be assembled. Handling is discussed in Chap. 3.51 and the reader is directed there to understand those issues. This chapter will concentrate on the remaining aspects.

The basic tenants of designing for assembly for large parts still apply for the micro size products. The most fundamental elements of design for assembly are:

- Reduce the number of components to the smallest possible number
- Design parts and fixtures to capture mating components (as opposed to having to position and hold for assembly)

To this list for micro parts, designers need to consider the following approaches also:

- Design for pop-up assembly
- Design for self-assembly
- Take advantage of the micro forces described in Chap. 3.51
- Utilize some of the advanced manufacturing processes which incorporate several of these issues

McGraw-Hill's *Design for Manufacturability Handbook* (edited by James G. Bralla), a predecessor to this handbook, provides a number of design concepts that incorporate the basic tenants of assemblies of any size and a number of *illustrative examples* that we do not repeat here, but we do list some of the more common approaches particularly amenable to micro components.

- Incorporate hinges rather than extra pieces.
- Design spring-like features rather than use springs (cantilevers bend at a predetermined rate just like springs).
- Use interlocking features to hold components.
- Incorporate guides and covers.
- Design parts to be self-aligning.
- Use processes that provide both mechanical and electrical features in the same process.
- Use bent tabs or crimping rather than screw-like components.
- Build in identification rather than adding on.
- Use press fits where possible.
- Design so that components cannot be misassembled.
- Eliminate the need for adjustments.
- Use funnel-like openings (or countersinks) for easy pin-in hole insertion.
- Design grasping features on components.
- Eliminate the need to hold parts while assembling.
- Eliminate the potential for parts to entangle when bagged together.
- Use families of similar parts where possible to facilitate pickup and placement.

Fortunately, several of the advanced manufacturing processes described in this handbook provide new approaches that were not available in past years and that make several of what once were considered "way-out" ideas absolutely possible. Specifically, Chaps. 3.44, 3.45, 3.46, 3.47, and 3.51 describe processes that make many of these ideas practical.

MICA Freeform (Chap. 3.44) provides moving assemblies without the need of any external assembly. Micro stereolithography (Chap. 3.45) allows designers to design exceptionally complex polymer configurations without having to worry about joining them together. Laminated films of metal combined with micro milling tools also result in completed assemblies of almost any shape.[1] The layer-by-layer approach builds the assembly as it goes.

MEMS manufacturing (Chap. 3.46) and printed circuit MEMS (PC-MEMS) already have demonstrated production capabilities to make pop-up assemblies that require no or little extra assembly work than popping up the uniquely designed flat configurations into 3D shapes and releasing them from their substrate or tabs to other similar components.

[1]All of the defined additive processes in this handbook provide similar conceptual advantages, but at present only micro stereolithography is uniquely positioned to provide truly micro size components. The laminated sheet process combined with micro milling can do that, but is more applicable to slightly larger sizes than millimeter-size parts.

Origami (Chap. 3.47), which focuses on how bend lines can be used to restructure 2D configurations into 3D components, has already demonstrated how it can be used for modern manufacturing of micro structures. Combined with MEMS production processes it can be very effective.

Handling approaches (Chap. 3.51) similarly is a key element of assembly that makes modern advanced manufacturing of micro parts a reality.

In addition, processes like micro laser welding (Chap. 3.37), laser soldering (Chap. 1.8), and other laser-assisted processes, micro resistance welding (Chap. 3.36), ultrasonic micro bonding (Chap. 3.40), and micro molding (Chaps. 3.22 and 3.23), in combination with or in addition to unique micro tools that heat a single part to cause some form of fusing (like soldering), have provided many avenues to more successful assembly. The presence of several advanced modeling or simulation computer programs similarly provides effective and timely enhanced design of thermal, electrical, mechanical, and production in both static and dynamic conditions.

Four elements for micro design deserve particular note:

- Positioning
- Joining
- Pop-up designing
- Self-assembly
- Biomedical issues

Positioning

Micro parts typically require unusual levels of precision positioning. In some instances, the requirement is for placement within 1 μm (40 μin.). Positioning, as a result, involves robots, telerobots, or other high-precision automation. Robots alone, however, may not be the total solution. For many, if not most, applications, robots are partnered with high-precision grippers, microscopes, vision systems, and sensors. When this is the case, designers need to assess the need for vision while assembling and how the product design can be designed to facilitate that aspect.

In other words, how conducive is the product design to automated or semiautomated assembly? If the part has to be seen to be assembled, are the shapes and colors amenable to vision? The manufacturing engineer has to determine how to set up the assembly operation and equipment, but the designer's choices will influence the difficulty faced in actual assembly. Once again, early discussions between the designer and the manufacturing engineer will facilitate assembly. Figure 3.52.1 illustrates the congestion of one micro assembly setup (Zhang, 2008).

Today, grippers are capable of picking up nanometer-size items and positioning to nanometer levels and some can do that in seconds. Others require more time. The sensors employed include identifying and controlling insertion forces, temperatures, voltage, current, and location.

Joining

Physically attaching one component to another, in general, is much slower than for more conventional size parts if conventional approaches are used. Screws are not

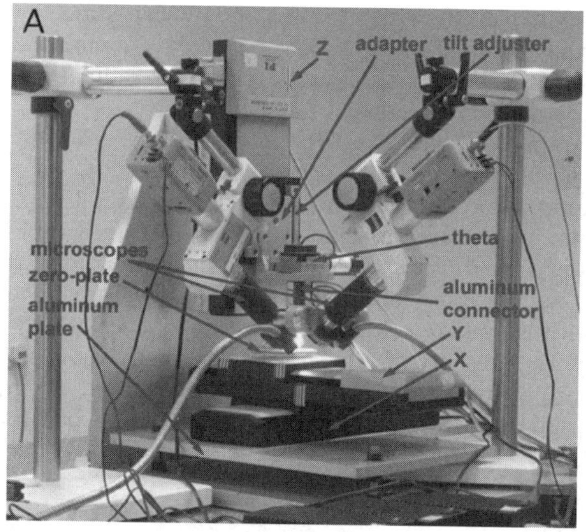

Figure 3.52.1 Congestion of one micro assembly setup. (Courtesy E. Burdett, Imperial College, London.)

usually available for the truly minute parts. Pining, "riveting," press fits, and bonding of some nature are the common joining processes. Pining implies both inserting pins (normally round ones) to position parts, and locking the pins in. The pins can be cold headed slightly to prevent vertical motion, or tabs can be bent to lock in parts. Both approaches pin the parts in place. The thermal approaches of micro laser welding (Chap. 3.37), laser soldering (Chap. 1.8), micro resistance welding (Chap. 3.36), and ultrasonic micro bonding (Chap. 3.40) are the most common joining processes for permanent bonding, although adhesives (Chap. 3.41) are also used. Each of those chapters describes the abilities for micro component joining. As mentioned earlier some of the grippers can also provide thermal bonding.

The use of thermal systems requires the components and nearby components to not be functionally affected by the bonding. Most of the micro thermal processes can carefully control the thermal aspects, but some components are particularly sensitive.

One of the unique aspects of perfectly smooth and clean metals is that when placed together with moderate pressure, or even no pressure, the two clean surfaces will atomically bond to each other if of the same material. It is not clear how strong this bond is, but it is a strong bond.

While the microscopic forces mentioned at the beginning of Chap. 3.51 greatly hinder handling and assembly, they can be useful in some situations if planned for. Van der Waals, electrostatic force, and surface tension can hold items in place while other assembly functions are performed. It does require some ingenuity, however, to employ them since they hate to let go.

Contamination

Most micro assembly occurs in clean rooms, but assembly itself can generate particle contamination. Welding can produce microscopic particles and fumes that land on the assembly. Friction fits can release or generate particles. Tooling/part interactions can release particles. Designs that eliminate these particle-producing contaminants have more assurance of particle-free assemblies. Even the pop-up design assemblies mentioned in the next section can release unseen particles. As a last consideration on this aspect, designers need to provide a way in conjunction with manufacturing engineers to assure that once assembled there are no stray particles in the assembly.

Pop-Up Design

One of the most significant advanced approaches to micro design is the use of pop-up features of parts. This approach makes use of hinges built in as part of the production process, which allow assemblers to simply pop up supports from parts made in a 2D plane. No additional components are needed to space parts or to hold pieces together. There is nothing to misapply and no component tracking is required. The 3D part is made entirely or mostly from components all built in a single piece having flexible hinges. Flexible circuit boards made almost 50 years ago represent one of the earliest implementations of the idea and that approach is directly viable and in fact highly needed today for micro assemblies.

More than simply being flexible, however, pop-up design often involves an ability to bend at least to 90 degrees and then lock-in to assure necessary support. In that sense the design element would be pop-up latching. One of the most successful applications is the design for a MEMS micro mirror which consists of two actuating electrodes, two pop-up supports for the mirror, supporting beams for the mirror, and the mirror itself (Keating, 2000). Made of polysilicon and metallic electrodes by MEMS manufacturing (surface micro machining), it consists of 10-μm-wide (0.0004-in.) pop-up beams and a total length of 400 μm (0.016 in.). Some components are only 3 μm (0.0001 in.) thick. It is assembled by simply applying pressure to one side of the supports, causing them to rotate to a vertical position. It locks into place when a rectangular tab falls into its mating rectangular hole. After assembly the embedded electrical circuits cause the mirror to rotate as a function of applied voltage.

Pop-up designs are not limited to MEMS production, however. Walking robot components have been made from laminated layers of Kapton and carbon fiber, resins, and acrylic adhesive with intermediate machining—all that occurring as a part of a hybrid sheet laminating—process. Pop-up designs may provide a total structure or they may provide just part of the final assembly, but their use simplifies assembly.

Self-Assembly

For certain materials self-assembly is possible. Chapter 3.47 provides some examples. While not a universal solution because of configuration, materials, and function, it is an area that designers need to consider. It truly has a very limited potential, but is viable for some designs. In addition to self-assembly, the ideas provided for bend or hinge areas are indispensable for pop-up assemblies.

Biomedical Issues

Assembly issues for biomedical components are beyond the goal of this chapter, but some key issues are discussed. First, the term "biomedical" encompasses many aspects from treatment to insertion in the body, to providing living cells. All of this brings a variety of biological issues as well as regulatory ones. At the same time some of these are important for conventional manufacturing and assembly in those environments, but perhaps not to the detail that others must be considered.

For most biomedical applications, the materials in the process must be non-antigenic, noncarcinogenic, nontoxic, nonteratogenic and have high cell/tissue biocompatibility. Many also have unique external shape and internal structure requirements. In addition, the design must prevent unwanted cell growth yet also in many applications desire both cell attachment and in some instances bioresorbable by newly grown cells.

In some materials, the buildup of mesh-like scaffolds provides yet another approach to assembly—bonding layers together which self-adhere as they are placed together. SU-8[2] and micro stereolithography parts from acrylic resin, for example, will tend to stick together when moderately pushed together. No extra bonding is required (see Zhang, 2008).

Shape Memory Alloys

Many of the advanced technology assemblies employ shape memory alloys, and many of the grippers handling micro parts similarly employ these materials. They can be configured for a variety of thermal conditions, physical motions, and even gripping forces. These materials can be a significant asset to assembly operations either as a component to be assembled or as a tool to perform the assembly.

Several references are included in the reference section of this chapter to allow further exploration of the insights provide in the chapter. The assembly of micro components can be a major challenge if not considered carefully before design release.

References

Baisch, A. and Robert J. Wood, Pop-up assembly of a quadrupedal ambulatory micro-Robot, *Proceedings of 2013 IEEE/RSJ International Conference on Intelligent Robots and Systems*, 2013, pp. 1518–1524 (doi:http://dx.doi.org/10.1109/IROS.2013.6696550).

Bralla, J. G., Design for assembly, *Design for Manufacturability Handbook*, 2nd ed., McGraw-Hill, New York, 1999, pp. 7.3–7.33.

Keating, D. J., Y. He, N. Finch, and J. Marchetti, Numerical simulation of micro-assembly of MEMS devices and post assembly electromechanical actuation, *Proceedings of SPIE 4228, Design, Modeling, and Simulation in Microelectronics*, October 24, 2000 (doi:10.1117/12.405440).

Saitou, K. and M. J. Jakiela, Design of a self closing compliant "mouse trap" for micro assembly, *ASME Dynamic Systems and Control Division Publication DSC*, vol. 59, pp. 421–426.

Tietje, C., Design for microassembly—A methodology for product design and process selection, *Proceedings of the IEEE International Symposium on Assembly and Manufacturing*, Ann Arbor, Michigan, 2007, pp. 178–183.

Zhang, H., E. Burdet, A. N. Poo, and D. W. Hutmacher, Microassembly fabrication of tissue engineering scaffolds with customized design, *IEEE Transactions on Automation Science and Engineering*, vol. 5, No. 3, July 2008, pp. 446–453.

[2]SU-8 is an epoxy-based negative resist, which, when cured, is resistant to solvents, acids, and bases. Because it is thermally and mechanically stable, it is used to make permanent micro structures.

Index

Note: Page numbers followed by *f* denote figures; by *t*, tables; by *n*, footnotes.

M

N

O

P

Q

R